Annals of Mathematics Studies

Number 124

Exponential Sums and Differential Equations

by

Nicholas M. Katz

PRINCETON UNIVERSITY PRESS

PRINCETON, NEW JERSEY
1990

Princeton University Press books are printed on acid-free paper, and meet the guidelines for permanence and durability of the Committee on Production Guidelines for Book Longevity of the Council on Library Resources

Printed in the United States of America
by Princeton University Press, 41 William Street
Princeton, New Jersey

Library of Congress Cataloging-in-Publication Data

Katz, Nicholas M., 1943–
Exponential sums and differential equations / by Nicholas M. Katz
p. cm. — (Annals of mathematics studies ; no. 124)
Includes bibliographical references.
ISBN 0-691-08598-6 (cloth)
ISBN 0-691-08599-4 (pbk.)
1. Exponential sums. 2. Differential equations. I. Title.
II. Series.
QA246.7.K38 1990
512'.73—dc20 90-34934

Contents

Leitfaden

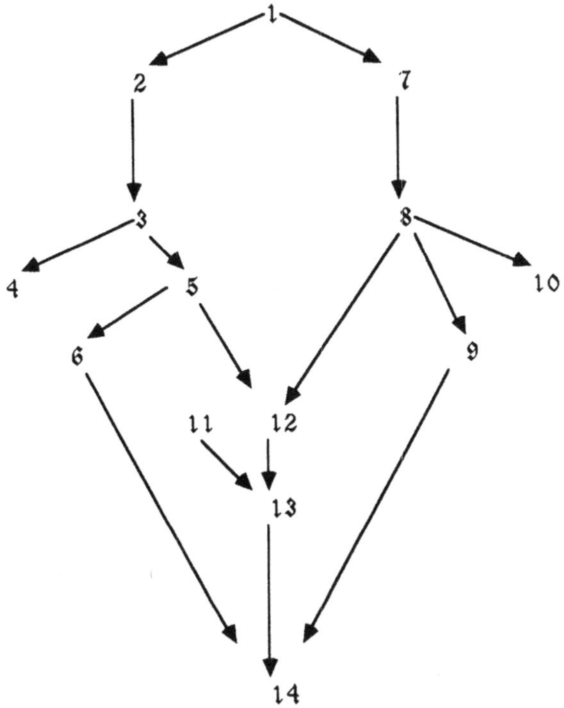

Introduction

This book is concerned with two areas of mathematics, at first sight disjoint, and with some of the analogies and interactions between them. These areas are the theory of linear differential equations in one complex variable with polynomial coefficients and the theory of one-parameter families of exponential sums over finite fields.

The simplest example of an exponential sum over a finite field is this: take a prime number p, a polynomial f(X) in $\mathbb{Z}[X]$, and form the sum

$$\sum_{x \text{ in } \mathbb{F}_p} \exp(2\pi i f(x)/p).$$

By letting the polynomial f(X) vary in a one-parameter family $f_t(X) \in \mathbb{Z}[t, X]$, e.g., the family $f_t(X) := tf(X)$, one is led to one-parameter families of exponential sums.

The above exponential **sum** is formally analogous to a complex path **integral**

$$\int e^{f(x)}dx.$$

And if we let the polynomial f(X) vary in a one parameter family $f_t(X) \in \mathbb{Z}[t, X]$, e.g., the family $f_t(X) := tf(X)$, the resulting integral, e.g.,

$$\int e^{tf(x)}dx,$$

formally satisfies a differential equation with respect to the variable t.

There are four basic questions about this concrete situation with which the book is concerned:

(1) For given p, how do the above exponential sums vary as t varies? Is there a "Sato-Tate law" which governs their distribution, and if so what is it? Thanks to Deligne, we know that under mild hypotheses there is such a Sato-Tate law, and that it is in turn governed by a certain complex semisimple algebraic group $G_{geom,p}$, the "geometric monodromy group" attached to our situation in characteristic p. So this question is essentially "What is $G_{geom,p}$?".

(2) How does the answer to (1) depend upon p? How does $G_{geom,p}$ depend upon p? In practice, one finds that whenever one can compute $G_{geom,p}$, its identity component, and often $G_{geom,p}$ itself, is independent of p >> 0. Can one prove this in general?

(3) What is the differential galois group G_{gal} of the the differential equation satisfied by the integral?

(4) What is the relation of G_{gal} to the common value of the groups $G_{geom,p}$ (or of their identity components) for $p >> 0$? It turns out in practice that when one can compute G_{gal}, and when G_{gal} turns out to be semisimple, then for $p >> 0$ the groups $G_{geom,p}$ are all equal to G_{gal}. Can one prove this in general?

There are of course more general sorts of exponential sums, and we will deal with them systematically in the course of the book. The reader should keep in mind that already the simple ones above illustrate the essential phenomena and contain the essential difficulties.

The book is arranged in four parts, in a diamond pattern of logical dependence

$$\begin{array}{ccc} & \text{Part I} & \\ \swarrow & & \searrow \\ \text{Part II} & & \text{Part III} \\ \searrow & & \swarrow \\ & \text{Part IV.} & \end{array}$$

Part I (Chapter 1): results from representation theory

Part II (Chapters 2,3,4,5,6): results about differential equations and their differential galois groups G_{gal}.

Part III (Chapters 7,8,9,10) : results about one-parameter families of exponential sums and their geometric monodromy groups G_{geom}.

Part IV (Chapters 11,12,13,14): comparison theorems relating G_{gal} and G_{geom} of suitably "corresponding" situations.

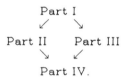

We have tried to strike a balance of emphasis between the underlying general theory and its application to concrete problems, in such a way that the two complement each other, with the applications serving both to illustrate and to motivate the general theory. The reader will judge how well we have succeeded.

Parts II and III especially are written in this spirit of "applied mathematics"; there is a strong emphasis on the **effective calculation** of the groups G_{gal} and G_{geom} respectively, when one is given a concrete differential equation, or a concrete one-parameter family of exponential sums.

The effective calculations in Parts II and III ultimately rely on the general representation-theoretic results of Part I. However, in order to be able to bring these results to bear, we make essential use of some recent developments in the theory of differential equations and in the theory of one-parameter families of exponential sums.

In the case of differential equations, there are four essential ingredients. The first is the theory of the "slopes", or "breaks", of a differential equation on an open curve at one of the points at infinity. The second is the general theory of holonomic $\mathcal{D}$-modules on curves, especially the theory of the "middle extension" and the structure theory of irreducible $\mathcal{D}$-modules. The third is the theory of the Fourier Transform of $\mathcal{D}$-modules on $\mathbb{A}^1$. The fourth is the idea of the (derived category) convolution of holonomic $\mathcal{D}$-modules on both the additive group $\mathbb{A}^1$ and on the multiplicative group $\mathbb{G}_m$. [There is also a natural notion of convolution of holonomic $\mathcal{D}$-modules on elliptic curves, which seems well worth exploring.]

What happens in the case of one-parameter families of exponential sums? Roughly speaking, studying a "one parameter family" means studying a lisse ℓ-adic sheaf on an open curve over a field of postitve characteristc $p \neq \ell$. In this case as well there are four essential ingredients, which are closely analogous to the $\mathcal{D}$-module ingredients discussed above. The first is the the theory of "breaks" (in the sense of the upper-numbering filtration) of ℓ-adic representations of inertia groups at the points at infinity. This theory was in fact the inspiration for its D.E. namesake. The second is the theory of perverse ℓ-adic sheaves on curves, especially the structure theory of irreducibles. This theory is analogous to the theory of holonomic $\mathcal{D}$-modules on curves over $\mathbb{C}$. The third is the theory of the ℓ-adic Fourier Transform for perverse sheaves on $\mathbb{A}^1$ over a field of positive characteristic $p \neq \ell$. In the ℓ-adic case, we have much more precise information about Fourier Transform than we do in the $\mathcal{D}$-module case, thanks to Laumon's "principle of stationary phase", which effectively determines the local monodromy of a Fourier Transform. [On the other hand, in the $\mathcal{D}$-module case we can "write down" the Fourier Transform of a $\mathcal{D}$-module (just interchange x and d/dx and change a sign) and then stare at it.] The fourth is the idea of convolution, entirely analogous to what is done in the $\mathcal{D}$-module context (and the inspiration for it).

Thus it is not surprising that there are remarkable similarities between the results obtained in Part II and those obtained in Part III. One of our guiding principles was to illustrate systematically these similarities by working out as completely as possible the theory of the generalized hypergeometric differential equation in Part II, and developing in Part III the analogous theory of ℓ-adic hypergeometric sheaves in characteristic p. Roughly speaking, we show that the differential galois group G_{gal} for a hypergeometric differential equation is essentially "the same" as the geometric monodromy group G_{geom} for a "corresponding" ℓ-adic hypergeometric sheaf in sufficiently large characteristic p. We also give, in both the $\mathcal{D}$-module and ℓ-adic contexts, intrinsic characterizations of irreducible hypergeometrics among all holonomic $\mathcal{D}$-modules (resp. perverse sheaves) on $\mathbb{G}_m$. [We also note that the "hypergeometric sums" which are the traces of Frobenius at closed points of our ℓ-adic hypergeometric sheaves are already known in combinatorics (cf. [Gre]) as "hypergeometric functions over finite fields".]

What is the "explanation" for these remarkable similarities? Is there some general theory of "exponential sums over $\mathbb{Z}$" which explains them? To the extent that most of the concrete calculations of both G_{gal} and of G_{geom} depend in analogous ways upon the same representation-theoretic results of Part I, one might think that these similarities are simply an instance of the anthropic fallacy, and that the true similarity between the $\mathcal{D}$-module and ℓ-adic cases we consider is that they are the only ones where we can compute what is going on. However, the results of Part IV show that this is not the case; they give an intrinsic conceptual explanation for at least some of these similarities.

The main result of Part IV is a comparison theorem which compares, for an object which lives on $\mathbb{A}^1$ over $\mathbb{Z}$ in a suitably strong sense, the differential galois group G_{gal} of the $\mathcal{D}$-module Fourier Transform of its $\mathbb{C}$-fibre to the geometric monodromy group G_{geom} of the ℓ-adic Fourier Transforms of its $\mathbb{F}_p$-fibres, as p runs over the primes. In contrast to Parts II and III, where we prove statements of the form "X = Y" by explicitly computing both X and Y and noting that they are equal, the main result of Part IV is of the form "X = Y", but it is proven by proving, as it were, that "X - Y = 0", **without** evaluating either X or Y.

In their most concrete instance, the results of Part IV provide affirmative answers to the questions (2) and (4) raised at the beginning of the introduction in the special case of the family $f_t(X) := tf(X)$. But already the same questions (2) and (4) for the case of a general one-parameter family of polynomial functions on $\mathbb{A}^1$ remain beyond the scope of our present knowledge.

We believe that the results of Part IV are in fact just the "tip of the iceberg" of a general theory, as yet inexistant, of "exponential sums over $\mathbb{Z}$". We refer the interested reader to [Ka-ES, last few pages] for some entirely speculative discussion of the possible categorical setting for such a theory, and for a conjectural statement of a general comparison theorem which would provide affirmative answers to the questions (2) and (4) for the case of a general one-parameter family.

We now turn to a brief discussion of some of the other open problems suggested by the results in this book.

That one obtains the classical groups in their standard representations both as G_{gal}'s of simple-to-write-down differential equations, and as G_{geom}'s for easy-to-write-down corresponding families of exponential sums in all characteristics $p >> 0$ is perhaps not surprising. That one obtains G_2 in its seven dimensional irreducible representation in similarly concrete fashion is perhaps less expected. Can one obtain all the exceptional groups so concretely? Can one obtain interesting finite simple groups this way?

The results of Chapter 6 show that a number of explicit differential equations on $\mathbb{A}^1$ have "diophantine meaning", including those of the form

$$(\partial)^n - x^m, \quad \partial := d/dx.$$

But what if any is the diophantine meaning of equations of the general form

$$P_n(\partial) - Q_m(x),$$

where P_n and Q_m are constant coefficient polynomials in one variable?

Similary, what if any is the the diophantine meaning of differential equations on an elliptic curve $\mathbb{C}/L$ of the form

$$(d/dz)^n - \wp^{(m)}(z; L),$$

where d/dz is the invariant derivation, and where $\wp^{(m)}(z; L)$ is the m'th derivative of the Weierstrass $\wp$-function $\wp(z; L)$? And what is the differential galois group of such an equation? Same questions for equations of the more general form

$$P_n(d/dz) - (\text{an elliptic function with poles only in } L),$$

where P_n is a constant coefficient polynomial in one variable.

Our results on G_{geom} for ℓ-adic hypergeometric sheaves in characteristic p were only valid for large p. For small p, can G_{geom} be an "interesting" group? In any case, what is it?

The ℓ-adic hypergeometrics we construct in characteristic p live naturally on $\mathbb{G}_m$. Up to a multiplicative translation and a shift, they are precisely those of the form $R\varphi_!\mathcal{F}$, for some integer $n \geq 0$, some homomorphism of tori

$$\varphi : (\mathbb{G}_m)^n \to \mathbb{G}_m,$$

and some lisse rank one sheaf $\mathcal{F}$ on the source $(\mathbb{G}_m)^n$ of the form

$$\mathcal{F} := \mathcal{L}_{\psi(\Sigma a_i x_i)} \otimes (\otimes_i \mathcal{L}_{\chi_i(x_i)}).$$

There is an obvious generalization of this sort of object on $\mathbb{G}_m$ to a class of "hypergeometric objects" on tori T of arbitrary dimension. One considers $R\varphi_!\mathcal{F}$, for some integer $n \geq 0$, some homomorphism of tori

$$\varphi : (\mathbb{G}_m)^n \to T,$$

and $\mathcal{F}$ on the source as above. This class of "hypergeometrics" on tori is stable by ! convolution, by external product, and by ! direct image by homomorphisms of tori, and for $\mathbb{G}_m$ gives back the original notion. One can also pursue the obvious holonomic $\mathcal{D}$-module analogue of this generalization, obtaining hypergeometric holonomic $\mathcal{D}$-modules on tori of arbitrary dimension over $\mathbb{C}$. This notion of hypergeometric in several variables can be viewed as an algebraic incarnation of the classical definitions [Er, 5.8] of hypergeometric functions of several variables as inverse Mellin Transforms of monomials in Γ-functions. What is the relation between this point of view and the current work of the Gelfand school [Gel] on the general theory of hypergeometric functions?

My interest in the generalized hypergeometric differential equation as a beautiful "test case" for the study and calculation of differential galois groups was aroused by the paper [B-B-H] of Beukers, Brownawell, and Heckman. Many of the results of the book were worked out with the collaboration of Ofer Gabber, whose contribution cannot be overestimated. It was Bill Messing who first posed the crucial question "If the groups $G_{geom,p}$ are independent of p >> 0, what is an a priori description of the group to which they are all equal?".

It is a pleasure to acknowledge the support of the John Simon Guggenheim Memorial Foundation, the I.H.E.S., the University of Paris at Orsay, the National Science Foundation, and Princeton University during the writing of this book. It is also a pleasure to thank Benji Fisher for his meticulous proofreading of the entire manuscript, and his many helpful comments, corrections, and suggestions.

I respectfully dedicate this book to my teacher Bernard Dwork, who discovered the intimate relations between the p-adic theory of classical differential equations and the p-adic variation with parameters of zeta and L-functions. In particular, he was the first person to understand (cf [Dw]) that classical differential equations with **irregular** singularities had deep meaning in arithmetic algebraic geometry. (The prevailing dogma held that only equations with regular singular points should have meaning.) Indeed, he showed that in many cases the p-adic variation with parameters of exponential sums was controlled by the p-adic theory of precisely the differential equations with irregular singularities whose G_{gal}'s play such a crucial role in this book.

CHAPTER 1
Results from Representation Theory

Throughout this chapter, we work over an algebraically closed field $\mathbb{C}$ of characteristic zero. We fix an integer $n \geq 2$, and denote by V an n-dimensional $\mathbb{C}$-vector space. We suppose given a Lie-subalgebra $\mathcal{G}$ of End(V) which is semisimple and which acts irreducibly on V. We first give a fundamental "torus trick" (Theorem 1.0) of Ofer Gabber which is extremely useful in diverse contexts. We then give a sequence of criteria (Theorems 1.1-1.6), some classical and some new, which insure that $\mathcal{G}$ is either $\mathcal{SL}(V)$ or $\mathcal{SO}(V)$ or (for dimV even) $\mathcal{SP}(V)$ **or** that dimV is 7,8, or 9 (and we give the list of possible $\mathcal{G}$'s for these). The basic tool in most of these proofs is that of "chains" of weights in representations; I am indebted to Ofer Gabber for having explained to me both this method and most of the criteria discussed below.

Theorem 1.0 (Gabber). Let $\mathcal{G}$ be a semisimple Lie-subalgebra of End(V) which acts irreducibly on V. Suppose that a diagonal subgroup K of GL(V) normalizes $\mathcal{G}$. Let $\chi_1, \ldots, \chi_n$ be the n characters of K defined by the diagonal matrix coefficients; i.e., $k = \mathrm{Diag}(\chi_1(k), \ldots, \chi_1(k))$ for k in K. Consider the "torus" $\mathcal{T}$ in End(V) consisting of those diagonal matrices $\mathrm{Diag}(X_1, \ldots, X_n)$ whose entries satisfy the conditions

$$\Sigma X_i = 0$$

$$X_i - X_j = X_k - X_m \text{ whenever } \chi_i/\chi_j = \chi_k/\chi_m \text{ on K.}$$

Then $\mathcal{T}$ lies in $\mathcal{G}$.

proof Consider the action of K on End(V) by conjugation. By assumption $\mathcal{G}$ is stable under this action. For any character ρ of K, the ρ-eigenspaces of $\mathcal{G}$ and of End(V) are related by $\mathcal{G}(\rho) = \mathcal{G} \cap \mathrm{End}(V)(\rho)$. Because K is diagonal, we have

$$\mathcal{G} = \oplus_\rho \mathcal{G}(\rho), \quad \mathrm{End}(V) = \oplus_\rho \mathrm{End}(V)(\rho).$$

Now in End(V), the line $E_{i,j}$ of matrices whose only possibly nonzero entry is in the (i,j) place transforms under K by χ_i/χ_j. Therefore $\mathrm{End}(V)(\rho) = 0$ unless $\rho = \chi_i/\chi_j$ for some (i,j), and for such ρ we have

$$\mathrm{End}(V)(\rho) = \oplus_{i,j \text{ such that } \rho = \chi_i/\chi_j} E_{i,j}.$$

Now consider a diagonal matrix $X = \mathrm{Diag}(X_1, \ldots, X_n)$. Acting on End(V), ad(X) stabilizes each line $E_{i,j}$, and multiplies it by $X_i - X_j$. So if X satisfies the condition

$$X_i - X_j = X_k - X_m \text{ whenever } \chi_i/\chi_j = \chi_k/\chi_m \text{ on K}$$

then for any ρ with $\mathrm{End}(V)(\rho) \neq 0$, ad(X) acts on $\mathrm{End}(V)(\rho)$ by a scalar X_ρ (namely $X_\rho := X_i - X_j$ for any (i,j) with $\rho = \chi_i/\chi_j$). Therefore ad(X) maps $\mathcal{G}(\rho)$ to itself (since ad(X) maps **every** subspace of End(V)(ρ) to itself), and hence ad(X) maps $\mathcal{G}$ to itself. Because ad(X) is a derivation of End(V), it must be a derivation of the subalgebra $\mathcal{G}$. Because $\mathcal{G}$ is semisimple, every derivation of $\mathcal{G}$ is inner, so there exists an element Y in $\mathcal{G}$ such that, on $\mathcal{G}$, ad(X) = ad(Y); this means precisely that X-Y in End(V) is an element which commutes with $\mathcal{G}$. Since $\mathcal{G}$ acts irreducibly on V, X - Y is necessarily a scalar. This scalar is necessarily (1/n)trace(X-Y). Because Y is in the semisimple $\mathcal{G}$, trace(Y) = 0. So if trace(X) = 0 in addition, then X - Y = 0, whence X lies in $\mathcal{G}$. QED

   In the following discussion of "recognition criteria" for the standard representations of the classical groups, we will sometimes abbreviate as "std" the standard representation of $\mathcal{SL}$(V), $\mathcal{SP}$(V), $\mathcal{SO}$(V) on V.

Theorem 1.1 (Kostant [Kos]) Let $\mathcal{G}$ be a semisimple Lie-subalgebra of End(V) which acts irreducibly on V. Suppose that with respect to some basis of V, $\mathcal{G}$ contains the diagonal matrix h := Diag(n-1, -1,..., -1). Then $\mathcal{G}$ is $\mathcal{SL}$(V).

Theorem 1.2 (Kostant [Kos], Zarhin [Za-WS, A.2.1]) Let $\mathcal{G}$ be a semisimple Lie-subalgebra of End(V) which acts irreducibly on V. Suppose that with respect to some basis of V, $\mathcal{G}$ contains the diagonal matrix h := Diag(1, 0, ..., 0, -1). Then $\mathcal{G}$ is either $\mathcal{SL}$(V) or $\mathcal{SO}$(V) or (for dimV even) $\mathcal{SP}$(V).

Theorem 1.3 (Gabber) Let $\mathcal{G}$ be a semisimple Lie-subalgebra of End(V) which acts irreducibly on V. Suppose that with respect to some basis of V, $\mathcal{G}$ contains the "G$_2$ torus" consisting of all diagonal matrices of the form
$$h(x,y) := \quad \text{Diag}(x+y, x, y, 0,...,0, -y, -x, -x-y).$$
Then $\mathcal{G}$ is either $\mathcal{SL}$(V) or $\mathcal{SO}$(V) or (for dimV even) $\mathcal{SP}$(V) or we have one of the following exceptional cases:

   n=7: $\mathcal{G}$ = Lie(G$_2$) in the 7-dim'l representation of G$_2$
   n=8: $\mathcal{G}$ = Lie(SO(7)) in the 8-dim'l spin representation
         $\mathcal{G}$ = Lie(SL(3)) in the adjoint representation
         $\mathcal{G}$ = Lie(SL(2)×SL(2)×SL(2)) in std⊗std⊗std
         $\mathcal{G}$ = Lie(SL(2)×Sp(4)) in std⊗std
         $\mathcal{G}$ = Lie(SL(2)×SL(4)) in std⊗std
   n=9: $\mathcal{G}$ = Lie(SL(3)×SL(3)) in std⊗std.

In fact, one uses the full strength of having a G_2 torus only to take care of the non-simple cases; the simple case is handled by

Theorem 1.4 (Gabber) Let $\mathfrak{g}$ be a semisimple Lie-subalgebra of End(V) which acts irreducibly on V. Suppose that $\mathfrak{g}$ is simple and that with respect to some basis of V, $\mathfrak{g}$ contains the diagonal matrix $h := \mathrm{Diag}(1,1,0,\ldots,0,-1,-1)$. Then $\mathfrak{g}$ is either $\mathfrak{sl}(V)$ or $\mathfrak{so}(V)$ or (for dimV even) $\mathfrak{sp}(V)$, or we have one of the following exceptional cases :

$n=7$: $\mathfrak{g}$ = Lie(G_2) in the 7-dim'l representation of G_2
$n=8$: $\mathfrak{g}$ = Lie(SO(7)) in the 8-dim'l spin representation
$\mathfrak{g}$ = Lie(SL(3)) in the adjoint representation.

Remark 1.4.1 If in Theorem 1.4 we no longer assume that $\mathfrak{g}$ is simple, the non-simple possibilities are the image of $\mathfrak{sl}(3)\times\mathfrak{sl}(3)$ in $\mathrm{std}_3\otimes\mathrm{std}_3$, and, for every $k\geq 2$, the image of

$\mathfrak{sl}(2)\times(\mathfrak{sl}(k)$ or $\mathfrak{so}(k)$ or (for k even)$\mathfrak{sp}(k))$ in $\mathrm{std}_2\otimes\mathrm{std}_k$.

Remark 1.4.2 In the simple case, there is an asymptotic result [Za-LS,Thm. 6] of Zarhin which gives Theorem 1.4 as soon as the dimension of the representation is sufficiently large. Zarhin proves that if $\mathfrak{g}$ is a simple irreducible subalgebra of End(V) and if there exists a semisimple element h in $\mathfrak{g}$ which has dim(h(V)) = d, then $\mathfrak{g}$ is $\mathfrak{sl}$, $\mathfrak{sp}$ or $\mathfrak{so}$ provided that $\dim V > 72d^2$. This given Theorem 1.4 (whose h has d= 4) as soon as $\dim V > 72\times 16$.

Theorem 1.5 (Kazhdan-Margulis, Gabber, Beukers-Heckman [B-H]) Let $\mathfrak{g}$ be a semisimple Lie-subalgebra of End(V) which acts irreducibly on V. Suppose that $\mathfrak{g}$ is normalized by a pseudo-reflection γ in GL(V). Then $\mathfrak{g}$ is either $\mathfrak{sl}(V)$ or $\mathfrak{so}(V)$ or (for dimV even) $\mathfrak{sp}(V)$. Moreover,

if $\det\gamma \neq \pm 1$, then $\mathfrak{g} = \mathfrak{sl}(V)$;
if $\det\gamma = +1$, then $\mathfrak{g} = \mathfrak{sl}(V)$ or (for dimV even) $\mathfrak{sp}(V)$;
if $\det\gamma = -1$, then $\mathfrak{g} = \mathfrak{sl}(V)$ or $\mathfrak{so}(V)$.

Theorem 1.6 (Gabber) Let $\mathfrak{g}$ be a semisimple Lie-subalgebra of End(V) which acts irreducibly on V. Suppose that dimV is a prime p. Then $\mathfrak{g}$ is either $\mathfrak{sl}(2)$ in $\mathrm{Sym}^{p-1}(\mathrm{std})$, or $\mathfrak{sl}(V)$ or $\mathfrak{so}(V)$ or, if n=7, possibly Lie(G_2) in the seven-dimensional irreducible representation of G_2.

1.7 The proofs

Notice that in Theorems 1.1, 1.2, 1.3, 1.4, 1.5, 1.6 the cases listed can be easily checked to have the property in question; the problem is to show that these are the **only** cases. In doing this we will make explicit use of classification, via the Bourbaki tables. Notice also that the " $\mathfrak{g}$ simple" case of Theorem 1.3 is a trivial consequence of Theorem 1.4.

Before beginning the proofs of Theorems 1.1 to 1.6, we need to review some basic facts, which are certainly well-known to the experts but for which we do not know a convenient explicit reference.

Lemma 1.7.1 Let $\mathfrak{g}$ be a semisimple Lie subalgebra of End(V), $\mathcal{H}$ a Cartan subalgebra of $\mathfrak{g}$, and h in $\mathcal{H}$ an element with rational eigenvalues. Then there exists a Weyl chamber such that for the corresponding notion of positive root, we have $\alpha(h) \geq 0$ for any positive root α.

proof The $\mathbb{Q}$-dual of the $\mathbb{Q}$-span of the roots is a $\mathbb{Q}$-form $\mathcal{H}_{\mathbb{Q}}$ of $\mathcal{H}$, and h lies in $\mathcal{H}_{\mathbb{Q}}$. Picking a vector-space basis of $\mathcal{H}_{\mathbb{Q}}$ which starts with h, we get a lexicographic order on the $\mathbb{Q}$-span of the roots, so a notion of positive root, such that if α is a positive root, then $\alpha(h) \geq 0$. One knows that this notion of positive root is the one associated to some Weyl chamber ([Bour L6], 1, 7, Cor 2 of Prop 20). QED

Suppose that $(\mathfrak{g}, \mathcal{H})$ is a split semisimple Lie algebra over $\mathbb{C}$, and that we have chosen a Weyl chamber. Thus we may speak of positive roots, simple roots, et cetera. We denote by w_0 the unique element of the Weyl group which interchanges positive and negative roots. Recall the notion of a "chain" between two weights μ and λ of a finite-dimensional representation M of $\mathfrak{g}$; this is a sequence of weights of M starting with μ, ending at λ, such that each successive weight in the sequence after the first is obtained from the previous one by subtracting a simple root. We will make essential use of the fact that

Lemma 1.7.2 Any two weights μ and λ of M such that $\mu - \lambda$ is a nontrivial sum $\Sigma n_\alpha \alpha$ of simple roots with integral coefficients $n_\alpha \geq 0$ can be joined by a chain.

proof The proof is by induction on Σn_α, the case $\Sigma n_\alpha = 1$ being trivial. In terms of a W-invariant inner product on the $\mathbb{Q}$-span of the roots, we write

$$0 < (\mu-\lambda, \mu-\lambda) = (\mu-\lambda, \Sigma n_\alpha \alpha) = \Sigma n_\alpha(\mu-\lambda,\alpha).$$

Since all the $n_\alpha \geq 0$ and at least one is >0, we see that for some simple α we must have $(\mu-\lambda,\alpha)>0$, i.e., $(\mu,\alpha) > (\lambda,\alpha)$. Therefore either $(\mu,\alpha) > 0$ or $0 > (\lambda,\alpha)$. Now for any W-invariant inner product, we have, for any root α and any σ in the $\mathbb{Q}$-span of the roots

$$2(\sigma,\alpha)/(\alpha,\alpha) = \sigma(H_\alpha).$$

Thus we have that either $\mu(H_\alpha)>0$ or $\lambda(H_\alpha)<0$. In the first case $\mu-\alpha$ is also a weight, and in the second case $\lambda + \alpha$ is also a weight ([Bour L8], 7, 2, Prop 3(i), page 124). Now by induction on Σn_α we can pass from $\mu-\alpha$ to λ, or from μ to $\lambda+\alpha$. QED

Corollary 1.7.2.1 Let M be an irreducible representation of $\mathfrak{g}$. There exists a chain going from the highest weight to any given weight of M, and there exists a chain from any given weight to the lowest weight.

Given an irreducible representation M of $\mathfrak{g}$, denote by λ and ν its highest and lowest weights respectively. For any chain $\lambda=\lambda_1, \lambda_2,\ldots,\lambda_n=\nu$ from λ to ν, with successive drops the simple roots $\alpha_i := \lambda_i - \lambda_{i+1}$ for i = 1,..,n-1, we have

$$\lambda - \nu = \Sigma_{i=1,\ldots,n-1}\, \alpha_i = \Sigma n_\alpha \alpha.$$

This shows that the multiplicity n_α with which a given simple root α occurs as drops in such a chain is independent of the chain. Moreover, the length of any such chain is $1 + \Sigma n_\alpha$.

Lemma 1.7.3 For a faithful irreducible representation M of $\mathfrak{g}$, every simple root occurs as a drop in every chain from highest weight to lowest.

proof If M is faithful, then every root, and in particular every simple root β, is a difference of two weights of M , say σ and τ: $\sigma = \tau - \beta$. So if we take a chain from λ to τ and concatenate to it a chain from σ to ν we get a chain from λ to ν in which β occurs as a drop; therefore $n_\beta > 0$ and hence β occurs as a drop in **every** chain from λ to ν. QED

Using ([Bour L8], 7, 2, Prop 3(i), page 124), one sees

Lemma 1.7.4 For a fundamental representation M of $\mathfrak{g}$, with highest weight ω_α, **either**

(1) α is an isolated point of the Dynkin diagram, the representation is

the standard representation of the corresponding $\mathcal{SL}(2)$ factor, and the only chain from highest weight to lowest is ω_α, $\omega_\alpha - \alpha$.

or

(2) α is not isolated, every chain from highest weight to lowest begins

$$\omega_\alpha, \quad \omega_\alpha - \alpha, \quad \omega_\alpha - \alpha - \beta,$$

where β is any simple root adjacent to α in the Dynkin diagram, and every chain ends

$$\gamma - w_0(\alpha) + w_0(\omega_\alpha), \quad -w_0(\alpha) + w_0(\omega_\alpha), \quad w_0(\omega_\alpha),$$

where γ is any simple root adjacent to $-w_0(\alpha)$ in the Dynkin diagram.

(1.7.5) We now take up the proofs of Theorems 1.1, 1.2, 1.3, 1.4. We will consider successively the three cases

$\mathcal{G}$ non-simple

$\mathcal{G}$ simple but V not fundamental

$\mathcal{G}$ simple and V fundamental.

We begin by considering the non-simple case. In Theorems 1.1, 1.2, we are given a nonzero diagonal element h which has an eigenvalue of multiplicity one on V. Therefore if $V = V_1 \otimes V_2$ with each factor of dimension at least two, then V_1 cannot be the trivial representation of h (otherwise all its weights in V would occur with multiplicity divisible by $\dim V_1$). By symmetry, h is also non-trivial on V_2.

In Theorem 1.3, we are given commuting diagonal elements h(1,0) and h(0,1), each of which has an eigenvalue of multiplicity two. Therefore if $V = V_1 \otimes V_2$ and $\dim V_1 \geq 3$, then each of these h's is nontrivial on V_1 (for otherwise all its weights in V would occur with multiplicity divisible by $\dim V_1$). If $\dim V_1 = 2$, then the sum h(1,0) + h(0,1) is nontrivial on V_1 (since it has an eigenvalue with multiplicity one) and hence at least one of the h's is nontrivial on V_1. Since $\dim V_2 = \dim V/2 \geq 3$, both h's are nontrivial on V_2. Therefore at least one of the two elements h(1,0) or h(0,1) must be nontrivial on both V_1 and on V_2.

Thus if $V = V_1 \otimes V_2$ in Theorems 1.1, 1.2, or 1.3, we have a diagonal element h of $\mathcal{G}$ which acts nontrivially on both V_1 and on V_2. This h is $X \otimes 1 + 1 \otimes Y$ with both X and Y non-trivial; both X and Y have at least two distinct eigenvalues (because they are semisimple and their traces are zero). Let $\mathcal{X}$ resp. $\mathcal{Y}$ denote the finite subset of $\mathbb{Q}$ consisting of the distinct eigenvalues of X resp. Y. Then the set of distinct eigenvalues of $h = X \otimes 1 + 1 \otimes Y$ is the set $\mathcal{X} + \mathcal{Y} := \{ x+y \mid x \text{ in } \mathcal{X}, y \text{ in } \mathcal{Y}\}$. Then by the $\mathrm{Card}(\mathcal{X} + \mathcal{Y}) \geq \mathrm{Card}\mathcal{X} + \mathrm{Card}\mathcal{Y} - 1$ inequality (valid in any $\mathbb{Q}$-vector space), and the fact that $\mathrm{Card}\mathcal{X}$ and $\mathrm{Card}\mathcal{Y}$ are both ≥ 2, we arrive at a contradiction in Theorem 1.1 (where h has only two distinct eigenvalues) and in the other cases we see that X and Y each have exactly two distinct eigenvalues. Thus we have $X = \mathrm{Diag}(a,\dots,a,b,\dots,b)$ and $Y = \mathrm{Diag}(c,\dots,c,d,\dots,d)$; with $a>b$, a with multiplicity A, b with multiplicity B and $c>d$, c with multiplicity C, d with multiplicity D.

To discuss Theorems 1.2 and 1.3 simultaneously, we introduce the parameter k such that the nonzero eigenvalues of $h := X \otimes 1 + 1 \otimes Y$ are 1 and -1, each with multiplicity k. Thus k=1 is Theorem 1.2, and k=2 is Theorem 1.3. The highest weight in $X \otimes 1 + 1 \otimes Y$ is a+c, hence AC=k. Similarly we have BD=k. Thus $A \leq k$ with equality iff C=1, and $B \leq k$ with equality iff D=1. Thus $\dim V_1 = A + B \leq 2k$, with equality iff $\dim V_2 = C+D = 2$. So either we have (2 dim'l)$\otimes$(2k dim'l) or both V_i have dimension in [3, 2k-1]. For k=1 this means (2dim)$\otimes$(2dim), so the standard representation of $\mathcal{SO}(4) \approx \mathcal{SL}(2) \times \mathcal{SL}(2)$. For k=2, this means that either we have (2 dim, Diag(1/2,-1/2))$\otimes$(4 dim, Diag(1/2,1/2,-1/2,-1/2)) or (3dim)$\otimes$(3dim). This last case can only be $\mathcal{SL}(3) \times \mathcal{SL}(3)$ in (std or its dual)$\otimes$(std. or its dual). In the penultimate case, the only possibility for the four dimensional faithful representation is one of the three classical groups $\mathcal{SL}(4)$, $\mathcal{SP}(4)$, or $\mathcal{SO}(4)$ in std. or its dual (in the representation $\mathrm{Symm}^3(\mathrm{std}_2)$ of $\mathcal{SL}(2)$, the element Diag(1,1,-1,-1) is not in the image of $\mathcal{SL}(2)$).

We now turn to Theorems 1.1, 1.2, 1.4 in the case when $\mathfrak{g}$ is simple, but where V is not fundamental. We pick a maximal torus containing the given element h and a Weyl chamber such that $\alpha(h) \geq 0$ for all positive roots α.

In Theorem 1.1 this case does not arise (V is automatically fundamental if h has only two distinct eigenvalues). In Theorems 1.2, 1.4, where the element h has exactly three eigenvalues, if V is not fundamental then its highest weight is a sum of two fundamental weights, in both of whose representations h has exactly two eigenvalues. [For suppose that the highest weight of V is the sum of two dominant weights λ and λ'. If $\mathcal{H}$ and $\mathcal{H}'$ are the weights occurring in V_λ and in $V_{\lambda'}$ respectively, then the set of weights of $V_{\lambda+\lambda'}$ is $\mathcal{H} + \mathcal{H}'$ ([Bour L8], 7, 4, Prop. 10), hence the same is true of the sets of h-weights. Denote by $\mathcal{X} := h(\mathcal{H})$, $\mathcal{Y} := h(\mathcal{H}')$ these sets of h-weights, and apply the inequality

$$\mathrm{Card}\,(\mathcal{X} + \mathcal{Y}) + 1 \geq \mathrm{Card}(\mathcal{X}) + \mathrm{Card}(\mathcal{Y}).$$

Because h is diagonal $\neq 0$ and the representations are faithful ($\mathfrak{g}$ being simple), we have Card($\mathcal{X}$), Card($\mathcal{Y}$) both ≥ 2. If h has only two distinct eigenvalues on V, then Card ($\mathcal{X}$ + $\mathcal{Y}$) = 2, and we have a contradiction, whence V was fundamental. If h has three distinct eigenvalues on V, then Card ($\mathcal{X}$ + $\mathcal{Y}$) = 3, and hence Card($\mathcal{X}$) = Card($\mathcal{Y}$) = 2, whence by the preceding argument both λ and λ' must be fundamental weights.]

We now use the additional fact that if if $\mathcal{W}_\lambda$ denotes the **set of weights with multiplicity** of V_λ then $\lambda' + \mathcal{W}_\lambda$ occurs in $\mathcal{W}_{\lambda+\lambda'}$ ([Bour L8], 7, exc. 21). Passing to the opposite (w_0) Weyl chamber, if we denote by $\lambda''$the **lowest** weight in $V_{\lambda'}$, it follows that $\lambda'' + \mathcal{W}_\lambda$ occurs in $\mathcal{W}_{\lambda+\lambda'}$. (In fact, the argument suggested in the Bourbaki exercise shows that if ξ is **any** weight of $V_{\lambda'}$, then $\xi + \mathcal{W}_\lambda$ occurs in $\mathcal{W}_{\lambda+\lambda'}$.) Using this, we can bound the pairs of fundamentals λ and λ' for which an h in $\mathfrak{g}$ acts on $V_{\lambda+\lambda'}$ as

Diag(1,...,1 repeated k times,..some 0's.., -1,...,-1 repeated k times).

Let X and Y be the diagonal matrices of trace zero with exactly two eigenvalues (necessarily rational) by which h acts on V_λ and on $V_{\lambda'}$ respectively; say X = (a rep A times, b rep B times) with $a>0>b$, and Y = (c rep C times, d rep D times) with $c>0>d$. The highest weight space of $V_{\lambda'}$ must have h-weight c (rather than d). Because

(h-weight a on V_λ)$\otimes$(highest wt λ')

appears weightwise **with multiplicity** in $V_{\lambda+\lambda'}$, we conclude that the h-weight a+c occurs with multiplicity $\geq$ A in $V_{\lambda+\lambda'}$, whence $A \leq k$. We claim that in fact $A \leq k-1$ if $C \geq 2$. For if $C \geq 2$, there is a second h-positive weight in $V_{\lambda'}$, and taking a chain from λ' to it we find an h-positive weight of $V_{\lambda'}$ of the form $\lambda' - \mu$ for μ a simple root. Now use the fact that

$$(\text{any wt of } V_\lambda) + (\text{any wt of } V_{\lambda'})$$

occurs as a weight in $V_{\lambda+\lambda'}$. We claim that among the weights

$$(\text{a wt where } h=a \text{ on } V_\lambda) + (\lambda' - \mu),$$

there is at least one which is **not** of the form

$$(\text{a wt where } h=a \text{ on } V_\lambda) + \lambda',$$

for if this were not the case then the set of weights occuring in (h=a on V_λ) would be closed under subtracting μ, which is absurd. Thus we have

$A \leq k$, and $A \leq k-1$ if $C > 1$.

Similarly arguing with the lowest weight, we obtain

$B \le k$, and $B \le k-1$ if $D > 1$.

Putting this together, we see that $A+B \le 2k$, with equality if and only if $C=D=1$. Thus either (i) one of our fundamentals is two-dimensional and the other has dimension 2k, or (ii)both have dimension $3 \le \dim \le 2k-1$.

For k=1, only the first case is possible, and it is the case of $\mathfrak{so}(3)$ in its standard representation.

For k=2, neither of the fundamentals has dimension 2 ($\mathfrak{g}$ would be $\mathfrak{sl}(2)$, which doesn't have a fundamental of dimension 2k for $k \ge 2$) so each of the two fundamentals is three-dimensional, each with a Diag(x,x,y) element, so by a trivial case of Kostant's theorem each is a fundamental representation of $\mathfrak{sl}(3)$. So the two possibilities are Symm^2(std. rep. or its dual of $\mathfrak{sl}(3)$), which by inspection does not contain a G_2 torus, or even a Diag(1,1,0,0,-1,-1), and the adjoint representation of $\mathfrak{sl}(3)$, which does by inspection contain a G_2 torus.

We now turn to the case when $\mathfrak{g}$ is simple and V is fundamental. Again we pick a Cartan subalgebra containing h and a Weyl chamber such that if β is a positive root, then $\beta(h) \ge 0$. We denote by w_0 the unique element of the Weyl group which interchanges positive and negative roots. Because V is fundamental, its highest weight λ is ω_α for some simple root α, and its lowest weight ν is $w_0(\omega_\alpha) = -\omega_{-w_0(\alpha)}$. By our choice of Weyl chamber we have

$$\lambda(h) \ \ge \ \text{any h-eigenvalue on V} \ \ge \ \nu(h),$$

so

$$\lambda(h) = \text{the highest h-weight} = n-1 \text{ in Thm 1}, =1 \text{ in Thm's 2,4},$$
$$\nu(h) = \text{the lowest h-weight} = -1.$$

Pick a chain of weights $\lambda = \lambda_1, \lambda_2, \ldots, \lambda_d = \nu$ from the highest to the lowest, and consider the sequence $\lambda_i(h)$ of h-weights. This is a non-increasing sequence of integers, whose successive drops are the integers $\alpha_i(h)$, where $\alpha_i := \lambda_i - \lambda_{i+1}$.

In Theorem 1.1, this sequence is n-1,-1,...,-1, and hence $\alpha(h)=n$ (from the first step ω_α to $\omega_\alpha - \alpha$); after this first drop the h-weight stays -1, so $\beta(h) = 0$ for β simple, $\beta \ne \alpha$. So in Theorem 1 the situation is

$$\boldsymbol{\omega_\alpha - w_0\omega_\alpha = \alpha + \textbf{a sum of the other basic roots};}$$

the only case is (A_n, std or its dual) as may be checked from the Planches in [Bour L6]. This concludes the proof of Theorem 1.1.

In Theorem 1.2, either $\mathfrak{g}$ is $\mathfrak{sl}(2)$ or every chain from top to bottom has at least three terms ($\ge 1 +$ rank, since every simple root occurs) so has h-weights 1,0,...,0,-1. Looking at the beginning and ending steps we see

$$\omega_\alpha - w_0\omega_\alpha = \alpha + (-w_0)\alpha + \text{other basic roots.}$$

If $\alpha \neq (-w_0)\alpha$, then this is the Theorem 1.1 situation above, while if $\alpha = (-w_0)\alpha$ then

$$\omega_\alpha = \alpha + \text{others}, \ \alpha = (-w_0)\alpha.$$

This occurs only for $B_{\geq 3}$, $C_{\geq 2}$, $D_{\geq 4}$, std., again by checking the Planches in [Bour L6]. This concludes the proof of Theorem 1.2.

We now consider the situation of Theorem 1.4, where the element h is Diag(1,1,0,...,0,-1,-1). Since the highest weight λ and the lowest weight ν each occur with multiplicity one, we see that there is a unique weight (say λ_2) other than the highest weight for which $\lambda_2(h) = 1$, and this weight occurs with multiplicity one. Similarly there is a unique weight λ_{d-1} other than the lowest one with $\lambda_{d-1}(h) = -1$. Taking a chain passing through λ_2, we see by looking at the h-weights that λ_2 must be the immediate successor of λ; similarly, looking at the h-weights in a chain passing through λ_{d-1} shows that its immediate successor must be ν. Therefore (by Lemma 4) **every** chain has the form $\lambda, \lambda_2, ???, \lambda_{d-1}, \nu$.

If $\lambda, \lambda_2, \lambda_{d-1}, \nu$ is a chain, then there are no other weights in V (otherwise some chain would exhibit it, and such a chain would have intermediate terms; therefore $\lambda_2 - \lambda_{d-1}$ would be the sum of two or more simple roots, and hence could not be a simple root itself). As these weights occur with multiplicity one we see that **dimV = 4; among fundamental representations of simple $\mathfrak{g}$'s, the only possibilities are $\mathfrak{sL}(V)$ and $\mathfrak{sP}(V)$ (this last is spin for $\mathfrak{sO}(5)$).**

In case $\lambda, \lambda_2, \lambda_{d-1}, \nu$ is not a chain, then every chain has the form $\lambda, \lambda_2, ???, \lambda_{d-1}, \nu$ with at least one ? term. In this chain, the sequence of h-weights is 1,1,0,...,0,-1,-1 with at least one intermediate zero. Therefore the simple roots β have $\beta(h) = 0$ or 1, and in writing $\lambda - \nu$ as $\Sigma n_\beta \beta$, either there are exactly two simple roots β with $\beta(h)=1$, each with coefficient $n_\beta = 1$, or there is a unique one, with coefficient $n_\beta = 2$. But we have seen (Lemma 1.7.4) that any simple root adjacent to α passes from second to third place in some chain, and similarly any neighbor of $-w_0(\alpha)$ passes from third last to second last in some chain. So any neighbor β of α occurs in $\lambda - \nu$, as does any neighbor of $-w_0(\alpha)$.

This means there are at most two distinct simple roots β, γ with $\beta(h) = 1 = \gamma(h)$. If there are precisely two, then

$\omega_\alpha - w_0\omega_\alpha = \beta + \gamma +$ others,

β adajcent to α,

γ adjacent to $-w_0(\alpha)$, and

{simple roots adj to α} $\cup$ {simple roots adj to $-w_0(\alpha)$} = $\{\beta, \gamma\}$.

If there is only one such, then it is the unique neighbor of α **and** the unique neighbor of $-w_0(\alpha)$, and we have

$\omega_\alpha - w_0\omega_\alpha = 2\beta +$ others,

β the unique neighbor of α,

β the unique neighbor of $-w_0(\alpha)$.

At this point one must patiently check the Bourbaki tables! To do this easily, we break up into four cases, depending on whether $\beta=\gamma$ or not, and on whether or not α is self dual.

If $\beta=\gamma$, then both α and $-w_0(\alpha)$ have unique neighbors, so both are extreme points of the Dynkin diagram. If α and $-w_0(\alpha)$ coincide, then α is a self-dual extreme point and we have

$\omega_\alpha = \beta +$ others, α a self-dual extreme pt, β its unique nbr.

Among the exceptional groups, inspection shows that only **G_2, ω_1** qualifies. The classical self-dual extreme points are none for $A_{\ell\geq 2}$, ω_1 and ω_ℓ for B_ℓ and C_ℓ, ω_1 for D_{odd}, ω_1 and the two spins for D_{even}. For **ω_1 in B,C,D** the coef of α_2 is 1, so these occur. For ω_ℓ in B_ℓ the coef of $\alpha_{\ell-1}$ is $(\ell-1)/2$, which is 1 only for $\ell=3$, corresponding to the **8-dim spin rep of $\mathfrak{so}(7)$**. For ω_ℓ in C_ℓ, the coef of $\alpha_{\ell-1}$ is $\ell-1$, only 1 for $\ell=2$, corresponding to the **standard rep of $\mathfrak{so}(5)$**. For either spin in $D_{even\,\ell}$ the coef of $\alpha_{\ell-2}$ is $(\ell-2)/2$, which is 1 only for $\ell=4$, in which case by triality the image of the rep. is **$\mathfrak{so}(8)$ in its standard representation.**

If α and $-w_0(\alpha)$ are distinct in the case ($\beta=\gamma$), then β has two extreme point neighbors, so one end of the diagram looks like the end of the $D_{\ell\geq 3}$ series, and ℓ is odd because the two ends α and $-w_0(\alpha)$ are distinct. This means we can only have one of the spin representations, but for these the sum $\omega_{\ell-1} + \omega_\ell$ contains $\alpha_{\ell-2}$ with coefficient $\ell-2$, which is never 2 since ℓ is odd. **So this case does not exist.**

If we are in the case $\beta\neq\gamma$, and if β is the unique neighbor of α, then α is extreme and hence so is $-w_0(\alpha)$; similarly if γ is the unique neighbor of $-w_0(\alpha)$ then $-w_0(\alpha)$ is extreme and hence so is α. This then means that both α and $-w_0(\alpha)$ are extreme, and have different neighbors, so α cannot be selfdual and

$$\omega_\alpha - w_0\omega_\alpha = \beta + \gamma + \text{others, } \alpha \text{ nonselfdual extreme.}$$

The possibilities here for nonselfdual extremes are only A_ℓ, ω_1 and ω_ℓ; D_{odd}, spin; E_6, ω_1 and ω_6. The D case is ruled out by the requirement that the unique neighbors of the duals be distinct. The E_6 case is ruled out because $\omega_1 + \omega_6$ has all coef's ≥ 2. The case of the **standard representation** ω_1 **of** A_ℓ **and its dual** ω_ℓ is fine for $\ell \geq 4$ (since $\omega_1 + \omega_\ell = \alpha_1 + \alpha_2 + \dots + \alpha_\ell$).

If $\beta \neq \gamma$, we see as above that if α (and hence $-w_0(\alpha)$) is **not** extreme, each of α and $-w_0(\alpha)$ has both β and γ as neighbors. As there are no closed diamonds in the diagram, it must be that $\alpha = -w_0(\alpha)$; then α has exactly two neighbors β and γ and

$$\omega_\alpha = 1/2(\beta + \gamma) + \text{others, } \alpha \text{ selfdual with nbrs } \beta, \gamma$$

Again by inspection the exceptional groups are ruled out (no ω_α has 1/2 as any coef.). In the A series, the only selfdual is ω_k for A_{2k-1}, and this only works if k=3, corresponding to the **standard representation of $\mathcal{SO}(6)$**. In the B and C series, no ω_α has two coef.'s 1/2. In the D series, those with exactly two neighbors are ω_2 through $\omega_{\ell-3}$, and none of these has any coef. 1/2. This concludes the proof of Theorem 1.4, and with it, Theorem 1.3. QED

(1.7.6) We now turn to the proof of Theorem 1.5. We begin with the case when $\mathcal{G}$ is normalized by a reflection γ. If the automorphism $Ad(\gamma)$ of the corresponding connected semisimple group G in GL(V) is inner, then G contains the diagonal (in a suitable basis of V) matrix

$$A := \text{Diag}(-a, a, a, \dots, a)$$

for some nonzero a. (Therefore $\mathcal{G} \neq \mathcal{SP}(V)$ if dimV ≥ 4, because the group Sp(V) does not contain Diag(-a, a,...,a); in Sp(2d), the 2d eigenvalues of any element can be grouped into d pairs of inverses.) Taking a maximal torus T of G which contains A, and fixing some Weyl chamber, we see that the (-a) eigenspace of A is a multiplicity one weight space for T, say with weight τ. Therefore there exists a simple root α with $\alpha(A)=-1$ (any simple root α that takes us to or from τ in a chain of weights which passes through τ). Now consider the corresponding $\mathcal{SL}(2)$ for this simple root. For any weight space V^λ, $X_\alpha V^\lambda$ is in $V^{\lambda+\alpha}$ and $X_{-\alpha} V^\lambda$ is in $V^{\lambda-\alpha}$. So if we write $V = V_a \oplus V_{-a}$ as the sum of the n-1 dimensional a-eigenspace for A and of the one dimensional (-a)-eigenspace for A, then X_α maps V_a to V_{-a}, and it maps V_{-a} to V_a. Therefore

$$\mathrm{Image}(X_\alpha) = X_\alpha V_a \oplus X_\alpha V_{-a} \subset V_{-a} \oplus X_\alpha V_{-a}$$

has dimension at most two.

If dim Image(X_α) ≤ 1, then dim Image(X_α) = 1 since the representation is faithful. So the $\mathfrak{sl}(2)$-representation we have is the direct sum of the standard one and a trivial one, and so the element H_α is Diag(1,-1,0,...,0). Now apply Theorem 1.2.

If dim Image(X_α) = 2, then $X_\alpha V_a = V_{-a}$ and $X_\alpha V_{-a} \neq 0$, so $(X_\alpha)^2 \neq 0$. Since dim Image(X_α) = 2, when we write our $\mathfrak{sl}(2)$ representation as a direct sum of Symm^{n_i}(std)'s, we have $\Sigma n_i = 2$. Since $(X_\alpha)^2 \neq 0$, we cannot have std$\oplus$std$\oplus$triv, so we must have Symm^2(std)$\oplus$triv, and H_α is Diag(2,-2,0,...,0). Now apply Theorem 1.2.

Next we consider the case when $\mathfrak{g}$ is normalized by a reflection γ, but the automorphism σ = Ad(γ) is not inner. (Here $\mathfrak{sp}$ is also ruled out, because every automorphism of it is inner.) Any automorphism σ of the corresponding connected semisimple group G stabilizes some Borel subgroup B (look at the coherent cohomology $H^i(G/B, \mathcal{O})$ of G/B, which vanishes except for $H^0 = \mathbb{C}$, and use the Lefschetz fixed point formula). If σ is an involution we claim it stabilizes some maximal torus T in B. For if we pick one maximal torus T in B, then σ(T) is another maximal torus in the same Borel, so it is conjugate to T by an element u of the unipotent radical U of B: $\sigma T = uTu^{-1}$. Moreover, the element u in U is uniquely determined, since inside B, T is its own normalizer. We need to find an element v in U such that $\sigma(vTv^{-1}) = vTv^{-1}$,i.e., an element v in U that satisfies $\sigma(v)u = v$. Since $\sigma^2 = 1$, we know $\sigma^2(T) = T$ and hence $\sigma(u)u = 1$, or $u^{-1} = \sigma u$. So if we write $u = \mu^2$ with μ in U, then $u^{-1} = \sigma u$ gives $\mu^{-2} = (\sigma\mu)^2$, so both μ^{-1} and $\sigma\mu$ are square roots in U of the same element. By the uniqueness of square roots in U we have $\mu^{-1} = \sigma\mu$. Then $(\sigma\mu)u = \mu^{-1}u = \mu$, so $\mu T \mu^{-1}$ is the desired σ-stable maximal torus.

Once σ stabilizes a pair (B, T), it induces an automorphism of the Dynkin diagram. This automorphism is necessarily nontrivial (lest σ be inner) and of order two, so a nontrivial involution of the Dynkin diagram. Consider the action of σ on the weight spaces of our representation; it must permute them nontrivially (otherwise it fixes the weights, so fixes pointwise the $\mathbb{Q}$-span of the weights, so fixes all roots, so would be inner). This means that the matrix of the reflection γ is a permutation-block shaped matrix. But the eigenvalues of a permutation-block shaped matrix with cycles of length d_i are the various d_i'th roots of the eigenvalues of the what the d_i'th power induces on any block in the orbit.

Since γ has eigenvalues $\{-1,1,1,...,1\}$, a set which contain no clumps of size d_i which are principal homogenous under multiplication by the d_i'th roots of unity with $d_i > 1$ except possibly for a single one of size $d_i = 2$, we conclude that σ permutes precisely two weight spaces, each of which is of multiplicity one, and stabilizes all the others. Consider the $\mathbb{Q}$-span S of those weights fixed by σ; S must be a proper subspace of the $\mathbb{Q}$-span of the roots (because σ does not fix all the roots, not being inner), and hence we can find a 1-parameter torus in T whose h kills S. This h, acting on V, acts as zero on the sum of all the non-permuted weight spaces, which is of codimension 2 in V. By faithfulness of V, h must act on V as $\mathrm{Diag}(x,-x,0,...,0)$ with $x\neq 0$. Now apply Theorem 1.2.

If $\mathfrak{g}$ is normalized by a unipotent pseudoreflection u, then its logarithm $N := \log(u)$ lies in $\mathfrak{g}$ (because every derivation of $\mathfrak{g}$ is inner, compare the proof of Theorem 0). As endomorphism of V, N has rank one. Using Jacobson-Morosov to complete N to an $\mathfrak{sl}(2)$-triple (N, h, ?) in $\mathfrak{g}$, we get a semisimple element h in $\mathfrak{g}$ which in a suitable basis of V is $\mathrm{Diag}(1,0,...,0,-1)$. Again apply Theorem 2.We have $\mathfrak{g} \neq \mathfrak{so}(V)$ simply because $\mathfrak{so}(V)$ doesn't contain any nilpotent N of rank one.

If $\mathfrak{g}$ is normalized by a pseudoreflection γ whose determinant ξ is not ± 1, then in a suitable basis of V, γ is $\mathrm{Diag}(\xi,1,...,1)$. Because $\xi \neq \xi^{-1}$, Gabber's Theorem 0 shows that $\mathfrak{g}$ contains $\mathrm{Diag}(n-1, -1, ..,-1)$, whence $\mathfrak{g}$ is $\mathfrak{sl}(V)$ by Theorem 1.1. This concludes the proof of Theorem 1.5. QED

(1.7.7) We now turn to the proof of Theorem 1.6 on prime-dimensional representations. First of all, $\mathfrak{g}$ must be simple since dimV is prime. If $\mathfrak{g}$ has rank one, we have the $\mathfrak{sl}(2)$ possibility. Assume now that $\mathfrak{g}$ has rank at least two. Pick a Cartan subalgebra $\mathcal{H}$, a Weyl chamber, et cetera. The Weyl dimension formula for dimV in terms of the highest weight λ of V and $\rho := \Sigma\mathrm{fd\ wts} = (1/2)\Sigma$ pos rts is

$$\dim V = \Pi_{\text{pos roots } \alpha} \; [(\lambda + \rho, H_\alpha)/ (\rho, H_\alpha)].$$

The two key observations are that for $\mathfrak{g}$ simple, and V irreduible with highest weight λ,

(1) the largest single term in the formula is the term $(\lambda+\rho, H_{\mathrm{highest}})$,

where $H_{highest}$ is the highest root of (i.e., highest weight of the adjoint representation of the algebra corrsponding to) the dual root system, and this term is **strictly** larger than any of the other terms.

(2) We have the inequality

$$(\lambda+\rho, H_{highest}) \leq \dim V.$$

If we grant both of these points, the first of which is obvious, and the second of which will be proven below, then for $\dim V = p$, we find that $(\lambda+\rho, H_{highest}) = \dim V$ (otherwise every single term in the formula is $< p$ and hence prime to p). We will then examine those irreducible representations V of simple $\mathfrak{g}$'s for which

$$(\lambda+\rho, H_{highest}) = \dim V.$$

To prove (2), consider the highest weight of the restriction of V to the principal $\mathfrak{sl}(2)$ in $\mathfrak{g}$; this is one less than the dimension of its biggest irreducible, so we have

$$\dim V \geq 1 + \text{highest weight of the principal } \mathfrak{sl}(2).$$

In virtue of ([Bour L8], 11, 4,Prop. 8 (i) and 7, 5, Lemma 2), the highest weight under the principal $\mathfrak{sl}(2)$ is

$$(\lambda, \text{the element } h^0 \text{ in } \mathcal{H} \text{ such that } \beta(h^0)=2 \text{ for all simple } \beta),$$

and $h^0 = \Sigma_{pos\ \alpha} H_\alpha$ ("the sum of the positive roots is twice the sum of the fundamental weights" for the dual root system). Thus

$$\begin{aligned}\dim V &\geq 1 + \text{highest weight of the principal } \mathfrak{sl}(2)\\ &= 1 + (\lambda, \Sigma_{pos\ \alpha} H_\alpha).\end{aligned}$$

Now pick some simple α such that $\lambda(H_\alpha) > 0$, and pick a chain which runs from H_α up to the highest dual root $H_{highest}$; we get

$$\begin{aligned}&1 + (\lambda, \Sigma_{pos\ \alpha} H_\alpha)\\ &= 1 + (\lambda, H_{highest}) + \Sigma_{\alpha\ pos,\ H_\alpha\ not\ highest} (\lambda, H_\alpha)\\ &\geq 1 + (\lambda, H_{highest}) + \Sigma_{chain\ omitting\ H_{highest}} (\lambda, H_{in\ chain})\\ &\geq 1 + (\lambda+\rho, H_{highest}) - (\rho, H_{highest}) + \Sigma_{chain\ omitting\ H_{highest}} (\lambda, H_{in\ chain}).\end{aligned}$$

Now we will establish that in fact

$$1 + \Sigma_{chain\ omitting\ H_{highest}} (\lambda, H_{in\ chain}) \geq (\rho, H_{highest}).$$

Writing $H_{highest} = \Sigma_{simple\ \beta}\, n_\beta H_\beta$ as a sum of simple dual roots, and recalling that ρ is the sum of the fundamental weights, we see that

$$(\rho, H_{\text{highest}}) = \Sigma n_\beta = \text{the "length" of } H_{\text{highest}},$$

so what is to be proved is

$$1 + \Sigma_{\text{chain omitting } H_{\text{highest}}} (\lambda, H_{\text{in chain}}) \geq \text{length of } H_{\text{highest}}.$$

This inequality is obvious, because the length of any chain from the simple root H_α to $H_{\text{highest}} = \Sigma_{\text{simple } \beta}\, n_\beta H_\beta$ is obviously the length Σn_β of H_{highest}, and α is chosen so that every term $(\lambda, H_{\text{in chain}})$ is at least one. Thus we have established the inequalities

$$\begin{aligned}
\dim V &\geq 1 + \text{highest weight of the principal } \mathcal{SL}(2)\\
&= 1 + (\lambda, \Sigma_{\text{pos } \alpha} H_\alpha) = 1 + (\lambda, H_{\text{highest}}) + \Sigma_{\alpha \text{ pos,not highest}} (\lambda, H_\alpha)\\
&\geq 1 + (\lambda, H_{\text{highest}}) + \Sigma_{\text{chain omitting } H_{\text{highest}}} (\lambda, H_{\text{in chain}})\\
&\geq (\lambda, H_{\text{highest}}) + \text{length of } H_{\text{highest}}\\
&= (\lambda + \rho, H_{\text{highest}}).
\end{aligned}$$

Now suppose we have the equality

$$\dim V = (\lambda + \rho, H_{\text{highest}}).$$

Then we see, from the top end of the inequality, that V is irreducible when restricted to the principal $\mathcal{SL}(2)$. By [Ka-GKM, 11.6], the only possibilities outside $\mathcal{SL}(2)$ are $\mathcal{SL}$, $\mathcal{SP}$, and $\mathcal{SO}$(odd) in their standard representations and Lie(G_2) in its seven-dimensional representation. QED

1.8 Appendix: Direct sums and tensor products

Let G_i, $i=1,2$, be connected semisimple groups over $\mathbb{C}$ whose Lie algebras are simple, and let $\rho_i: G_i \to GL(V_i)$, $i=1,2$, be faithful irreducible representations of the G_i on finite-dimensional $\mathbb{C}$-spaces V_i. We say that (G_1, V_1) and (G_2, V_2) are **Goursat-adapted** if either

(a) the Lie algebras of G_1 and G_2 are not isomorphic, or

(b) for any isomorphism γ : Lie(G_1) $\xrightarrow{\sim}$ Lie(G_2), at least one of the following two conditions holds:

(b1) there exists an isomorphism $A: V_1 \xrightarrow{\sim} V_2$ such that, viewing Lie(G_i)$\subset$End(V_i), we have $\gamma(X) = AXA^{-1}$ for every $X \in$ Lie(G_1). [If this A exists, then $g \mapsto AgA^{-1}$ defines an isomorphism from G_1 to G_2.]

(b2) there exists an isomorphism $A:(V_1)^* \tilde{\rightarrow} V_2$ from the dual of V_1 to V_2 such that, viewing $\mathrm{Lie}(G_i) \subset \mathrm{End}(V_i)$, we have $\gamma(X) = A(-X^t)A^{-1}$ for every $X \in \mathrm{Lie}(G_1)$, where $X^t \in \mathrm{End}((V_1)^*)$ denotes the intrinsic transpose of X.[If this A exists, then $g \mapsto Ag^{-t}A^{-1}$ defines an isomorphism from G_1 to G_2.]

Examples 1.8.1 If both (G_i, V_i) are isomorphic to (G, std_n) where G is one of the classical simple groups $SL(n \geq 3)$, $SO(\text{odd } n \geq 3)$, $Sp(\text{even } n \geq 2)$, $SO(\text{even } n \geq 6)$ and std_n is its standard n-dimensional representation ω_1, then only for $(SO(8), \mathrm{std}_8)$ does Goursat-adapted fail. Indeed, every automorphism of $SO(\text{odd } n \geq 3)$ or of $Sp(\text{even } n \geq 2)$ is inner, the only nontrivial outer automorphism of $SL(n \geq 3)$ is the Cartan involution $X \mapsto X^{-t}$, and every automorphism of $SO(\text{even } n \geq 6, n \neq 8)$ is induced by conjugation by an element of $O(n)$.

If both (G_i, V_i) are isomorphic to (G, V), where G is a connected semisimple irreducible subgroup of $GL(V)$ such that $\mathrm{Lie}(G)$ is simple and such that every automorphism of $\mathrm{Lie}(G)$ is inner (types B, C, F_4, E_7, E_8, G_2) then they are automatically Goursat-adapted, **whatever** the particular irreducible representation V.

Proposition 1.8.2 (Goursat-Kolchin-Ribet, [Kol], [Ri]) Let G be an algebraic group over $\mathbb{C}$, $n \geq 2$ an integer, and $\rho_i: G \rightarrow GL(V_i)$, $i = 1, \ldots, n$, a set of n finite-dimensional irreducible representations of G whose direct sum $\oplus_i V_i$ is faithful. For each i, let $G_i := \rho_i(G)$ be the image of G in $GL(V_i)$. Suppose that

(1) for each i, $G_i^{0,\mathrm{der}}$ operates irreducibly on V_i, and $\mathrm{Lie}(G_i^{0,\mathrm{der}})$ is simple.

(2) for $i \neq j$, $(G_i^{0,\mathrm{der}}, V_i)$ and $(G_j^{0,\mathrm{der}}, V_j)$ are **Goursat-adapted**.

(3) for $i \neq j$, and for any character χ of G, the representations ρ_i and $\chi \otimes \rho_j$ of G are not isomorphic.

(4) for $i \neq j$, and for any character χ of G, the representations $(\rho_i)^*$ and $\chi \otimes \rho_j$ of G are not isomorphic.

Then $G^{0,\mathrm{der}}$ is the subgroup $\prod G_i^{0,\mathrm{der}}$ of $\prod GL(V_i)$.

proof By definition, G maps onto each $G_i := \rho_i(G)$. Therefore G^0 maps onto each G_i^0, and $G^{0,\mathrm{der}}$ maps onto each $G_i^{0,\mathrm{der}}$. By faithfulness we have an a priori inclusion $G \subset \prod_i G_i$, so $G^{0,\mathrm{der}} \subset \prod G_i^{0,\mathrm{der}}$. Notice that G is reductive (it has a faithful completely reducible representation) and hence $G^{0,\mathrm{der}}$ is semisimple.

We first reduce to the case where $G = G^{0,\mathrm{der}}$ and $G_i = G_i^{0,\mathrm{der}}$ for each i. Replacing G by $G^{0,\mathrm{der}}$ does not change $G_i^{0,\mathrm{der}} = \rho_i(G)^{0,\mathrm{der}} = \rho_i(G^{0,\mathrm{der}})$. Each V_i is $G^{0,\mathrm{der}}$-irreducible, since it is $G_i^{0,\mathrm{der}}$-irreducible. If hypotheses (3) and (4) hold for G then they hold for $G^{0,\mathrm{der}}$ [Indeed if H is **any** normal subgroup of G, and if two H-irreducible representations V and W of G become isomorphic on H, then $\mathrm{Hom}_H(V, W)$ is a one-dimensional representation of G/H, and $\mathrm{Hom}_H(V, W) \otimes V \approx W$ as G-representations.]

So we may assume that $G = G^{0,\mathrm{der}}$ and $G_i = G_i^{0,\mathrm{der}}$ for each i. By Lie theory, it suffices to prove that $\mathrm{Lie}(G) = \prod \mathrm{Lie}(G_i)$ inside $\prod \mathrm{End}(V_i)$. Each $\mathrm{Lie}(G_i)$ being simple, it suffices by [Ri, pp. 790-791] to show that for any two indices $i \neq j$, Lie(G) maps onto $\mathrm{Lie}(G_i) \times \mathrm{Lie}(G_j)$. So (replacing G by $(\rho_i \times \rho_j)(G)$)we are reduced to the case n=2.

When n=2, Lie(G) is a Lie-subalgebra of the product $\mathrm{Lie}(G_1) \times \mathrm{Lie}(G_2)$ which maps onto each factor, so by Goursat's Lemma, either Lie(G) is the product $\mathrm{Lie}(G_1) \times \mathrm{Lie}(G_2)$, or it is the graph of an isomorphism $\gamma: \mathrm{Lie}(G_1) \xrightarrow{\sim} \mathrm{Lie}(G_2)$. In this second case, we will derive a contradiction, by using the Goursat-adaptedness, which tells us that either

(b1) there exists an isomorphism $A: V_1 \xrightarrow{\sim} V_2$ such that $\gamma(X) = AXA^{-1}$ for every $X \in \mathrm{Lie}(G_1)$, or

(b2) there exists an isomorphism $A: (V_1)^* \xrightarrow{\sim} V_2$ such that for every $X \in \mathrm{Lie}(G_1)$, $\gamma(X) = A(-X^t)A^{-1}$.

Since Lie(G) is the graph of γ, we conclude that G is the graph of an isomorphism from G_1 to G_2 either of the form $g \mapsto AgA^{-1}$, which contradicts (3), or of the form $g \mapsto A g^{-t} A^{-1}$, which contradicts (4). QED

The following standard lemma is stated for ease of reference.

Lemma 1.8.3 Let $G_1, \dots, G_r$ be connected semisimple linear algebraic groups over $\mathbb{C}$ whose Lie algebras $\mathfrak{g}_i$ are simple. Then

(1) if σ is any automorphism of $\prod \mathfrak{g}_i$, there exists a permutation $i \mapsto s(i)$ of the index set $\{1, \dots, r\}$ and isomorphisms $\sigma_{s(i),i} : \mathfrak{g}_{s(i)} \to \mathfrak{g}_i$ such that

$$\sigma(g_1, g_2, \dots, g_r) = (\sigma_{s(1),1}(g_{s(1)}), \sigma_{s(2),2}(g_{s(2)}), \dots, \sigma_{s(r),r}(g_{s(r)})).$$

(2) if σ is any automorphism of $\prod G_i$, there exists a permutation $i \mapsto s(i)$ of the index set $\{1, \dots, r\}$ and isomorphisms $\sigma_{s(i),i} : G_{s(i)} \to G_i$ such that

$$\sigma(g_1, g_2, \dots, g_r) = (\sigma_{s(1),1}(g_{s(1)}), \sigma_{s(2),2}(g_{s(2)}), \dots, \sigma_{s(r),r}(g_{s(r)})).$$

proof To prove (1), it suffices to note that the individual factors $\mathfrak{g}_i$ of the product $\mathfrak{g} := \prod \mathfrak{g}_i$ are intrinsically the minimal nonzero ideals of $\mathfrak{g}$, hence are necessarily permuted by any automorphism of $\mathfrak{g}$. To prove (2), denote by $\tilde{\sigma}$ the automorphism induced by σ on the universal covering $\prod \tilde{G}_i$, and by Lie(σ) the automorphism induced by σ on $\mathrm{Lie}(\prod G_i) = \prod \mathrm{Lie}(G_i)$. By (1) applied to Lie($\sigma$), and the equivalence of categories between connected simply connected complex Lie groups and finite dimensional Lie algebras over $\mathbb{C}$, we infer that there exists a permutation $i \mapsto s(i)$ of the index set $\{1, \dots, r\}$ and isomorphisms $\tilde{\sigma}_{s(i),i} : \tilde{G}_{s(i)} \to \tilde{G}_i$ such that

$$\tilde{\sigma}(\tilde{g}_1, \tilde{g}_2, \dots, \tilde{g}_r) = (\tilde{\sigma}_{s(1),1}(\tilde{g}_{s(1)}), \tilde{\sigma}_{s(2),2}(\tilde{g}_{s(2)}), \dots, \tilde{\sigma}_{s(r),r}(\tilde{g}_{s(r)})).$$

But as $\tilde{\sigma}$ is induced by the automorphism σ of $\prod G_i$, $\tilde{\sigma}$ maps the finite covering group $\mathrm{Ker}(\prod \tilde{G}_i \to \prod G_i) = \prod(\mathrm{Ker}(\tilde{G}_i \to G_i))$to itself. Therefore each $\tilde{\sigma}_{s(i),i} : \tilde{G}_{s(i)} \to \tilde{G}_i$ maps $\mathrm{Ker}(\tilde{G}_{s(i)} \to G_{s(i)})$ isomorphically to $\mathrm{Ker}(\tilde{G}_i \to G_i)$, whence each $\tilde{\sigma}_{s(i),i} : \tilde{G}_{s(i)} \to \tilde{G}_i$ descends to an isomorphism $\sigma_{s(i),i} : G_{s(i)} \to G_i$. That σ is

$$(g_1, g_2, \dots, g_r) \mapsto (\sigma_{s(1),1}(g_{s(1)}), \sigma_{s(2),2}(g_{s(2)}), \dots, \sigma_{s(r),r}(g_{s(r)}))$$

follows from the fact that this automorphism of the connected group $\prod G_i$ induces Lie(σ) on the Lie algebra. QED

Lemma 1.8.4 Let V be a $\mathbb{C}$-vector space of dimension $n \geq 2$, and let $G \subset GL(V)$ be one of the following groups:

$SL(n)$,

Sp(n) if n is even,
SO(n) if n is odd.

Let Γ denote the image of $G\times G\times\ldots\times G$ in $GL(V^{\otimes n})$. Then the normalizer of Γ in $GL(V^{\otimes n})$ is the semidirect product $\mathbb{G}_m\Gamma\ltimes S_n$, where S_n acts on $V^{\otimes n}$ by permuting the factors.

proof Suppose A in $GL(V^{\otimes n})$ normalizes Γ. The automorphism $\sigma := \mathrm{Ad}(A)$ of Γ lifts to an automorphism $\tilde{\sigma}$ of $G\times G\times\ldots\times G$ in each of the cases envisioned (for SL or Sp, $G\times G\times\ldots\times G$ is the universal covering of Γ, and for SO(odd), $G\times G\times\ldots\times G \approx \Gamma$). Any automorphism of $G\times G\times\ldots\times G$ is the composition of a permutation of the factors with an automorphism of the form $\prod\sigma_i : (g_i)_i \mapsto (\sigma_i(g_i))_i$, where for each i, σ_i is an automorphism of G.

If G is Sp or SO(odd), each σ_i is inner, say $\sigma_i = \mathrm{Int}(g_i)$. Then successively correcting Ad(A) by the permutation it induces of the factors and by $\mathrm{Int}(g_1, \ldots, g_n)$, we obtain an element of the centralizer of Γ in $GL(V^{\otimes n})$. But Γ acts irreducibly on $V^{\otimes n}$ (since G acts irreducibly on V), so this centralizer consists of the scalars $\mathbb{G}_m$.

If G is SL(V) with $n = \dim V \geq 3$, then an automorphism σ of G is inner if and only if the given representation ρ of G on V is equivalent to (i.e., has the same trace function as) $\rho^\sigma := \rho\circ\sigma$. Similarly, an automorphism $\prod\sigma_i : (g_i)_i \mapsto (\sigma_i(g_i))_i$ of $G\times G\times\ldots\times G$ is inner if and only if $\rho\otimes\rho\otimes\ldots\otimes\rho$ has the same trace function on $G\times G\times\ldots\times G$ as $(\rho\circ\sigma_1)\otimes\ldots\otimes(\rho\circ\sigma_n)$. The given representation $\rho^{\otimes n}$ of Γ on $V^{\otimes n}$ is the restriction to Γ of the standard representaion of $GL(V^{\otimes n})$, which is tautologically equivalent to its transform by Ad(A) for **any** A in $GL(V^{\otimes n})$. Applying this both to our A which normalizes Γ and to the permutation it induces, we see that Ad(A) is the composite of a permutation and of an inner automorphism of $G\times G\times\ldots\times G$, as above. QED

Lemma 1.8.5 Let $r \geq 2$, and pick r distinct integers n_i,

$$2 \leq n_1 < n_2 < \ldots < n_r.$$

For each i, let G_i be one of the groups

$SL(n_i)$,
$Sp(n_i)$ if n_i is even,
$SO(n_i)$ if n_i is odd ≥ 7,

SO(3) if $n_i = 3$ and no n_j is 2,

SO(5) if $n_i = 5$ and no n_j is 4.

and denote by G the image of $\prod G_i$ in $\otimes \mathrm{std}_{n_i}$. Then the normalizer of G in $GL(\otimes \mathrm{std}_{n_i})$ is $\mathbb{G}_m G$.

proof For each i, let ρ_i denote the standard representation std_{n_i} of G_i. If G_i is $SL(n_i)$, an automorphism σ_i of G_i is inner if and only if $(\rho_i)^{\sigma_i}$ is equivalent to ρ_i. For the other possibilities, every automorphism σ_i of G_i is inner. So in all cases an automorphism σ_i of G_i is inner if and only if $(\rho_i)^{\sigma_i}$ is equivalent to ρ_i.

Suppose A in $GL(\otimes \mathrm{std}_{n_i})$ normalizes G. Denote by σ the automorphism $\mathrm{Ad}(A) : g \mapsto AgA^{-1}$ of G it induces, and by $\mathrm{Lie}(\sigma)$ the automorphism σ induces of $\mathrm{Lie}(G) = \prod \mathrm{Lie}(G_i)$. Since the $\mathrm{Lie}(G_i)$ are pairwise nonisomorphic, $\mathrm{Lie}(\sigma) = \prod \mathrm{Lie}(\sigma)_i$ where $\mathrm{Lie}(\sigma)_i$ is an automorphism of $\mathrm{Lie}(G_i)$. Because G_i is either simply connected or adjoint, any automorphism $\mathrm{Lie}(\sigma)_i$ of its Lie algebra is of the form $\mathrm{Lie}(\sigma_i)$ for some automorphism σ_i of G_i. Therefore σ is induced by the automorphism $\tilde{\sigma} := \prod \sigma_i$ of $\prod G_i$. The representation $\rho := \otimes \rho_i$ of $\prod G_i$ is tautologically equivalent to $\rho^{\tilde{\sigma}}$ (by the intertwining operator A), and $\rho^{\tilde{\sigma}}$ is equivalent to $\otimes (\rho_i)^{\sigma_i}$. Therefore $\otimes \rho_i$ is equivalent to $\otimes (\rho_i)^{\sigma_i}$. Restricting both sides to the subgroup G_i of $\prod G_i$ and comparing characters, we see that ρ_i is equivalent to $(\rho_i)^{\sigma_i}$. Therefore each $\tilde{\sigma}_i$ is inner, $\tilde{\sigma} = \prod \sigma_i$ is inner, and hence $\sigma = \mathrm{Ad}(A)$ is inner. Since G acts irreducibly on $(\otimes \mathrm{std}_{n_i})$, we find $A \in \mathbb{G}_m G$, as required. QED

Lemma 1.8.6 Let $r \geq 2$, and pick r distinct integers n_i,

$$2 \leq n_1 < n_2 < \dots < n_r.$$

For each i, let G_i be one of the groups

$SL(n_i)$,

$Sp(n_i)$ if n_i is even,

$SO(n_i)$ if n_i is odd ≥ 7,

SO(3) if $n_i = 3$ and no n_j is 2,

SO(5) if $n_i = 5$ and no n_j is 4.

For each $i=1, \ldots, r$, pick an integer $m_i \geq 1$, denote by $(G_i)^{m_i}$ the m_i fold product $G_i \times \ldots \times G_i$, and denote by G the image of $\prod (G_i)^{m_i}$ in $GL(\otimes_i (std_{n_i})^{\otimes m_i})$. Then the normalizer of G in $GL(\otimes_i (std_{n_i})^{\otimes m_i})$ is the semidirect product $(\mathbb{G}_m G) \ltimes \prod_i S_{m_i}$, where $\prod_i S_{m_i}$ acts on $\otimes_i (std_{n_i})^{\otimes m_i}$ by permuting the factors.

proof Since the $Lie(G_i)$ are pairwise nonisomorphic, any automorphism of $Lie(G)$ is of the form $\prod_i$(an auto. of $Lie((G_i)^{m_i})$). Now G_i being either simply connected or adjoint, any automorphism of $Lie((G_i)^{m_i})$ is induced by an automorphism of $(G_i)^{m_i}$. Exactly as above, any automorphism of $(G_i)^{m_i}$ is the composition of a permutation of the factors and of an automorphism of the form $\prod_{j=1,\ldots,m_i} \sigma_{i,j}$, where $\sigma_{i,j}$ is an automorphism of G_i.

Suppose now that A in $GL(\otimes_i (std_{n_i})^{\otimes m_i})$ normalizes G. Modifying A by the element of $\prod_i S_{m_i}$ that $Ad(A)$ induces on the factors of each $Lie((G_i)^{m_i})$, we may suppose that there exist automorphisms $\sigma_{i,j}$ of G_i such that the product automorphism $\prod_{i,j} \sigma_{i,j}$ of $\prod_{i,j} G_i$ induces $Ad(A)$ on the Lie algebra. Denote by $\tilde{G}_i$ the universal covering of G_i, and by $\tilde{\sigma}_{i,j}$ the automorphism of $\tilde{G}_i$ induced by $\sigma_{i,j}$. Since $\prod_{i,j} \tilde{G}_i$ is connected, $\prod_{i,j} \tilde{\sigma}_{i,j}$ is the **unique** automorphism of $\prod_{i,j} \tilde{G}_i$ which induces $Ad(A)$ on its Lie algebra. But $\prod_{i,j} \tilde{G}_i$ is also the universal covering of G, so by uniqueness we conclude that $Ad(A)$ acting on G induces $\prod_{i,j} \tilde{\sigma}_{i,j}$ on its universal covering. Therefore $\prod_{i,j} \tilde{\sigma}_{i,j}$ stabilizes the two finite central subgroups of $\prod_{i,j} \tilde{G}_i$ corresponding to the two quotients $\prod_{i,j} G_i$ and G of $\prod_{i,j} \tilde{G}_i$. Therefore the action of $\prod_{i,j} \sigma_{i,j}$ on $\prod_{i,j} G_i$ is stable on the kernel of the canonical projection $\prod_{i,j} G_i \to G$, and induces $Ad(A)$ on G.

So if we denote by $\rho_{i,j}$ the standard representation of G_i, then the tensor product representation $\otimes_{i,j} \rho_{i,j}$ of $\prod_{i,j} G_i$ is equivalent to $\otimes_{i,j} (\rho_{i,j} \circ \sigma_{i,j})$. Just as above, we infer that each $\sigma_{i,j}$ is inner. QED

CHAPTER 2
D.E.'s and $\mathfrak{D}$-modules

The basic set-up and the Main D.E. Theorem

We will apply the representation-theoretic results of the last chapter to the calculation of some differential galois groups. Let us first recall the basic setup (cf [Ka-DGG, 1.1]).

(2.1) Let K be a field of characteristic zero, and X be a smooth geometrically connected separated K-scheme of finite type with X(K) nonempty, and ω a K-valued fibre functor on the category D.E.(X/K) (for instance ω ="fibre at x" for any K-valued point x of X). We denote by $\pi_1^{\mathrm{diff}}(X/K,\omega)$ the affine pro-algebraic K-group-scheme $Aut^{\otimes}(\omega)$. The fibre functor ω defines an equivalence of $\otimes$-categories

$$\text{D.E.}(X/K) \quad \approx \quad (\text{fin.-dim'l K-reps of } \pi_1^{\mathrm{diff}}(X/K,\omega)).$$

Given an object V in D.E.(X/K), denote by ⟨V⟩ the full subcategory of D.E.(X/K) whose objects are all subquotients of all finite direct sums of the objects $V^{\otimes n}\otimes(V^{\vee})^{\otimes m}$, all n,m $\geq$ 0. The restriction to ⟨V⟩ of ω is a fibre functor on ⟨V⟩. The K-group-scheme $Aut^{\otimes}(\omega \mid \langle V\rangle)$, denoted $G_{gal}(V, \omega)$, is by definition the differential galois group of V; it is a Zariski-closed subgroup of $GL(\omega(V))$. The restriction to ⟨V⟩ of the functor ω defines an equivalence of $\otimes$-categories

$$\langle V\rangle \quad \approx \quad (\text{fin-dim K-reps of } G_{gal}(V, \omega)).$$

If we view V as a representation ρ_V of $\pi_1^{\mathrm{diff}}(X/K,\omega)$ on $\omega(V)$, then $G_{gal}(V, \omega)$ is none other than the image under ρ_V of $\pi_1^{\mathrm{diff}}(X/K,\omega)$ in $GL(\omega(V))$.

2.2 Torsors and Lifting Problems

(2.2.0) What about the interpretation of homomorphisms of $\pi_1^{\mathrm{diff}}(X/K,\omega)$ to a linear algebraic group H over K other that GL(n)? There is a "general nonsense" interpretation of homomorphisms $\varphi : \pi_1^{\mathrm{diff}}(X, x) \to H$ as (isomorphism classes of) triples (P, $\square$, $P_x \approx H$) consisting of a right (etale) H-torsor P on X, an H-equivariant integrable connection $\square$ on P as X-scheme, and a trivialization of P_x as right H-torsor. Let us briefly sketch this interpretation. Suppose we begin with the homomorphism $\varphi : \pi_1^{\mathrm{diff}}(X, x) \to H$. View H as a closed subgroupscheme of some SL(N). Denote by A the coordinate ring of H, and filter it as K-vector space by the finite dimensional subspaces A_n := the functions on H which are the restrictions of polynomials of degree $\leq$ n in the functions $X_{i,j}$ (:= (i,j)'th matrix coefficient) on the ambient SL(N). Then A_n is stable by both the right and left translation actions of H on its coordinate ring. Via the left action, and the given homomorphism $f : \pi_1^{\mathrm{diff}}(X, x) \to H$, we may view A_n as a finite-dimensional representation of $\pi_1^{\mathrm{diff}}(X, x)$. By the main theorem of Tannakien categories, this representation corresponds to a D.E. $(\mathbb{A}_n, \square)$ on X together with an isomorphism $\omega_x(\mathbb{A}_n) \approx A_n$. The right action of H on A_n commutes with its left action, so it commutes with the left action of $\pi_1^{\mathrm{diff}}(X, x)$, and thus gives a horizontal right action of H on $\mathbb{A}_n$. The multiplication maps $A_n \otimes_K A_m \to A_{n+m}$ are both right and left H-equivariant, so we get horizontal multiplication maps $\mathbb{A}_n \otimes_K \mathbb{A}_m \to \mathbb{A}_{n+m}$ which are right H-equivariant. Thus if we define $\mathbb{A}$ to be the direct limit of the $\mathbb{A}_n$ (via the inclusions $\mathbb{A}_n \to \mathbb{A}_m$ corresponding to $A_n \to A_m$), then $P := \mathit{Spec}(\mathbb{A})$ is an affine X-scheme endowed with a right action of H, an integrable connection $\square$ for which this H-action is horizontal, and whose fibre over x is given as $P_x := \mathrm{Spec}(A) := H$. It remains to verify that P is in fact a right H-torsor on X_{et}. Indeed, P is faithfully flat over X [each $\mathbb{A}_n$ is a locally free $\mathcal{O}_X$-module of finite rank, and the inclusions $\mathbb{A}_n \to \mathbb{A}_{n+m}$ being horizontal, have cokernels which are $\mathcal{O}_X$-locally free as well; taking n=0 and passing to the limit over m, we see that $\mathcal{O}_X \to \mathbb{A}$ has $\mathcal{O}_X$-flat cokernel, so for any O_X-module $\mathcal{M}$, the map $\mathcal{M} \to \mathcal{M} \otimes_{\mathcal{O}_X} \mathbb{A}$ is injective]. To show that the map $H \times_K P \to P \times_X P$, $(h,p) \mapsto (p, ph)$ is an isomorphism of X-schemes, it suffices that the corresponding map $\mathbb{A} \otimes_{\mathcal{O}_X} \mathbb{A} \to A \otimes_K \mathbb{A}$ be an isomorphism of $\mathcal{O}_X$-modules. But this is a horizontal map of ind-objects of the category D.E.(X/K), whose fibre over x is an isomorphism, so it is an isomorphism. Thus P is a right H-torsor on X_{fpqc}, and as H is smooth over K, P is consequently a right H-torsor on X_{et} as well.

In the opposite direction, suppose we start with data $(P, \square, P_x \approx H)$. Let $\mathbb{A}$ denote the sheaf of $\mathcal{O}_X$-algebras whose *Spec* is P. Because P as right H-torsor is etale locally trivial, we see by descent that there exists a unique filtration of $\mathbb{A}$ by locally free $\mathcal{O}_X$-modules $\mathbb{A}_n$ of finite rank, which, after any etale $E \to X$ such that $P_E \approx H \times_K E$, corresponds to the filtration of $A \otimes_K \mathcal{O}_E$ by the $A_n \otimes_K \mathcal{O}_E$. The connection

$\square$ on P is an integrable connection on $\mathbb{A}$ for which the multiplication map and the right action of H are both horizontal. Looking at $\square$ after etale localization, we see that the $\mathbb{A}_n$ are horizontal. We are given an isomorphism $\mathbb{A}(x) \approx A$ compatible with the right H-actions on both $\mathbb{A}$ and A, so the tautological action of $\pi_1^{\mathrm{diff}}(X, x)$ on A respects both its ring structure and the right H-torsor structure of Spec(A):= H. Therefore the action of $\pi_1^{\mathrm{diff}}(X, x)$ on A must be through left translations by elements of H; this is the required homomorphism φ of $\pi_1^{\mathrm{diff}}(X, x)$ to H. It is clear that these two constructions are mutually inverse.

(2.2.1) Let $\rho : G \to H$ be a finite etale homomorphism of linear algebraic groups over K, whose kernel Γ is a finite etale central subgroup of G (e.g., $SL(n) \to PGL(n)$). Suppose we are given a homomorphism

$$\varphi: \pi_1^{\mathrm{diff}}(X, x) \to H$$

of algebraic groups over K. We look for homomorphisms

$$\tilde{\varphi} : \pi_1^{\mathrm{diff}}(X, x) \to G$$

with $\varphi = \rho\tilde{\varphi}$, and we call such a $\tilde{\varphi}$ a lift of φ.

Proposition 2.2.2 The obstruction to the existence of a lifting of φ lies in $H^2(X_{et}, \Gamma)$, and (if a lifting exists) the indeterminacy in a lifting is the group $H^1(X_{et}, \Gamma)$.

proof This is best seen in terms of the associated torsors with connection.

Using this interpretation, we argue as follows. Suppose we are given a right G-torsor P on X. Denote by ρP the right H-torsor on X gotten from P by the change of structural group $\rho: G \to H$. Given a right G-equivariant (resp. integrable) connection $\square$ on P, there is a natural right H-equivariant (resp. integrable) connection $\rho\square$ on ρP. We claim that the construction $\square \mapsto \rho\square$ is a bijection; in other words, right H-equivariant (resp. integrable) connections ρP lift uniquely to P. [To see this unicity of lifting, it suffices by etale descent to treat the case when P is the trivial torsor $G \times_K X$ on an affine X. Denote by A the coordinate ring of G. Then a right G-equivariant connection on $G \times_K X$, i.e., on $A \otimes_K \mathcal{O}_X$, is a rule which to every K-linear derivation ∂ of $\mathcal{O}_X$ to itself assigns a K-linear derivation $\square(\partial)$ of $A \otimes_K \mathcal{O}_X$ to itself which prolongs ∂

and which is right G-equivariant; to give such $\square(\partial)$ it suffices to specify the right G-equivariant derivation of A to the A-module $A\otimes_K\mathcal{O}_X$ given by $a \mapsto \square(\partial)(a\otimes 1)$, and this is none other than an element $D_{\square(\partial)}$ of $\mathrm{Lie}(G)\otimes_K\mathcal{O}_X$ (where Lie(G) is viewed as the right G-equivariant K-linear derivations of A to A). The mapping $\partial \mapsto D_{\square(\partial)}$ is $\mathcal{O}_X$-linear, so all in all our connection on P is an element ω of $\mathrm{Lie}(G)\otimes_K\Omega^1{}_{X/K}$. The induced connection on $\rho P = H\times_K X = \mathrm{Spec}(B\otimes_K\mathcal{O}_X)$ is $\partial \mapsto \square(\partial)$ restricted to the subalgebra $B\otimes_K\mathcal{O}_X$ of $A\otimes_K\mathcal{O}_X$. Since A is finite etale over B, this restriction is the same element ω viewed in $\mathrm{Lie}(H)\otimes_K\Omega^1{}_{X/K}$, when we identify $\mathrm{Lie}(G)\approx\mathrm{Lie}(H)$ by ρ. The connection is integrable if and only if for any pair of commuting derivations ∂_i, $i=1,2$, of $\mathcal{O}_X$ the derivations $\square(\partial_i)$ of $A\otimes_K\mathcal{O}_X$ commute, or equivalently if their restrictions to $B\otimes_K\mathcal{O}_X$ commute.]

So the problem of lifting φ to a $\tilde{\varphi}$ is that of lifting a given H-torsor P (which happens to have an integrable connection) to a G-torsor $\tilde{P}$ (which will then have a unique connection which lifts the given one). In terms of the short exact sequence on X_{et} of **algebraic groups/X**

$$1 \to \Gamma \to G \to H \to 1,$$

the cohomology sequence

$$H^1(X_{et}, \Gamma) \to H^1(X_{et}, G) \to H^1(X_{et}, H) \to H^2(X_{et}, \Gamma)$$

shows that the obstruction to lifting (the isomorphism class of) a given H-torsor P is in $H^2(X_{et}, \Gamma)$, and the indeterminacy in lifting it is in $H^1(X_{et}, \Gamma)$. QED

Corollary 2.2.2.1 If K is algebraically closed and if X/K is an open curve, then liftings exist.

(2.2.3) Suppose now that K is $\mathbb{C}$. The exact tensor functor $V \mapsto V^{an}$ from D.E.$(X/\mathbb{C})$ to D.E.$(X^{an}) \approx \mathrm{Rep}(\pi_1{}^{top}(X^{an}, x))$ defines a homomorphism $\iota : \pi_1{}^{top} \to \pi_1{}^{diff}$ from the topological π_1 of the complex manifold X^{an} to $\pi_1{}^{diff}(X, x)$. With respect to this homomorphism $\iota : \pi_1{}^{top} \to \pi_1{}^{diff}$, we have the following variant of the lifting problem:

Let $\varphi_{top} := \varphi\circ\iota$ be the restriction of φ to $\pi_1{}^{top}$. We look for liftings

$\tilde{\varphi}_{top}$ of φ_{top}, i.e., for homomorphisms $\tilde{\varphi}_{top} : \pi_1{}^{top}(X^{an}, x) \to G$ with $\varphi_{top} = \rho\tilde{\varphi}_{top}$.

Of course, if $\tilde{\varphi}$ is a lift of φ, then $(\tilde{\varphi})_{top} := \tilde{\varphi}\circ\iota$ is a lift of φ_{top}.

Proposition 2.2.4 When $K = \mathbb{C}$, the construction $\tilde{\varphi} \mapsto (\tilde{\varphi})_{top} := \tilde{\varphi}\circ\iota$ defines a bijection {lifts of φ} $\approx$ {lifts of φ_{top}}.

proof The analogous short exact sequence on X_{an} of **algebraic groups**

$$1 \to \Gamma \to G \to H \to 1,$$

and its cohomology sequence

$$H^1(X^{an}, \Gamma) \to H^1(X^{an}, G) \to H^1(X^{an}, H) \to H^2(X^{an}, \Gamma)$$

analyses the problem of lifting P^{an}; here the obstruction is in $H^2(X^{an}, \Gamma)$, and the indeterminacy in $H^1(X^{an}, \Gamma)$. By the comparison of etale and classical cohomology with finite coefficients, we have

$$H^i(X_{et}, \Gamma) \cong H^i(X^{an}, \Gamma) \text{ for all i.}$$

Therefore the liftings of the H-torsor P to a G-torsor $\tilde{P}$ are equivalent (by the functor $\tilde{P} \mapsto (\tilde{P})^{an}$) to the liftings of P^{an} to a G-torsor on X^{an}.

When P^{an} has an integrable connection, its sheaf of germs of horizontal sections (in its coordinate ring) is a principal $H(\mathbb{C})$-bundle on X^{an}, and this construction is an equivalence of categories

{right H-torsors on X^{an} with H-equivariant integrable connection} $\approx$

$\approx$ {principal right $H(\mathbb{C})$-bundles on X^{an}}

Now look at the exact sequence on X^{an} of constant groups

$$1 \to \Gamma \to G(\mathbb{C}) \to H(\mathbb{C}) \to 1.$$

The cohomology sequence here gives the obstruction and indeterminacy for lifting corresponding principal $H(\mathbb{C})$-bundle to a principal $G(\mathbb{C})$-bundle as again being the **same** elements in $H^2(X^{an}, \Gamma)$, and $H^1(X^{an}, \Gamma)$. But this lifting problem is that of lifting the map φ_{top}. Thus we find the asserted equivalences. QED

Corollary 2.2.4.1 Suppose that the topological fundamental group $\pi_1{}^{top}(X^{an}, x)$ of X^{an} is a free group (e.g., X an open curve or an Artin good neighborhood). Let $\rho : G \to H$ be a surjective homomorphism of linear algebraic groups over $\mathbb{C}$, whose kernel Γ is a finite central subgroup of G. Then any homomorphism $\varphi: \pi_1{}^{diff}(X, x) \to H$ of algebraic groups over $\mathbb{C}$ lifts to a homomorphism $\tilde{\varphi} : \pi_1{}^{diff}(X, x) \to G$

with $\varphi = \rho\tilde{\varphi}$.

Remark 2.2.4.2 Here is a slightly more cohomological formulation. Given a field K of characteristic zero, and a smooth X/K, we have the "crystalline-etale site" of X/K, noted $(X/K)_{\text{crys-et}}$. Its objects are the pairs $(U/X, \iota: U \to T)$ consisting of an etale X-scheme U/X and a closed K-immersion ι of U into a K-scheme T such that U is defined in T by a nilpotent ideal of $\mathcal{O}_T$. The morphisms are the obvious commutative diagrams, and a family of objects $(U_i/X, \iota: U_i \to T_i)$ over $(U/X, \iota: U \to T)$ is a covering if and only if the $T_i \to T$ are an etale covering of T.

Given any smooth K-groupscheme G, we get a sheaf, still noted G, on $(X/K)_{\text{crys-et}}$ by the rule $(U/X, \iota: U \to T) \mapsto G(T)$. It is essentially tautological that a right G-torsor on $(X/K)_{\text{crys-et}}$ is the same as a right G-torsor on X_{et} endowed with a right G-equivariant integrable connection. Now suppose that X(K) is nonempty. In view of the above discussion, we see that the set $H^1((X/K)_{\text{crys-et}}, G)$ of isomorphism classes of right G-torsors on $(X/K)_{\text{crys-et}}$, is none other than the quotient set $\text{Hom}_{K\text{-gpsch}}(\pi_1{}^{\text{diff}}(X, x), G)$ modulo the conjugation action of G (this action because we have not specified the trivialization over x). In other words, we have

$$H^1((X/K)_{\text{crys-et}}, G) \approx \text{Hom}_{K\text{-gpsch}}(\pi_1{}^{\text{diff}}(X, x), G)/G,$$

so that $\pi_1{}^{\text{diff}}(X, x)$ is a kind of "fundamental group" of $(X/K)_{\text{crys-et}}$.

The point is that if we have $\rho: G \to H$ as above with finite etale central kernel Γ, then we can use the standard cohomological setup on $(X/K)_{\text{crys-et}}$ to investigate our lifting problem. Consider the exact sequence of sheaves on $(X/K)_{\text{crys-et}}$

$$1 \to \Gamma \to G \to H \to 1,$$

which gives rise to an exact sequence of cohomology

$$H^1((X/K)_{\text{crys-et}}, \Gamma) \to H^1((X/K)_{\text{crys-et}}, G) \to$$
$$\to H^1((X/K)_{\text{crys-et}}, H) \to H^2((X/K)_{\text{crys-et}}, \Gamma).$$

By its very construction the site $(X/K)_{\text{crys-et}}$ maps to (both the usual crystalline site $(X/K)_{\text{crys}}$, and to) the etale site X_{et}. The Cech-Alexander calculation (cf. [Gro-CDR, 5.5], [Bert, V, 1.2]) of the cohomology shows that for any **etale** (resp. and commutative) K-groupscheme Γ, the canonical maps

$$H^i((X/K)_{\text{crys-et}}, \Gamma) \to H^i(X_{\text{et}}, \Gamma)$$

are isomorphisms for i=0,1 (resp. for all i). On the other hand for the smooth groups G and H, the canonical maps

$$H^1((X/K)_{crys-et}, G \text{ (resp. H)}) \to H^1(X_{et}, G \text{ (resp. H)})$$

correspond to the map "forget the connection". Thus we find once again that the obstruction and indeterminacy in our lifting problem lie in $H^2(X_{et}, \Gamma)$ and $H^1(X_{et}, \Gamma)$ respectively, and are the same as the obstruction and the indeterminacy in the lifting problem for the underlying "naked" torsors without connection.

2.3 Relation to Transcendence

We now turn to a brief discussion of the relation between the differential galois group of a D.E. and the transcendence properties of its power series solutions. This material is "well-known", indeed it was a large part of the basic motivation for the classical differential galois theory, but it does not seem to be written down anywhere in the Tannakian context (however cf. [De-CT, 9] for a Tannakian proof of the general existence of Picard-Vessiot extensions).

Proposition 2.3.1 Let V be a D.E. on X/K, $G := G_{gal}(V, x)$ its differential galois group, $v \in V_x$, and $\hat{v} \in V\otimes(\mathcal{O}_{X,x})^\wedge$ be the corresponding horizontal section. Suppose that K is algebraically closed. The the transcendence degree over K(X) of the coefficients (with repect to any K(X)-basis of $V\otimes_{\mathcal{O}_X}K(X)$) of $\hat{v}$ is the K-dimension of (the closure in V_x of) the G-orbit Gv.

proof Inside $V^\vee$, the annihilator of $\hat{v}$ is a horizontal submodule W, so it corresponds to a G-stable subspace W_x of $(V_x)^\vee$ which lies in the annihilator of v. Being G-stable, W_x must annihilate the entire G-orbit of v, so W_x is contained in the annihilator S_x of Gv; this S_x is a G-stable subspace of $(V_x)^\vee$, so it corresponds to a sub-D.E. S of $V^\vee$. Since S is horizontal, and S_x annihilates v, S annihilates $\hat{v}$. Thus $S \subset W$, and as $W_x \subset S_x$ we have S = W. Therefore the K-dimension of the space S_x of K-linear forms on V_x which annihilate Gv is the same as the $\mathcal{O}_X$-rank of W, i.e., the same as the K(X)-dimension of the space $W\otimes_{\mathcal{O}_X}K(X)$ of K(X)-linear forms on $V\otimes_{\mathcal{O}_X}K(X)$ which annihilate $\hat{v}$.

Applying this equality of dimensions to the situation

$\oplus_{j \le n}\mathrm{Symm}^j(v) \in \oplus_{j \le n}\mathrm{Symm}^j(V)_x$, for all n, we obtain equality of the corresponding Hilbert polynomials, whence the asserted equality of dimensions. QED

Corollary 2.3.1.1 Suppose K algebraically closed. The transcendence degree over K(X) of the $(\mathrm{rank}(V))^2$ matrix coefficients of any fundamental solution matrix at x is the dimension of G_{gal}.

proof Simply apply the above result to the internal hom D.E. $W := \mathcal{H}om(V_x \otimes_K \mathcal{O}_X, V)$, whose fibre W_x is $\mathrm{End}(V_x)$ with G acting by right translation. Fundamental solution matrices at x are precisely those horizontal $\hat{w}$'s in $W \otimes (\mathcal{O}_{X,x})^\wedge$ whose value w at x lies in $GL(V_x)$. Since G acts freely on $GL(V_x)$, the orbit dimension is dimG. QED

Remark 2.3.2 Here is a slightly more precise version of the above numerical result. Consider the right G-torsor P on X corresponding to the homomorphism $\pi_1{}^{\mathrm{diff}}(X, x) \to G$ which "classifies" the D.E. V. An a priori description of it is this. Consider the subcategory $\langle V \rangle$ of D.E.(X/K). On it we have two obvious $\mathcal{O}_X$-valued fiber functors, namely

$$\omega_x : W \mapsto W_x \otimes_K \mathcal{O}_X,$$
$$\omega_{id} : W \mapsto W \text{ as } \mathcal{O}_X\text{-module.}$$

It is essentially tautological that P is the right G-torsor $\mathcal{I}som^{\otimes, \langle V \rangle}(\omega_x, \omega_{id})$. The horizontal sections of P over $(\mathcal{O}_{X,x})^\wedge$ are precisely the set (actually a right G(K) torsor) S of those fundamental solution matrices $\hat{w}$ which have the following property: for every "construction of linear algebra" Constr(V), and every sub-D.E. W of Constr(V), the induced horizontal section Constr(w) of $\mathcal{I}som(\mathrm{Constr}(V)_x \otimes_K \mathcal{O}_X, \mathrm{Constr}(V))$ maps $W_x \otimes_K \mathcal{O}_X$ to W. We claim that if G(K) is Zariski dense in G (e.g., if K is algebraically closed), then this set S is Zariski dense in P. Indeed, this is more general nonsense:

Suppose that G(K) is Zariski dense in G. Then for any right G-torsor P on X with right G-equivariant integrable connection, the set (actually a right G(K) torsor) S of its horizontal sections over $(\mathcal{O}_{X,x})^\wedge$ is Zariski dense in P. [proof: the annihilator ideal I of S in $\mathbb{A}$ is horizontal, the union of the D.E.'s $I \cap \mathbb{A}_n$, so determined by its fibre at x. But at x it annihilates all the K-rational points G(K) of $G \approx P_x$, and as these are Zariski dense I = 0.]

Corollary 2.3.2.1 Suppose K is algebraically closed. The algebraic group G acts transitively on $V_x - \{0\}$ if and only if for every nonzero $v \in V_x$, the corresponding horizontal section $\hat{v}$ has its n=rank(V) coefficients algebraically independent over K(X).
proof If G acts transitively, then for each nonzero v, the orbit Gv is n-dimensional. Conversely, let $v \neq 0$. If Gv is n-dimensional, then Gv must be all of $V_x - \{0\}$, simply because Gv is constructible. QED

2.4 Behavior of G_{gal} under Specialization

Let K be a field of characteristic zero, t an indeterminate, R the ring K[[t]], X/R a smooth separated R-scheme of finite type with geometrically connected fibres, and $x \in X(R)$. Suppose we are given a locally free $\mathcal{O}_X$-module V of finite rank n together with an integrable connection $\nabla : V \to V \otimes_{\mathcal{O}_X} \Omega^1{}_{X/R}$ relative to the base R. Let us denote by η: R $\to$ K((t)) the inclusion (generic point of Spec(R)), and by s: R $\to$ K the specialization map $t \mapsto 0$ (special point of Spec(R)). Via these extensions of scalars, we obtain D.E.'s $V(\eta)$ on $X_\eta/K((t))$ and V(s) on X_s/K, and rational points $x_\eta \in X_\eta(K((t)))$ and $x_s \in X_s(K)$. Thus we can speak of the differential galois groups $G_{gal}(V(\eta), x_\eta) \subset GL(V(\eta)_{x_\eta})$ and $G_{gal}(V(s), x_s) \subset GL(V(s)_{x_s})$

Specialization Theorem 2.4.1 (Ofer Gabber) Let G/R be the closed R-flat subgroupscheme of $GL(V_x) \approx GL(n)/R$ obtained as the schematic closure of $G_{gal}(V(\eta), x_\eta)$, and let G_s denote its special fibre. Then there is natural inclusion $G_{gal}(V(s), x_s) \subset G_s$ (inside $GL(V(s)_{x_s}) \approx GL(n)/K$).

proof Denote by A the coordinate ring of $GL(V_x)$, by $A_d \subset A$ the R-submodule of those functions which are the restrictions of polynomials of degree $\leq$ d in the n^2 matrix coefficients $X_{i,j}$ and the function $1/\det(X_{i,j})$, and by $I_\eta \subset A_\eta$ the ideal defining $G_{gal}(V(\eta), x_\eta)$ in $GL(V(\eta))$. Then the schematic closure G of $G_{gal}(V(\eta), x_\eta)$ in $GL(V_x)$ is defined by the ideal $I_\eta \cap A$. Since A is noetherian, this ideal is generated by $I_\eta \cap A_d$ for sufficiently large d.

For each d, there is a natural "construction of linear algebra" $\mathbb{A}_d := \mathrm{Constr}_d(V)$ whose pullback $(\mathbb{A}_d)_x$ by the section x is A_d. Since the intersection $I_\eta \cap (A_d)_\eta$ is tautologically $G_{gal}(V(\eta), x_\eta)$-stable, it makes sense to speak of the horizontal submodule M_d of $\mathbb{A}_d(\eta)$ whose x_η-fibre is $I_\eta \cap (A_d)_\eta$. Let $\mathcal{M}_d$ denote the intersection $M_d \cap \mathbb{A}_d$ inside $\mathbb{A}_d(\eta)$. Then $\mathcal{M}_d$ is visibly horizontal, R-flat, and $\mathcal{O}_X$-coherent. Hence it is $\mathcal{O}_X$-locally free, as results from

Lemma 2.4.2 Let X/R as above. Let $(\mathcal{M}, \square)$ be an $\mathcal{O}_X$-coherent module $\mathcal{M}$ together with an integrable connection $\square: \mathcal{M} \to \mathcal{M} \otimes_{\mathcal{O}_X} \Omega^1{}_{X/R}$ relative to the base R. Then

(1) $\mathcal{M}$ is $\mathcal{O}_X$-locally free if (and only if) it is R-flat.

(2) If there exists a section $x \in X(R)$, then $\mathcal{M}$ is $\mathcal{O}_X$-locally free if (and only if) it $\mathcal{M}_x := x^*\mathcal{M}$ is R-flat.

proof of Lemma The "only if" is trivial, since X/R is flat. Because $\mathcal{M}$ is $\mathcal{O}_X$-coherent and is endowed with an integrable connection, it is automatically locally free on X_η. Suppose that $\mathcal{M}$ is R-flat. To prove (1), it suffices to show that $\mathcal{M}$ is locally free over the local ring $\mathcal{O}_{X,p}$ of X at every closed point p of the special fibre. Since finite extensions of K are harmless, we may assume that p is K-rational. By faithful flatness of the completion, it suffices to treat the formal case, where X is the spec of $R[[x_1, \ldots, x_n]]$. But for any noetherian $\mathbb{Q}$-algebra R, the functor "horizontal sections", $\mathcal{M} \mapsto \mathcal{M}^\square$ defines an equivalence of categories

{coherent $R[[x_1, \ldots, x_n]]$-modules with integrable connection over R} ≈

≈ {coherent R-modules},

whose inverse functor is

$$M \mapsto M[[x_1, \ldots, x_n]] \text{ with the trivial connection } 1 \otimes d.$$

From this explicit description of the inverse, it is obvious that for R local we have

$\mathcal{M}$ is R-flat $\Leftrightarrow$ $\mathcal{M}^\square$ is R-flat $\Leftrightarrow$ $\mathcal{M}^\square$ is R-free $\Leftrightarrow$ $\mathcal{M}$ is $R[[x_1, \ldots, x_n]]$-free.

Supose that there exists a section $x \in X(R)$, and denote by p the point x_s. Working at p as above, we see from the explicit description that $\mathcal{M}^\square$ is R-isomorphic to $x^*\mathcal{M}$. So if $x^*\mathcal{M}$ is R-flat, then $\mathcal{M}$ is R-flat in a neighborhood of p in X. But the locus of non R-flatness of $\mathcal{M}$ is the support of Ker(Left(t): $\mathcal{M} \to \mathcal{M}$); but this is a coherent $\mathcal{O}_{X_s}$-module with integrable connection, i.e., a D.E. on X_s/K, so it is $\mathcal{O}_{X_s}$-locally free; as it vanishes near p, it must be zero. QED

By the definition of $\mathcal{M}_d \subset \mathbb{A}_d$, the quotient $\mathbb{A}_d/\mathcal{M}_d$ is R-flat.

Therefore it too is $\mathcal{O}_X$-locally free; in other words, $\mathcal{M}_d$ is **locally a direct factor** of $\mathbb{A}_d$. Pulling back by the section x, we find that $(\mathcal{M}_d)_x$ is a direct factor of $(\mathbb{A}_d)_x. = A_d$. As the generic fibre of $(\mathcal{M}_d)_x$ is $I_\eta \cap (A_d)_\eta$, we conclude that $(\mathcal{M}_d)_x$ is $I_\eta \cap A_d$

Now consider the special fibre G_s of G. It is defined by the ideal $(I_\eta \cap A)_s$ of A_s. Since $I_\eta \cap A$ is the union of the $I_\eta \cap A_d$, and each $I_\eta \cap A_d$ is a direct factor of A_d, we see that $(I_\eta \cap A)_s$ is the union of the $(I_\eta \cap A_d)_s$. But $(I_\eta \cap A_d)_s$ is the x_s-fibre of $(\mathcal{M}_d)(s)$. Since $\mathcal{M}_d$ is a locally direct factor of $\mathbb{A}_d$, $(\mathcal{M}_d)(s)$ is a sub-D.E. of $(\mathbb{A}_d)(s)$, and therefore its x_s-fibre $(I_\eta \cap A_d)_s$ is a $G_{gal}(V(s), x_s)$-stable subspace of $(\mathbb{A}_d)_{x_s} = (A_d)_s$. Therefore the entire ideal $(I_\eta \cap A)_s$ of A_s is $G_{gal}(V(s), x_s)$-stable. As this ideal kills the identity element in GL(n), it kills all of $G_{gal}(V(s), x_s)$. This means precisely that $G_{gal}(V(s), x_s) \subset G_s$. QED

Corollary 2.4.3.1 Hypotheses and notations as in the specialization theorem 2.4.1 above, we have the inequality of dimensions

$$\dim_K(G_{gal}(V(s), x_s)) \leq \dim_{K((t))}(G_{gal}(V(\eta), x_\eta)).$$

By successive specialization, we find

Corollary 2.4.3.2 Let K be a field of characteristic zero, R a smooth geometrically connected affine K-algebra or a power series ring in finitely many variables over K, X/R a smooth separated R-scheme of finite type with geometrically connected fibres, and $x \in X(R)$. Suppose given a locally free $\mathcal{O}_X$-module V of finite rank n with an integrable connection $\square : V \to V \otimes_{\mathcal{O}_X} \Omega^1_{X/R}$ relative to R. Let η be the generic point of Spec(R), and s any closed point of Spec(R). Then we have the inequality of dimensions

$$\dim_{K(s)}(G_{gal}(V(s), x_s)) \leq \dim_{K(\eta)}(G_{gal}(V(\eta), x_\eta)).$$

Remark 2.4.4 Although the differential galois group G_{gal} "decreases under specialization", it is not true that G_{gal} is algebraically constructible in a family. Consider for example over the ground ring $R := \mathbb{C}[t]$ the scheme $X := (\mathbb{G}_m)_R = \mathrm{Spec}(R[x, x^{-1}])$, on X the rank one $\mathcal{O}_X$-module $\mathcal{O}_X$ with the connection (relative to R) given by $\square(D)(f) = D(f) + tf$, where D is xd/dx. For any irrational value of t, G_{gal} is $\mathbb{G}_m$, while if t is a rational number in lowest terms a/b, then G_{gal} is μ_b.

2.5 Specialization of Morphisms

Let K be a field of characteristic zero, R a discrete valuation ring with residue field K, uniformizing parameter π, and fraction field L. Denote by η: R → L the inclusion (generic point of Spec(R)), and by s: R → R/πR = K the specialization map (special point of Spec(R)).Let X/R be a smooth separated R-scheme of finite type with geometrically connected fibres. Given a locally free $\mathcal{O}_X$-module V of finite rank n together with an integrable connection $\square : V \to V \otimes_{\mathcal{O}_X} \Omega^1_{X/R}$ relative to the base R, we obtain D.E.'s V(η) on X_η/L and V(s) on X_s/K by restriction to the two fibres of X/R.

Proposition 2.5.1 On X/R as above, let V and W be two locally free $\mathcal{O}_X$-modules of finite rank together with integrable connections relative to the base R. If $\mathrm{Hom}_{\mathrm{D.E.}(X_\eta/L)}(V(\eta), W(\eta))$ is nonzero, then $\mathrm{Hom}_{\mathrm{D.E.}(X_s/K)}(V(s), W(s))$ is nonzero.

proof Suppose that φ: V(η) → W(η) is a nonzero horizontal morphism. Since V and W are locally free of finite rank, $\mathrm{Hom}_{\mathcal{O}_X}(V, W)$ is a torsion free R-module and $\mathrm{Hom}_{\mathcal{O}_X}(V, W)[1/\pi] \cong \mathrm{Hom}_{\mathcal{O}_{X_\eta}}(V(\eta), W(\eta))$. So multiplying φ by a suitable power of π, we may suppose that $\varphi(V) \subset W$ and $\varphi(V) \not\subset \pi W$. This "new" φ is still nonzero and horizontal (π is a "constant") on X_η/L. Therefore φ is horizontal as a map V → W on X/R (because the obstruction to its horizontality lies in the torsion free R-module $\mathrm{Hom}_{\mathcal{O}_X}(V, W \otimes_{\mathcal{O}_X} \Omega^1_{X/R})$) . Its special fibre φ_s is therefore a horizontal map V(s) → W(s) on X_s/K, which by construction is nonzero. QED

One variant of this gives a sort of "Brauer theory":

Variant 2.5.2 (O. Gabber) On X/R as above, let V and W be two locally free $\mathcal{O}_X$-modules of finite rank together with integrable connections relative to the base R. Denote by $V(s)^{ss}$ and by $W(s)^{ss}$ the semisimplifications of V(s) and of W(s) respectively in the category D.E.(X_s/K). If there exists an isomorphism V(η) ≈ W(η) as D.E.'s on X_η/L, then there exists an isomorphism $V(s)^{ss} \approx W(s)^{ss}$ as D.E.'s on X_s/K.

proof Let $\varphi: V(\eta) \xrightarrow{\sim} W(\eta)$ and $\psi: W(\eta) \xrightarrow{\sim} V(\eta)$ be inverse isomorphisms. View V (resp. W) as an $\mathcal{O}_X$-submodule of $V(\eta)$ (resp. $W(\eta)$). Then for n >> 0, $\pi^n\varphi(V) \subset W$, and $\pi^n\psi(W) \subset V$. Now identify $V(\eta)$ to $W(\eta)$ by means of φ and ψ; V and W then appear as horizontal $\mathcal{O}_X$-submodules of their common generic fibre $V(\eta) = W(\eta)$ with

$$(\pi^n)V \subset W, \ (\pi^n)W \subset V.$$

For each pair of integers (a,b), let us denote

$$M(a,b) := \pi^a V + \pi^b W, \text{ an } \mathcal{O}_X\text{-submodule of } V(\eta) = W(\eta).$$

This M(a,b) is $\mathcal{O}_X$-coherent (a quotient of $V \oplus W$), horizontal, and R-flat (being a submodule of $V(\eta) = W(\eta)$), so by 2.4.2 M(a,b) is $\mathcal{O}_X$-locally free. Its generic fibre $M(a,b)(\eta)$ is $V(\eta) = W(\eta)$. Clearly we have

$$V = M(0,n), \ W = M(n,0).$$

So it suffices to show that the isomorphism class of $M(a,b)(s)^{ss}$ is independent of (a,b). For this it suffices to compare (a,b) to both (a+1,b) and to (a,b+1). By symmetry, it suffices to compare (a,b) to, say, (a+1,b). We have

$$\pi M(a+1, b) \subset \pi M(a,b) = M(a+1, b+1) \subset M(a+1,b) \subset M(a,b) .$$

Thus we are reduced to treating universally the case in which

$$\pi W \subset \pi V \subset W \subset V.$$

In this case we have short exact sequences of DE.'s on X_s/K

$$0 \to W/\pi V \to V/\pi V \to V/W \to 0,$$

$$\begin{array}{c} 0 \to \pi V/\pi W \to W/\pi W \to W/\pi V \to 0, \\ \uparrow\wr\ \pi \qquad\qquad\qquad\qquad\qquad \\ V/W \qquad\qquad\qquad\qquad\qquad \end{array}$$

which show that $V(s) := V/\pi V$ has the same semisimplification as $W(s) := W/\pi W$, namely $(W/\pi V)^{ss} \oplus (V/W)^{ss}$. QED

In the case of several parameters, we have only the weaker

Theorem 2.5.3 (Specialization of (Iso)morphisms) Let K be a field of characteristic zero, R a smooth geometrically connected affine K-algebra or a power series ring in finitely many variables over K, X/R a smooth separated R-scheme of finite type with geometrically connected fibres. Suppose given locally free $\mathcal{O}_X$-modules V and W of finite rank with integrable connections relative to R. Let η be the generic point of Spec(R), and s any closed point of Spec(R). Then

(1) If $\mathrm{Hom}_{D.E.(X_\eta/K(\eta))}(V(\eta), W(\eta))$ is nonzero, then $\mathrm{Hom}_{D.E.(X_s/K(s))}(V(s), W(s))$ is nonzero.

(2)If there exists an isomorphism $V(\eta) \approx W(\eta)$ as D.E.'s on $X_\eta/K(\eta)$, and if at least one of $V(s)$ or $W(s)$ is irreducible as D.E. on X_s/K, then there exists an isomorphism $V(s) \approx W(s)$ as D.E.'s on $X_s/K(s)$.

proof Assertion (1) is immediate from the Proposition above by successive specialization. For (2), notice that if $V(\eta) \approx W(\eta)$ then V and W have the same rank, and hence $V(s)$ and $W(s)$ have the same rank. So if either $V(s)$ or $W(s)$ is irreducible as D.E., any nonzero map between them as D.E.'s is an isomorphism. QED

For the rest of this chapter we will take $K = \mathbb{C}$ unless there is explicit mention to the contrary.

2.6 Direct Sums and Tensor Products

Proposition 2.6.1 (Goursat-Kolchin-Ribet, [Kol], [Ri]) Suppose that V_1, ..., V_n are $n \geq 2$ D.E.'s on $X/\mathbb{C}$, ω a $\mathbb{C}$-valued fibre functor on D.E.$(X/\mathbb{C})$. Suppose that for each i, V_i has rank $n_i \geq 2$, and that its differential galois group $G_i := G_{gal}(V_i, \omega) \subset GL(n_i)$ has $G_i^{0,der}$ one of the groups

- $SL(n_i)$, any $n_i \geq 2$,
- $Sp(n_i)$, any even $n_i \geq 4$,
- $SO(n_i)$, $n_i = 7$ or any $n_i \geq 9$,
- $SO(3)$, if $n_i = 3$ and no $n_j = 2$,
- $SO(5)$, if $n_i = 5$ and no $n_j = 4$,
- $SO(6)$, if $n_i = 6$ and no $n_j = 4$,
- $G_2 \subset SO(7)$, if $n_i = 7$,
- $Spin(7) \subset SO(8)$ if $n_i = 8$, and no $n_j = 7$.

Suppose that for all $i \neq j$, and all rank one D.E.'s L on X, there exist no isomorphism from V_i to either $L \otimes V_j$ or to $L \otimes (V_j^{\vee})$. Then the differential galois group G of $\oplus V_i$ has $G^{0,der} = \prod G_i^{0,der}$, and (consequently) that of $\otimes V_i$ has $G^{0,der}$ = the image of $\prod G_i^{0,der}$ in $\otimes \mathrm{std}_{n_i}$.

proof This is an immediate application of 1.8.2, taking $G := G_{gal}(\oplus V_i, \omega)$ and $\rho_i :=$ the action of G on $\omega(V_i)$, since we have eliminated SO(8), the nonsimple SO(4), the nonsemisimple SO(2), and the repetitions of isomorphism classes of simple Lie algebras $A_1=B_1$, $B_2=C_2$, $A_3=D_3$. QED

2.7 A Basic Trichotomy

We say that V is **irreducible** on X if it is irreducible as an object of D.E.$(X/\mathbb{C})$. This is equivalent to saying that $\omega(V)$ is an irreducible representation of $G_{gal}(V, \omega)$, or equivalently that it is an irreducible representation of $\pi_1^{diff}(X/\mathbb{C},\omega)$.

In analogy with the case of ℓ-adic sheaves (cf [Ka-MG, Part I]), we say that V is **Lie-irreducible** if $\omega(V)$ is an irreducible representation of the identity component $(G_{gal}(V, \omega))^0$ of $G_{gal}(V, \omega)$. In view of [Ka-DGG,1.2.5.4 and 1.4.4], V is Lie-irreducible if and only if for every connected finite etale covering $\pi:Y\rightarrow X$, the inverse image $\pi^*(V)$ on Y is irreducible on Y. Clearly V is Lie-irreducible if and only if its restriction to some, or to every, finite etale connected covering is Lie-irreducible.

Rigidity Lemma 2.7.1 Suppose that V and W are Lie-irreducible D.E.'s on $X/\mathbb{C}$, and that $\pi: Y \rightarrow X$ is a finite etale connected galois covering on which $\pi^*V \approx \pi^*W$. Then there exists a rank one D.E. L on X with π^*L trivial and an isomorphism $W \approx V\otimes L$ on X.

proof Denote by G the π_1^{diff} of X, and by H that of Y. Then H is a normal subgroup of G, and V and W are two representations of G whose restrictions to H are both irreducible and isomorphic to each other. But whenever H is a normal subgroup of G, two representations V and W of G whose restrictions to H are both isomorphic and irreducible differ by the character $\mathrm{Hom}_H(V, W)$ of G/H, i.e., $V\otimes\mathrm{Hom}_H(V, W) \approx W$. QED

We say that V is **induced** if there exists a connected finite etale covering $\pi:Y\rightarrow X$ of degree $d\geq 2$ and an object W in D.E.$(Y/\mathbb{C})$ such that V $\approx \pi_*(W)$; equivalently(cf [Ka-DGG, 1.4.6, 1.4.7], V is induced if and only if the representation $\omega(V)$ of $G_{gal}(V, \omega)$ is induced from a representation of an open subgroup H of $G_{gal}(V, \omega)$ of finite index $d\geq 2$.

Proposition 2.7.2 (compare Prop. 1 of [Ka-MG]) Suppose that the topological fundamental group of the complex manifold X^{an} is a free group (e.g.,X an open curve). Then for any irreducible object V in D.E.$(X/\mathbb{C})$, either V is induced, or V is Lie-irreducible, or there exists a divisor $d\geq 2$ of the rank n of V,and a factorization of V as a tensor product $V = W\otimes K$ where W is Lie-irreducible of rank n/d, and where K is an irreducible of rank d which becomes trivial on a finite etale covering of X. In this last case, the pair (W, K) is unique up to replacing it by $(W\otimes\mathcal{L}, K\otimes\mathcal{L}^\vee)$ with $\mathcal{L}$ of rank-one and of finite order (i.e., $\mathcal{L}$ corresponds to a character of finite order of $\pi_1^{\mathrm{diff}}(X/\mathbb{C},\omega)$).

proof Let us denote by G the differential galois group $G_{\mathrm{gal}}(V, \omega)$, and by G^0 its identity component. Because G^0 is a normal subgroup of G, we have the customary dichotomy: an irreducible representation M of G is either isotypical when restricted to G^0, or M is induced from a representation of a proper subgroup H of G which contains G^0. Apply this to $M=\omega(V)$: if either M is irreducible on G^0, or if M is induced, there is nothing further to prove.

The troublesome case is that in which the restriction of M to G^0 is isotypical but not irreducible, say $M \approx$ d copies of an irreducible representation M_0 of G^0 with $d \geq 2$. In terms of differential equations, this means there exists a connected finite etale galois covering of X, say $\pi:Y\to X$ with galois group H, such that on Y

$$\pi^*(V) \approx d\geq 2 \text{ copies of a Lie-irreducible } W \text{ on } Y.$$

By Jordan-Holder theory, the isomorphism class of W must be H-invariant (in the sense that for every h in H, there exists an isomorphism of h^*W with W). We now apply the following

Lemma 2.7.3 Suppose that the topological fundamental group of the complex manifold X^{an} is a free group (e.g.,X an open curve). Let $\pi:Y\to X$ be a connected finite etale galois covering of X with galois group H, and W an irreducible object of $D.E.(Y/\mathbb{C})$ whose isomorphism class is H-invariant. Then there exists a connected finite etale covering $p : Z\to Y$

$$Z \xrightarrow{p} Y \xrightarrow{\pi} X$$

such that Z is finite etale galois over X and such that $p^*(W)$ on Z descends to an object **W** on X. Moreover, **W** is Lie irreducible on X if W is Lie-irreducible on Y.

proof of Lemma For each h in H, choose an isomorphism $A(h):W \cong h^*W$. For each g in H, the pullback $g^*(A(h))$ is an isomorphism from g^*W to g^*h^*W. If $A(hg)$ were equal to $g^*(A(h))\circ A(g)$ for all pairs (g,h) of elements of H, we could interpret our choice of A(h)'s as descent data for W relative to the covering π, and our W would descend to X. At worst there exists a $\mathbb{C}^\times$ factor a(h,g) with

$$a(h,g) = (A(hg))^{-1}\, g^*(A(h))\circ A(g),$$

simply because the right hand side is an automorphism of the irreducible object W. This a(h,g) is a two-cocycle on H with values in $\mathbb{C}^\times$ (with trivial H-action). If its cohomology class were trivial, say $a(h,g) = b(hg)/b(h)b(g)$ for some $\mathbb{C}^\times$-valued function b on H, then $h \mapsto b(h)A(h)$ is descent data, and W descends to X.

Because $H^2(H, \mathbb{C}^\times)$ is killed by N := Card(H), the Kummer sequence shows that every element of $H^2(H, \mathbb{C}^\times)$ is in the image of $H^2(H, \mu_N(\mathbb{C}))$. So we may correct the choice of the A(h)'s by scalar factors so that a(h,g) lies in $\mu_N(\mathbb{C})$.

Because the topological fundamental group π_1 of X^{an} is free, any subgroup F of π_1 is also free. So for any finite abelian coefficient group A (e.g., $\mu_N(\mathbb{C})$) with trivial action of π_1, and any normal subgroup F of finite index in π_1, the Hochschild-Serre spectral sequence

$$E_2^{a,b} = H^a(\pi_1/F, H^b(F, A)) \Rightarrow H^{a+b}(\pi_1, A)$$

has $E_2^{a,b} = 0$ for $b \neq 0,1$. Any element of $H^1(F, A)$ dies when restricted to a smaller normal F of finite index, so the direct limit over all normal F's of finite index of these spectral sequences has $E_2^{a,b} = 0$ for $b \neq 0$. Thus

$$H^*(\pi_1, A) = \lim_{\text{finite quotients H of } \pi_1} H^*(H, A).$$

For any $i \geq 2$, we have $H^i(\pi_1, A) = 0$, so in particular the direct limit

$$\lim_{\text{finite quotients H of } \pi_1} H^2(H, A) \text{ must vanish.}$$

Therefore any given element in $H^2(H, \mu_N(\mathbb{C}))$ dies in $H^2(\tilde{H}, \mu_N(\mathbb{C}))$ for some larger finite quotient $\tilde{H}$ of π_1. The covering Z of X defined by such an $\tilde{H}$ sits in a diagram

$$Z \xrightarrow{p} Y \xrightarrow{\pi} X.$$

When we pull back W from Y to Z, and denote by $\tilde{h} \mapsto h$ the canonical projection of $\tilde{H}$ onto H, the dying means precisely that the choice of isomorphisms on Z

$$A(\tilde{h}) := p^*(A(h)) : p^*W \to \tilde{h}^*p^*W = p^*h^*W$$

can be corrected by invertible scalars to give descent data on p^*W for the covering $Z \to X$. Finally, if W is Lie-irreducible on Y, then p^*W is Lie-irreducible on Z, so any descent W of p^*W to X is itself Lie-irreducible. QED

We now return to the proof of 2.7.2. Applying the lemma to W on Y,we see that at the expense of enlarging the covering group H, there exists a Lie-irreducible **W** on X and a connected finite etale galois covering $\pi:Y\to X$ with galois group H such that on Y

$$\pi^*V \approx \text{d copies of } \pi^*\mathbf{W}.$$

Denote by $Hom_Y(\mathbf{W}, V)$ the subobject of the internal hom object $Hom(\mathbf{W}, V)$ on X obtained by descending through π the $\mathcal{O}_Y$-span of the global horizontal sections of the internal hom $Hom(\pi^*\mathbf{W}, \pi^*V)$ ($= \pi^*Hom(\mathbf{W}, V)$) on Y. [Because **W** and V are each irreducible on X, their internal hom $Hom(\mathbf{W}, V)$ is completely reducible on X (being a representation of the reductive group $G_{gal}(\mathbf{W}\oplus V, \omega)$), and so $\pi^*Hom(\mathbf{W}, V)$ is completely reducible on Y. We are descending its trivial isotypical component.] We have a canonical morphism of D.E.'s on X

$$\mathbf{W}\otimes Hom_Y(\mathbf{W}, V) \to V, \quad (\mathbf{w}, \varphi) \mapsto \varphi(\mathbf{w})$$

which becomes an isomorphism on Y, so must already be an isomorphism on X.

This is the desired factorization of V as $W\otimes K$, with W Lie-irreducible of rank n/d and K of rank d becoming trivial on a finite etale covering of X (K must be irreducible because $W\otimes K$ is). To see its essential uniqueness, suppose that $W_1\otimes K_1$ were another. Pulling back to a sufficiently small connected finite etale galois covering Z of X, V becomes isomorphic to d copies of the Lie-irreducible W, and to d_1 copies of the Lie-irreducible W_1. By Jordan-Holder theory, we must have $d = d_1$, and the two Lie-irreducible representations W and W_1 of $\pi_1^{\text{diff}}(X/\mathbb{C},\omega)$ become isomorphic irreducibles on the open normal subgroup $\pi_1^{\text{diff}}(Z/\mathbb{C},\omega)$. Therefore W and W_1 are projectively equivalent as representations of $\pi_1^{\text{diff}}(X/\mathbb{C},\omega)$, and hence $W_1 \approx W\otimes\mathcal{L}$ for some character of $\pi_1^{\text{diff}}(X/\mathbb{C},\omega)$) which is trivial on $\pi_1^{\text{diff}}(Z/\mathbb{C},\omega)$. QED

Remarks 2.7.3 (1) The two non-Lie-irreducible cases of the Proposition are not mutually exclusive. Indeed, if W is Lie-irreducible, and if K is an irreducible which is induced from a K_0 on a connected finite etale $\pi:Y\to X$ such that K_0 itself becomes trivial on a finite etale covering of

Y, then $W \otimes K$ is also the induction $\pi_*(\pi^*(W) \otimes K_0)$. Of course this is a reflection of the fact that already for a finite group G with a normal subgroup H, the "dichotomy" for irreducibles of G to be either induced or H-isotypical is not a true dichotomy; for H = {e}, every representation is H-isotypical.
(2) One could also give a proof of this proposition which is analogous to the proof of Prop. 1 of [Ka-MG], by using the fact (2.2.4.1) that when the topological π_1 is a free group, any projective representation of π_1^{diff} can be linearized.

Corollary 2.7.4 (compare [Ka-MG], Cor.3) Suppose that X is an open curve, and that V is an irreducible object of D.E.(X/ℂ). Suppose that the rank of V is a prime p. If det(V) is of finite order, then V is either Lie-irreducible or induced or it becomes trivial on a finite etale covering.
proof Indeed if V is neither induced nor Lie-irreducible, it is a tensor product $W \otimes K$ of a Lie-irreducible W of rank one and of an irreducible K of rank p which becomes trivial on a finite etale covering. So $\det(V) \approx W^{\otimes p} \otimes \det(K)$, with det(K) of finite order. So if det(V) is of finite order, W is of finite order, whence $V = W \otimes K$ is trivial on a finite etale covering. QED

Corollary 2.7.5 Suppose that X is an open curve, with complete nonsingular model $\overline{X}$, and that V is an irreducible object of D.E.(X/ℂ). At each point at infinity $\infty_i \in \overline{X} - X$, let the slopes of V, written in lowest terms, be the rational numbers $a_{i,j}/b_{i,j}$, with multiplicities $n_{i,j}b_{i,j}$. Suppose that $\gcd_{i,j}(\text{all } n_{i,j}) = 1$. Then V is either induced or V is Lie-irreducible.
proof If V is neither induced nor Lie-irreducible, then for some integer $d \geq 2$ we have a factorization of V as $W \otimes K$, where K is a rank d object which becomes trivial on a finite etale covering of X. Such a K is entirely of slope zero at any ∞_i (cf [Ka-DGG, 2.6.2]). Therefore the slopes with multiplicity of V at ∞_i are the slopes with multiplicity of W at ∞_i, repeated d times. In particular, d divides every $n_{i,j}$. QED
Remark 2.7.6 This slope criterion for "induced or Lie-irreducible" is very easy to verify when it applies. The problem comes after, in deciding which of the two cases one is in. Only on $\mathbb{G}_m$, where "induced" is necessarily "Kummer induced", does one know a manageable sufficient conditions that a D.E. V not be induced (namely that there

exist no integer $d \geq 2$ such that both $V \mid I_0$ and $V \mid I_\infty$ are induced from the unique subgroups of index d in I_0 and I_∞ respectively). Ignoring this problem for a moment, we state a quite general result.

2.8 The Main D.E. Theorem

Main D.E. Theorem 2.8.1 Suppose that X is an open curve, with complete nonsingular model $\overline{X}$, and that V is a Lie-irreducible object of D.E.$(X/\mathbb{C})$ of rank n. Suppose that at some point at infinity $\infty \in \overline{X} - X$, the highest slope of V, written a/b in lowest terms, is > 0 and occurs with multiplicity b. Let $G \subset GL(\omega(V))$ denote the differential galois group of V, G^0 its identity component, and $G^{0,der}$ the commutator subgroup of G^0. Then G^0 is equal either to $G^{0,der}$ or to $\mathbb{G}_m G^{0,der}$, and the list of possible $G^{0,der}$ is given by:

(1) If b is odd, $G^{0,der}$ is $SL(\omega(V))$; if b=1, then G is $GL(\omega(V))$.

(2) If b is even, then either $G^{0,der}$ is $SL(\omega(v))$ or $SO(\omega(v))$ or (if n is even) $SP(\omega(v))$, or b=6, n=7,8 or 9, and $G^{0,der}$ is one of

n=7: the image of G_2 in its 7-dim'l irreducible representation
n=8: the image of Spin(7) in the 8-dim'l spin representation
the image of SL(3) in the adjoint representation
the image of $SL(2) \times SL(2) \times SL(2)$ in $std \otimes std \otimes std$
the image of $SL(2) \times Sp(4)$ in $std \otimes std$
the image of $SL(2) \times SL(4)$ in $std \otimes std$
n=9: the image of $SL(3) \times SL(3)$ in $std \otimes std$.

proof Since V is irreducible, G^0 is reductive; its connected center $Z^0 = (Z(G^0))^0$ is a torus, its derived group $G^{0,der}$ is semisimple, and $G^0 = Z^0 G^{0,der}$. Because V is Lie-irreducible, G^0 acts irreducibly on $\omega(V)$, and therefore its center $Z := Z(G^0)$ acts as scalars. Since $\omega(V)$ is a faithful representation of G^0, it follows that Z and a fortiori Z^0 are contained in the scalars. Since Z^0 is a torus, either Z^0 is trivial or it is $\mathbb{G}_m$. Because $G^0 = Z^0 G^{0,der}$ acts irreducibly on $\omega(V)$, already $G^{0,der}$ acts irreducibly on it. Therefore $\mathfrak{g} := \mathrm{Lie}(G^{0,der})$ is a semisimple Lie-subalgebra of $\mathrm{End}(\omega(V))$ which acts irreducibly on $\omega(V)$. By its very construction, $\mathfrak{g}$ is normalized by **any** subgroup K of G.

We now use the slope hypothesis that at some point at infinity ∞ the highest slope is a/b in lowest terms and its multiplicity is b to construct a diagonal subgroup K of G, to which we will then apply Gabber's "torus trick" Theorem 0.

As a representation of I_∞, V is the direct sum

$$V = V_{a/b} \oplus V_{< a/b} = (\text{slope } a/b, \text{ rank } b) \oplus (\text{all slopes} < a/b).$$

In order to describe the representation $V_{a/b}$ of I_∞ explicitly, fix a uniformizing parameter $1/x$ at ∞. This identifies the ∞-adic completion of the function field of X with the Laurent series field $K:=\mathbb{C}((1/x))$. Fix a b'th root t of x, and denote by K_b the Laurent series field $\mathbb{C}((1/t))$. In this setting, I_∞ is the local differential galois group $I(K/\mathbb{C})$, and $I(K_b/\mathbb{C})$ is its unique closed subgroup of index b (cf [Ka-DGG, 2.6.3]). The representation $V_{a/b}$ of $I(K/\mathbb{C})$ is irreducible [Ka-DGG, 2.5.9.2], so it is the direct image from K_b of a rank-one D.E. of the form $(K_b, td/dt + P_a(t))$, where $P_a(t)$ is a polynomial in t of degree a (cf [Ka-DGG, proof of 2.6.6]). Therefore after pullback to K_b we have

$$V_{a/b}\otimes K_b = \oplus_{\zeta \in \mu_b} (K_b, td/dt + P_a(\zeta t))$$

$$V_{< a/b}\otimes K_b = \text{all slopes} < a.$$

At the expense of scaling the parameter $1/x$, we may assume that $P_a(t)$ is monic, say $t^a + f_{<a}(t)$, with $f_{<a}(t)$ a polynomial of degree strictly less than a. As D.E. on K_b, we have

$$(K_b, td/dt + P_a(\zeta t)) = (K_b, td/dt + (\zeta t)^a)\otimes(K_b, td/dt + f_{<a}(\zeta t)).$$

We will now analyse V as representation of the upper numbering subgroup $(I(K_b/\mathbb{C}))^{(a)}$ of $I(K_b/\mathbb{C})$. Because $(K_b, td/dt + f_{<a}(\zeta t))$ has slope $< a$, it is trivial as a character of $(I(K_b/\mathbb{C}))^{(a)}$. Therefore V as representation of the upper numbering subgroup $(I(K_b/\mathbb{C}))^{(a)}$ of $I(K_b/\mathbb{C})$ is given by

$$V \mid (I(K_b/\mathbb{C}))^{(a)} \approx \oplus_{\zeta \in \mu_b} (K_b, td/dt + \zeta^a t^a) \oplus (\text{trivial of rank } n-b).$$

Because $\gcd(a, b) = 1$, as ζ runs over $\mu_b(\mathbb{C})$ the ζ^a's are just a permutaion of the ζ's, so we may rewrite this as

$$V \mid (I(K_b/\mathbb{C}))^{(a)} \approx \oplus_{\zeta \in \mu_b} (K_b, td/dt + \zeta t^a) \quad \oplus \quad (\text{trivial of rank } n-b).$$

For each ξ in $\mathbb{C}$ we denote by

$$\chi_\xi := \text{the character of } (I(K_b/\mathbb{C}))^{(a)} \text{ given by } (K_b, td/dt + \xi t^a).$$

The key observation is that for ξ, ν in $\mathbb{C}$ we have

$$\chi_\xi \chi_\nu = \chi_{\xi+\nu},$$

χ_ξ is trivial on $(I(K_b/\mathbb{C}))^{(a)}$ iff $\xi = 0$.

Let

$K :=$ the image of $(I(K_b/\mathbb{C}))^{(a)}$ in G.

Then K is a diagonal subgroup of G, and the diagonal entries of K are the n characters

the b characters χ_ξ as ξ runs over $\mu_b(\mathbb{C})$,

n-b repetitions of the trivial character χ_0.

We now apply the "torus trick" 1.0 to K. If b=n, we must find all relations of the form

$\chi_\alpha/\chi_\beta = \chi_\gamma/\chi_\delta$ on K, where α, β, γ, δ lie in $\mu_b(\mathbb{C})$,

or equivalently all relations of the form

Rel(b = n)($\mathbb{C}$) $\alpha - \beta = \gamma - \delta$, where α, β, γ, δ lie in $\mu_b(\mathbb{C})$.

If $b < n$, then we must find all relations of the form

Rel(b < n)($\mathbb{C}$) $\alpha - \beta = \gamma - \delta$, where α, β, γ, δ lie in $\mu_b(\mathbb{C}) \cup \{0\}$.

Lemma 2.8.2 If α, β, γ, δ are complex numbers of absolute value one which satisfy $\alpha - \beta = \gamma - \delta$, then we are in one of the following three cases:

(1) $\alpha = \beta$ and $\gamma = \delta$
(2) $\alpha = \gamma$ and $\beta = \delta$
(3) $\alpha = -\delta$ and $\beta = -\gamma$.

proof Here is a geometric argument. Suppose that we are not in case (1). Then the line segment $\alpha \rightarrow \beta$ is an oriented chord of the unit circle, and $\gamma \rightarrow \delta$ is a parallel oriented chord of the same circle having the same length. So either the two chords coincide (this is case (2)) or they are symmetric with respect to the unique diameter to which they are both parallel (this is case (3)).QED

Corollary 2.8.2.1 If α, β, γ, δ in $\mu_b(\mathbb{C})$ satisfy $\alpha - \beta = \gamma - \delta$, then either

(1) $\alpha = \beta$ and $\gamma = \delta$
or (2) $\alpha = \gamma$ and $\beta = \delta$
or b is even and (3) $\alpha = -\delta$ and $\beta = -\gamma$.

Lemma 2.8.3 If α, β, γ are complex numbers of absolute value one which satisfy $\alpha - \beta = \gamma$, then α/β is a primitive sixth root of unity, and γ/β is $(\alpha/\beta)^2$.

proof Applying complex conjugation, $\overline{\alpha} - \overline{\beta} = \overline{\gamma}$. Since α, β, γ are on the unit circle, we can rewrite this as $\alpha^{-1} - \beta^{-1} = \gamma^{-1}$. Then

$$(\alpha-\beta)(\alpha^{-1} - \beta^{-1}) = \gamma\gamma^{-1} = 1,$$

which simplifies to $(\alpha/\beta) + (\alpha/\beta)^{-1} = 1$, i.e. to $(\alpha/\beta)^2 - (\alpha/\beta) + 1 = 0$. Therefore α/β is a primitive sixth root of unity, and $\gamma/\beta = (\alpha/\beta) - 1$ is equal to $(\alpha/\beta)^2$. QED

Corollary 2.8.3.1 If α, β, γ in $\mu_b(\mathbb{C})$ satisfy $\alpha-\beta = \pm\gamma$, then 6 divides b, α/β is a primitive sixth root of unity, and $\pm\gamma/\beta$ is $(\alpha/\beta)^2$.

Using these two corollaries, we can compute **explicitly** the torus $\mathcal{T}$ which Theorem 1.0 assures us lies in $\mathcal{G}$.

If b=n, then the torus $\mathcal{T}$ is obviously contained in the set $\mathcal{S}_b$ of all diagonal matrices (X_ζ) of trace zero whose entries are indexed by the b'th roots of unity in $\mathbb{C}$. The relations defining $\mathcal{T}$ in $\mathcal{S}_b$ are

$$X_\alpha - X_\beta = X_\gamma - X_\delta \text{ whenever } \alpha, \beta, \gamma, \delta \in \mu_b(\mathbb{C}) \text{ and } \alpha - \beta = \gamma - \delta.$$

By 2.8.2.1, these relations are of three types:

(1) $X_\alpha - X_\alpha = X_\gamma - X_\gamma$ for all α,γ

or (2) $X_\alpha - X_\beta = X_\alpha - X_\beta$ for all a,β

or if b is even (3) $X_\alpha - X_\beta = X_{-\beta} - X_{-\alpha}$ for all α,β.

Of these, only type (3) relations impose any conditions, and these are

if b is even, $X_\alpha + X_{-\alpha} = X_\beta + X_{-\beta}$ for all α,β.

Since the trace is zero on $\mathcal{S}_b$, the common value of $X_\alpha + X_{-\alpha}$ can only be zero, and so type (3) relations are equivlent to

if b is even, $X_\alpha + X_{-\alpha} = 0$ for all α.

Case(b=n, b odd) $\mathcal{T}$ is all of $\mathcal{S}_b$; thus $\mathcal{T}$ contains Diag(n-1, -1,...,-1), whence $\mathcal{G}$ is $\mathcal{SL}(\omega(V))$ by Theorem 1.1.

Case (b=n, b even) $\mathcal{T}$ consists of those elements of $\mathcal{S}_b$ whose entries satisfy $X_\zeta + X_{-\zeta} = 0$ for every ζ in $\mu_b(\mathbb{C})$. In particular, $\mathcal{T}$ contains Diag(1,-1,0,...,0), and so $\mathcal{G}$ is $\mathcal{SL}(\omega(V))$ or $\mathcal{SO}(\omega(V))$ or $\mathcal{SP}(\omega(V))$ by Theorem 1.2.

If b < n, $\mathcal{T}$ is obviously contained in the set $\mathcal{S}_{b,n}$ of all diagonal matrices of trace zero of the form

(X_ζ's indexed by ζ in $\mu_b(\mathbb{C})$,X_0 repeated n-b times).

[Use the observation that in applying the torus trick, whenever two of the characters χ_i and χ_j of K are equal, then the corresponding entries X_i and X_j are equal, simply because $\chi_i/\chi_j = \chi_i/\chi_i$ on K; apply this to the n-b trivial characters of K to see that "their" entry X_0 is repeated n-b times.]

The relations defining $\mathcal{T}$ in $\mathcal{S}_{b,n}$ are

$$X_\alpha - X_\beta = X_\gamma - X_\delta$$

whenever α, β, γ, δ in $\mu_b(\mathbb{C}) \cup \{0\}$ and $\alpha - \beta = \gamma - \delta$.

To analyse these relations, it is best to distinguish cases according to how many of α, β, γ, δ are nonzero. When all four are nonzero, we are in the (b=n) case above; we get conditions on $\mathcal{T}$ only for b even, in which case the relations imposed are

(if b even) $\qquad X_\alpha + X_{-\alpha} = X_\beta + X_{-\beta}$ for all α,β in $\mu_b(\mathbb{C})$.

If exactly one of α, β, γ, δ is zero, say γ or δ, we get relations

$$X_\alpha - X_\beta = X_\gamma - X_0 \text{ if } \alpha,\beta,\gamma \text{ in } \mu_b(\mathbb{C}) \text{ have } \alpha-\beta=\gamma,$$

$$X_\alpha - X_\beta = X_0 - X_\delta \text{ if } \alpha,\beta,\delta \text{ in } \mu_b(\mathbb{C}) \text{ have } \alpha-\beta=-\delta.$$

By 2.8.3.1, such relations exist only if $6|b$, in which case α/β is a primitive sixth root of unity and γ/β (resp. $-\delta/\beta$) is $(\alpha/\beta)^2$.

If exactly two of α, β, γ, δ vanish, we get either a trivial relation

$$X_\alpha - X_\alpha = X_0 - X_0 \text{ if } \alpha \text{ in } \mu_b(\mathbb{C})$$

or, for b even, the nontrivial relation

(if b even) $\qquad X_\alpha - X_0 = X_0 - X_{-\alpha}$ if α in $\mu_b(\mathbb{C})$

which we rewrite as

(if b even) $\qquad X_\alpha + X_{-\alpha} = 2X_0$ if α in $\mu_b(\mathbb{C})$.

Since the trace vanishes on $\mathcal{S}_{n,b}$, we see that in fact

(if b even) $\qquad X_\alpha + X_{-\alpha} = 0 = X_0$ if α in $\mu_b(\mathbb{C})$.

If exactly three of α, β, γ, δ vanish, there are no relations, and if all four vanish there is only the trivial relation $X_0 - X_0 = X_0 - X_0$.

Case (b < n, b odd) $\mathcal{T}$ is all of $\mathcal{S}_{b,n}$; $\mathcal{T}$ contains $\mathrm{Diag}(n-1,-1,\ldots,-1)$, and so $\mathcal{G}$ is $\mathcal{SL}(\omega(V))$ by Theorem 1.1.

Case (b < n, b even not divisible by 6) $\mathcal{T}$ consists of those elements of $\mathcal{S}_{b,n}$ whose entries satisfy $X_0 = 0$, and $X_\zeta + X_{-\zeta} = 0$ for every ζ in $\mu_b(\mathbb{C})$. In particular, $\mathcal{T}$ contains $\mathrm{Diag}(1,-1,0,\ldots,0)$, and so $\mathcal{G}$ is $\mathcal{SL}(\omega(V))$ or $\mathcal{SO}(\omega(V))$ or (if n is even) $\mathcal{SP}(\omega(V))$ by Theorem 1.2.

Case(b < n, 6 divides b) In this case, it is most convenient to choose coset representatives α's for $\mu_b(\mathbb{C})$ modulo $\mu_6(\mathbb{C})$, and a primitive sixth root of unity ζ. Then $\mathcal{T}$ consists of those elements of $\mathcal{S}_{b,n}$ which satisfy $X_0=0$, and, for each coset representative α, the relations

$$X_\alpha - X_{\alpha\zeta} + X_{\alpha\zeta^2} = 0,$$

$$X_\alpha + X_{-\alpha} = 0,$$

$$X_{\alpha\zeta} + X_{-\alpha\zeta} = 0,$$

$X_{\alpha\,\xi}2 + X_{-\alpha\,\xi}2 = 0$.

In this case, the set of elements of $\mathcal{T}$ with $X_\xi = 0$ whenever $\xi^6 \neq 1$ is a G_2 torus. So by Theorem 1.3, either $\mathcal{G}$ is $\mathcal{SL}(\omega(V))$ or $\mathcal{SO}(\omega(V))$ or (if n is even) $\mathcal{SP}(\omega(V))$, or n is 7,8, or 9 (and so b =6, since 6|b and b < n) and $\mathcal{G}$ is one of the exceptional cases of Theorem 1.3. This concludes the proof of the Main D.E. Theorem 2.8.1, except for the fact that if b= 1, then G is $GL(\omega(V))$. But in this case, the highest slope a/b > 0, which occurs with multiplicity b= 1, is the slope of det(V), so det(V) is necessarily of infinite order. QED

2.9 Generalities on $\mathcal{D}$-modules on curves

We now turn to the explicit examination of some D.E.'s on $\mathbb{A}^1$ and on open sets of $\mathbb{A}^1$. Recall (cf. [Ka-DGG, 1.2.5]) that the differential galois group of a D.E. on $\mathbb{A}^1$ is always connected, so any irreducible V in D.E.$(\mathbb{A}^1/\mathbb{C})$ is automatically Lie-irreducible. In general, the theory of $\mathcal{D}$-modules provides us with reasonable sufficient conditions for the irreducibility of D.E.'s on open sets of $\mathbb{A}^1$ (though not for their Lie-irreducibility; this seems a much more difficult problem).

It will be convenient first to recall some of the basic facts about $\mathcal{D}$-modules on open curves. Thus let $X/\mathbb{C}$ be a nonempty smooth connected affine curve with coordinate ring $\mathcal{O} = \mathcal{O}_X$. For simplicity of exposition we assume that the invertible $\mathcal{O}$-module $\mathrm{Der}_\mathbb{C}(\mathcal{O}, \mathcal{O})$ is $\mathcal{O}$-free, and we pick an $\mathcal{O}$-basis ∂ of it . [For example, if X is an open set of $\mathbb{A}^1 := \mathrm{Spec}(\mathbb{C}[x])$, then d/dx is such a basis; if X is an open set of $\mathbb{G}_m$, then xd/dx is such a basis; if X is an open set of an elliptic curve E : $y^2 = f_3(x)$, then yd/dx is such a basis.]

We denote by $\mathcal{D} = \mathcal{D}_X$ the ring $\mathcal{O}[\partial]$ of all differential operators on X. The adjoint L^* of an element $L = \Sigma f_i \partial^i$ of $\mathcal{D}$ is defined by $L^* = \Sigma (-\partial)^i f_i$. The map $L \mapsto L^*$ is a ring isomorphism of $\mathcal{D}$ to the opposite ring $\mathcal{D}^{\mathrm{opp}}$, whose "square" is the identity.

Attached to an operator L in $\mathcal{D}$ is the left $\mathcal{D}$-module $\mathcal{D}/\mathcal{D}L$; it is holonomic so long as $L \neq 0$. Attached to any point α in $X(\mathbb{C})$ is the "delta-module supported at α", the holonomic left $\mathcal{D}$-module $\delta_\alpha := \mathcal{D}/\mathcal{D}I_\alpha$, where $I_\alpha \subset \mathcal{O}$ is the ideal of functions which vanish at α.

Recall that if $\mathcal{M}$ is any holonomic left $\mathcal{D}$-module, then its intrinsic dual $\mathcal{M}^\vee$ is the right $\mathcal{D}$-module $\mathrm{Ext}^1{}_{\mathrm{left}\mathcal{D}\text{-mod}}(\mathcal{M},\mathcal{D})$, where the right $\mathcal{D}$-module structure is via the right structure of the second argument. If $\mathcal{M}$ is $\mathcal{D}/\mathcal{D}L$, its intrinsic dual is $\mathcal{D}/L\mathcal{D}$ (calculate the ext using the resolution $\mathrm{Right}(L):\mathcal{D}\to\mathcal{D}$ of $\mathcal{M}$). Formation of this intrinsic dual is an exact equivalence of categories from the category of holonomic left $\mathcal{D}$-modules to that of holonomic right $\mathcal{D}$-modules; it is involutive in the sense that its inverse is $\mathcal{N}\mapsto\mathrm{Ext}^1{}_{\mathrm{right}\mathcal{D}\text{-mod}}(\mathcal{N},\mathcal{D})$.

Recall how one passes from right $\mathcal{D}$-modules back to left ones, by using the $\mathcal{O}$-invertible **right** $\mathcal{D}$-module $\omega:=\Omega^1{}_{X/\mathbb{C}}$. In terms of the chosen ∂, ω is $\mathcal{D}/\partial\mathcal{D}$. For any two right $\mathcal{D}$-modules R_1 and R_2, the $\mathcal{O}$-module $\mathrm{Hom}_{\mathcal{O}}(R_1, R_2)$ carries a canonical structure of left $\mathcal{D}$-module for which $(\partial\varphi)(r_1) = \varphi(r_1\partial) - (\varphi(r_1))\partial$; applying this with $R_1 = \omega$ and a variable $R=R_2$, we obtain a left $\mathcal{D}$-module $R_{\mathrm{left}}:=\mathrm{Hom}_{\mathcal{O}}(\omega,R)$. If R is $\mathcal{D}/L\mathcal{D}$, then R_{left} is $\mathcal{D}/\mathcal{D}L^*$. Another way of describing the functor $R\mapsto R_{\mathrm{left}}$ is to view right $\mathcal{D}$-modules R as left $\mathcal{D}^{\mathrm{opp}}$-modules, and then to use the adjoint isomorphism $L\mapsto L^*$ to identify $\mathcal{D}$ to $\mathcal{D}^{\mathrm{opp}}$.

If we apply the construction $R\mapsto R_{\mathrm{left}}$ to $\mathcal{M}^\vee$ we obtain a left $\mathcal{D}$-module $\mathcal{M}^*:= (\mathcal{M}^\vee)_{\mathrm{left}}$, called the adjoint of $\mathcal{M}$. Formation of this adjoint is a contravariant involution of the category of holonomic left $\mathcal{D}$-modules, which commutes with Zariski (indeed, with etale) localization on X. If $\mathcal{M}$ is $\mathcal{D}/\mathcal{D}L$, its adjoint $\mathcal{M}^*$ is $\mathcal{D}/\mathcal{D}L^*$. The delta module δ_α is its own adjoint.

For purposes of later globalization, it is important to keep in mind that the notion of the adjoint $\mathcal{M}^*$ of a holonomic left $\mathcal{D}$-module $\mathcal{M}$ is an intrinsic one which does not depend on the auxiliary choice of ∂, namely it is $\mathcal{M}\mapsto\mathrm{Hom}_{\mathcal{O}}(\omega, \mathrm{Ext}^1{}_{\mathcal{D}}(\mathcal{M},\mathcal{D}))$. On the other hand, the notion of the adjoint L^* of an operator L in $\mathcal{D}$ does depend on the choice of ∂; only the associated $\mathcal{D}$-module $\mathcal{D}/\mathcal{D}L^* \approx (\mathcal{D}/\mathcal{D}L)^*$ is intrinsic.

Suppose that U is a nonempty open set of X, $j:U\to X$ the inclusion. There is a natural inverse image functor j^* from $\mathcal{D}$-modules on X to those on U, namely $\mathcal{M}\mapsto\mathcal{D}_U\otimes_{\mathcal{D}}\mathcal{M} := j^*\mathcal{M}$, via the canonical inclusion of rings $\mathcal{D}\to\mathcal{D}_U$.There is a natural direct image functor j_* from left $\mathcal{D}$-modules on U to those on X, which is right adjoint to j^*: given V on U, j_*V is the $\mathcal{D}$-module on X obtained by using the canonical inclusion of rings $\mathcal{D} \to \mathcal{D}_U$ to view the $\mathcal{D}_U$-module V as a $\mathcal{D}$-module. By Bernstein's theorem ([Ber], [Bor]), if V is holonomic on U then j_*V is holonomic on X. The restriction of j_*V to U is just V again. However, there is a "better" prolongation of a holonomic V on U to a holonomic on X, the "middle extension" $j_{!*}(V)$. It is defined as follows. On U, form the adjoint V^* of V and take its direct image $j_*(V^*)$; its adjoint $(j_*(V^*))^*$ is called $j_!(V)$. This $j_!(V)$ is also a prolongation of V (because formation of the adjoint commutes with Zariski localization), so adjunction applied to the resulting isomorphism $j^*j_!(V) \approx V$ gives a canonical map $j_!(V) \to j_*V$, whose image is defined to be $j_{!*}(V)$.

More generally, for any holonomic $\mathcal{N}$ on X whose restriction to U is V, from $j^*\mathcal{N} \approx V$, we get by adjunction $\mathcal{N} \to j_*V$. If $\mathcal{N}_1 \to \mathcal{N}_2$ is any map of such prolongations of V which is the identity over U, then by functoriality of the adjunction map the commutative diagram

$$\begin{array}{ccc} j^*\mathcal{N}_1 & \to & j^*\mathcal{N}_2 \\ & \searrow \ \swarrow & \\ & V & \end{array} \quad \text{gives the commutative diagram} \quad \begin{array}{ccc} \mathcal{N}_1 & \to & \mathcal{N}_2. \\ & \searrow \ \swarrow & \\ & j_*V. & \end{array}$$

For example, if we take for $\mathcal{N}$ the module j_*V, the adjunction map is the identity. For the prolongation $j_!V$ of V, the adjunction map is the canonical map $j_!(V) \to j_*V$ above. Therefore if we have any prolongation $\mathcal{N}$ of V which sits in $j_!V \to \mathcal{N} \to j_*V$ then the adjunction maps sit in

$$\begin{array}{ccccc} j_!V & \to & \mathcal{N} & \to & j_*V \\ & \text{can}\searrow & \downarrow & \swarrow = & \\ & & j_*V, & & \end{array}$$

and the rightmost arrow down is the identity, while the first one is the canonical map $j_!(V) \to j_*V$. In other words, to identify the middle extension ,i.e.,the image of $j_!(V) \to j_*V$, we have only to find an $\mathcal{N}$ which extends V and which is simultaneously a quotient of $j_!(V)$ and a subobject ofj_*V. Using this, we can easily prove

Lemma 2.9.1 (characterization of middle extensions) Given a holonomic left $\mathcal{D}$-module $\mathcal{M}$ on X, a nonempty open set U of X and $j{:}U \to X$ the inclusion, then $\mathcal{M} \approx j_{!*}(j^*\mathcal{M})$ by an isomorphism which is the identity on U if and only if $\mathcal{M}$ satisfies

$$\mathrm{Hom}_{\mathcal{D}}(\mathcal{M}, \delta_\alpha) = 0 = \mathrm{Hom}_{\mathcal{D}}(\delta_\alpha, \mathcal{M}) \text{ for every } \alpha \text{ in } X-U,$$

or equivalently (by duality), if and only if $\mathcal{M}$ satisfies

$$\mathrm{Hom}_{\mathcal{D}}(\mathcal{M}, \delta_\alpha) = 0 = \mathrm{Hom}_{\mathcal{D}}(\mathcal{M}^*, \delta_\alpha) \text{ for every } \alpha \text{ in } X-U.$$

or equivalently (by duality), if and only if $\mathcal{M}$ satisfies

$$\mathrm{Hom}_{\mathcal{D}}(\delta_\alpha, \mathcal{M}) = 0 = \mathrm{Hom}_{\mathcal{D}}(\delta_\alpha, \mathcal{M}^*) \text{ for every } \alpha \text{ in } X-U.$$

proof.We denote $j^*\mathcal{M}$ by V. Notice that j_*V has no nonzero $\mathcal{O}$-torsion outside of U, while any δ-module consists entirely of $\mathcal{O}$-torsion. Thus we obviously have $\mathrm{Hom}_{\mathcal{D}}(\delta_\alpha, j_*(V))=0$ for any α in X-U. Applying this to V^*, we see that $\mathrm{Hom}_{\mathcal{D}}(\delta_\alpha, j_*(V^*))=0$ for any α in X-U, so by duality $\mathrm{Hom}_{\mathcal{D}}(j_!(V),\delta_\alpha)=0$ for any α in X-U. Since $j_{!*}(V)$ is both a subobject of $j_*(V)$ and a quotient of $j_!(V)$, we have

$$\mathrm{Hom}_{\mathcal{D}}(\delta_\alpha, j_{!*}(V)) = 0 = \mathrm{Hom}_{\mathcal{D}}(j_{!*}(V), \delta_\alpha) \text{ for any } \alpha \text{ in} X-U.$$

Now suppose we are given any holonomic $\mathcal{N}$ on X whose restriction to U is V, and which satisfies the two conditions

$$\mathrm{Hom}_{\mathcal{D}}(\delta_\alpha, \mathcal{N}) = 0 = \mathrm{Hom}_{\mathcal{D}}(\delta_\alpha, \mathcal{N}^*) \text{ for every } \alpha \text{ in } X-U.$$

We claim that $\mathcal{N}$ is necessarily$j_{!*}(V)$. From the given isomorphism $j^*\mathcal{N} \to V$ we get by adjunction a map $\mathcal{N} \to j_*V$. This map is injective (because being an isomorphism on U its kernel can only be a successive extension of δ-modules supported in X-U, but in view of $0 = \mathrm{Hom}_{\mathcal{D}}(\delta_\alpha,\mathcal{N})$ for α in X-U, $\mathcal{N}$ contains no δ-modules supported in X-U). Similarly, the adjoint $\mathcal{N}^*$ restricts to V^*, and the natural map of adjunction $\mathcal{N}^* \to j_*(V^*)$ is injective (its kernel is punctual; now use the vanishing of $\mathrm{Hom}_{\mathcal{D}}(\delta_\alpha,\mathcal{N}^*)$ for α in X-U). So by duality $\mathcal{N}$ is a quotient of $j_!(V)$, as well as a subobject of j_*V. As explained above, this completes the proof.QED

Corollary 2.9.1.1 If the holonomic left $\mathcal{D}$-module $\mathcal{M}$ on X lies in D.E.(X/$\mathbb{C}$), then for any nonempty open set $j:U \to X$, $\mathcal{M} \approx j_{!*}(j^*\mathcal{M})$.

proof For $\mathcal{M}$ in D.E.(X/$\mathbb{C}$), viewed as coherent locally free $\mathcal{O}$-module with integrable connection, the adjoint $\mathcal{M}^*$ is also in D.E.(X/$\mathbb{C}$), being the dual $\mathcal{O}$-module with the dual connection. [To see this, we may Zariski localize and suppose $\mathcal{M}$ is $\mathcal{O}$-free of rank n, with connection matrix $A \in M_n(\mathcal{O})$, so $\mathcal{M} \approx \mathcal{D}^n/\mathcal{D}^n(\partial - A)$, whence $\mathrm{Ext}^1{}_{\mathrm{left}\mathcal{D}\text{-mod}}(\mathcal{M},\mathcal{D})$

is $\mathcal{D}^n/(\partial - A)\mathcal{D}^n$, and so $\mathcal{M}^*$ is $\mathcal{D}^n/\mathcal{D}^n(-\partial - A^t)$.] Therefore both $\mathcal{M}$ and $\mathcal{M}^*$ are torsion-free $\mathcal{O}$-modules, so $\mathrm{Hom}_{\mathcal{D}}(\delta_\alpha, \mathcal{M}) = 0 = \mathrm{Hom}_{\mathcal{D}}(\delta_\alpha, \mathcal{M}^*)$ for every α in X-U. QED

Corollary 2.9.1.2 Given a nonempty open set $j: U \to X$, formation of the middle extension $j_{!*}$ commutes with formation of the adjoint.

proof Indeed, given a holonomic left $\mathcal{D}$-module V on U, its middle extension $\mathcal{M}$ satisfies $\mathrm{Hom}_{\mathcal{D}}(\delta_\alpha, \mathcal{M}) = 0 = \mathrm{Hom}_{\mathcal{D}}(\delta_\alpha, \mathcal{M}^*)$ for every α in X-U; as this condition is symmetric in $\mathcal{M}$ and $\mathcal{M}^*$, we see from the characterization of middle extensions that $\mathcal{M}^* \approx j_{!*}(j^*(\mathcal{M}^*))$. But $j^*(\mathcal{M}^*)$ is canonically $(j^*(\mathcal{M}))^* = V^*$, whence $\mathcal{M}^* \approx j_{!*}(V^*)$. QED

Corollary 2.9.1.3 Given a nonempty open set $j: U \to X$, and $\mathcal{M}, \mathcal{N}$ holonomic on U, the restriction map defines an isomorphism of Hom groups

$$\mathrm{Hom}_{\mathcal{D}\text{-mod on } X}(j_{!*}\mathcal{M}, j_{!*}\mathcal{N}) \approx \mathrm{Hom}_{\mathcal{D}\text{-mod on } U}(\mathcal{M}, \mathcal{N}).$$

proof The quotient $j_*\mathcal{N}/j_{!*}\mathcal{N}$ is punctual with support in X - U, so we have $\mathrm{Hom}_{\mathcal{D}\text{-mod on } X}(j_{!*}\mathcal{M}, j_{!*}\mathcal{N}) = \mathrm{Hom}_{\mathcal{D}\text{-mod on } X}(j_{!*}\mathcal{M}, j_*\mathcal{N}) =$ (by adjunction) $= \mathrm{Hom}_{\mathcal{D}\text{-mod on } U}(j^*j_{!*}\mathcal{M}, \mathcal{N}) = \mathrm{Hom}_{\mathcal{D}\text{-mod on } U}(\mathcal{M}, \mathcal{N})$. QED

Lemma 2.9.2 Let α in $X(\mathbb{C})$, and $j : X-\{\alpha\} \to X$ the inclusion. Suppose that L is a nonzero element of $\mathcal{D}$. The following conditions are equivalent:

(1) the natural map

$$\mathcal{D}/\mathcal{D}L \to j_*(j^*(\mathcal{D}/\mathcal{D}L))$$

is an isomorphism.

(2) L^* operates bijectively on the delta-module δ_α.

proof The question is Zariski local around α in X, so by shrinking down we may assume that the ideal defining α in X is principal, with generator denoted x. By definition we have $j_*(j^*(\mathcal{D}/\mathcal{D}L)) = (\mathcal{D}/\mathcal{D}L)[1/x]$, so (1) is equivalent to the statement that the operator Left(x): $\lambda \mapsto x\lambda$ is bijective on $\mathcal{D}/\mathcal{D}L$.

To say that Left(x) is injective on $\mathcal{D}/\mathcal{D}L$ is to say that if a,b in $\mathcal{D}$ satisfy $xa = bL$ in $\mathcal{D}$, then there exists c in $\mathcal{D}$ such that $a = cL$; if this c exists, then $xcL = bL$ in $\mathcal{D}$, so $b = xc$. Read backwards, this is precisely the condition that Right(L):$\lambda \mapsto \lambda L$ is injective on $\mathcal{D}/x\mathcal{D}$.

To say that Left(x) is surjective on $\mathcal{D}/\mathcal{D}L$ is to say that for any a in $\mathcal{D}$ there exist b and c in $\mathcal{D}$ such that $a=xb+cL$ in $\mathcal{D}$. This is precisely the condition that Right(L) is surjective on $\mathcal{D}/x\mathcal{D}$.

Therefore Left(x) is bijective on $\mathcal{D}/\mathcal{D}L$ if and only if Right(L) is bijective on $\mathcal{D}/x\mathcal{D}$. Passing from right modules to left, this is in turn the same as saying that Left(L^*) is bijective on $\mathcal{D}/\mathcal{D}x := \delta_\alpha$. Thus (1) and (2) are equivalent.QED

Lemma 2.9.3 Let α in X($\mathbb{C}$), and $j : X-\{\alpha\} \to X$ the inclusion. Suppose that L is a nonzero element of $\mathcal{D}$. The following conditions are equivalent:

(1) the natural map

$$j_!(j^*(\mathcal{D}/\mathcal{D}L)) \to \mathcal{D}/\mathcal{D}L$$

is an isomorphism.

(2) L operates bijectively on the delta-module δ_α.

proof The map (1) is the dual of the map (1) of the preceding Lemma with L replaced by L^*.QED

Lemma 2.9.4 Let α in X($\mathbb{C}$),and $j : X-\{\alpha\} \to X$ the inclusion. Choose a formal uniformizing parameter x at α, i.e., an isomorphism $\mathbb{C}[[x]] \approx (\mathcal{O}_{X,\alpha})^\wedge$. Let L be a nonzero element of $\mathcal{D}$ of degree $n \ge 0$ in ∂, which satisfies the following condition (*):

(*) viewed in $\mathbb{C}[[x]]\otimes_{\mathcal{O}}\mathcal{D} \approx \mathbb{C}[[x]][d/dx]$, L lies in the subring $\mathbb{C}[[x]][xd/dx]$, say

$$L = \Sigma_{i\ge 0}\, x^i P_i(xd/dx),$$

where the $\{P_i(t)\}_{i\ge 0}$ are a sequence of polynomials in $\mathbb{C}[t]$ of degree $\le n$. Then the following conditions are equivalent:

(1) L and L^* both operate injectively on δ_α.

(2) L and L^* both operate bijectively on δ_α.

(3) The "indicial polynomial" $P_0(t)$ has no zeroes in $\mathbb{Z}$.

(4) $\mathcal{D}/\mathcal{D}L \approx j_{!*}j^*(\mathcal{D}/\mathcal{D}L)$.

(5) $j_!(j^*(\mathcal{D}/\mathcal{D}L)) \approx \mathcal{D}/\mathcal{D}L \approx j_*(j^*(\mathcal{D}/\mathcal{D}L))$.

proof We have already seen that (1)$\Leftrightarrow$(4) and that (2)$\Leftrightarrow$(5), and (2)$\Rightarrow$(1) is trivial. So it remains to show, under the hypothesis (*) made on L, that (1)⇒(2)⇔(3). Let us denote by I the ideal which defines α in X. L acts on $\mathcal{O}_{X-\{\alpha\}} = \cup_n I^{-n}$. Intrinsically, the hypothesis (*) is that L, acting on $\mathcal{O}_{X-\{\alpha\}}$, maps every power I^n of I to itself. Thus for every $n\in\mathbb{Z}$, L induces a $\mathbb{C}$-linear endomorphism $gr_n(L)$ of the one-dimensional $\mathbb{C}$-space I^n/I^{n+1}. This endomorphism $gr_n(L)$ is none other than multiplication by $P_0(n)$.

This shows both that (*) is independent of the choice of formal parameter x, and that, when it holds, the condition (3) is also independent of this choice. This allows us to choose the formal parameter x in a convenient way. We will adopt its choice to the derivation ∂ used in the explicit definition of the adjoint,by requiring that $\partial(x)=1$. [This is clearly possible, since for any initial choice of formal parameter x, ∂ is $f(x)d/dx$ for some unit $f(x)$ in $\mathbb{C}[[x]]$. The required parameter is then $\int_0^x dt/f(t)$.] With this choice of x, the adjoint of $xd/dx = x\partial$ is $-\partial x = -1-x\partial = -1-xd/dx$, and so the formal expansion of the adjoint L* is

$$L^* = \Sigma_{i\geq 0} P_i(-1-xd/dx)x^i = \Sigma_{i\geq 0} x^i P_i(-1-i-xd/dx).$$

Thus L* also satisfies (*), and its indicial polynomial is $P_0(-1-t)$.

Now consider the delta-module $\delta_\alpha := \mathcal{D}/\mathcal{D}I$; it is isomorphic to $\mathbb{C}((x))/\mathbb{C}[[x]]$, by the $\mathcal{D}$-linear map $1 \mapsto 1/x$. By the hypothesis (*), each of the finite-dimensional subspaces $F_{-n} := x^{-n}\mathbb{C}[[x]]/\mathbb{C}[[x]]$, $n\geq 1$, is stable by L (resp. L*); as δ_α is their union, we see that L (resp. L*) is injective on δ_α if and only if it is injective on each F_{-n}. Since F_{-n} is finite-dimensional, L (resp. L*) is injective on F_{-n} if and only if it bijective on F_{-n}. Thus if L (resp. L*) is injective on δ_α, it is bijective on each F_{-n}, so surjective on δ_α and hence bijective on δ_α. Thus (1)⇒(2).

It remains to see that (2)⇔(3). Since L (resp. L*) is stable on each F_{-n}, and induces multiplication by $P_0(-n)$ (resp. $P_0(-1+n)$) on F_{-n}/F_{1-n}, we see that L (resp. L*) is bijective on F_{-n} if and only if $P_0(-t)$ (resp. $P_0(-1+t)$) has no zeroes in $\{1,2,...,n\}$. Thus L and L* are both bijective on δ_α if and only if $P_0(t)$ has no zeroes in $\mathbb{Z}$. QED

Remark 2.9.4.1 The proof as given shows that one has the slightly more precise

Lemma 2.9.5 Hypotheses and notations as above, the following four conditions are equivalent:

(1) L^* (resp. L) operates injectively on δ_α.

(2) L^* (resp. L) operates bijectively on δ_α.

(3) The "indicial polynomial" $P_0(t)$ has no zeroes in $\mathbb{Z}_{<0}$ (resp. in $\mathbb{Z}_{\geq 0}$).

(4) $j_!(j^*(\mathcal{D}/\mathcal{D}L)) \approx \mathcal{D}/\mathcal{D}L$ (resp. $\mathcal{D}/\mathcal{D}L \approx j_*(j^*(\mathcal{D}/\mathcal{D}L))$.

Corollary 2.9.5.1 Let $j: \mathbb{G}_m \to \mathbb{A}^1$ the inclusion, $\partial := d/dx$, $D := x\partial$. Then

(a) $j_!j^*\mathcal{O} \approx \mathcal{D}/\mathcal{D}x\partial = \mathcal{D}/\mathcal{D}D$.

(b) $j_*j^*\mathcal{O} \approx \mathcal{D}/\mathcal{D}\partial x = \mathcal{D}/\mathcal{D}(D + 1)$.

proof On $\mathbb{G}_m$, both $\mathcal{D}/\mathcal{D}D$ and $\mathcal{D}/\mathcal{D}(D + 1)$ are isomorphic to $j^*\mathcal{O} = \mathbb{C}[x, x^{-1}]$, by the $\mathcal{D}$-linear maps $1 \mapsto 1$ and $1 \mapsto 1/x$ respectively. So (a) and (b) result from the above lemma's (3) $\Leftrightarrow$ (4), applied to the operators D and D + 1 respectively. QED

(2.9.6) One knows that in the category of holonomic left $\mathcal{D}$-modules, every object is of finite length, and that the irreducibles are of two kinds:

(1)for each α in $X(\mathbb{C})$, the delta-module δ_α is irreducible.

(2)for each nonempty open U in X, with $j:U\to X$ the inclusion, and each irreducible object V in D.E.(U/$\mathbb{C}$), $j_{!*}(V)$ is irreducible.

Recall that for an object V of D.E.(U/$\mathbb{C}$), any subobject N of V as holonomic (or even as $\mathcal{O}$-quasicoherent) $\mathcal{D}$-module is itself an object of D.E.(U/$\mathbb{C}$), simply because D.E.(U/$\mathbb{C}$) is precisely the category of $\mathcal{O}$-coherent $\mathcal{D}$-modules. This means that for an object V of D.E.(U/$\mathbb{C}$), the notions of "irreducible as D.E." and of "irreducible as holonomic $\mathcal{D}$-module" coincide. Recall also that if an object V in D.E.(U/$\mathbb{C}$) is irreducible, then its restriction to any nonempty open set U' of U remains irreducible in D.E.(U'/$\mathbb{C}$) ("birational invariance of the differential galois group, cf [Ka-CAT, 4.2]).

Therefore, given an irreducible $\mathcal{M}$ on X which is not a delta-module, then for any nonempty open $j:U\to X$ such that $j^*\mathcal{M}$ lies in D.E.(U/$\mathbb{C}$), $j^*\mathcal{M}$ is irreducible in D.E.(U/$\mathbb{C}$) and $\mathcal{M}$ is $j_{!*}(j^*\mathcal{M})$.

Thus we find

Corollary 2.9.6.1 Let $\mathcal{M}$ be holonomic left $\mathcal{D}$-module on X whose support is not punctual. Then $\mathcal{M}$ is irreducible if and only if there exists a nonempty open $j:U \to X$ such that

$j^*\mathcal{M}$ is an irreducible object in D.E.$(U/\mathbb{C})$ and $\mathcal{M} \approx j_{!*}j^*\mathcal{M}$.

Moreover, if this condition holds for some U, then it holds for any U such that $\mathcal{M}|U$ is in D.E.$(U/\mathbb{C})$.

Corollary 2.9.6.2 Let f, g $\in \mathcal{O}$, with $f \neq 0$. The first order operator $L := f\partial + g$ has $\mathcal{D}/\mathcal{D}L$ an irreducible $\mathcal{D}$-module on X if and only if the following conditions hold:
(1) at every simple zero α of f, the ratio $g(\alpha)/(\partial f)(\alpha)$ is not in $\mathbb{Z}$.
(2) at every multiple zero α of f, $g(\alpha) \neq 0$.
proof On the open set U where f is invertible, we have a rank one D.E., which is automatically irreducible. We must show that $\mathcal{D}/\mathcal{D}L$ is a middle extension from U. At a zero α of f, choose a formal parameter x with $\partial x = 1$. Formally at α, ∂ is d/dx and so the operator L is $f(x)d/dx + g(x) = ((\partial f)(\alpha) + \text{higher terms})xd/dx + g(x)$. Therefore 2.9.4 applies. The indicial polynomial at α is $P_0(t) = (\partial f)(\alpha)t + g(\alpha)$, so the conditions (1) and (2) just amount to requiring $P_0(t)$ to have no roots in $\mathbb{Z}$. QED

Lemma 2.9.7 (Pochammer) Let $\alpha \in X(\mathbb{C})$, $U := X - \{\alpha\}$ $j:U \to X$ the inclusion. Choose a formal parameter x at α, i.e., an isomorphism $\mathbb{C}[[x]] \approx (\mathcal{O}_{X,\alpha})^{\wedge}$. Let $L := \Sigma f_i \partial^i$ be a nonzero element of $\mathcal{D}$ of degree $n \geq 1$ in ∂, whose leading coefficient f_n has a simple zero at α, and is invertible on $U := X - \{\alpha\}$. Let $\mathcal{M} := \mathcal{D}/\mathcal{D}L$. Then
(1) $j^*\mathcal{M}$ and $j^*\mathcal{M}^*$ each lie in D.E.$(U/\mathbb{C})$, and as D.E. on U each has a regular singular point at α.
(2) if the formal parameter x is convergent, i.e., if $x \in \mathcal{O}_{X^{an},\alpha}$, every solution of $L\varphi = 0$ (resp. of $L^*\varphi = 0$) in $\mathbb{C}((x))$ is convergent in a punctured (classical) neighborhood of 0 in $\mathbb{C}$.
(3) the equations $L\varphi = 0$ and $L^*\varphi = 0$ have the same number $\geq n-1$ of $\mathbb{C}$-linearly independent solutions in $\mathbb{C}((x))$, i.e.,

$$\dim_{\mathbb{C}}\mathrm{Hom}_{\mathcal{D}}(\mathcal{M}, \mathbb{C}((x))) = \dim_{\mathbb{C}}\mathrm{Hom}_{\mathcal{D}}(\mathcal{M}^*, \mathbb{C}((x))) \geq n-1.$$

Moreover, at least one of $\mathcal{M}$ or $\mathcal{M}^*$ has $\dim_{\mathbb{C}}\mathrm{Hom}_{\mathcal{D}}(\mathcal{M}, \mathbb{C}[[x]]) = n-1$.

(4) If $\mathcal{M} \approx j_{!*}(j^*\mathcal{M})$, e.g., if $\mathcal{M}$ is irreducible, then every solution of $L\varphi = 0$ (resp. of $L^*\varphi = 0$) in $\mathbb{C}((x))$ lies in $\mathbb{C}[[x]]$, and

$$\dim_{\mathbb{C}}\mathrm{Hom}_{\mathcal{D}}(\mathcal{M}, \mathbb{C}((x))) = \dim_{\mathbb{C}}\mathrm{Hom}_{\mathcal{D}}(\mathcal{M}^*, \mathbb{C}((x))) = n-1,$$

i.e., local monodromy around α is a pseudoreflection.

proof Notice first that the hypothesis on L is also satisfied by L^*; its leading coefficient is $(-1)^n f_n$. Because f_n is invertible on $U:=X - \{\alpha\}$, both $j^*\mathcal{M}$ and $j^*\mathcal{M}^*$ lie in $\mathrm{D.E.}(U/\mathbb{C})$, and as D.E.'s on U both visibly (Fuch's criterion) have a regular singular point at α. This proves (1), and (1)$\Rightarrow$(2). Let us denote $X^{an} - \{\alpha\}$ by $\mathcal{U}$, and denote by $\mathcal{L}$ and $\mathcal{L}^*$ the dual local systems on $\mathcal{U}$ of germs of holomorphic solutions of L and L^* respectively. Because $j^*\mathcal{M}$ and $j^*\mathcal{M}^*$ both have a regular singularity at α, their spaces of $\mathbb{C}((x))$-valued solutions are the invariants of "local monodromy around α" in these dual local systems. Looking at the Jordan normal form of local monodromy around α on $\mathcal{L}$, we see that it and its contragredient have equal-dimensional spaces of invariants. This proves the "equal dimension" part of (3).

To prove the rest of (3), we argue as follows. The dimensions in question depend on what happens over $(\mathcal{O}_{X,\alpha})^\wedge \approx \mathbb{C}[[x]]$ and over $\mathcal{O}_{X - \{\alpha\}}\otimes(\mathcal{O}_{X,\alpha})^\wedge \approx \mathbb{C}((x))$. Therefore we may and will choose the formal parameter x so that ∂ is d/dx. Then in $\mathbb{C}[[x]][\partial]$, we can multiply L by a unit $u(x)$ in $\mathbb{C}[[x]]$ so that it is of the form

$$u(x)L = x\partial^n + \text{ lower terms in } \partial, \text{ coef's in } \mathbb{C}[[x]],$$

$$= x\partial^n + (\beta + \text{ higher terms in } x\,)\partial^{n-1} + \Sigma_{j<n-1} f_j\partial^j, \quad f_j \in \mathbb{C}[[x]].$$

One readily computes that $(-1)^n u(x)L^*$ is of the form

$$x\partial^n + (n-\beta + \text{ higher terms in } x)\partial^{n-1} + \Sigma_{j<n-1} g_j\partial^j, \quad g_j \in \mathbb{C}[[x]].$$

Let us admit for a moment

(*) if β is not in $\mathbb{Z}_{\leq 0}$, then $\dim_{\mathbb{C}}\mathrm{Hom}_{\mathcal{D}}(\mathcal{M}, \mathbb{C}[[x]]) = n-1$.

Then we may complete the proof as follows. At the expense of interchanging L and L^* we may suppose that β is not in $\mathbb{Z}_{<0}$. Then by (*) we trivially have $\dim_{\mathbb{C}}\mathrm{Hom}_{\mathcal{D}}(\mathcal{M}, \mathbb{C}((x))) \geq n-1$. This proves (3). Finally, if $\mathcal{M} \approx j_{!*}(\mathcal{M})$, then $\mathrm{Hom}_{\mathcal{D}}(\mathcal{M}, \delta_\alpha) = \mathrm{Hom}_{\mathcal{D}}(\mathcal{M}^*, \delta_\alpha) = 0$, so (4) follows from (*) and (3) by applying the functors $\mathrm{Hom}_{\mathcal{D}}(\mathcal{M}, ?)$ and $\mathrm{Hom}_{\mathcal{D}}(\mathcal{M}^*, ?)$ to the short exact sequence of $\mathcal{D}$-modules

$$0 \to \mathbb{C}[[x]] \to \mathbb{C}((x)) \to \delta_\alpha \to 0.$$

It remains to prove (*). Now $\mathrm{Hom}_{\mathcal{D}}(\mathcal{M}, \mathbb{C}[[x]])$ is precisely the kernel of L on $\mathbb{C}[[x]]$. Because $\mathbb{C}[[x]]$ is an integral domain, this kernel is the same for L and for $(x^{n-1}u(x))L$. This operator is readily seen to lie in the subring $\mathbb{C}[[x]][xd/dx]$ of $\mathbb{C}[[x]][d/dx]$, and its expansion (cf 2.9.4)

$$(x^{n-1}u(x))L = \Sigma_{i\geq 0}\, x^i P_i(xd/dx)$$

has $P_0(T) = T(T-1)(T-2)......(T-(n-2))(T-(n-1-\beta))$, as one sees using the identity

$$x^k(d/dx)^k = (xd/dx)(xd/dx - 1)(xd/dx - 2)......(xd/dx - (k-1)).$$

Therefore

$(x^{n-1}u(x))L$ acts stably on each ideal of $\mathbb{C}[[x]]$, and, because $(n-1-\beta)$ is not in $\mathbb{Z}_{\geq n-1}$, it acts bijectively on $(x^{n-1})\mathbb{C}[[x]]$. The snake lemma for the short exact sequence

$$0 \to (x^{n-1})\mathbb{C}[[x]] \to \mathbb{C}[[x]] \to \mathbb{C}[[x]]/(x^{n-1})\mathbb{C}[[x]] \to 0$$

then shows that $(x^{n-1}u(x))L$ has isomorphic kernels on $\mathbb{C}[[x]]$ and on $\mathbb{C}[[x]]/(x^{n-1})\mathbb{C}[[x]]$. But

$$(x^{n-1}u(x))L = (x^{n-1})(\text{ an endomorphism of } \mathbb{C}[[x]]\ )$$

so it kills $\mathbb{C}[[x]]/(x^{n-1})\mathbb{C}[[x]]$. This concludes the proof of (*). QED

Remark 2.9.7.1 The indicial polynomial of $x^{n-1}u(x)L$ at α has roots $0,1,2,...,n-2$, and $n-1-\beta$, while that of $x^{n-1}u(x)L^*$ has roots $0,1,2,...,n-2$, and $\beta-1$. So if either β is a noninteger or if β lies in $\{1,2,...,n-1\}$, then $\mathcal{M} \approx j_{!*}(\mathcal{M})$. [For then both $x^{n-1}u(x)L$ and $x^{n-1}u(x)L^*$ act bijectively on δ_α, and hence L and L^* are injective on δ_α.] In any case, if $\mathcal{M} \approx j_{!*}(\mathcal{M})$, then the determinant of the pseudoreflection which is its local monodromy at α is $\exp(2\pi i\beta)$.

Proposition 2.9.8 Let $\alpha \in X(\mathbb{C})$, $U := X-\{\alpha\}$ $j:U \to X$ the inclusion. Choose a formal parameter x at α, i.e., an isomorphism $\mathbb{C}[[x]] \approx (\mathcal{O}_{X,\alpha})^\wedge$. Given a holonomic $\mathcal{M}$ on U, denote by Soln_α the finite-dimensional $\mathbb{C}$-vector space $\mathrm{Soln}_\alpha := \mathrm{Hom}_{\mathcal{D}}(\mathcal{M}, (\mathcal{O}_{X,\alpha})^\wedge[1/x]) = \mathrm{Hom}_{\mathcal{D}}(\mathcal{M}, \mathbb{C}((x))) = \mathrm{Hom}_{\mathcal{D}}(\mathcal{M}\otimes_{\mathcal{O}}\mathbb{C}((x)), \mathbb{C}((x)))$ of its formal meromorphic solutions at α. Consider the tautological short exact sequence on X

$$0 \to j_{!*}\mathcal{M} \to j_*\mathcal{M} \to j_*\mathcal{M}/j_{!*}\mathcal{M} \to 0.$$

The quotient $j_*\mathcal{M}/j_{!*}\mathcal{M}$ is the punctual $\mathcal{D}$-module

$$j_*\mathcal{M}/j_{!*}\mathcal{M} \approx \delta_\alpha \otimes_{\mathbb{C}} \mathrm{Hom}_{\mathbb{C}}(\mathrm{Soln}_\alpha, \mathbb{C}).$$

proof The question is Zariski local on X around α, and independent of the choice of the uniformizing parameter x. So we may and will assume that x is a function on X with a simple zero at α and no other zeroes. Admit for a moment the following assertion (*):

(*) $$\mathrm{Hom}_{\mathcal{D}}(j_*\mathcal{M}, (\mathcal{O}_{X,\alpha})^\wedge) = 0 = \mathrm{Ext}^1{}_{\mathcal{D}}(j_*\mathcal{M}, (\mathcal{O}_{X,\alpha})^\wedge).$$

In the short exact sequence

$$0 \to j_{!*}\mathcal{M} \to j_*\mathcal{M} \to j_*\mathcal{M}/j_{!*}\mathcal{M} \to 0,$$

the quotient $j_*\mathcal{M}/j_{!*}\mathcal{M}$ is holonomic and supported in α, so necessarily of the form $\delta_\alpha \otimes_{\mathbb{C}} V$ for some finite-dimensional $\mathbb{C}$-space V.

Apply the functor $\mathrm{Hom}_{\mathcal{D}}(\,?, (\mathcal{O}_{X,\alpha})^\wedge)$ and look at the long exact cohomology sequence for this exact sequence. In virue of (*), the coboundary induces an isomorphism

$$\mathrm{Hom}_{\mathcal{D}}(j_{!*}\mathcal{M}, (\mathcal{O}_{X,\alpha})^\wedge) \approx \mathrm{Ext}^1{}_{\mathcal{D}}(j_*\mathcal{M}/j_{!*}\mathcal{M}, (\mathcal{O}_{X,\alpha})^\wedge)$$
$$\approx \mathrm{Ext}^1{}_{\mathcal{D}}(\delta_\alpha \otimes_{\mathbb{C}} V, (\mathcal{O}_{X,\alpha})^\wedge)$$
$$\approx \mathrm{Hom}_{\mathbb{C}}(V, \mathbb{C}) \otimes_{\mathbb{C}} \mathrm{Ext}^1{}_{\mathcal{D}}(\delta_\alpha, (\mathcal{O}_{X,\alpha})^\wedge).$$

The same consideration with $\mathcal{M}$ replaced by the trivial $\mathcal{D}$-module $\mathcal{O}$ shows that $\mathrm{Ext}^1{}_{\mathcal{D}}(\delta_\alpha, (\mathcal{O}_{X,\alpha})^\wedge)$ is canonically $\mathbb{C}$. Therefore

$$\mathrm{Hom}_{\mathcal{D}}(j_{!*}\mathcal{M}, (\mathcal{O}_{X,\alpha})^\wedge) \approx \mathrm{Hom}_{\mathbb{C}}(V, \mathbb{C}).$$

From the short exact sequence

$$0 \to (\mathcal{O}_{X,\alpha})^\wedge \to (\mathcal{O}_{X,\alpha})^\wedge[1/x] \to \delta_\alpha \to 0$$

and the vanishing of $\mathrm{Hom}_{\mathcal{D}}(j_{!*}\mathcal{M}, \delta_\alpha)$ we obtain

$$\mathrm{Hom}_{\mathcal{D}}(j_{!*}\mathcal{M}, (\mathcal{O}_{X,\alpha})^\wedge) \approx \mathrm{Hom}_{\mathcal{D}}(j_{!*}\mathcal{M}, (\mathcal{O}_{X,\alpha})^\wedge[1/x]),$$

(by adjunction) $\approx \mathrm{Hom}_{\mathcal{D}}(\mathcal{M}, (\mathcal{O}_{X,\alpha})^\wedge[1/x]) := \mathrm{Soln}_\alpha$.

Thus we find $\mathrm{Soln}_\alpha \approx \mathrm{Hom}_{\mathbb{C}}(V, \mathbb{C})$, as required.

It remains to prove the assertion (*). The vanishing of $\mathrm{Hom}_{\mathcal{D}}(j_*\mathcal{M}, (\mathcal{O}_{X,\alpha})^\wedge)$ is obvious, for already $\mathrm{Hom}_{\mathcal{O}}(j_*\mathcal{M}, (\mathcal{O}_{X,\alpha})^\wedge)=0$, simply because every element of the source $j_*\mathcal{M}$ is infintely x-divisible, while no nonzero element of $(\mathcal{O}_{X,\alpha})^\wedge \approx \mathbb{C}[[x]]$ is infinitely x-divisible.

To prove the vanishing of $\mathrm{Ext}^1{}_{\mathcal{D}}(j_*\mathcal{M}, (\mathcal{O}_{X,\alpha})^\wedge)$, we will use the unique x-divisibility of $j_*\mathcal{M}$ to show that any such extension splits (the splitting is unique if it exists, because $\mathrm{Hom}_{\mathcal{D}}(j_*\mathcal{M}, (\mathcal{O}_{X,\alpha})^\wedge) = 0$). Thus suppose we have any short exact sequence of $\mathcal{D}$-modules

$$0 \to (\mathcal{O}_{X,\alpha})^\wedge \to A \xrightarrow{\pi} B \to 0$$

in which B is a $\mathcal{D}[1/x]$-module. Denote by $C \subset A$ the intersection $C := \bigcap_{n \geq 0} x^n A$. Clearly C is a $\mathcal{D}$-submodule of A (as $\partial(x^n A) \subset x^{n-1}A$ for $n \geq 1$), and $C \cap (\mathcal{O}_{X,\alpha})^\wedge = 0$ (for if $f \in (\mathcal{O}_{X,\alpha})^\wedge$ lies in $x^n A$, say $f = x^n a$, then $0 = \pi(f) = x^n \pi(a)$ in B, so $\pi(a) = 0$, so $a \in (\mathcal{O}_{X,\alpha})^\wedge$, and so we find $f \in x^n (\mathcal{O}_{X,\alpha})^\wedge$). To split π, it suffices to show that π maps C onto B. (For $\pi|C : C \to B$ is automatically injective, as $C \cap (\mathcal{O}_{X,\alpha})^\wedge = 0$.)

Given an element β of B, choose for each $n \geq 0$ an element $\alpha_n \in A$ which lifts $x^{-n}\beta$. For each $n \geq 0$, let $f_n := \alpha_n - x\alpha_{n+1}$. Then $\pi(f_n) = 0$, so $f_n \in (\mathcal{O}_{X,\alpha})^\wedge$. The series $\Sigma_{n \geq 0}\, x^n f_n$ converges in $(\mathcal{O}_{X,\alpha})^\wedge$, say to F. Now define new liftings $\gamma_n \in A$ of the $x^{-n}\beta$ by $\gamma_n := \alpha_n - F$. With this choice, the differences $c_n := \gamma_n - x\gamma_{n+1}$ are $c_n = f_n - (1-x)F$, so $\Sigma_{n \geq 0}\, x^n c_n = 0$. For each $n \geq 0$, define $C_n \in (\mathcal{O}_{X,\alpha})^\wedge$ to be $C_n := \Sigma_{i \geq 0}\, x^i c_{i+n}$. Then

$$\gamma_0 - x^{n+1}\gamma_{n+1} = \Sigma_{i=0,\ldots,n}\, x^i c_i = -\Sigma_{i \geq n+1}\, x^i c_i = -x^{n+1} C_{n+1},$$

and so $\gamma_0 = x^{n+1}(\gamma_{n+1} - C_{n+1})$ lies in $x^{n+1}A$ for every $n \geq 0$, and hence $\gamma_0 \in C$ is a lifting of β to C. QED

This Proposition leads immediately to the following $\mathcal{D}$-module complement to Deligne's Euler-Poincare formula [De-ED,II, 6.21], which was suggested to me by Ofer Gabber. Recall that for a holonomic $\mathcal{D}$-module $\mathcal{M}$ on X, we define

$$\chi(X, \mathcal{M}) := \chi(H^*{}_{DR}(X, \mathcal{M})) = \Sigma(-1)^i \dim_{\mathbb{C}} H^i{}_{DR}(X, \mathcal{M}).$$

Corollary 2.9.8.1 Let $j : U \to X$ be the inclusion of a nonempty open set. Let $\mathcal{M}$ be a holonomic $\mathcal{D}$-module on U. For each $\alpha \in X-U$, denote by Soln_α the finite-dimensional $\mathbb{C}$-vector space of formal meromorphic solutions of $\mathcal{M}$ at α. Then

$$\chi(X, j_{!*}\mathcal{M}) = \chi(U, \mathcal{M}) + \Sigma_{\alpha \in X-U} \dim_{\mathbb{C}} \mathrm{Soln}_\alpha.$$

proof Because $\mathcal{M}$ is $\mathcal{O}$-quasicoherent, $H_{DR}(U, \mathcal{M}) \approx H_{DR}(X, j_*\mathcal{M})$, and so $\chi(U, \mathcal{M}) = \chi(X, j_*\mathcal{M})$. The short exact sequence on X

$$0 \to j_{!*}\mathcal{M} \to j_*\mathcal{M} \to j_*\mathcal{M} / j_{!*}\mathcal{M} \to 0$$

has

$$j_*\mathcal{M} / j_{!*}\mathcal{M} \approx \oplus_{\alpha \in X-U} \delta_\alpha \otimes_{\mathbb{C}} \mathrm{Hom}_{\mathbb{C}}(\mathrm{Soln}_\alpha, \mathbb{C}),$$

so

$$\chi(X, j_{!*}\mathcal{M}) = \chi(X, j_*\mathcal{M}) - \Sigma_{\alpha \in X-U} \dim_{\mathbb{C}}(\mathrm{Soln}_\alpha)\chi(X, \delta_\alpha).$$

But $\chi(X, \delta_\alpha) = -1$ (the map $d/dx : \mathbb{C}((x))/\mathbb{C}[[x]] \to \mathbb{C}((x))/\mathbb{C}[[x]]$ is visibly injective with one-dimensional cokernel). QED

(2.9.8.2) Denote by $\overline{X}$ the complete nonsingulat model of X. For each x in $\overline{X}$, denote by $\mathrm{Irr}_x(\mathcal{M})$ the irregularity (sum of the slopes with multiplicity) of $\mathcal{M}$ at x. Deligne's formula asserts that if $\mathcal{M}$ is a D.E. on X, i.e., if $\mathcal{M}$ is $\mathcal{O}$-locally free of finite rank, then

$$\chi(X, \mathcal{M}) = \mathrm{rank}_{\mathcal{O}}(\mathcal{M})\chi(X) - \Sigma_{x \in \overline{X}-X} \mathrm{Irr}_x(\mathcal{M}),$$

where $\chi(X) := 2-2g(\overline{X}) - \mathrm{Card}(\overline{X} - X)$ is the topological Euler characteristic of X. Combining this with the above corollary, we obtain

Theorem 2.9.9 (Deligne, Gabber) Let $j : U \to X$ be the inclusion of a nonempty open set, $\overline{X}$ the complete nonsingular model of X. Suppose $\mathcal{M}$ is a D.E. on U. For each $\alpha \in X-U$ define integers

$$\mathrm{drop}_\alpha := \mathrm{rank}(\mathcal{M}) - \dim_{\mathbb{C}}\mathrm{Soln}_\alpha,$$

$$\mathrm{totdrop}_\alpha := \mathrm{Irr}_\alpha(\mathcal{M}) + \mathrm{drop}_\alpha.$$

These integers are nonnegative, and

$$\chi(X, j_{!*}\mathcal{M}) = \mathrm{rank}_{\mathcal{O}}(\mathcal{M})\chi(X) - \Sigma_{x \in \overline{X}-X} \mathrm{Irr}_x(\mathcal{M}) - \Sigma_{\alpha \in X-U} \mathrm{totdrop}_\alpha.$$

proof That $\mathrm{drop}_\alpha \geq 0$ is the fact that a rank n D.E. on $\mathbb{C}((x))$ has a solution space of dimension at most n. The irregularity is by definition nonnegative. The χ-formula is a trivial concatenation of the previous corollary with Deligne's formula for $\chi(U, \mathcal{M})$. QED

Lemma 2.9.10 Let α in $X(\mathbb{C})$, $U = X - \{\alpha\}$, $j: U \to X$ the inclusion, $\mathcal{M}$ a D.E. on U of rank $r \geq 1$. Then the following conditions are equivalent:

(1) $j_{!*}\mathcal{M}$ is a D.E. on X.

(2) $\mathrm{totdrop}_\alpha = 0$.

(3) $\mathrm{drop}_\alpha = 0$.

proof Pick a formal parameter x at α. The quantities $\mathrm{totdrop}_\alpha$ and drop_α depend only on $\mathcal{M} \otimes_{\mathcal{O}} \mathbb{C}((x))$. If $\mathcal{N} := j_{!*}\mathcal{M}$ is a D.E. on X, then

$\mathcal{N}\otimes_{\mathcal{O}}\mathbb{C}[[x]]$ is spanned by its horizontal sections, so $\mathcal{N}\otimes_{\mathcal{O}}\mathbb{C}[[x]] \approx (\mathbb{C}[[x]])^r$. As $\mathcal{M} = j^*\mathcal{N}$, $\mathcal{M}\otimes_{\mathcal{O}}\mathbb{C}((x)) \approx (\mathbb{C}((x)))^r$, so obviously (1)$\Rightarrow$(2)$\Rightarrow$(3).

If $\mathrm{drop}_\alpha = 0$, then $\mathcal{M}\otimes_{\mathcal{O}}\mathbb{C}((x)) \approx (\mathbb{C}((x)))^r$. It suffices to extend $\mathcal{M}$ to a D.E. $\mathcal{N}$ on X, i.e., to a $\mathcal{D}$-module $\mathcal{N}$ on X which is a locally free $\mathcal{O}$-module of rank r (for then $j_{!*}\mathcal{M} = j_{!*}j^*\mathcal{N} = \mathcal{N}$ by 2.9.1.1). Now $\mathcal{O}$-locally free extensions of $\mathcal{M}$ as $\mathcal{O}$-module to X are in bijective correspondence with $\mathbb{C}[[x]]$-lattices in $\mathcal{M}\otimes_{\mathcal{O}}\mathbb{C}((x))$, and the corresponding locally free extension is $\mathcal{D}$-stable (inside $j_*\mathcal{M}$) if and only if its $\mathbb{C}[[x]]$-lattice is $\mathcal{D}$-stable (inside $\mathcal{M}\otimes_{\mathcal{O}}\mathbb{C}((x))$). Since $\mathcal{M}\otimes_{\mathcal{O}}\mathbb{C}((x)) \approx (\mathbb{C}((x)))^r$, we have only to take for $\mathbb{C}[[x]]$-lattice $(\mathbb{C}[[x]])^r \subset (\mathbb{C}((x)))^r$ to produce the required $\mathcal{N}$. QED

Corollary 2.9.10.1 Let α in $X(\mathbb{C})$, $U = X - \{\alpha\}$, $j:U\to X$ the inclusion, $\mathcal{M}$ a holonomic $\mathcal{D}$-module on X such that $j^*\mathcal{M}$ is a D.E. on U of rank $r \geq 1$. Suppose x is a function on X which has a simple zero at α and which is invertible on U. Then $\mathcal{M}$ is a D.E. on X if and only if the following three conditions hold:

(1) the map Left(x) : $\mathcal{M} \to \mathcal{M}$ is injective, i.e., $\mathrm{Hom}_{\mathcal{D}}(\delta_\alpha, \mathcal{M}) = 0$.

(2) the map Left(x) : $\mathcal{M}^* \to \mathcal{M}^*$ is injective, i.e., $\mathrm{Hom}_{\mathcal{D}}(\delta_\alpha, \mathcal{M}^*) = 0$.

(3) $\dim_{\mathbb{C}}(\mathcal{M}/x\mathcal{M}) = r$, or equivalently

(3 bis) The function on $X(\mathbb{C})$ given by

$$\beta \mapsto \dim_{\mathbb{C}}(\mathcal{M}/I_\beta\mathcal{M}),\ I_\beta := \text{the ideal sheaf of } \beta,$$

is constant.

proof The conditions listed are trivially necessary. To show that they are sufficient, we argue as follows.

The first two conditions together imply

$$\mathcal{M} \approx j_{!*}(j^*\mathcal{M}),$$

in virtue of 2.9.1. Using this, 2.9.8 gives a short exact sequence

$$0 \to \mathcal{M} \to j_*j^*\mathcal{M} \to \delta_\alpha\otimes_{\mathbb{C}}\mathrm{Hom}_{\mathbb{C}}(\mathrm{Soln}_\alpha, \mathbb{C}) \to 0.$$

Now apply the snake lemma to the endomorphism Left(x) of this short exact sequence. Since Left(x) is bijective on $j_*j^*\mathcal{M}$, the coboundary defines an isomorphism

$$\mathrm{Hom}_{\mathbb{C}}(\mathrm{Soln}_\alpha, \mathbb{C}) \approx \mathcal{M}/x\mathcal{M}.$$

So by (3), we see that $\dim_{\mathbb{C}}\mathrm{Soln}_\alpha = r$. By 2.9.10,(1) $\Leftrightarrow$ (3), we conclude that $j_{!*}(j^*\mathcal{M})$ is a D.E. on X, whence $\mathcal{M} \approx j_{!*}(j^*\mathcal{M})$ is a D.E. on X. QED

(2.9.11) We next recall what the general global duality theorem (for coherent $\mathcal{D}$-modules with respect to projective morphisms) gives in our situation. Thus suppose $\mathcal{M}$ is a holonomic left $\mathcal{D}$-module on the smooth connected curve X. The de Rham cohomology groups $H^i_{DR}(X, \mathcal{M})$ can also be described as the global Ext groups $Ext^i_{\mathcal{D}}(\mathcal{O}, \mathcal{M})$. Passage to adjoints gives $Ext^i_{\mathcal{D}}(\mathcal{O}, \mathcal{M}) \approx Ext^i_{\mathcal{D}}(\mathcal{M}^*, \mathcal{O})$. We have a natural pairing

$$Ext^i_{\mathcal{D}}(\mathcal{O}, \mathcal{M}^*) \times Ext^j_{\mathcal{D}}(\mathcal{M}^*, \mathcal{O}) \to Ext^{i+j}_{\mathcal{D}}(\mathcal{O}, \mathcal{O}),$$

which via the above isomorphisms becomes a pairing

$$H^i_{DR}(X, \mathcal{M}^*) \times H^j_{DR}(X, \mathcal{M}) \to H^{i+j}_{DR}(X, \mathcal{O}) := H^{i+j}_{DR}(X).$$

The global duality theorem (cf. [Ber], [Bor]) asserts that if $X = \overline{X}$ is a **complete** nonsingular connected curve, then for any holonomic $\mathcal{M}$ on $\overline{X}$, the pairings

$$H^i_{DR}(\overline{X}, \mathcal{M}^*) \times H^{2-i}_{DR}(\overline{X}, \mathcal{M}) \to H^2_{DR}(\overline{X}) \approx \mathbb{C}$$

are perfect dualities of finite-dimensional $\mathbb{C}$-vector spaces.

To conclude this section, we give elementary Euler characteristic formulas for the special case of $\mathcal{D}$-modules of the form $\mathcal{D}/\mathcal{D}L$ on $\mathbb{A}^1$ and on $\mathbb{G}_m$.

Lemma 2.9.12 On $\mathbb{A}^1$ with parameter x, write ∂ for d/dx, and consider a nonzero operator $L := \Sigma a_{i,j}x^i\partial^j$. Define the integer $d = d(L)$ by

$$d := \max(\, i-j \mid a_{i,j} \neq 0).$$

Then $\chi(\mathbb{A}^1, \mathcal{D}/\mathcal{D}L) = -d$.

proof We have

$$\chi(\mathbb{A}^1, \mathcal{D}/\mathcal{D}L) := \chi(Ext_{\mathcal{D}}(\mathcal{O}, \mathcal{D}/\mathcal{D}L)) = \chi(Ext_{\mathcal{D}}(\mathcal{D}/\mathcal{D}L^*, \mathcal{O})),$$

$$= \dim(Ker) - \dim(Coker) \text{ for the map } L^* : \mathbb{C}[x] \to \mathbb{C}[x].$$

Now $L^* = \Sigma a_{i,j}(-\partial)^j x^i$, and each operator $(-\partial)^j x^i$ is homogeneous of degree i-j when it acts on the graded ring $\mathbb{C}[x]$. Moreover, the associated graded map

$$(-\partial)^j x^i : (\text{degree } n) \to (\text{degree } n+i-j)$$

is given by a nonzero polynomial $P_{i,j}(n)$ in n of degree j, namely

$$(-\partial)^j x^i(x^n) = P_{i,j}(n)x^{n+i-j}, \text{ where } P_{i,j}(t) = (-1)^j \prod_{k=0,\dots,j-1}(t+i-k).$$

So by definition of the integer d, there exists a nonzero polynomial $P(t)$ (namely $P := \Sigma_{i-j=d}\, a_{i,j}P_{i,j}$) such that L^* acting on $\mathbb{C}[x]$ maps (degree $\le$ n) to (degree $\le$ n+d), and induces $x^n \mapsto P(n)x^{n+d}$ on the associated graded. So if we denote by K_n the complex

$$L^*: (\text{degree} \le n) \to (\text{degree} \le n+d),$$

then the inclusion of K_n into K_{n+1} is a quasiisomorphism if $P(n+1) \neq 0$. The direct limit K_∞ of the K_n is the complex which calculates the Ext groups, so if n is larger than any integer zero of P we have

$$H^*{}_{DR}(\mathbb{A}^1, \mathcal{D}/\mathcal{D}L) = H^*(K_\infty) = H^*(K_n).$$

But for n large enough that $n+d \ge 0$ we have

$$\chi(K_n) = \chi[L^*: (\text{degree} \le n) \to (\text{degree} \le n+d)] = -d. \quad \text{QED}$$

Lemma 2.9.13 On $\mathbb{G}_m$ with parameter x, write D for xd/dx, and consider a nonzero operator $L := \Sigma x^i P_i(D)$. Define integers a, b, d by

$$a := \max(\, i \mid P_i \neq 0), \quad b := \min(\, i \mid P_i \neq 0), \quad d := a-b.$$

Then $\chi(\mathbb{G}_m, \mathcal{D}/\mathcal{D}L) = -d$.

proof. The proof is exactly analogous to that given above, taking for K_n the subcomplex

$$L^* : (-n \le \text{degree} \le n) \to (b-n \le \text{degree} \le a+n)$$

of the complex $L^* : \mathbb{C}[x, x^{-1}] \to \mathbb{C}[x, x^{-1}]$. QED

2.10 Some equations on $\mathbb{A}^1$, with a transition to $\mathbb{G}_m$

(2.10.0) We now turn to the special case where X is $\mathbb{A}^1$. We will write ∂ for d/dx. Thus $\mathcal{O}$ is the polynomial ring $\mathbb{C}[x]$ and $\mathcal{D}$ is the Weyl algebra $\mathcal{O}[\partial] = \mathbb{C}[x,\partial]$.The Fourier Transform FT(L) of an element $L = \Sigma f_i(x)\partial^i$ of $\mathcal{D}$ is defined by

$$FT(L) = \Sigma\, f_i(\partial)(-x)^i.$$

The map $L \mapsto FT(L)$ is a ring isomorphism of $\mathcal{D}$ with itself, whose square is $[-1]^*$:

$$\text{for } L = \Sigma f_i(x)\partial^i, \quad FT(FT(L)) = [-1]^*L = \Sigma f_i(-x)(-\partial)^i.$$

Notice that FT and adjoint nearly commute: one has

$$(FT(L))^* = [-1]^*(FT(L^*)) = FT([-1]^*(L^*)).$$

Given a left (resp. left holonomic) $\mathcal{D}$-module $\mathcal{M}$ on $\mathbb{A}^1$, its Fourier

Transform FT($\mathcal{M}$) is the left (resp. left holonomic) $\mathcal{D}$-module $\mathcal{D}\otimes_{\mathcal{D}}\mathcal{M}$, where in forming the tensor product the leftmost $\mathcal{D}$ is viewed as a right $\mathcal{D}$-module by the ring isomorphism FT:$\mathcal{D}\to\mathcal{D}$. If $\mathcal{M}$ is $\mathcal{D}/\mathcal{D}L$, then FT($\mathcal{M}$) is $\mathcal{D}/\mathcal{D}FT(L)$.

Example 2.10.1 (1) Let j: $\mathbb{G}_m \to \mathbb{A}^1$ the inclusion. Then

$$FT(j_!j^*\mathcal{O}) \approx j_*j^*\mathcal{O},\ FT(j_*j^*\mathcal{O}) \approx j_!j^*\mathcal{O}.$$

Indeed, by 2.9.5.1, $j_!j^*\mathcal{O} \approx \mathcal{D}/\mathcal{D}x\partial$, and $j_*j^*\mathcal{O} \approx \mathcal{D}/\mathcal{D}\partial x$. Visibly we have $FT(x\partial) = -\partial x$, $FT(\partial x) = -x\partial$.

Example 2.10.1 (2) For α in $\mathbb{C}$, the delta module δ_α is $\mathcal{D}/\mathcal{D}(x-\alpha)$. Its FT is $\mathcal{D}/\mathcal{D}(\partial-\alpha)$, which is isomorphic to the $\mathcal{D}$-module $e^{\alpha x}\mathbb{C}[x]$ by means of the $\mathcal{D}$-linear map $1\mapsto e^{\alpha x}$.

(2.10.2) Thus $\mathcal{M}\mapsto FT(\mathcal{M})$ is an exact autoequivalence of the category of left (resp. left holonomic) $\mathcal{D}$-modules. If we iterate FT, we find

$$FT(FT(\mathcal{M})) \approx [-1]^*(\mathcal{M}).$$

A key point for later applications is the apparently trivial consequence that a holonomic left $\mathcal{D}$-module $\mathcal{M}$ is irreducible if and only if FT($\mathcal{M}$) is irreducible. Here is a simple illustration :

Theorem 2.10.3 Let $P{:=}P_n(x) = \Sigma\, p_ix^i$ and $Q{:=}Q_m(x) = \Sigma\, q_ix^i$ be nonzero polynomials in $\mathbb{C}[x]$, of degrees n and m respectively, and suppose that

(1) if α is a simple root of Q, then $P(\alpha)/Q'(\alpha)$ is not in $\mathbb{Z}$.

(2) if α is a multiple root of Q, then $P(\alpha)\neq 0$.

Denote by L the operator $L := P(\partial) + xQ(\partial)$, $\mathcal{M} := \mathcal{D}/\mathcal{D}L$. Then

(1) $\mathcal{M}$ is an irreducible $\mathcal{D}$-module on $\mathbb{A}^1$.

(2) If $n > m$, $\mathcal{M}$ is a Lie-irreducible object of D.E.($\mathbb{A}^1/\mathbb{C}$), whose largest slope at ∞ is $(n+1-m)/(n-m)$, with multiplicity $n-m$.

(3a) If $n \le m$, then for $\alpha := -p_m/q_m$, $\mathcal{M} \mid \mathbb{A}^1 - \{\alpha\}$ is an irreducible object of D.E.($\mathbb{A}^1 - \{\alpha\}/\mathbb{C}$). Its local monodromy at α is a pseudoreflection of determinant $\exp(-2\pi i\beta)$, where β is given by

$$\beta := (p_{m-1}q_m - p_mq_{m-1})/(q_m)^2.$$

(3b) If $n \le m$ and if either $m\neq 2$ or $\exp(2\pi i\beta)\neq -1$, then $\mathcal{M} \mid \mathbb{A}^1 - \{\alpha\}$ is Lie-irreducible.

proof The operator L is the FT of the first order operator $-\partial Q(x) + P(x) = -Q(x)\partial + P(x) - Q'(x)$ to which we apply 2.9.6.2 to get the irreducibility (1). (2) is obvious from the definition of L, and the fact that Lie-irreducibility on $\mathbb{A}^1$ results from irreducibility(1). (3a) is just the spelling out of 2.9.7 (4) and 2.9.7.1. For (3b), we argue as follows (cf. [Ka-Pi],Cor.6 and Criterion 7). By additive translation, we may suppose that $\alpha=0$, so $\mathbb{A}^1 - \{\alpha\}$ is $\mathbb{G}_m$. But an irreducible D.E. on $\mathbb{G}_m$ is either Lie-irreducible or it is Kummer induced. If it is Kummer induced and regular singular at zero, then its exponents at zero in $\mathbb{C}/\mathbb{Z}$ are Kummer induced. The exponents mod $\mathbb{Z}$ are { 0 repeated $m-1$ times, $-\beta$}, which are visibly not Kummer induced unless $m=2$ and $\beta \equiv 1/2$ mod $\mathbb{Z}$. QED

Remark 2.10.3.1 Another way to state the hypotheses on P and Q is to say that P and Q are relatively prime and that the partial fraction expression of P/Q is

$$P/Q = \Sigma\, \lambda_i/(x - \alpha_i) + g'(x),$$

where $g(x)$ in $\mathbb{C}(x)$ blows up at those α_i for which λ_i is in $\mathbb{Z}$.

Notice that $L = P(\partial) + xQ(\partial)$ is the FT of $P(x) - \partial Q(x)$, annihilator of

$$(1/Q(x))\exp(\textstyle\int(P/Q)(t)dt) = (1/Q(x))(\prod(x - \alpha_i)^{\lambda_i})e^{g(x)},$$

and $L^* = P(-\partial) + Q(-\partial)x$ is the FT of $P(-x) - Q(-x)\partial$, annihilator of

$$\exp(\textstyle\int(P/Q)(-t)dt) = (\prod(x + \alpha_i)^{-\lambda_i})e^{-g(-x)}.$$

Conversely, if we begin with a function of the form $(\prod(x - \alpha_i)^{\lambda_i})e^{g(x)}$ with $g(x)$ a rational function which blows up at those α_i for which λ_i is in $\mathbb{Z}$, we recover P and Q by writing $P/Q = \Sigma\, \lambda_i/(x - \alpha_i) + g'(x)$ with $(P,Q)=1$. If $g=0$, then $n=m-1$; otherwise $m-1-n = \mathrm{ord}_\infty(g)$. We will see below that already the sequence of functions $x^{-1/2}\exp(-x^n/n)$ leads to some surprises.

Theorem 2.10.4 Let $P:=P_n(x) = \Sigma\, p_i x^i$ and $Q:=Q_m(x) = \Sigma\, q_i x^i$ be nonzero polynomials in $\mathbb{C}[x]$, of degrees n and m respectively, and suppose that

(1) if α is a simple root of Q, then $P(\alpha)/Q'(\alpha)$ is not in $\mathbb{Z}$.

(2) if α is a multiple root of Q, then $P(\alpha) \neq 0$.

Suppose $n > m$. The differential galois group G of $P(\partial) + xQ(\partial)$ on $\mathbb{A}^1$ is connected and reductive. If $p_{n-1} = q_{n-1} = 0$, then $G = G^{0,\mathrm{der}}$; otherwise $G = \mathbb{G}_m G^{0,\mathrm{der}}$. The possibilities for $G^{0,\mathrm{der}}$ are given by:

(1) If n-m is odd, $G^{0,der}$ is SL(n); if n-m=1, then G is GL(n).
(2) If n-m is even, then either $G^{0\ der}$ is SL(n) or SO(n) or (if n is even) SP(n), or n-m=6, n=7,8 or 9, and $G^{0,der}$ is one of

n=7: the image of G_2 in its 7-dim'l irreducible representation
n=8: the image of Spin(7) in the 8-dim'l spin representation
the image of SL(3) in the adjoint representation
the image of SL(2)×SL(2)×SL(2) in std⊗std⊗std
the image of SL(2)×Sp(4) in std⊗std
the image of SL(2)×SL(4) in std⊗std
n=9: the image of SL(3)×SL(3) in std⊗std.

proof In view of 2.10.3(2) above, this is just the Main D.E. Theorem 2.8.1 on $\mathbb{A}^1$, with a/b = (n+1-m)/(n-m), together with the remark that on $\mathbb{A}^1$ one has detG = {1} or $\mathbb{G}_m$, and detG = {1} if and only if the coefficient of ∂^{n-1} vanishes. QED

To give a concrete illustration of this theory, let us compute G for the operator $\partial^n - x\partial - 1/2$, whose FT defines $x^{-1/2}\exp(-(-x)^n/n)$.

Theorem 2.10.5 The differential galois group G of $\partial^n - x\partial - 1/2$ on $\mathbb{A}^1$ is
GL(2) for n=2,
SL(n) for n even ≥ 4,
SO(n) for n≠7 odd ≥ 3,
G_2 for n=7.

proof This is an instance of the above theorem with $P(x) = x^n - 1/2$, Q(x) = x, m=1. For n even, n-m = n-1 is odd, and so G is SL(n) or GL(n); looking at the ∂^{n-1} term, we see that G is inside SL iff n > 2. If n is odd, this operator is self-adjoint (up to a sign), and as n is odd the resulting autoduality is necessarily symmetric. Therefore G is inside SO(n) for n≥3 odd; in view of the limited possibilities for G, it must be SO(n) except for n=7, where the (only) other possibility is G_2. That it is G_2 in this case results from the following

G_2 Theorem 2.10.6 For any polynomial f in $\mathbb{C}[x]$ of degree k prime to 6, the differential galois group G of $\partial^7 - f\partial - (1/2)f'$ on $\mathbb{A}^1$ is G_2.

proof We first prove that the D.E.on $\mathbb{A}^1$

$$\mathcal{M} := \mathcal{D}/\mathcal{D}L, \quad L := \partial^7 - f\partial - (1/2)f'$$

is irreducible. Its ∞-slopes are 1 + (k/6) with multiplicity six and one slope 0. Since (k, 6)=1 by hypothesis, the I_∞-representation is the direct sum of an irreducible of dimension 6 and a tame character. So if $\mathcal{M}$ is reducible on $\mathbb{A}^1$, its Jordan-Holder constituents must be an irreducible D.E. $\mathcal{N}$ on $\mathbb{A}^1$ of rank six and a rank one D.E. $\mathcal{L}$ on $\mathbb{A}^1$ which is regular singular at ∞, and therefore isomorphic to the trivial $\mathfrak{D}$-module $\mathcal{O}$. So either $\mathcal{O}$ is a quotient of $\mathcal{M}$, or it is a subobject. Since $\mathcal{M}$ is self-adjoint, and $\mathcal{N}$ and $\mathcal{O}$ are nonisomorphic irreducibles, $\mathcal{M} \approx \mathcal{N} \oplus \mathcal{O}$. Therefore $\mathcal{O}$ is a quotient of $\mathcal{M}$. This means that the equation $L\varphi=0$ has nonzero solutions in $\mathbb{C}[x]$. But L acts injectively on $\mathbb{C}[x]$; indeed if $f = Ax^k + \ldots$, then L maps x^d + lower terms to $(-d - (1/2)k)Ax^{d+k-1}$ + lower terms, and as k is odd (being prime to 6), $(-d - (1/2)k)$ is nonzero for all $d \in \mathbb{Z}$. Therefore $\mathcal{M}$ is irreducible.

Once $\mathcal{M}$ is irreducible, it is Lie-irreducible (we are on $\mathbb{A}^1$). Its ∞-slopes qualify it for the Main D.E. Theorem. Because it is self-adjoint, the only possibilities are G_2 or $SO(7)$. It thus suffices to rule out $SO(7)$. Because $\Lambda^3(\mathrm{std}_7)$ is irreducible for $SO(7)$, it suffices to show that $\mathcal{M}$ has a nonzero horizontal section in $\Lambda^3\mathcal{M}$. We can view $\mathcal{M}$ as the free $\mathcal{O}$-module with basis $e_0, \ldots, e_6$, where ∂ acts by

$$\partial e_i = e_{i+1} \text{ for } i=0,1,\ldots,5 \qquad \partial e_6 = (1/2)f' e_0 + f e_1.$$

Then one readily verifies that in $\Lambda^3\mathcal{M}$ the element

$$e_0\wedge e_4\wedge e_5 + e_2\wedge e_3\wedge e_4 + 2e_1\wedge e_2\wedge e_6 - e_1\wedge e_3\wedge e_5 - e_0\wedge e_3\wedge e_6 - fe_0\wedge e_1\wedge e_2$$

is killed by ∂. QED

Theorem 2.10.7 Let $P:=P_n(x) = \Sigma\, p_i x^i$ and $Q:=Q_m(x) = \Sigma\, q_i x^i$ be nonzero polynomials in $\mathbb{C}[x]$, of degrees n and m respectively, and suppose that

(1) if α is a simple root of Q, then $P(\alpha)/Q'(\alpha)$ is not in $\mathbb{Z}$.

(2) if α is a multiple root of Q, then $P(\alpha)\neq 0$.

Suppose $m \geq n$. Define

$$\alpha := -p_m/q_m, \qquad \beta := (p_{m-1}q_m - p_m q_{m-1})/(q_m)^2.$$

Suppose that either $m \geq 3$ or $m=2$ and $\exp(2\pi i\beta) \neq -1$. Then the differential galois group G of $xQ(\partial) + P(\partial)$ on $\mathbb{A}^1 - \{\alpha\}$ is reductive. If $q_{m-1} = 0$ and $\beta \in \mathbb{Q}$, then $G^0 = G^{0,\mathrm{der}}$ and detG is the cyclic subgroup of $\mathbb{G}_m$ generated by $\exp(2\pi i\beta)$; otherwise $G^0 = \mathbb{G}_m G^{0,\mathrm{der}}$. The group $G^{0,\mathrm{der}}$ is either $SL(m)$ or $SO(m)$ or (if m is even) $Sp(m)$. Moreover,

(1) if $\exp(2\pi i\beta) \neq \pm 1$, $G^{0,\mathrm{der}} = SL(m)$;
(2) if $\exp(2\pi i\beta) = +1$, $G = G^0$ and $G^{0,\mathrm{der}} = SL(m)$ or (for m even) $Sp(m)$;
(3) if $\exp(2\pi i\beta) = -1$, $G^{0,\mathrm{der}} = SL(m)$ or $SO(m)$.

proof For any Lie-irreducible D.E., G is reductive, $G^{0,\mathrm{der}}$ is semisimple and irreducible in its given representation, and $G^0 = G^{0,\mathrm{der}}$ or $\mathbb{G}_m G^{0,\mathrm{der}}$, depending on whether detG is finite or not. Here detG is the differential galois group of $\partial + (q_{m-1}/q_m) + \beta/(x - \alpha)$ on $\mathbb{A}^1 - \{\alpha\}$. For any first order D.E. on any nonempty open set of $\mathbb{P}^1$, the differential galois group is finite if and only if all the singularities are regular and at each the exponent is rational. So detG is finite if and only if $q_{m-1} = 0$ (so that ∞ is regular singular) and $\beta \in \mathbb{Q}$, in which case it is the cyclic subgroup of $\mathbb{G}_m$ generated by $\exp(2\pi i\beta)$.

The local monodromy around α is a pseudoreflection γ of determinant $\exp(-2\pi i\beta)$. As G contains the monodromy group (cf. [Ka-DGG], 1.2.2.1), G and hence $G^{0,\mathrm{der}}$ are normalized by γ. The result follows from the Pseudoreflection Theorem 1.5, except for the connectedness of G when $\exp(2\pi i\beta) = +1$. If $\exp(2\pi i\beta) = +1$, then γ is unipotent, and, as as we are on $\mathbb{A}^1 - \{\alpha\}$, whose π_1 is generated by local monodromy around α, G_{mono} is connected, whence G is connected (cf. [Ka-DGG]1.2.5). QED

(2.10.8) We now return to the general properties of Fourier Transform. We recall for later use the "Fourier integral" interpretation (cf. [Ka-Lau, 7.1.4, 7.5]) of $FT(\mathcal{M})$ as $\int \mathcal{M}(x)e^{xy}dx$; one takes on $\mathbb{A}^2 := \mathrm{Spec}(\mathbb{C}[x,y])$ the $\mathcal{D}$-module $pr_1{}^*(\mathcal{M})$ (this is the term $\mathcal{M}(x)$ in the integral), one tensors it over $\mathcal{O}$ with the $\mathcal{D}$-module $\mathcal{D}/\mathcal{D}(\partial_x - y, \partial_y - x)$ [which is the $\mathcal{D}$-module $e^{xy}\mathbb{C}[x,y]$ by means of the $\mathcal{D}$-linear map $1 \mapsto e^{xy}$] (this is the term e^{xy} in the integral) and one takes the relative $H^1{}_{DR}$ for the map $pr_2 : \mathbb{A}^2 \to \mathbb{A}^1$ (this is the meaning of $\int dy$; the other $H^i{}_{DR}$ vanish).

(2.10.9) Given a $\mathcal{D}$-module $\mathcal{M}$ on $\mathbb{A}^1$, and $\alpha \in \mathbb{C}$, we denote by $\mathcal{M} \otimes e^{\alpha x}$ the $\mathcal{D}$-module $\mathcal{M} \otimes_{\mathcal{O}} (\mathcal{D}/\mathcal{D}(\partial - \alpha))$. An alternate description is this. For each $\alpha \in \mathbb{C}$, there is a $\mathbb{C}$-linear ring automorphism A_α of $\mathcal{D}$ which sends $x \mapsto x$ and which sends $\partial \mapsto \partial - \alpha$. (This automorphism is the Fourier Transform of the automorphism of $\mathcal{D}$ induced by the automorphism $x \mapsto x-\alpha$ of $\mathbb{A}^1$.) One could also describe $\mathcal{M} \otimes e^{\alpha x}$ as the $\mathcal{D}$-module $\mathcal{D} \otimes_{\mathcal{D}} \mathcal{M}$ obtained from $\mathcal{M}$ by the extension of scalars $A_\alpha : \mathcal{D} \to \mathcal{D}$.

Lemma 2.10.10 If $\mathcal{M}$ is a non-punctual holonomic $\mathcal{D}$-module on $\mathbb{A}^1$, all of whose slopes at ∞ are ≤ 1, there exists an $\alpha \in \mathbb{C}$ such that $\mathcal{M} \otimes e^{\alpha x}$ has some ∞-slope < 1

proof If $\mathcal{M}$ has some slope < 1, there is nothing to prove. If $\mathcal{M}$ has all slopes 1, we use the following local

Break Depression Lemma 2.10.11 (compare [Ka-GKM, 8.5.7.1]) Let $n \geq 1$ be an integer, and V a nonzero D.E. on $\mathbb{C}((1/x))$, all of whose slopes $= n$. Then there exists an $\alpha \in \mathbb{C}$ such that $V \otimes e^{\alpha x^n}$ has some ∞-slope $< n$.

proof This is obvious from Levelt's structure theorem (cf. [Ka-DGG, 2.2.2]). For the assertion is invariant under extension of scalars from $\mathbb{C}((1/x))$ to $\mathbb{C}((1/x^{1/m}))$ (with n replaced by nm). But after such an extension, any D.E. becomes a successive extension of rank one D.E.'s. So we reduce to the case when V is rank one and slope n, so isomorphic to the D.E. for $x^\delta e^{P(x)}$ where P(x) is a polynomial of degree n, and the assertion is obvious; take for $-\alpha$ the leading coefficient of P(x). QED

Lemma 2.10.12 Let $\mathcal{M}$ be an irreducible holonomic $\mathcal{D}$-module on $\mathbb{A}^1$, all of whose slopes at ∞ are ≤ 1. If $\mathcal{M}$ is in D.E.($\mathbb{A}^1/\mathbb{C}$), then $\mathcal{M}$ is the $\mathcal{D}$-module $\mathcal{D}/\mathcal{D}(\partial - \alpha)$ corresponding to $e^{\alpha x}$ for some $\alpha \in \mathbb{C}$.

proof Twisting by a suitable $e^{\alpha x}$, we reduce to the case where all ∞-slopes of $\mathcal{M}$ are ≤ 1, and at least one slope is < 1. Therefore its irregularity $\mathrm{Irr}_\infty(\mathcal{M})$ is $<$ rank($\mathcal{M}$). By Deligne's Euler-Poincare formula for a D.E. $\mathcal{M}$ on $\mathbb{A}^1/\mathbb{C}$,

$$\dim H^0{}_{DR}(\mathbb{A}^1/\mathbb{C}, \mathcal{M}) - \dim H^1{}_{DR}(\mathbb{A}^1/\mathbb{C}, \mathcal{M}) = \mathrm{rank}(\mathcal{M}) - \mathrm{Irr}_\infty(\mathcal{M})$$

is strictly positive. Therefore $H^0{}_{DR}(\mathbb{A}^1/\mathbb{C}, \mathcal{M}) = \mathrm{Hom}_{\mathcal{D}}(\mathcal{O}, \mathcal{M})$ is nonzero, and this contradicts the irreducibility of $\mathcal{M}$ unless $\mathcal{M}$ is $\mathcal{O}$ itself. QED

Corollary 2.10.12.1 Let $\mathcal{M}$ be a holonomic $\mathcal{D}$-module on $\mathbb{A}^1$, all of whose slopes at ∞ are < 1. If $\mathcal{M}$ is in D.E.($\mathbb{A}^1/\mathbb{C}$), then $\mathcal{M}$ is the trivial $\mathcal{D}$-module $H^0{}_{DR}(\mathbb{A}^1/\mathbb{C}, \mathcal{M}) \otimes_{\mathbb{C}} \mathcal{O}$.

proof By the previous Lemma, $\mathcal{M}$ is a successive extension of trivials.

Since $\mathrm{Ext}^1_{\mathcal{D}}(\mathcal{O}, \mathcal{O}) = H^1_{DR}(\mathbb{A}^1/\mathbb{C}, \mathcal{O}) = 0$, the extensions are themselves trivial. QED

Corollary 2.10.12.2 Let $\mathcal{M}$ be a holonomic $\mathcal{D}$-module on $\mathbb{A}^1$, all of whose slopes at ∞ are ≤ 1. If $\mathcal{M}$ is in D.E.($\mathbb{A}^1/\mathbb{C}$), then $\mathcal{M}$ is the direct sum $\oplus_\alpha H^0_{DR}(\mathbb{A}^1/\mathbb{C}, \mathcal{M}\otimes e^{-\alpha x})\otimes_{\mathbb{C}}(\mathcal{D}/\mathcal{D}(\partial-\alpha))$.

proof Again by the previous Lemma, $\mathcal{M}$ is a successive extension of the $\mathcal{D}/\mathcal{D}(\partial-\alpha)$'s, and again the extensions are trivial because

$$\mathrm{Ext}^1_{\mathcal{D}}(\mathcal{D}/\mathcal{D}(\partial-\alpha),\mathcal{D}/\mathcal{D}(\partial-\beta)) = H^1_{DR}(\mathbb{A}^1/\mathbb{C}, \mathcal{D}/\mathcal{D}(\partial+\alpha-\beta)) = 0.\ \text{QED}$$

Proposition 2.10.13 Suppose that $L = \Sigma f_i(x)\partial^i$ is a monic polynomial in ∂ of degree $n \ge 1$ with coefficients $f_i(x)$ in $\mathbb{C}[x]$. Suppose that the highest slope at ∞ of L is a/b in lowest terms, with multiplicity b, and $a/b > 1$. Then $\mathcal{M} := \mathcal{D}/\mathcal{D}L$ is irreducible on $\mathbb{A}^1$ if any of the following conditions holds:

(1) Every other slope λ of L at ∞ satisfies $0 < \lambda < 1$.

(2) Every other slope λ of L at ∞ satisfies $\lambda < 1$, and neither L nor L^* has any nonzero polynomial solutions.

(3) Every other slope λ of L at ∞ satisfies $\lambda \le 1$, and neither L nor L^* has any nonzero solutions in $e^{\alpha x}\mathbb{C}[x]$ for any α in $\mathbb{C}$.

(4) Every other slope λ of L at ∞ satisfies $\lambda \le 1$, and FT(L) is a middle extension, i.e., for $j:U\to\mathbb{A}^1$ the inclusion of any nonempty open set on which $\mathrm{FT}(\mathcal{M})=\mathcal{D}/\mathcal{D}\mathrm{FT}(L)$ is a D.E., $\mathrm{FT}(\mathcal{M})\approx j_{!*}(j^*\mathrm{FT}(\mathcal{M}))$.

proof The slope hypotheses assure in the break decomposition of $\mathcal{M}$ as I_∞-representation, $\mathcal{M} \approx \mathcal{M}_{a/b}\oplus\mathcal{M}_{\le 1}$, the term $\mathcal{M}_{a/b}$ is I_∞-irreducible. Therefore if $\mathcal{M}$ is reducible, it has either a nonzero subobject or a nonzero quotient $\mathcal{N}$ all of whose ∞-slopes are ≤ 1. At the expense of switching $\mathcal{M}$ and $\mathcal{M}^*$, we may assume the existence of such a quotient $\mathcal{N}$. In view of 2.10.12, $\mathcal{N}$ itself has a quotient $e^{\alpha x}\mathbb{C}[x]$ for some α. The sufficiency of (1), (2), (3) is now obvious, and (3)$\Leftrightarrow$(4) by Fourier Transform. QED

Examples 2.10.14 In the examples below, P_n and Q_m denote polynomials in $\mathbb{C}[x]$ of degrees n and m respectively.

example of (1): $L = P_n(\partial) + x\partial^m$ with $2 \le m < n$ and $P(0)\neq 0$; its slopes are $1 + 1/(n-m)$ with mult. $n-m$, and $1 - (1/m)$ with mult. m.

example of (2) : $L = P_n(\partial) + Q_m(x\partial)$ with $m < n$, $(n,m)=1$, $P_n(0) = 0$, Q_m has no zeroes in $\mathbb{Z}$; its slopes are $1 + m/(n-m)$ with mult. $n-m$, 0 with mult. m. Because $P(0)=0$, $L(x^d + \text{lower terms}) = Q_m(d)x^d + \text{lower terms}$, so L is injective on $\mathbb{C}[x]$ if $Q_m(x)$ has no zeroes in $\mathbb{Z}_{\geq 0}$. The adjoint L^* is $P_n(-\partial) + Q_m(-1-x\partial)$, which will be injective on $\mathbb{C}[x]$ if $Q_m(-1-x)$ has no zeroes is $\mathbb{Z}_{\geq 0}$.

example of (3) via (4): $L = P_n(\partial) + xQ_m(\partial)$, $n > m$, $(n,m) = 1$, where at simple roots of Q, P/Q' has non-$\mathbb{Z}$ values, and P is nonzero at multiple roots of Q. Let x^d be the highest power of x which divides Q_m. The slopes are $1 + 1/(n-m)$ with mult. $n-m$, 1 with multiplicity $m-d$, and, if $d \neq 0$, $1 - 1/d$ with multiplicity d. This is the case we have already discussed at some length, without the extra hypotheses on n and m. Notice that taking Q to be ∂^m gives back our example of (1).

So only example (2) is really new. An alternate and more fruitful approach to it comes by noticing that $L := P_n(\partial) + Q_m(x\partial)$ is the Fourier transform of $K := Q_m(-1-xd/dx) + P_n(x)$, a "Kloosterman operator of bidegree (m,n)" in the terminolgy of [Ka-DGG, 4.4]. The ∞-slopes of K are all n/m, so $\mathcal{D}/\mathcal{D}K$ is irreducible on $\mathbb{G}_m$ (because $(n,m)=1$), and on $\mathbb{A}^1$ it is the middle extension of $(\mathcal{D}/\mathcal{D}K) \mid \mathbb{G}_m$ (because Q has no zeroes in $\mathbb{Z}$). So in fact $L = P_n(\partial) + Q_m(x\partial)$ defines an irreducible $\mathcal{D}$-module on $\mathbb{A}^1$ whether or not $n > m$.

If $m > n$, L gives a D.E. on $\mathbb{G}_m$, which is regular singular at ∞ and whose 0-slopes are $n/(m-n)$ with multiplicity $m-n$, and 0 with multiplicity n. It is Lie-irreducible as well. Indeed, if $(\mathcal{D}/\mathcal{D}L) \mid \mathbb{G}_m$ were Kummer induced of degree d, then d would divide the multiplicity of each 0-slope, and these multiplicities ($m-n$ and n) are relatively prime.

Applying the Main D.E. Theorem 2.8.1, we find

Theorem 2.10.15 Let $P := P_n(x) = \Sigma\, p_i x^i$ and $Q := Q_m(x) = \Sigma\, q_i x^i$ be nonzero polynomials in $\mathbb{C}[x]$, of degrees n and $m \neq n$ respectively, and suppose that

(1) $Q_m(x)$ has no roots in $\mathbb{Z}$, and $P_0(0)=0$.

(2) $(n,m)=1$.

Denote by G the differential galois group G of $P(\partial) + Q(x\partial)$ on $\mathbb{A}^1$ (if $n > m$) or on $\mathbb{G}_m$ (if $m > n$) and define $N := \max(n,m)$. Then

(case $n > m$) G is connected. If $p_{n-1} = q_{n-1} = 0$, then $G = G^{0,\mathrm{der}}$; otherwise $G = \mathbb{G}_m G^{0,\mathrm{der}}$.

(case $m > n$) If $m-n=1$, then G is $GL(m)$. If $m-n \geq 2$ and $q_{m-1}/q_m \in \mathbb{Q}$, then $G^0 = G^{0,\mathrm{der}}$ and $\det G$ is $\langle \exp(2\pi i q_{m-1}/q_m) \rangle$; otherwise $G = \mathbb{G}_m G^{0,\mathrm{der}}$.

In both cases, the possibilities for $G^{0,\mathrm{der}}$ are given by:

(1) If $n-m$ is odd, $G^{0,\mathrm{der}}$ is $SL(N)$; if $|n-m|=1$, then G is $GL(N)$.

(2) If $n-m$ is even, then $G^{0\ \mathrm{der}}$ is $SL(N)$ or $SO(N)$ or (if N is even) $SP(N)$, or $|n-m| = 6$, $N=7,8$ or 9, and $G^{0,\mathrm{der}}$ is one of

$N=7$: the image of G_2 in its 7-dim'l irreducible representation

$N=8$: the image of Spin(7) in the 8-dim'l spin representation
the image of $SL(3)$ in the adjoint representation
the image of $SL(2)\times SL(2)\times SL(2)$ in $\mathrm{std}\otimes\mathrm{std}\otimes\mathrm{std}$
the image of $SL(2)\times Sp(4)$ in $\mathrm{std}\otimes\mathrm{std}$
the image of $SL(2)\times SL(4)$ in $\mathrm{std}\otimes\mathrm{std}$

$N=9$: the image of $SL(3)\times SL(3)$ in $\mathrm{std}\otimes\mathrm{std}$.

Location of the singularities of a Fourier Transform

We conclude this section with a general result on the location of the singularities of a Fourier Transform. Recall that any holonomic $\mathcal{D}$-module $\mathcal{M}$ on $\mathbb{A}^1$ is a D.E. on some dense open set of $\mathbb{A}^1$. Therefore it makes sense to speak of the ∞-slopes of $\mathcal{M}$.

Theorem 2.10.16 (compare [Ka-GKM, 8.5.8]) Let $\mathcal{M}$ be a holonomic $\mathcal{D}$-module on $\mathbb{A}^1$. Then

(1) $FT\mathcal{M}$ is a D.E. on $\mathbb{G}_m$ if and only if $\mathcal{M}$ has no ∞-slope $=1$.

(2) $FT\mathcal{M}$ is a D.E. near α if and only if $\mathcal{M}\otimes e^{\alpha x}$ has all ∞-slopes ≥ 1.

(3) $FT\mathcal{M}$ is a D.E. near α if and only if the function $\mathbb{C} \to \mathbb{Z}$

$$\beta \mapsto \mathrm{Irr}_\infty(\mathcal{M}\otimes e^{\beta x})$$

is constant in a neighborhood of α.

proof We first show that (2) implies (1). Indeed, if $\mathcal{M}$ has no ∞-slope $=1$, then for any $\alpha \neq 0$, $\mathcal{M}\otimes e^{\alpha x}$ has all ∞-slopes ≥ 1 (cf. [Ka-DGG, 2.2.11.4]). Conversely, if $\mathcal{M}$ has some ∞-slope $=1$, apply the "slope decomposition" [Ka-DGG, 2.3.4] to $\mathcal{M}\otimes_{\mathcal{O}}\mathbb{C}((1/x))$ and then apply the Break Depression Lemma 2.10.11 to its "slope $=1$" part to produce an α such that $\mathcal{M}\otimes e^{\alpha x}$ has some ∞-slopes < 1.

To prove (2) and (3), we may, by additive translation, reduce to the case $\alpha = 0$. Thus we must show that $\mathrm{FT}\mathcal{M}$ is a D.E. near zero if and only if $\mathcal{M}$ has all ∞-slopes ≥ 1. By Levelt's structure theorem [Le-JD], $\mathcal{M}$ has all ∞-slopes ≥ 1 if and only if the function

$$\mathbb{C} \rightarrow \mathbb{Z}$$
$$\alpha \mapsto \mathrm{Irr}_{\infty}(\mathcal{M}\otimes e^{\alpha x})$$

is constant in a neighborhood of $\alpha = 0$. [Indeed, if $\mathcal{M}$ has ∞-slopes $\lambda_1, \ldots, \lambda_r$, then $\mathrm{Irr}_{\infty}(\mathcal{M}) := \Sigma_i \lambda_i$, while for all but finitely many $\alpha \neq 0$, $\mathrm{Irr}_{\infty}(\mathcal{M}\otimes e^{\alpha x}) = \Sigma_i \max(\lambda_i, 1)$.] So (2) and (3) are equivalent.

We may further reduce to the case in which $\mathcal{M}$ is irreducible. For in any case $\mathcal{M}$ is a successive extension of finitely many irreducibles $\mathcal{M}_i$, and $\mathrm{FT}\mathcal{M}$ is a successive extension of the finitely many irreducibles $\mathrm{FT}\mathcal{M}_i$. Now $\mathrm{FT}\mathcal{M}$ is a D.E. near zero (i.e., is $\mathcal{O}$-coherent near zero) if and only if each of its Jordan Holder constituents $\mathrm{FT}\mathcal{M}_i$ is a D.E. near zero (i.e., is $\mathcal{O}$-coherent near zero). Clearly the condition (2), that $\mathcal{M}$ have all ∞-slopes ≥ 1, holds for $\mathcal{M}$ if and only if it holds for each $\mathcal{M}_i$. Thus it suffices to treat the case when $\mathcal{M}$ is irreducible.

We first check by hand a few special cases.

If $\mathcal{M}$ is a delta module δ_{α}, then it has no ∞-slopes, and $\mathrm{FT}\mathcal{M} \approx \mathcal{D}/\mathcal{D}(\partial-\alpha)$ is a D.E. on all of $\mathbb{A}1$.

If $\mathcal{M}$ is the constant $\mathcal{D}$-module $\mathcal{O} \approx \mathcal{D}/\mathcal{D}\partial$, whose unique ∞-slope is 0, then $\mathrm{FT}\mathcal{M}$ is δ_0, which is not a D.E. near zero.

If $\mathcal{M}$ is the $\mathcal{D}$-module $\mathcal{D}/\mathcal{D}(\partial+\alpha) \approx e^{-\alpha x}\mathbb{C}[x]$ with $\alpha \neq 0$, whose unique ∞-slope is 1, then $\mathrm{FT}\mathcal{M}$ is δ_{α}, which is a D.E. near zero.

Suppose now that $\mathcal{M}$ is irreducible, not punctual, and not of the form $\mathcal{D}/\mathcal{D}(\partial+\alpha)$ for any α in $\mathbb{C}$. For every α in $\mathbb{C}$ we have

$$H^0{}_{DR}(\mathbb{A}^1, \mathcal{M}\otimes e^{\alpha x}) = \mathrm{Hom}_{\mathcal{D}}(\mathcal{D}/\mathcal{D}(\partial+\alpha), \mathcal{M}) = 0.$$

The adjoint $\mathcal{M}^*$ of $\mathcal{M}$ is of the same type, as is its pullback by $x \mapsto -x$, $[-1]^*\mathcal{M}^*$, so we also have

$$H^0_{DR}(\mathbb{A}^1, [-1]^*\mathcal{M}^*\otimes e^{\alpha x}) = \mathrm{Hom}_{\mathcal{D}}(\mathcal{D}/\mathcal{D}(\partial+\alpha), [-1]^*\mathcal{M}^*) = 0.$$

If we denote by $U = \mathbb{A}^1 - S$ a nonempty open set of $\mathbb{A}^1$ on which $\mathcal{M}$ is a D.E., then $\mathcal{M} \approx j_{!*}j^*\mathcal{M}$ (since $\mathcal{M}$ is irreducible nonpunctual). The Euler-Poincare formula for $\mathcal{M} \approx j_{!*}j^*\mathcal{M}$ on $\mathbb{A}^1$ gives

$$\chi(\mathbb{A}^1, \mathcal{M}\otimes e^{\alpha x}) =$$
$$= \mathrm{rank}_{\mathcal{O}_U}(j^*\mathcal{M}\otimes e^{\alpha x}) - \mathrm{Irr}_\infty(\mathcal{M}\otimes e^{\alpha x}) - \Sigma_{\beta\in S}\,\mathrm{totdrop}_\beta(\mathcal{M}\otimes e^{\alpha x}).$$

Because $e^{\alpha x}$ is a rank one D.E. on all of $\mathbb{A}^1$, it is formally trivial at each point $\beta \in S$, so we can rewrite the above formula as

$$\chi(\mathbb{A}^1, \mathcal{M}\otimes e^{\alpha x}) = \mathrm{rank}_{\mathcal{O}_U}(j^*\mathcal{M}) - \mathrm{Irr}_\infty(\mathcal{M}\otimes e^{\alpha x}) - \Sigma_{\beta\in S}\,\mathrm{totdrop}_\beta(\mathcal{M})$$
$$= (\text{function of } \mathcal{M} \text{ alone}) - \mathrm{Irr}_\infty(\mathcal{M}\otimes e^{\alpha x}).$$

We now express this information in terms of the De Rham complexes of $\mathcal{M}\otimes e^{\alpha x}$, and of $[-1]^*\mathcal{M}^*\otimes e^{\alpha x}$, which we view as the two-term complexes (in degrees 0 and 1)

$$\mathcal{M} \xrightarrow{\partial+\alpha} \mathcal{M}, \qquad [-1]^*\mathcal{M}^* \xrightarrow{\partial+\alpha} [-1]^*\mathcal{M}^*.$$

We are told that both of these arrows are injective for all α in $\mathbb{C}$, and that

$$\dim_{\mathbb{C}}(\mathcal{M}/(\partial+\alpha)\mathcal{M}) = \mathrm{Irr}_\infty(\mathcal{M}\otimes e^{\alpha x}) - (\text{function of } \mathcal{M} \text{ alone}).$$

In terms of the Fourier Transforms $FT\mathcal{M}$ and $FT([-1]^*\mathcal{M}^*) \approx (FT\mathcal{M})^*$, this says precisely that the two arrows

$$FT\mathcal{M} \xrightarrow{-x+\alpha} FT\mathcal{M}, \qquad (FT\mathcal{M})^* \xrightarrow{-x+\alpha} (FT\mathcal{M})^*.$$

are both injective for all α in $\mathbb{C}$, and that

$$\dim_{\mathbb{C}}(FT\mathcal{M}/(x-\alpha)FT\mathcal{M}) = \mathrm{Irr}_\infty(\mathcal{M}\otimes e^{\alpha x}) - (\text{function of } \mathcal{M} \text{ alone}).$$

According to 2.9.10.1, $FT\mathcal{M}$ is a D.E. near zero if and only if the following three conditions hold:

(1) the map $\mathrm{Left}(x) : FT\mathcal{M} \to FT\mathcal{M}$ is injective.

(2) the map $\mathrm{Left}(x) : (FT\mathcal{M})^* \to (FT\mathcal{M})^*$ is injective.

(3bis) the function $\alpha \mapsto \dim_{\mathbb{C}}(FT\mathcal{M}/(x - \alpha)FT\mathcal{M})$ is constant near $\alpha = 0$.

Thus for $\mathcal{M}$ is irreducible, not punctual, and not of the form $\mathcal{D}/\mathcal{D}(\partial+\alpha)$ for any α in $\mathbb{C}$, $FT\mathcal{M}$ is a D.E. near zero if and only if the function $\alpha \mapsto \mathrm{Irr}_\infty(\mathcal{M}\otimes e^{\alpha x})$ is constant for α near zero. QED

2.11 Systematic study of equations on $\mathbb{G}_m$

(2.11.0) We now turn to the systematic study of equations on $\mathbb{G}_m := \mathrm{Spec}(\mathbb{C}[x,x^{-1}])$. We write

$$\partial := d/dx, \qquad D := xd/dx.$$

The ring $\mathcal{D}_{\mathbb{G}_m}$ on $\mathbb{G}_m$ is $\mathbb{C}[x,x^{-1},D] = \mathbb{C}[x,x^{-1},\partial] = \mathcal{D}_{\mathbb{A}^1}[1/x]$. The multiplicative inversion "inv" on $\mathbb{G}_m$ interchanges x and x^{-1}, and sends D to -D. Because we will sometimes want to think of $\mathbb{G}_m$ as sitting in $\mathbb{A}^1$, we will use the induced notion of adjoint $\partial \mapsto -\partial$, which sends D to $-1-D$. Every element of $\mathcal{D}_{\mathbb{G}_m}$ is the (right or left) product of a power of x, which is a unit in $\mathcal{D}_{\mathbb{G}_m}$, and of an element of the subring $\mathbb{C}[x,D]$. We will generally write elements of $\mathcal{D}_{\mathbb{G}_m}$ in the form $\Sigma\, x^i P_i(D)$, where the $P_i(t)$ are polynomials in $\mathbb{C}[t]$.

Given a $\mathcal{D}$-module $\mathcal{M}$ on $\mathbb{G}_m$ and an $\alpha \in \mathbb{C}$, we denote by $\mathcal{M}\otimes x^\alpha$ the $\mathcal{D}$-module $\mathcal{M}\otimes_{\mathcal{O}}(\mathcal{D}/\mathcal{D}(D-\alpha))$. Notice that $\mathcal{D}/\mathcal{D}(D-\alpha) \approx x^\alpha\mathbb{C}[x,x^{-1}]$ by the $\mathcal{D}$-linear map $1 \mapsto x^\alpha$. For each $\alpha \in \mathbb{C}$, there is a $\mathbb{C}$-linear ring automorphism B_α of $\mathcal{D}$ which sends $x \mapsto x$ and which sends $D \mapsto D-\alpha$. One could also describe $\mathcal{M}\otimes x^\alpha$ as the $\mathcal{D}$-module obtained from $\mathcal{M}$ by this extension of scalars $B_\alpha : \mathcal{D} \to \mathcal{D}$.

Lemma 2.11.1 Let $\mathcal{M}$ be a holonomic $\mathcal{D}$-module on $\mathbb{G}_m$. Then for any $\alpha \in \mathbb{C}$, $\chi(\mathbb{G}_m, \mathcal{M}\otimes x^\alpha) = \chi(\mathbb{G}_m, \mathcal{M})$.

proof This is immediate from the Euler-Poincare formula, because $x^\alpha\mathbb{C}[x,x^{-1}]$ is a rank one D.E. on $\mathbb{G}_m$ of slope zero at 0 and ∞. QED

Lemma 2.11.2 Let V be a nonzero object of $\mathrm{D.E.}(\mathbb{G}_m/\mathbb{C})$ all of whose slopes at both 0 and ∞ are zero. Then V is a successive extension of objects $x^\alpha\mathbb{C}[x,x^{-1}]$.

proof Since V is regular singular at both 0 and ∞, it is uniquely determined by the monodromy representation it gives of the topological $\pi_1(\mathbb{G}_m) \approx \mathbb{Z}$. As $\mathbb{Z}$ is abelian, V is a successive extension of rank one D.E.'s on $\mathbb{G}_m$ which are regular singular at 0 and ∞. So we are reduced to the case when V is $\mathcal{O}$-invertible. As $\mathcal{O} = \mathbb{C}[x,x^{-1}]$ is a principal ideal domain, V is free of rank one over $\mathcal{O}$, say $V = \mathbb{C}[x,x^{-1}]e_0$ with $De_0 = fe_0$, so $V \approx \mathcal{D}/\mathcal{D}(D-f)$ for some $f \in \mathbb{C}[x,x^{-1}]$. In terms of f, the slopes of V at 0 and ∞ are $\max(0, -\mathrm{ord}_0(f))$ and $\max(0, -\mathrm{ord}_\infty(f))$ respectively. As V has slope zero at both 0 and ∞, f is constant. QED

Here is a formal version of this, in a form which inductively isolates the entire slope zero part from the rest.

Factorization Lemma 2.11.3(cf. [Ma]) Consider an element L of $\mathbb{C}[[x]][D]$ of the form

$$L = \Sigma_{i \geq 0}\, x^i A_i(D)$$

where the $A_i(t)$ are polynomials in $\mathbb{C}[t]$ of uniformly bounded degree. Put $N := \sup_i(\deg A_i)$. Suppose that A_0 is nonzero and that we are given a factorization of A_0,

$$A_0(t) = P(t)Q(t), \text{ such that for all } n \geq 1, \quad \gcd(P(t+n), Q(t)) = 1.$$

Denote by M the degree of P. Then in $\mathbb{C}[[x]][D]$ there exists a unique factorization of L

$$L = (\, \Sigma_{i \geq 0}\, x^i P_i(D)\,) \cdot (\, \Sigma_{i \geq 0}\, x^i Q_i(D)\,)$$

such that

$P_0 = P$, $\deg(P_i) < M$ for $i > 0$

$Q_0 = Q$, $\deg(Q_i) \leq N - M$ for $i > 0$, with equality if $\deg(A_i) = N$.

proof If such a factorization exists, then equating like powers of x and using the commutation relation $P_i(D)x^j = x^j P_i(D+j)$, we find

$$A_n(t) = \Sigma_{i+j=n} P_i(t+j)Q_j(t) =$$
$$= P(t+n)Q_n(t) + P_n(t)Q(t) + \Sigma_{i+j=n,\ 1 \leq i,j \leq n-1} P_i(t+j)Q_j(t).$$

If we rewrite this as

$$A_n(t) - \Sigma_{i+j=n,\ 1 \leq i,j \leq n-1} P_i(t+j)Q_j(t) = P(t+n)Q_n(t) + P_n(t)Q(t),$$

we see that P_n and Q_n are obtained inductively by using the relative primality of $P(t+n)$ and $Q(t)$ to write the left hand side as lying in the ideal $(P(t+n), Q(t))$ they generate, in such a way that the coefficient P_n of Q is of lowest possible degree. This shows the unicity, and also gives an inductive proof of existence. QED

Isomorphism Lemma 2.11.4 (cf. [Ma]) Suppose that $P(t)$ in $\mathbb{C}[t]$ is a polynomial of degree $N \geq 1$ which satisfies

$$\text{for all } n \geq 1, \quad \gcd(P(t+n), P(t)) = 1.$$

Suppose that $L = \Sigma_{i \geq 0}\, x^i P_i(D)$ in $\mathbb{C}[[x]][D]$ satisfies

$$P_0 = P, \text{ and for } i \geq 1,\ \deg P_i < N.$$

Then there is a unique isomorphism of $\mathbb{C}[[x]][D]$-modules $\mathbb{C}[[x]][D]/\mathbb{C}[[x]][D]L \approx \mathbb{C}[[x]][D]/\mathbb{C}[[x]][D]P(D)$ which modulo (x) is the identity on $\mathbb{C}[D]/\mathbb{C}[D]P(D)$.
In particular, the two $\mathfrak{D}$-modules on $\mathbb{C}((x))$, $\mathbb{C}((x))[D]/\mathbb{C}((x))[D]L$ and $\mathbb{C}((x))[D]/\mathbb{C}((x))[D]P(D)$ are isomorphic.

This is a special case of

Isomorphism Lemma bis 2.11.5 (cf. [Ma]) Suppose that V is a finite-dimensional $\mathbb{C}$-vector space, and that $\mathcal{V} := V\otimes_{\mathbb{C}}\mathbb{C}[[x]]$ is endowed with a structure of $\mathbb{C}[[x]][D]$-module. Write the action of D on elements of V:

$$D(v) = \Sigma_{i\geq 0}\, x^i A_i(v), \text{ for unique elements } A_i \text{ in } \mathrm{End}_{\mathbb{C}}(V).$$

Suppose that the distinct eigenvalues of A_0 are incongruent mod $\mathbb{Z}$. Denote by $\mathcal{V}_1$ the $\mathbb{C}[[x]][D]$-module whose underlying $\mathbb{C}[[x]]$-module is $V\otimes_{\mathbb{C}}\mathbb{C}[[x]]$ but where $D(v) = A_0(v)$ for v in V. Then there exists a unique isomorphism of $\mathbb{C}[[x]][D]$-modules from $\mathcal{V}_1$ to $\mathcal{V}$ which is the identity modulo (x).

proof We must show that there is a unique sequence of elements $\{B_i\}_{i\geq 0}$ in $\mathrm{End}(V)$ such that $B_0=1$ and such that in $\mathcal{V}$, we have

$$D(\Sigma_{i\geq 0}x^iB_i(v)) = \Sigma_{i\geq 0}x^iB_i(A_0(v)).$$

The desired isomorphism is then the unique $\mathbb{C}[[x]]$-linear one which maps $v\mapsto\Sigma_{i\geq 0}x^iB_i(v)$. Expanding the left side we find

$$D(\Sigma_{i\geq 0}\, x^iB_i(v)) = \Sigma_{i\geq 0}\, ix^iB_i(v) + \Sigma_{i\geq 0}x^i\Sigma_{j\geq 0}\, x^jA_jB_i(v).$$

Comparing coefficients of like powers of x, we find, for each $n \geq 1$,

$B_nA_0 = nB_n + A_0B_n + \Sigma_{j<n} A_{n-j}B_j$, which we rewrite

$-[A_0, B_n] - nB_n = \Sigma_{j<n} A_{n-j}B_j$.

But the hypothesis on the eigenvalues of A_0 is precisely that the endomorphism $\mathrm{ad}(A_0)$ of $\mathrm{End}(V)$ has no nonzero eigenvalue in $\mathbb{Z}$. Therefore this last equation allows us to solve uniqely for B_n inductively. QED

Ext Lemma 2.11.6 (cf. [Ma]) Suppose that V and W are finite-dimensional $\mathbb{C}$-vector spaces, and that $\mathcal{V} := V\otimes_{\mathbb{C}}\mathbb{C}[[x]]$ and $\mathcal{W} := W\otimes_{\mathbb{C}}\mathbb{C}[[x]]$ are endowed with the structure of $\mathbb{C}[[x]][D]$-modules. Write the actions of D on elements of V and W:

$$D(v) = \Sigma_{i\geq 0}\, x^iA_i(v), \text{ for unique elements } A_i \text{ in } \mathrm{End}_{\mathbb{C}}(V),$$

$$D(w) = \Sigma_{i\geq 0}\ x^i B_i(w), \text{ for unique elements } B_i \text{ in } \mathrm{End}_{\mathbb{C}}(W).$$

(1) If $\dim(V) > 0$, and $\det(T - A_0|V) = \prod(T - \alpha_i)$, then $\mathcal{V}$ is a successive extension of the objects $\mathbb{C}[[x]][D]/\mathbb{C}[[x]][D](D - \alpha_i)$.

(2) Suppose that A_0 and B_0 have no common eigenvalues mod $\mathbb{Z}$. Then

$$\mathrm{Hom}_{\mathbb{C}((x))[D]}(\mathcal{V}[1/x], \mathcal{W}[1/x]) = 0 = \mathrm{Ext}_{\mathbb{C}((x))[D]}(\mathcal{V}[1/x], \mathcal{W}[1/x]).$$

proof (1) Intrinsically, $\mathcal{V}$ is a $\mathbb{C}[[x]][D]$ module which is $\mathbb{C}[[x]]$-free of some finite rank $r \geq 1$, V is the r-dimensional $\mathbb{C}$-space $\mathcal{V}/x\mathcal{V}$, and A_0 in $\mathrm{End}_{\mathbb{C}}(V)$ is the induced action of D on $\mathcal{V}/x\mathcal{V}$. To prove (1), it suffices (by induction on r) to exhibit an eigenvalue α of A_0 and an element v_∞ of $\mathcal{V}$ such that both of the following conditions hold:

$$v_\infty \bmod x\mathcal{V} \text{ is nonzero in } V,$$

$$Dv_\infty = \alpha v_\infty \text{ in } \mathcal{V}.$$

For such an element v_∞ defines an injective mapping of $\mathbb{C}[[x]][D]$-modules

$$\mathbb{C}[[x]][D]/\mathbb{C}[[x]][D](D - \alpha) \to \mathcal{V}$$

whose cokernel is $\mathbb{C}[[x]]$-free of rank r-1. To do this, pick for α any eigenvalue of A_0 whose real part $\mathrm{Re}(\alpha)$ is minimal, and any nonzero eigenvector v_{00} in V with eigenvalue α: $A_0 v_{00} = \alpha v_{00}$. Because $\mathrm{Re}(\alpha)$ is minimal, for every integer $n \geq 1$, $\alpha - n$ is **not** an eigenvalue of A_0, and consequently $A_0 - \alpha + n$ is bijective on V, for every integer $n \geq 1$. Using this bijectivity, one shows that there exists a unique sequence of elements v_i in $\mathcal{V}/x^{i+1}\mathcal{V}$ which satisfy the three conditions

$$v_0 = v_{00},$$

$$v_{i+1} \equiv v_i \bmod x^{i+1}\mathcal{V},$$

$$Dv_i \equiv \alpha v_i \bmod x^{i+1}\mathcal{V}.$$

The inverse limit of the v_i is then the desired element v_∞.

(2) Applying (1) to both $\mathcal{V}$ and $\mathcal{W}$, we reduce immediately to the case where V and W are one-dimensional, and $D(v) = av$, $D(w) = bw$ with $a-b$ not in $\mathbb{Z}$. Then the Hom and Ext in question are just the kernel and cokernel of $D+b-a$ on $\mathbb{C}((x))$; as $a-b$ is not an integer, this map is bijective. QED

Corollary 2.11.7 Let $L = \Sigma_{i\geq 0}\ x^i A_i(D)$ be an element of $\mathbb{C}[[x]][D]$ of degree N in D, and suppose that A_0 is nonzero of degree M. Let V be the D.E. on $\mathbb{C}((x))$ given by L. Then in the the slope decomposition of V as $V_{\text{slope}=0} \oplus V_{\text{slope}>0}$, we have

$$\text{rank}(V_{\text{slope}=0}) = M.$$

If $M > 0$, and $A_0(t) = (\mathbb{C}^\times \text{ factor})\prod_\alpha (t - \alpha)^{n_\alpha}$, then $V_{\text{slope}=0}$ is a successive extension of the rank one objects $x^\alpha\mathbb{C}((x))$ with multiplicity n_α. Moreover, if the distinct zeros of $A_0(t)$ are incongruent mod $\mathbb{Z}$, then

$$V_{\text{slope}=0} \approx \mathbb{C}((x))[D]/\mathbb{C}((x))[D]A_0(D) \approx \oplus_\alpha \mathbb{C}((x))[D]/\mathbb{C}((x))[D](D-\alpha)^{n_\alpha}.$$

proof If $M = 0$, it is clear that all slopes are strictly positive. If $M > 0$, then by the Factorization Lemma, we may reduce to the case $N=M$ with the same A_0. If A_0 has its distinct zeroes incongruent mod $\mathbb{Z}$, apply the Isomorphism Lemma, and then the Ext Lemma to separate the roots. If not, successively apply the Factorization Lemma to the α's in increasing order of their real part, to express $V_{\text{slope}=0}$ as a successive extension of rank one equations of the form $D - f$, where $f=\alpha$ + higher terms in x. To each of these apply the Isomorphism Lemma. QED

To put this into perspective, recall that one always has

Formal Jordan Decomposition Lemma 2.11.8 Let V be any D.E. on $\mathbb{C}((x))$ which is entirely of slope zero. Pick any fundamental domain in $\mathbb{C}$ for $\mathbb{C}/\mathbb{Z}$. Then

(1) V is isomorphic to a direct sum of indecomposables of slope zero.

(2) Any indecomposable of slope zero is isomorphic to

$$\text{Loc}(\alpha, n_\alpha) := \mathbb{C}((x))[D]/\mathbb{C}((x))[D](D-\alpha)^{n_\alpha},$$

for some unique α in the chosen fundamental domain, and some integer $n_\alpha \geq 1$. We call such an indecomposable "of type α mod $\mathbb{Z}$".

(3) Given two indecomposables $\text{Loc}(\alpha, n_\alpha)$ and $\text{Loc}(\beta, m_\beta)$, we have $\text{Hom}_{\mathcal{D}}(\text{Loc}(\alpha, n_\alpha), \text{Loc}(\beta, m_\beta)) = 0$ unless $\alpha \equiv \beta$ mod $\mathbb{Z}$, in which case the Hom has dimension $\min(n_\alpha, m_\beta)$.

(4) Given an $\alpha\in\mathbb{C}$, the number of indecomposables in V of type α mod $\mathbb{Z}$ is the $\mathbb{C}$-dimension of the space

$$\text{Hom}_{\mathcal{D}}(V, x^\alpha\mathbb{C}((x)))=\text{Hom}_{\mathcal{D}}(V\otimes x^{-\alpha}, \mathbb{C}((x)))$$

of $\mathbb{C}((x))$-valued solutions of $V\otimes x^{-\alpha}$.

proof One knows that the restriction functor

$$\text{D.E.}(\mathbb{G}_m/\mathbb{C})_{\text{RS at } 0,\infty} \rightarrow \text{D.E.}(\mathbb{C}((x))/\mathbb{C})_{\text{slope}=0}$$

is an equivalence of categories. In view of the topological interpretation of the source as the category of representations of $\mathbb{Z}$, this lemma amounts to Jordan normal form. QED

We now return to the global theory on $\mathbb{G}_m$.

Proposition 2.11.9 Let (d,n,m) be a triple of nonnegative integers with $d \geq 1$, $n+m \neq 0$ and $\gcd(d,n-m)=1$. Suppose the operator $L = \Sigma\, x^i P_i(D)$ satisfies the following three conditions:

(a) $P_i = 0$ except for $i=0,\ldots,d$; P_0 and P_d are nonzero.

(b) $\deg P_0 = n$, $\deg P_d = m$, and $\min(n,m) \geq \deg P_i$ for $0<i<d$.

(c) P_0 and P_d have no common zeroes mod $\mathbb{Z}$ (i.e., if α is a zero of P_0 and β is a zero of P_d, then $\alpha-\beta$ is not in $\mathbb{Z}$).

Put $\mathcal{M} := \mathcal{D}/\mathcal{D}L$. Then

(1) If $n \neq m$, $\mathcal{M}$ is an irreducible object of $\mathrm{D.E.}(\mathbb{G}_m/\mathbb{C})$. If $n>m$ (resp. if $m>n$), $\mathcal{M}$ is regular singular at 0 (resp. ∞), and at ∞ (resp. 0) its slopes are $d/|n-m|$ with multiplicity $|n-m|$ and 0 with multiplicity $\min(n,m)$. It is Lie-irreducible unless there exists a divisor $D > 1$ of $\gcd(n,m)$ such that both the roots mod $\mathbb{Z}$ of P_0 and the roots mod $\mathbb{Z}$ of P_d are Kummer-induced of degree D.

(2) If $n=m$, (so $d=1$) write $P_i(x) = \Sigma p_{i,j}x^j$ for $i=0,1$ and define

$\alpha = -p_{0,n}/p_{1,n}$ $U := \mathbb{G}_m - \{\alpha\}$, $j\colon U \to \mathbb{G}_m$ the inclusion.

Then $j^*\mathcal{M}$ is an irreducible object of $\mathrm{D.E.}(U/\mathbb{C})$ and $\mathcal{M} \approx j_{!*}j^*\mathcal{M}$. Local monodromy around α is a pseudoreflection.

proof The adjoint of an L which satisfies (a), (b), (c) is another one, with the same (d,n,m), as is, for any $\lambda \in \mathbb{C}^\times$, its pullback by multiplicative translation $T_\lambda\colon x \mapsto \lambda x$. And if L is such an operator, then $x^d \mathrm{inv}^*(L)$ is one of type (d,m,n).

So to prove (1) it suffices (by an inversion) to treat the case $n>m$. Then L is (up to a $\mathbb{C}^\times$ factor) monic in D of degree n, so $\mathcal{M}$ is certainly a D.E. on $\mathbb{G}_m$. The calculation of the slopes at zero and ∞ is immediate from (b). Because $\gcd(d, n-m) = 1$ by hypothesis, as I_∞-representation $\mathcal{M}$ is the direct sum of an irreducible of rank $n-m$ and of a tame I_∞-representation of rank m. So if $m=0$, $\mathcal{M}$ is already I_∞-irreducible. If $m>0$ and $\mathcal{M}$ is reducible, it has either a nonzero subobject or a nonzero quotient $\mathcal{N}$ in $\mathrm{D.E.}(\mathbb{G}_m/\mathbb{C})$ all of whose slopes at both zero and ∞ are 0.

Replacing $\mathcal{M}$ by its adjoint if necessary, we may assume that it has such a quotient. But such an $\mathcal{N}$ is itself a successive extension of $\mathcal{D}$-modules of the form $x^{\alpha}\mathbb{C}[x,x^{-1}]$, so M would admit some $x^{\alpha}\mathbb{C}[x,x^{-1}]$ as a quotient. Concretely, this means that L kills some nonzero element of $x^{\alpha}\mathbb{C}[x,x^{-1}]$, say $x^{\alpha}f(x)$ where $f = \Sigma_{i=a,\ldots,b}\,\lambda_i x^i$ is a Laurent polynomial of bidegree (a,b). Looking at the highest order term in $L(x^{\alpha}f)=0$, we see that $P_d(\alpha+b)=0$; looking at lowest order terms we see that $P_0(\alpha+a)=0$, contradicting (c). Thus $\mathcal{M}$ is irreducible.

If $\mathcal{M}$ is not Lie-irreducible, it is Kummer induced of some degree $D > 1$. Looking at the slope decompositions of the I_0- and I_∞-representations, we see that every slope=λ component of each decomposition is itself Kummer induced of degree D. Therefore D must divide the multiplicities with which any of the slopes occurs, whence D divides gcd(n,m). Looking at the semisimplification of the slope=0 part at zero (resp. ∞), we see that the roots mod $\mathbb{Z}$ of P_0 (resp. P_d) are Kummer induced of degree D. This concludes the proof of (1).

We now turn to the proof of (2). It is clear that $j^*\mathcal{M}$ is a D.E. on U, which is regular singular at 0, α, and ∞. Let us admit temporarily that $\mathcal{M}\approx j_{!*}j^*\mathcal{M}$. Then from Pochammer's Lemma 2.9.7 we see that local monodromy around α is a pseudoreflection. So if $j^*\mathcal{M}$ is not irreducible, it has either a nonzero subobject or a nonzero quotient $\mathcal{N}$ in D.E.(U/$\mathbb{C}$) which is regular singular at $0,\alpha,\infty$ and whose local monodromy at α is **trivial**. Such an $\mathcal{N}$ is a successive extension of $j^*(x^{\alpha}\mathbb{C}[x,x^{-1}])$'s, and hence, at the expense of replacing $\mathcal{M}$ by its adjoint, we may assume that $j^*\mathcal{M}$ has a quotient $j^*(x^{\alpha}\mathbb{C}[x,x^{-1}])$. This means that $\mathcal{M}\approx j_{!*}j^*\mathcal{M}$ has a nonzero map to $j_{!*}j^*(x^{\alpha}\mathbb{C}[x,x^{-1}]) = x^{\alpha}\mathbb{C}[x,x^{-1}]$, which means precisely that L kills a nonzero element of $x^{\alpha}\mathbb{C}[x,x^{-1}]$. Exactly as above, this contradicts (c), and so establishes the irreducibility of $j^*\mathcal{M}$. Again using $\mathcal{M}\approx j_{!*}j^*\mathcal{M}$, we find that $\mathcal{M}$ itself is an irreducible $\mathcal{D}$-module on $\mathbb{G}_m$.

Thus it remains only to establish that $\mathcal{M}\approx j_{!*}j^*\mathcal{M}$ in case (2). This means showing that both L and L^* act injectively on the delta-module δ_{α}. As the hypotheses are self-adjoint, it suffices to prove this for L. At the expense of a multiplicative translation, we may assume that $\alpha=1$. We write

$L = P(D) - xQ(D)$ with both P, Q of degree $n \geq 1$.

Passing to the formal parameter t at $\alpha = 1$ such that $x = e^t$, our operator becomes

$L = P(d/dt) - e^t Q(d/dt)$,

and we must show that this operates injectively on $\mathcal{S}_0 = \mathbb{C}((t))/\mathbb{C}[[t]]$. This is the Fourier Transform of the equivalent problem of showing that, denoting by Sub = exp(-d/dt) the endomorphism of $\mathbb{C}[t]$ given by $t \mapsto t-1$, the operator $P(t) - \mathrm{Sub}\circ Q(t)$ is injective on $\mathbb{C}[t]$. But if f(t) is a nonzero polynomial which satisfies

$P(t)f(t) = Q(t-1)f(t-1)$, i.e., $f(t)/f(t-1) = Q(t-1)/P(t)$,

then for each $k \geq 2$ we have

$f(t)/f(t-k) = [Q(t-1)Q(t-2)\ldots Q(t-k)]/[P(t)P(t-1)\ldots P(t-(k-1))]$.

By hypothesis (c), the right hand fraction is in lowest terms, which implies that f(t-k) has at least nk zeroes. Therefore no such nonzero f exists. QED

Applying the Main D.E. Theorem 2.8.1 in the case $n \neq m$, (we will take up the case $n = m$ further on, in 3.5, 3.5.8) we obtain

Theorem 2.11.10 Let (d,n,m) be a triple of nonnegative integers with $d \geq 1$, $n \neq m$ and $\gcd(d,n-m) = 1$. Suppose the operator $L := \sum x^i P_i(D)$ satisfies the following four conditions:

(a) $P_i = 0$ except for $i = 0,\ldots,d$; P_0 and P_d are nonzero.

(b) $\deg P_0 = n$, $\deg P_d = m$, and $\min(n,m) \geq \deg P_i$ for $0 < i < d$.

(c) P_0 and P_d have no common zeroes mod $\mathbb{Z}$ (i.e., if α is a zero of P_0 and β is a zero of P_d, then $\alpha - \beta$ is not in $\mathbb{Z}$).

(d) There exists no divisor $D > 1$ of gcd(n,m) such that both the roots mod $\mathbb{Z}$ of P_0 and of P_d are Kummer induced of degree D.

Let $N := \max(n,m)$ be the order of L. Then the differential galois group G of $\mathcal{M} := \mathcal{D}/\mathcal{D}L$ on $\mathbb{G}_m$ is reductive. If $\det\mathcal{M}$ is of finite order then $G^0 = G^{0,\mathrm{der}}$; otherwise $G^0 = \mathbb{G}_m G^{0,\mathrm{der}}$. The possibilities for $G^{0,\mathrm{der}}$ are given by:

(1) If $|n-m|$ is odd, $G^{0,\mathrm{der}}$ is SL(N); if $|n-m| = 1$, then G is GL(N).

(2) If $|n-m|$ is even, then $G^{0,\mathrm{der}}$ is SL(N) or SO(N) or (if N is even) SP(N), or $|n-m| = 6$, N=7,8 or 9, and $G^{0,\mathrm{der}}$ is one of

N=7: the image of G_2 in its 7-dim'l irreducible representation

N=8: the image of Spin(7) in the 8-dim'l spin representation

the image of SL(3) in the adjoint representation

the image of $SL(2)\times SL(2)\times SL(2)$ in $std\otimes std\otimes std$
the image of $SL(2)\times Sp(4)$ in $std\otimes std$
the image of $SL(2)\times SL(4)$ in $std\otimes std$
$N=9$: the image of $SL(3)\times SL(3)$ in $std\otimes std$.

CHAPTER 3
The Generalized Hypergeometric Equation

The generalized hypergeometric equation

We now turn to the detailed study of the case d=1, which is the case of the classical "generalized hypergeometric equation". My interest in this case was aroused in 1986 by the paper [B-B-H] of Beukers, Brownawell, and Heckman, concerned with the case d=1, $n \neq m$ (cf. also the recent paper [B-H] of Beukers and Heckman devoted to the case d=1,n=m).

(3.1) Let $(n,m) \neq (0,0)$ be a pair of nonnegative integers. Given nonzero polynomials $P := P_n$, $Q := Q_m$ in $\mathbb{C}[t]$ of degrees n and m respectively, we define the hypergeometric operator Hyp(P,Q) in $\mathcal{D}$ to be

$$\mathrm{Hyp}(P,Q) := P(D) - xQ(D).$$

We call it a hypergeometric operator of type (n,m). We define the hypergeometric $\mathcal{D}$-module $\mathcal{H}(P, Q)$ on $\mathbb{G}_m$ by

$$\mathcal{H}(P, Q) := \mathcal{D}/\mathcal{D}\mathrm{Hyp}(P,Q).$$

Of course this $\mathcal{D}$-module does not change if we multiply Hyp(P,Q) by a $\mathbb{C}^\times$ factor. This permits us when convenient (which it is not always) to suppose that Q is monic.

It will sometimes be convenient to have a notation which makes explicit the factorizations of P and Q. Suppose

$$P(t) = p\Pi(t - \alpha_i), \text{ with } p \in \mathbb{C}^\times,$$

$$Q(t) = q\Pi(t - \beta_j), \text{ with } q \in \mathbb{C}^\times,$$

$$\lambda := p/q.$$

In terms of the the n roots (with multiplicity) α_i of P, the m roots (with multiplicity) β_j of Q, and the scaling factor λ in $\mathbb{C}^\times$, we define

$$\mathrm{Hyp}_\lambda(\alpha_i\text{'s}; \beta_j\text{'s}) := \lambda\Pi(D - \alpha_i) - x\Pi(D - \beta_j) = (1/q)\mathrm{Hyp}(P,Q),$$

$$\mathcal{H}_\lambda(\alpha_i\text{'s}; \beta_j\text{'s}) := \mathcal{D}/\mathcal{D}\mathrm{Hyp}_\lambda(\alpha_i\text{'s}; \beta_j\text{'s}) = \mathcal{D}/\mathcal{D}\mathrm{Hyp}(P,Q).$$

The effect of the operation $\mathcal{M} \mapsto \mathcal{M} \otimes x^\gamma$ is particularly easy to see:

$$\mathcal{H}_\lambda(\alpha_i\text{'s}; \beta_j\text{'s}) \otimes x^\gamma \approx \mathcal{H}_\lambda(\gamma+\alpha_i\text{'s}; \gamma+\beta_j\text{'s}),$$

as is the behaviour under multiplicative translation:

$$[x \mapsto \mu x]^* \mathcal{H}_\lambda(\alpha_i\text{'s}; \beta_j\text{'s}) \approx \mathcal{H}_{\lambda/\mu}(\alpha_i\text{'s}; \beta_j\text{'s}),$$

$$[x \mapsto \mu x]_* \mathcal{H}_\lambda(\alpha_i\text{'s}; \beta_j\text{'s}) \approx \mathcal{H}_{\lambda\mu}(\alpha_i\text{'s}; \beta_j\text{'s}).$$

The behavior of the operators under multiplicative inversion and passage to adjoint is given by

$$\mathrm{inv}^*\mathrm{Hyp}(P(t), Q(t)) = \mathrm{inv}_*\mathrm{Hyp}(P(t), Q(t)) = (-1/x)\mathrm{Hyp}(Q(-t), P(-t))$$

$$\mathrm{Hyp}(P(t), Q(t))^* = \mathrm{Hyp}(P(-1-t), Q(-2-t)).$$

In the λ, α, β notation this gives

$$\mathrm{inv}^*\mathcal{H}_\lambda(\alpha_i\text{'s}; \beta_j\text{'s}) \approx \mathrm{inv}_*\mathcal{H}_\lambda(\alpha_i\text{'s}; \beta_j\text{'s}) \approx \mathcal{H}_{(-1)^{n+m}/\lambda}(-\beta_j\text{'s}; -\alpha_i\text{'s})$$

$$\mathcal{H}_\lambda(\alpha_i\text{'s}; \beta_j\text{'s})^* \approx \mathcal{H}_{\lambda(-1)^{n+m}}(-1-\alpha_i\text{'s}; -2-\beta_j\text{'s}).$$

Proposition 3.2 Suppose that $\mathcal{H}_\lambda(\alpha_i\text{'s}; \beta_j\text{'s}) = \mathcal{H}(P,Q)$ is an irreducible $\mathcal{D}$-module on $\mathbb{G}_m$. Then

(1) For fixed λ, the isomorphism class of $\mathcal{H}_\lambda(\alpha_i\text{'s}; \beta_j\text{'s})$ depends only on the α_i and the β_j mod $\mathbb{Z}$.

(2) P and Q have no common zeroes mod $\mathbb{Z}$, i.e., for any root α_i of P and any root β_j of Q, $\alpha_i - \beta_j$ is not in $\mathbb{Z}$.

proof (1) We must show that if $\mathcal{H}_\lambda(\alpha_i\text{'s}; \beta_j\text{'s})$ is irreducible, then for any choice of $\tilde{\alpha}_i$'s; $\tilde{\beta}_j$'s which are congruent mod $\mathbb{Z}$ to the α_i's; β_j's, $\mathcal{H}_\lambda(\tilde{\alpha}_i\text{'s}; \tilde{\beta}_j\text{'s})$ is irreducible and isomorphic to $\mathcal{H}_\lambda(\alpha_i\text{'s}; \beta_j\text{'s})$. Using the fact that multiplicative inversion interchanges the role of α's and β's, and proceeding step by step, it suffices to treat the case when all α's and β's are equal to their respective $\tilde{\alpha}$'s and $\tilde{\beta}$'s except for α_1 and $\tilde{\alpha}_1$, which differ by 1. Passing to the adjoint if necessary, it suffices to treat the case where in addition $\tilde{\alpha}_1$ is $\alpha_1 - 1$. Writing simply α for α_1, the situation is that

$$\alpha \text{ is a root of P, say } P(D) = (D-\alpha)R(D).$$

The commutation relation

$$[(D-\alpha)R(D) - xQ(D)](D+1-\alpha) = (D-\alpha)[(D+1-\alpha)R(D) - xQ(D)]$$

shows that Right$(D+1-\alpha)$ defines a map of $\mathcal{D}$-modules

$$\mathrm{Right}(D+1-\alpha) : \mathcal{H}(P,Q) \to \mathcal{H}((D+1-\alpha)R(D), Q).$$

This map is nonzero (otherwise $D+1-\alpha \in \mathcal{D}[(D+1-\alpha)R(D) - xQ(D)]$; looking at degrees in D we infer that $D+1-\alpha = f(x)[(D+1-\alpha)R(D) - xQ(D)]$ for some $f(x)$ in $\mathbb{C}[x,x^{-1}]$. Looking at x-degrees now leads to a contradiction.). Since its source is irreducible, it is injective. If $n \neq m$, both source and target are $\mathcal{O}$-locally free $\mathcal{D}$-modules on $\mathbb{G}_m$ of the same rank $\max(n,m)$, so our map, being injective, is an isomorphism.

If $n=m$, then our map is an isomorphism on $\mathbb{G}_m - \{\lambda\}$, and it is injective on all of $\mathbb{G}_m$. So if it fails to be an isomorphism, its cokernel is a successive extension of delta-modules δ_λ. In particular, $(D+1-\alpha)R(D)$ -

$xQ(D)$ operates noninjectively on δ_λ. By a multiplicative translation, we may assume that $\lambda = 1$. Exactly as in the proof of 2.11.9 above, this noninjectivity then means that there exists a nonzero $f(t)$ in $\mathbb{C}[t]$ such that

$$(t+1-\alpha)R(t)f(t) = Q(t-1)f(t-1).$$

Multiply both sides by $(t-\alpha)$:

$$(t-\alpha)R(t)f(t)(t+1-\alpha) = Q(t-1)f(t-1)(t-\alpha),$$

which says precisely that $F(t) := f(t)(t+1-\alpha)$ satisfies

$$P(t)F(t) = Q(t-1)F(t-1),$$

which in turn says that $\mathrm{Hyp}(P,Q)$ acts noninjectively on δ_1. But this is not the case, because $\mathcal{H}(P,Q)$, being an irreducible $\mathcal{D}$-module on $\mathbb{G}_m$, is a middle extension across 1. This completes the proof of (1).

To prove (2), simply observe that in virtue of (1), any irreducible $\mathcal{H}_\lambda(\alpha_i\text{'s}; \beta_j\text{'s})$ is isomorphic to one all of whose α's and β's lie in any prechosen set of coset representatives for $\mathbb{C}/\mathbb{Z}$ (e.g., in $0 \le \mathrm{Re}(z) < 1$). For one of these, $\alpha_i - \beta_j \in \mathbb{Z}$ implies $\alpha_i = \beta_j$, whence $D - \alpha_i$ is a right divisor of $\mathrm{Hyp}_\lambda(\alpha_i\text{'s}; \beta_j\text{'s})$, contradicting irreducibility. QED

Corollary 3.2.1 $\mathcal{H} = \mathcal{H}_\lambda(\alpha_i\text{'s}; \beta_j\text{'s}) = \mathcal{H}(P,Q)$ is an irreducible $\mathcal{D}$-module on $\mathbb{G}_m$ if and only if P and Q have no common zeroes mod $\mathbb{Z}$.

proof This is immediate from 3.2 and 2.11.9. QED

Corollary 3.2.2 Suppose that $\mathcal{H} = \mathcal{H}_\lambda(\alpha_i\text{'s}; \beta_j\text{'s}) = \mathcal{H}(P,Q)$ is an irreducible hypergeometric $\mathcal{D}$-module on $\mathbb{G}_m$ of type (n,m) (i.e., $P = P_n$ and $Q = Q_m$ have no common zeroes mod $\mathbb{Z}$).

(1) The formal Jordan decomposition of $(\mathcal{H} \otimes_{\mathcal{O}} \mathbb{C}((x)))_{\text{slope}=0}$ is

$$(\mathcal{H} \otimes_{\mathcal{O}} \mathbb{C}((x)))_{\text{slope}=0} \approx \oplus_{0 \le \mathrm{Re}(\alpha) < 1} \mathbb{C}((x))[D]/\mathbb{C}((x))[D](D-\alpha)^{n_\alpha}$$

and n_α is the number of α_i which are congruent mod $\mathbb{Z}$ to α.

(2) The formal Jordan decomposition of $(\mathcal{H} \otimes_{\mathcal{O}} \mathbb{C}((1/x)))_{\text{slope}=0}$ is

$$(\mathcal{H} \otimes_{\mathcal{O}} \mathbb{C}((1/x)))_{\text{slope}=0} \approx \oplus_{0 \le \mathrm{Re}(\beta) < 1} \mathbb{C}((1/x))[D]/\mathbb{C}((1/x))[D](D-\beta)^{n_\beta}$$

and n_β is the number of β_j which are congruent mod $\mathbb{Z}$ to β.

(3) If $n \ge m$ (resp. if $m \ge n$) the local monodromy at the regular singular point 0 (resp. ∞) has eigenvalues with multiplicity $\{\exp(2\pi i\alpha)\}_{P(\alpha)=0}$ (resp. $\{\exp(-2\pi i\beta)\}_{Q(\beta)=0}$). It has a single Jordan block for each distinct eigenvalue, of size the multiplicity of that eigenvalue (i.e., the minimal polynomial is the characteristic polynomial).

proof To prove (1) and (2), use the irreducibility and the previous Proposition to reduce to treating the case when all the α_i and the β_j have real part in [0,1), in which case it follows immediately from 2.11.7.

To prove (3), we may, by inversion, suppose $n \geq m$. That the local monodromy around zero has the asserted eigenvalues is standard from the classical theory of the indicial polynomial at a regular singularity. Using the irreducibility, we may as above assume that all the α_i have real part in [0,1). Twisting by x^γ, it suffices to show that if 0 is the only integer root of P(t), then the unipotent part of the local monodromy consists of a single Jordan block. Because we are at a regular singularity, the number of unipotent Jordan blocks (which is always the dimension of the space of single-valued solutions in a punctured classical neighborhood of the singularity) is the dimension of the space of $\mathbb{C}((x))$-solutions. It follows from the shape of the formal decomposition that this dimension is one; alternately, if $\Sigma a_i x^i = x^d + \ldots$ is killed by $P(D) - xQ(D)$, then looking at the lowest degree term we see that $P(d)=0$, whence $d=0$ since 0 is the only integer root of P. The coefficients a_n are subject to the two term recursion

$$P(n)a_n = Q(n-1)a_{n-1}.$$

Since $P(0)=0$, all a_{neg} vanish, and the positive coefficients $a_{n>0}$ may be determined uniquely by this recurrence, since for $n > 0$ the factor P(n) does not vanish. QED

Lemma 3.3 Suppose that $\mathcal{H} := \mathcal{H}_\lambda(\alpha_i\text{'s}; \beta_j\text{'s})$ is a hypergeometric $\mathcal{D}$-module on $\mathbb{G}_m$ of type (n,m). Then its isomorphism class as $\mathcal{D}$-module on $\mathbb{G}_m$ determines the type (n,m), the set (with multiplicity) of the α_i mod $\mathbb{Z}$, the set (with multiplicity) of the β_j mod $\mathbb{Z}$, and, if either $n=m$ or if $\mathcal{H}$ is irreducible, the scalar $\lambda \in \mathbb{C}^\times$.

proof Denote by r the generic $\mathcal{O}$-rank of $\mathcal{H}$. At least one of zero or ∞ is a regular singularity. Performing a multiplicative inversion if necessary, we may assume that zero is a regular singularity. Then $n=r$, and m is the dimension of the slope zero part at ∞. The α_i mod $\mathbb{Z}$ and the β_j mod $\mathbb{Z}$ are determined by the formal Jordan decompositions of the slope zero parts at 0 and ∞ respectively.

If $n=m$, then λ is the "other" singularity of $\mathcal{H}$, i.e., it is the unique point α in $\mathbb{G}_m$ such that, denoting by $\mathcal{O}^\wedge$ the complete local ring at α, either $\mathcal{H}$ or its adjoint $\mathcal{H}^*$ has $\dim_{\mathbb{C}}\mathrm{Hom}_{\mathcal{D}}(\mathcal{H}, \mathcal{O}^\wedge) = n-1$ (by Pochammer's Lemma 2.9.7).

If $n \neq m$, we do not know any a priori description of λ, but we can prove its unicity as follows when $\mathcal{H}$ is irreducible. If λ is not unique, then for some $\mu \neq 1$ in $\mathbb{C}^\times$ there exists an isomorphism of $\mathcal{H}$ with its pullback by the map $x \mapsto \mu x$. Then each piece of the slope decomposition at ∞ is isomorphic to its pullback by this map. Since $(\mathcal{H} \otimes \mathbb{C}((1/x)))_{\mathrm{slope}>0}$ has irregularity **one**, this contradicts [Ka-DGG, 2.3.8]. QED

Remark 3.3.1 If we knew that the isomorphism class of the **semisimplification** of $\mathcal{H}$ as $\mathcal{D}$-module on $\mathbb{G}_m$ depended only on the data (α_i mod $\mathbb{Z}$, β_j mod $\mathbb{Z}$, λ), then at the last step we could replace "isomorphism" by "isomorphism of semisimplifications". This would still be adequate to show $\mu = 1$, using [Ka-DGG, 2.3.8].

Duality Recognition Theorem 3.4 Suppose that $\mathcal{H} := \mathcal{H}_\lambda(\alpha_i\text{'s}; \beta_j\text{'s})$ is an irreducible hypergeometric $\mathcal{D}$-module on $\mathbb{G}_m$ of type (n,m). In order that there exist an isomorphism of $\mathcal{H}$ with its adjoint, it is necessary and sufficient that the following three conditions hold:
(1) $n-m$ is even.
(2) there exists a permutation $i \mapsto i'$ of $[1,\ldots,n]$ such that $\alpha_{i'} + \alpha_i \in \mathbb{Z}$.
(3) there exists a permutation $j \mapsto j'$ of $[1,\ldots,m]$ such that $\beta_{j'} + \beta_j \in \mathbb{Z}$.
Moreover, if these conditions are satisfied, then the resulting autoduality of $\mathcal{H}$ (on the dense open set where $\mathcal{H}$ is a D.E.) is alternating if and only if

$$\max(n,m) \text{ is even, and } \gamma := \Sigma\beta_j - \Sigma\alpha_i \in \mathbb{Z};$$

otherwise (i.e., if $\max(n,m)$ is odd or if $\gamma \in 1/2 + \mathbb{Z}$) it is symmetric.
proof Indeed, the adjoint is $\mathcal{H}_{\lambda(-1)^{n+m}}(-1-\alpha_i\text{'s}; -2-\beta_j\text{'s})$, so the first assertion is obvious from 3.2 and 3.3. Because $\mathcal{H}$ is irreducible, it has at most one autoduality (up to a $\mathbb{C}^\times$ factor) $\langle x,y\rangle$, which is either alternating or symmetric.

If the generic rank $\max(n,m)$ of $\mathcal{H}$ is odd, the autoduality has no choice but to be symmetric. The only problem comes when $\max(n,m)$ is even. At the expense of an inversion, we may assume $n \geq m$.

If $n=m$, then local monodromy at λ is a pseudoreflection of determinant $\exp(2\pi i\gamma)$. If $\gamma \in \mathbb{Z}$, this is a unipotent pseudoreflection, and so $\langle x,y\rangle$ cannot be symmetric. [For denoting by N the log of this unipotent pseudoreflection, $\langle Nx, y\rangle + \langle x, Ny\rangle=0$, so if $\langle x,y\rangle=\langle y,x\rangle$, we find $\langle Nx,x\rangle=0$. Since N has one-dimensional image, say $\mathbb{C}e$, we have $\langle e,x\rangle=0$ if $Nx\neq 0$, and then for all x, whence $e=0$, contradiction.] Conversely, if $\gamma \in 1/2 + \mathbb{Z}$, then we have a true reflection, which lies in no symplectic group (in $Sp(2d)$, the eigenvalues of any element can be grouped into d pairs of inverses).

If $n > m$, and both n and $n-m$ are even, then $\det\mathcal{H}$ is a rank one D.E. on $\mathbb{G}_m$ which has slope zero at both 0 and ∞ (because all the ∞-slopes of $\mathcal{H}$ are $1/(n-m) < 1$). So $\det\mathcal{H}$ must be of the form $x^{\delta}\mathbb{C}[x,x^{-1}]$ for some δ. Looking at the slope decompositions of $\mathcal{H}$ at both 0 and ∞, and at the formal Jordan decompositions of the slope zero parts, we see that

$$\det(\mathcal{H}\otimes\mathbb{C}((x))) \approx x^{\alpha}\mathbb{C}((x)) \text{ for } \alpha:=\Sigma\alpha_i,$$

$$\det(\mathcal{H}\otimes\mathbb{C}((1/x))) \approx \det((\mathcal{H}\otimes\mathbb{C}((1/x)))_{\text{slope}>0})\otimes x^{\beta}\mathbb{C}((1/x)) \text{ for } \beta:=\Sigma\beta_j.$$

Comparing these local expressions for $\det\mathcal{H}$ with $x^{\delta}\mathbb{C}[x,x^{-1}]$, we see first that $\delta\equiv\alpha \mod \mathbb{Z}$ and then that

$$\det((\mathcal{H}\otimes\mathbb{C}((1/x)))_{\text{slope}>0}) \approx x^{-\gamma}\mathbb{C}((1/x)).$$

So $\det((\mathcal{H}\otimes\mathbb{C}((1/x)))_{\text{slope}>0})$ is either trivial or of order two, depending on whether $\gamma\in\mathbb{Z}$ or not.

On the other hand, our autoduality of $\mathcal{H}$ must induce an autoduality on $\mathcal{H}\otimes\mathbb{C}((1/x))$, which in turn induces an autoduality on each piece of the slope decomposition. As $(\mathcal{H}\otimes\mathbb{C}((1/x)))_{\text{slope}>0}$ is irreducible (because it has rank $n-m$ and all slopes $1/(n-m)$), it has at most one autoduality (up to a $\mathbb{C}^{\times}$ factor), say (x,y), and that autoduality has a sign (i.e., is either symmetric or alternating) which must be the **same** sign as that of our global one $\langle x,y\rangle$. So our sign rule follows from the following

Lemma 3.4.1 Let $d \geq 2$. Let W be a D.E. on $\mathbb{C}((1/x))$ of rank d all of whose slopes are $1/d$.

(1) The isomorphism class of W is determined up to a multiplicative translate by the isomorphism class of $\det(W)$.

(2)W is self dual if and only if d is even and $\det(W)^{\otimes 2}$ is trivial, and the duality is alternating if and only if $\det(W)$ is trivial.

proof Any D.E. W on $\mathbb{C}((1/x))$ of rank d all of whose slopes are $1/d$ is a multiplicative translate of one of the form $([d]_*\mathcal{L})\otimes x^{\lambda}$ where $\mathcal{L}$ is the rank one D.E. for e^x, and $\lambda \in\mathbb{C}$. Notice that the isomorphism class of

$([d]_*\mathcal{L})\otimes x^\lambda \approx [d]_*(\mathcal{L}\otimes x^{d\lambda})$ depends only on $d\lambda$ mod $\mathbb{Z}$. Now $\det(([d]_*\mathcal{L})\otimes x^\lambda) \approx \det([d]_*\mathcal{L})\otimes x^{d\lambda}$ visibly determines $d\lambda$ mod $\mathbb{Z}$. Because $[d]_*\mathcal{L}$ has all slopes < 1, its determinant is of the form $x^\mu\mathbb{C}((x))$, so the isomorphism class of $\det([d]_*\mathcal{L})$ is translation invariant. Therefore the isomorphism class of $\det(W)$ is translation invariant, and hence $\det(W)$ determines $d\lambda$ mod $\mathbb{Z}$. Therefore $\det(W)$ determines the isomorphism class of W itself, up to a multiplicative translation. This proves (1).

We next observe that if d is odd, then W cannot be self dual, because $W\otimes W$ has all slopes 1/d for odd d. [To see this, use the fact that $[d]^*W$ as representation of the upper-numbering subgroup $(I_\infty)^{(0+)}$ is, for some $a\in\mathbb{C}^\times$,

$$[d]^*W \mid (I_\infty)^{(0+)} \approx \oplus_{\zeta\in\mu_d}\chi_{a\zeta},$$

where $\chi_{a\zeta}$ is the character of $(I_\infty)^{(0+)}$ given by the rank one D.E. for $e^{a\zeta x}$.]

If W is self dual, so is $\det(W)$, whence $\det(W)^{\otimes 2}$ is trivial. Suppose now that $\det(W)^{\otimes 2}$ is trivial. We argue globally as follows. Consider a Kloosterman equation of even rank d,i.e., a hypergeometric equation of type (d,0), say $\mathcal{H}_\lambda(a_1,\dots, a_d; \varnothing)$. Since $d\geq 2$, its determinant is $x^a\mathbb{C}[x,x^{-1}]$ for $a=\Sigma a_i$, so over $\mathbb{C}((1/x))$ we obtain a W as above with $\det(W) \approx x^a\mathbb{C}((1/x))$. As λ varies over $\mathbb{C}^\times$, but the a_i remain fixed, we obtain all translates of this W. So it suffices to analyse the sign of the autoduality for the two particular Kloosterman equations of even rank d

$$\mathcal{H}_\lambda(0, 0, 1/(d-1),\dots, (d-2)/(d-1); \varnothing),$$

$$\mathcal{H}_\lambda(1/2, 0, 1/(d-1),\dots, (d-2)/(d-1); \varnothing).$$

Consider for each the d-1'st power of local monodromy at zero. For the first, it is a unipotent pseudoreflection (not in any O(d)), and for the second it is a true reflection (not in any Sp(d)). QED

Corollary 3.4.1.1 Hypotheses and notations as in the above lemma, denote by $\mathcal{L}$ the rank one D.E. for e^x. Then

(1) $\det([d]_*\mathcal{L}) \approx x^{(d-1)/2}\mathbb{C}((1/x))$.

(2) $W \approx$ a multiplicative translate of $[d]_*\mathcal{L}$ if and only if we have $\det(W) \approx x^{(d-1)/2}\mathbb{C}((1/x))$.

(3) If d is odd, then the dual of $[d]_*\mathcal{L}$ is $[x \mapsto -x]^*([d]_*\mathcal{L})$.

proof To prove (1) and (3), we use a global argument. $\mathcal{L}$ is $\mathcal{H}_1(0; \emptyset)$. As will be shown later in 3.5.6 (but with no circularity), we have the formula

$$[d]_*\mathcal{H}_1(0; \emptyset) \approx \mathcal{H}_1(0, 1/d, 2/d, \ldots, (d-1)/d; \emptyset).$$

Clearly $\det(\mathcal{H}_1(0, 1/d, 2/d, \ldots, (d-1)/d; \emptyset)) \approx x^\delta\mathbb{C}[x, x^{-1}]$, simply because its slope is 0 at zero and $\leq 1/d < 1$ at ∞, so also 0 at ∞. To evaluate δ, we look at local monodromy at zero; this shows $\delta = (d-1)/2$. Now looking over $\mathbb{C}((1/x))$ we get the asserted formula for $\det([d]_*\mathcal{L})$. That (1) $\Rightarrow$ (2) was proven as part (1) of the previous Lemma. To prove (3), notice that for d odd, the dual of $\mathcal{H}_1(0, 1/d, 2/d, \ldots, (d-1)/d; \emptyset)$ is its multiplicative translate by -1. QED

(3.5) We now study the Lie-irreducibility of the hypergeometric equation in the case n=m. Our analysis is a geometric version of the more group-theoretic one of [B-H, 5.8].

Given a pair (a,b) of strictly positive integers, and λ in $\mathbb{C}^\times$, consider the "Belyi polynomial"

$$\mathrm{Bel}_{a,b,\lambda}(x) := \lambda\mu_{a,b}x^a(1-x)^b, \text{ where } \mu_{a,b} := (a+b)^{a+b}/a^ab^b.$$

We call it the Belyi polynomial because of the brilliant use Belyi makes of it in [Bel, Part 4]. It is a morphism of degree n:=a+b from $\mathbb{P}^1$ to $\mathbb{P}^1$, which induces a finite etale covering

$$\mathbb{P}^1 - \{0, 1, a/(a+b), \infty\} \rightarrow \mathbb{P}^1 - \{0, \lambda, \infty\}$$

whose ramified fibres are

over 0: exactly two points; one with mult. a and one with mult. b,
over ∞: exactly one point,
over λ: exactly n-1 points.

We call this covering the Belyi covering of type (a,b). The covering defined by $1/\mathrm{Bel}_{a,b,\lambda^{-1}}(x)$ we call the inverse Belyi covering of type (a,b).

Lemma 3.5.1 Over $\mathbb{C}$, any finite etale connected covering $X \rightarrow \mathbb{P}^1 - \{0, \lambda, \infty\}$ of degree n such that the induced map of complete nonsingular models $\pi: \overline{X} \rightarrow \mathbb{P}^1$ has exactly n-1 points in the fibre over λ is isomorphic to either a Belyi covering or an inverse Belyi covering of type (a,b) for some partition of n =a+b as the sum of two strictly positive integers.

proof Since $\overline{X} - \pi^{-1}(0, \lambda, \infty)$ is finite etale over $\mathbb{P}^1 - \{0, \lambda, \infty\}$ of degree n, and we are in characteristic zero, the Euler characteristics multiply:

$$2 - 2g(\overline{X}) - \mathrm{Card}(\pi^{-1}(0)) - \mathrm{Card}(\pi^{-1}(\lambda)) - \mathrm{Card}(\pi^{-1}(\infty)) = -n.$$

By assumption, $\mathrm{Card}(\pi^{-1}(\lambda)) = n-1$, so we find

$2 - 2g(\overline{X}) - \mathrm{Card}(\pi^{-1}(0)) - \mathrm{Card}(\pi^{-1}(\infty)) = -1$, which we rewrite

$$2g(\overline{X}) + \mathrm{Card}(\pi^{-1}(0)) + \mathrm{Card}(\pi^{-1}(\infty)) = 3.$$

Since each of $\mathrm{Card}(\pi^{-1}(0))$ and $\mathrm{Card}(\pi^{-1}(\infty))$ is ≥ 1, and $g(\overline{X}) \geq 0$, we see that $g(\overline{X}) = 0$, and $\mathrm{Card}(\pi^{-1}(0)) + \mathrm{Card}(\pi^{-1}(\infty)) = 3$. After a multiplicative inversion on the base, we may assume that

$$\mathrm{Card}(\pi^{-1}(0)) = 2, \qquad \mathrm{Card}(\pi^{-1}(\infty)) = 1.$$

Since $\overline{X}$ is a noncanonical $\mathbb{P}^1$, we may decree that ∞ is the unique point over ∞, and that the two points over 0 are 0 and 1. That ∞ is the unique point over ∞ means that our covering is given by a polynomial of degree n, and looking at the fibre over 0 shows that its only zeroes are 0 and 1. So our covering is given by $\alpha x^a(1-x)^b$ for some α in $\mathbb{C}^\times$. The critical point $a/(a+b)$ is mapped to $\alpha/\mu_{a,b}$, so $\alpha/\mu_{a,b} = \lambda$. QED

Proposition 3.5.2 Suppose that $\mathcal{H} := \mathcal{H}_\lambda(\alpha_i\text{'s}; \beta_j\text{'s})$ is a hypergeometric $\mathcal{D}$-module on $\mathbb{G}_m$ of type (n,n). Suppose that the D.E. $\mathcal{H} \mid \mathbb{G}_m - \{\lambda\}$ is induced, i.e., it is the direct image of a D.E. $\mathcal{V}$ on a connected finite etale covering $\pi: X \to \mathbb{G}_m - \{\lambda\}$ of degree $d \geq 2$. Then either the covering π is isomorphic to a Kummer covering (i.e., the restriction to $\mathbb{G}_m - \{\lambda\}$ of the d-fold Kummer covering of $\mathbb{G}_m$ by itself) or d=n and the covering is isomorphic to either a Belyi covering or an inverse Belyi covering of type (a,b) for some partition of n =a+b as the sum of two strictly positive integers. Moreover, in the case of a Belyi or inverse Belyi covering, local monodromy of $\mathcal{H}$ around λ is a true reflection.

proof Suppose that the fibre of π over λ consists of points p_i with multiplicities e_i. Then $\mathcal{H} \otimes \mathbb{C}((x-\lambda))$ is the direct sum of the e_i-fold Kummer inductions of the $\mathcal{V} \otimes \mathbb{C}((x-p_i))$:

$$\mathcal{H} \otimes \mathbb{C}((x-\lambda)) \approx \oplus_i \; [e_i]_*(\mathcal{V} \otimes \mathbb{C}((x-p_i))).$$

Because $\mathcal{H} \otimes \mathbb{C}((x-\lambda))$ is of slope zero, with local monodromy a pseudoreflection, we see all the terms $[e_i]_*(\mathcal{V} \otimes \mathbb{C}((x-p_i)))$ are of slope zero, exactly one of them (say i=1) has local monodromy a pseudoreflection, and all the others have trivial local monodromy.

Now in order for $[e_1]_*(\mathcal{V} \otimes \mathbb{C}((x-p_1)))$ to have local monodromy a pseudoreflection, either $e_1 = 1$ and $\mathcal{V} \otimes \mathbb{C}((x-p_1))$ itself has local monodromy a pseudoreflecton, or $e_1 = 2$ and $\mathcal{V}$ is of rank one with trivial local monodromy around p_1. In this second case, notice that the pseudoreflection is a true reflection.

If $[e_i]_*(\mathcal{V}\otimes\mathbb{C}((x-p_i)))$ with $i \geq 2$ is of slope zero and has trivial local monodromy, then $e_i = 1$, and $\mathcal{V}$ is of slope zero at p_i and has trivial local monodromy around p_i.

If all the $e_i = 1$, then our covering is unramified over λ, so it is the restriction to $\mathbb{G}_m - \{\lambda\}$ of a d-fold connected finite etale covering of $\mathbb{G}_m$, necessarily the d-fold Kummer covering of $\mathbb{G}_m$ by itself.

If $e_1 = 2$ and all $e_{i\geq 2} = 1$, then $\mathcal{V}$ has rank one, whence $d=n$, and there are $n-1$ points in the fibre over λ, so our covering is either Belyi or inverse Belyi. QED

Belyi Recognition Lemma 3.5.3 Suppose that $\mathcal{H}:=\mathcal{H}_\lambda(\alpha_i\text{'s}; \beta_j\text{'s})$ is an irreducible hypergeometric $\mathcal{D}$-module on $\mathbb{G}_m$ of type (n,n). Then $\mathcal{H}$ is Belyi induced (resp. inverse Belyi induced) of type (a,b), with $a \geq 1$, $b \geq 1$, $n= a+b$, if and only if its exponents mod $\mathbb{Z}$ at 0 and ∞ are Belyi induced (resp. inverse Belyi induced) of type (a,b) in the following sense: there exist $A, B \in \mathbb{C}$ such that the sets of α_i's and of β_j's mod $\mathbb{Z}$ (with multiplicity) are given by

$\{\alpha_i\text{'s}\}$ (resp. $\{\beta_j\text{'s}\}$) $= \{(A+i)/a\}_{i=0,\ldots,a-1} \cup \{(B+j)/b\}_{j=0,\ldots,b-1}$ mod $\mathbb{Z}$.

$\{\beta_j\text{'s}\}$ (resp. $\{\alpha_i\text{'s}\}$) $= \{(A+B+k)/n\}_{k=0,\ldots,n-1}$ mod $\mathbb{Z}$.

proof By multiplicative inversion, it suffices to treat the case of Belyi induced. If $\mathcal{H}$ is Belyi induced, then the inducing equation is a rank one D.E. on $\mathbb{P}^1 - \{0, 1, \infty\}$ which has only regular singularities. Any such D.E. is of the form $(\mathcal{O}, d - Adx/x - Bdx/(x-1)) \approx \mathcal{H}_1(A; A+B)$ Looking at the exponents of its Belyi induction at 0 and ∞ gives the formulas for the α_i mod $\mathbb{Z}$ and the β_i mod $\mathbb{Z}$. So far we have not used the irreducibility of $\mathcal{H}$. This is needed only for the converse.

Conversely, suppose that the exponents of $\mathcal{H}:=\mathcal{H}_\lambda(\alpha_i\text{'s}; \beta_j\text{'s})$ are Belyi induced. We claim that $\mathcal{H}$ is the Belyi induction, say $\mathcal{V}$, of the corresponding $\mathcal{H}_1(A; A+B)$. We know that $\mathcal{V}$ is a D.E. on $\mathbb{G}_m - \{\lambda\}$ of order n with regular singular points at $0, \lambda, \infty$, whose local monodromy at λ is a true reflection, and whose exponents mod $\mathbb{Z}$ at 0 (resp. ∞) are the α_i (resp. the $-\beta_j$). Because $\mathcal{H}$ is irreducible, no α_i is a β_j mod $\mathbb{Z}$, and hence (exactly as in the proof of 2.11.9, (2)), we see that $\mathcal{V}$ is irreducible on $\mathbb{G}_m - \{\lambda\}$. We must show that $\mathcal{V}$ is isomorphic to $\mathcal{H}$ on

$\mathbb{G}_m - \{\lambda\}$. Because both $\mathcal{V}$ and $\mathcal{H}$ have regular singular points, it suffices to show that on $(\mathbb{G}_m - \{\lambda\})^{an}$ they give rise to isomorphic local systems. This is given by the following rigidity theorem 3.5.4. QED

The rigidity of the hypergeometric equation in the case n=m is given by the following theorem. In the case n=2 it goes back to Riemann (his "$\mathcal{P}$-scheme"; the point is that for n=2, we may always twist by some $(x-\lambda)^\gamma$ to make local monodromy around λ a pseudoreflection). Levelt gave a simple group-theoretic proof [Lev-HF] in the general case (cf [B-H, 3.5]). The proof we give below is due to Ofer Gabber.

Rigidity Theorem 3.5.4 Let $\mathcal{F}$ and $\mathcal{G}$ be two irreducible local systems on $(\mathbb{G}_m - \{\lambda\})^{an}$ of the same rank $n \geq 1$, and suppose that
(a) the local monodromies at λ of both $\mathcal{F}$ and $\mathcal{G}$ are pseudoreflections.
(b) $\mathcal{F}$ and $\mathcal{G}$ have the same characteristic polynomial of local monodromy at 0.
(c) $\mathcal{F}$ and $\mathcal{G}$ have the same characteristic polynomial of local monodromy at ∞.
Denote by $j: \mathbb{G}_m - \{\lambda\} \rightarrow \mathbb{P}^1$ the inclusion, and by $\mathit{Hom}(\mathcal{F},\mathcal{G})$ the internal hom local system $\mathcal{F}^\vee \otimes \mathcal{G}$ on $\mathbb{G}_m - \{\lambda\}$.Then
(1) $\chi(\mathbb{P}^1, j_*\mathit{Hom}(\mathcal{F},\mathcal{G})) = 2 > 0$.
(2) There exists an isomorphism $\mathcal{F} \approx \mathcal{G}$.

proof Let us first prove that (1) $\Rightarrow$ (2). Since $\chi = h^0 + h^2 - h^1$, the positivity forces at least one of h^0 or h^2 to be nonzero. By duality, $H^2(\mathbb{P}^1, j_*\mathit{Hom}(\mathcal{F},\mathcal{G}))$ is dual to $H^0(\mathbb{P}^1, j_*\mathit{Hom}(\mathcal{G},\mathcal{F}))$, so at the expense of interchanging $\mathcal{F}$ and $\mathcal{G}$, $H^0(\mathbb{P}^1, j_*\mathit{Hom}(\mathcal{F},\mathcal{G}))=H^0(\mathbb{G}_m-\{\lambda\},\mathit{Hom}(\mathcal{F},\mathcal{G}))$ $= \mathrm{Hom}(\mathcal{F},\mathcal{G})$ is nonzero. Since $\mathcal{F}$ and $\mathcal{G}$ are irreducible, any nonzero hom is an isomorphism. It remains to prove (1). For this it is convenient to give the following Lemma.

Lemma 3.5.5 Let $\mathcal{F}$ be an irreducible local system on $(\mathbb{G}_m - \{\lambda\})^{an}$ of rank $n \geq 2$, whose local monodromy at λ is a pseudoreflection. Then
(1) $\chi(\mathbb{G}_m, j_*\mathcal{F}|\mathbb{G}_m) = -1$.
(2) $H^i_c(\mathbb{G}_m, j_*\mathcal{F}|\mathbb{G}_m)$ vanishes for $i \neq 1$, and for i=1 it has dimension 1.

(3) the local monodromy of $\mathcal{F}$ at 0 (resp. ∞) has a single Jordan block for each distinct eigenvalue, i.e., its characteristic polynomial is its minimal polynomial. Moreover, if $\exp(2\pi i\delta)$ is an eigenvalue of local monodromy at 0 (resp. ∞) then $\exp(-2\pi i\delta)$ is not an eigenvalue of local monodromy at ∞ (resp. 0).

proof Let us denote by $h: \mathbb{G}_m - \{\lambda\} \to \mathbb{G}_m$ and by $k: \mathbb{G}_m \to \mathbb{P}^1$ the inclusions (so $j=kh$). We have an exact sequence on $\mathbb{G}_m$

$$0 \to h_!\mathcal{F} \to h_*\mathcal{F} \to h_*\mathcal{F}/h_!\mathcal{F} \to 0, \quad \text{and}$$

$h_*\mathcal{F}/h_!\mathcal{F}$ is the punctual sheaf $\mathcal{F}^{I_\lambda}$ at λ.

By hypothesis, $\mathcal{F}^{I_\lambda}$ has dimension n-1. But

$$\chi(\mathbb{G}_m, h_!\mathcal{F}) = \chi(\mathbb{G}_m - \{\lambda\}, \mathcal{F}) = \mathrm{rank}(\mathcal{F})\chi(\mathbb{G}_m - \{\lambda\}) = -n,$$

so (1) is obvious.

For (2), the irreducibility of $\mathcal{F}$ forces the vanishing of the H^2_c, and the fact that $j_*\mathcal{F}|\,\mathbb{G}_m$ has no punctual section on the affine curve $\mathbb{G}_m$ forces the vanishing of the H^0_c. That H^1_c has dimension 1 now results from (1).

For (3), consider the short exact sequence

$$0 \to k_!h_*\mathcal{F} \to k_*h_*\mathcal{F} \to \mathcal{F}^{I_0} \oplus \mathcal{F}^{I_\infty} \to 0.$$

The long exact cohomology sequence on $\mathbb{P}^1$ has $H^0(\mathbb{P}^1, k_*h_*\mathcal{F})=0$ by the irreducibility of $\mathcal{F}$, so the coboundary gives us an **injection**

$$0 \to \mathcal{F}^{I_0} \oplus \mathcal{F}^{I_\infty} \to H^1(\mathbb{P}^1, k_!h_*\mathcal{F}) = H^1_c(\mathbb{G}_m, j_*\mathcal{F}|\,\mathbb{G}_m) = \text{1-dim'l}.$$

Therefore at most one of $\mathcal{F}^{I_0}$ or $\mathcal{F}^{I_\infty}$ is nonzero, and if nonzero is one-dimensional. This means that if 1 is an eigenvalue of local monodromy at either 0 or ∞, then it occurs at only one of 0 or ∞, and local monodromy there has a single Jordan block which is unipotent. Applying this to all twists $\mathcal{F}\otimes x^\gamma$ of $\mathcal{F}$, we get (3). QED

We now return to the proof of the rigidity theorem. Let us denote by $\mathcal{K}$ the internal hom sheaf $\mathcal{H}om(\mathcal{F},\mathcal{G})$. The short exact sequence on $\mathbb{P}^1$

$$0 \to j_!\mathcal{K} \to j_*\mathcal{K} \to \mathcal{K}^{I_0} \oplus \mathcal{K}^{I_\infty} \oplus \mathcal{K}^{I_\lambda} \to 0$$

gives

$$\chi(\mathbb{P}^1, j_*\mathcal{K}) = -n^2 + \dim\mathcal{K}^{I_0} + \dim\mathcal{K}^{I_\infty} + \dim\mathcal{K}^{I_\lambda}.$$

Now for any of the missing points $p = 0, \infty$, or λ, $\mathcal{K}^{I_p}$ is the space $\mathrm{Hom}_{I_p}(\mathcal{F},\mathcal{G})$ of I_p-equivariant maps.

If p is 0 or ∞, then denoting by T the local monodromy, and by P(T) its characteristic polynomial, we have proven that both $\mathcal{F}$ and $\mathcal{G}$ as $\mathbb{C}[T]$ modules are isomorphic to $\mathbb{C}[T]/(P(T))$. So $\mathrm{Hom}_{I_p}(\mathcal{F},\mathcal{G})$ is just $\mathrm{Hom}_{\mathbb{C}[T]}(\mathbb{C}[T]/(P(T)), \mathbb{C}[T]/(P(T))) \approx \mathbb{C}[T]/(P(T))$, which is n-dimensional. So $\dim \mathcal{K}^{I_p} = n$ for both p=0 and $p=\infty$.

At λ, both local monodromies T are pseudoreflections. Their determinants are equal, because they are determined by the determinants of the local monodromies at 0 and ∞. If their common determinant is $\xi \neq 1$, then both $\mathcal{F}$ and $\mathcal{G}$ as $\mathbb{C}[T]$-modules are

$$\oplus_{n-1 \text{ copies}} \mathbb{C}[T]/(T-1) \;\oplus\; \mathbb{C}[T]/(T-\xi),$$

whose space of $\mathbb{C}[T]$-endomorphisms has dimension $(n-1)^2 + 1$. If their common determinant is 1, then both $\mathcal{F}$ and $\mathcal{G}$ as $\mathbb{C}[T]$-modules are

$$\oplus_{n-2 \text{ copies}} \mathbb{C}[T]/(T-1) \;\oplus\; \mathbb{C}[T]/(T-1)^2,$$

whose space of $\mathbb{C}[T]$-endomorphisms has dimension $(n-2)^2 + 2 + 2(n-2)$. So in both cases we find miraculously that $\dim \mathcal{K}^{I_\lambda} = n^2 - 2n + 2$. Adding up the contributions from 0, ∞, and λ we find $\chi(\mathbb{P}^1, j_*\mathcal{K}) = 2$. QED

Recall that this discussion of rigidity grew out of our desire to recognize which hypergeometrics of type (n,n) are induced. For the sake of completeness, we state the following

Kummer Recognition Lemma 3.5.6 Suppose that $\mathcal{H} := \mathcal{H}_\lambda(\alpha_i\text{'s}; \beta_j\text{'s})$ is an irreducible hypergeometric $\mathcal{D}$-module on $\mathbb{G}_m$ of type (n,m). Let $d \geq 2$ be a divisor of both n and m. Then the D.E. $\mathcal{H} \mid \mathbb{G}_m - \{\lambda\}$ is Kummer induced of degree d if and only if there exist $A_1, \ldots, A_{n/d}$ and $B_1, \ldots, B_{m/d}$ in $\mathbb{C}$ such that the sets of α_i's and of β_j's mod $\mathbb{Z}$ (with multiplicity) are given by

$$\{\alpha_i\text{'s}\} = \{ (A_i - j)/d \}_{i=1,\ldots,n/d;\ j=0,\ldots,d-1} \mod \mathbb{Z},$$

$$\{\beta_j\text{'s}\} = \{ (B_i + j)/d \}_{i=1,\ldots,m/d;\ j=0,\ldots,d-1} \mod \mathbb{Z}.$$

proof If $\mathcal{H}$ is Kummer induced of degree d, then looking at the effect of Kummer induction on the slope zero parts at 0 and ∞ shows that the α_i and the β_j mod $\mathbb{Z}$ are of the asserted form. Conversely, if the α_i and the β_j mod $\mathbb{Z}$ have the asserted form, then we may, by the irreducibility of $\mathcal{H}$, suppose that the α_i and the β_j are given exactly (not just mod $\mathbb{Z}$) by the above formulas (with the asymmetry in the sign of $\pm j$). Denote by $\mu \in \mathbb{C}^\times$ any solution of the equation

$$(\mu d^{n-m})^d = \lambda.$$

We will establish the following **Kummer Induction Formula**:

(3.5.6.1) $\mathcal{H}_\lambda(\alpha_i\text{'s}; \beta_j\text{'s}) \approx [d]_* \mathcal{H}_\mu(A_i\text{'s}; B_j\text{'s})$.

Notice first that $\mathcal{H}_\mu(A_i\text{'s}; B_j\text{'s})$ is irreducible, for if $A_i = B_j + r$ for some $r \in \mathbb{Z}$, then dividing by d we would find that some α_i is congruent to some β_j mod $\mathbb{Z}$, contradicting the irreducibility of $\mathcal{H}_\lambda(\alpha_i\text{'s}; \beta_j\text{'s})$. Notice next that by Frobenius reciprocity, $[d]_* \mathcal{H}_\mu(A_i\text{'s}; B_j\text{'s})$ is irreducible (on $\mathbb{G}_m - \{\lambda\}$ if n=m, on $\mathbb{G}_m$ otherwise), because

$$[d]^*[d]_* \mathcal{H}_\mu(A_i\text{'s}; B_j\text{'s}) \approx \oplus_{\zeta \in \mu_d} [x \mapsto \zeta x]^* \mathcal{H}_\mu(A_i\text{'s}; B_j\text{'s})$$
$$\approx \oplus_{\zeta \in \mu_d} \mathcal{H}_{\mu/\zeta}(A_i\text{'s}; B_j\text{'s})$$

is a direct sum of d pairwise-nonisomorphic irreducibles (on $\mathbb{G}_m - \{\mu\mu_d\}$ if n=m, on $\mathbb{G}_m$ otherwise).

So it suffices to construct a nonzero map of $\mathcal{D}$-modules from $\mathcal{H}_\lambda(\alpha_i\text{'s}; \beta_j\text{'s})$ to $[d]_* \mathcal{H}_\mu(A_i\text{'s}; B_j\text{'s})$. We will do this explicitly. It will be easier to see what is going on if we denote by P(t) and Q(t) the polynomials

$$P(t) := \mu \Pi_{i=1,\ldots,n/d}(T - A_i), \quad , Q(t) := \Pi_{j=1,\ldots,m/d}(T - B_j).$$

For each integer $k \geq 1$ we denote by $P_k(t)$ and $Q_k(t)$ the polynomials (note the asymmetry in the sign of $\pm j$)

$$P_k(t) := \Pi_{j=0,\ldots,k-1} P(t - j), \quad Q_k(t) := \Pi_{j=0,\ldots,k-1} Q(t + j),$$

and by $\mathrm{Hyp}_k(P, Q)$ the operator

$$\mathrm{Hyp}_k(P, Q) := P_k(D) - x^k Q_k(D).$$

Thus $\mathrm{Hyp}_k(P, Q)$ for k=1 is just the operator Hyp(P,Q) defining $\mathcal{H}_\mu(A_i\text{'s}; B_j\text{'s}) := \mathcal{D}/\mathcal{D}\mathrm{Hyp}(P,Q)$. One verifies easily by induction on k that when $\mathrm{Hyp}_k(P, Q)$ acts on the left $\mathcal{D}$-module $\mathcal{D}/\mathcal{D}\mathrm{Hyp}(P,Q)$, it kills the image of 1. Now the operator $\mathrm{Hyp}_k(P, Q)$ lies in the subring

$$\mathcal{D}_k := \mathbb{C}[x^k, x^{-k}, D = kD_k], \text{ where } D_k := x^k d/dx^k,$$

and so there is $\mathcal{D}_k$-linear map

$$\mathcal{D}_k/\mathcal{D}_k\mathrm{Hyp}_k(P,Q) \to \mathcal{D}/\mathcal{D}\mathrm{Hyp}(P,Q), \; 1 \mapsto 1.$$

This map is obviously nonzero. But for any $\mathcal{D}$-module $\mathcal{M}$, the k-fold Kummer induction $[k]_* \mathcal{M}$ is precisely $\mathcal{M}$ viewed as a $\mathcal{D}_k$-module. So we have constructed a nonzero map

$$\mathcal{D}_k/\mathcal{D}_k\mathrm{Hyp}_k(P,Q) \to [k]_*(\mathcal{D}/\mathcal{D}\mathrm{Hyp}(P,Q)) = [k]_* \mathcal{H}_\mu(A_i\text{'s}; B_j\text{'s}).$$

Write the operator $\mathrm{Hyp}_k(P,Q)$ in $\mathcal{D}_k$ in terms of $t := x^k$ and $D_k := td/dt$, and view $\mathcal{D}_k$ as isomorphic to $\mathcal{D}$ by $t \mapsto x$, $D_k \mapsto D$. Then for $k=d$, $\mathcal{D}_k/\mathcal{D}_k\mathrm{Hyp}_k(P,Q)$ is none other than $\mathcal{H}_\lambda(\alpha_i\text{'s}; \beta_j\text{'s})$. QED

Lemma 3.5.7 Suppose that $\mathcal{H}:=\mathcal{H}_\lambda(\alpha_i\text{'s}; \beta_j\text{'s})$ is an irreducible hypergeometric $\mathcal{D}$-module on $\mathbb{G}_m$ of type (n,n). Then either
(1) $\mathcal{H} \mid \mathbb{G}_m - \{\lambda\}$ is Lie-irreducible, or
(2) $\mathcal{H} \mid \mathbb{G}_m - \{\lambda\}$ is induced, or
(3) $\mathcal{H} \mid \mathbb{G}_m - \{\lambda\}$ is the tensor product $W \otimes K$ of a D.E. W of rank one with an irreducible D.E. K of rank n whose G_{gal} is finite. If in addition $\det\mathcal{H}$ is of finite order, then $\mathcal{H} \mid \mathbb{G}_m - \{\lambda\}$ itself has G_{gal} finite in case (3).

proof By 2.7.2 we know that if $\mathcal{H} \mid \mathbb{G}_m - \{\lambda\}$ is neither Lie-irreducible nor induced, it is $W \otimes K$ for some Lie-irreducible W of rank $d < n$ with $d|n$, and K some irreducible D.E. of rank $d' := n/d$ whose G_{gal} is finite. Therefore local monodromy at λ is of the form $A \otimes B$ with A in GL(d) and B in GL(d'). But if both d and d' are ≥ 2, no pseudoreflection can be of this form. Therefore $d=1$, as required. Then $\det\mathcal{H} \approx W^{\otimes n} \otimes \det K$, with $\det K$ of finite order, so W is of finite order if and only if $\det\mathcal{H}$ is of finite order; if it is, then $\mathcal{H} \approx W \otimes K$ itself has G_{gal} finite. QED

Theorem 3.5.8 ([B-H], 6.5]) Suppose that $\mathcal{H}:=\mathcal{H}_\lambda(\alpha_i\text{'s}; \beta_j\text{'s})$ is an irreducible hypergeometric $\mathcal{D}$-module on $\mathbb{G}_m$ of type (n,n) which is neither Kummer induced nor Belyi induced nor inverse Belyi induced. Denote by G the differential galois group of $\mathcal{H} \mid \mathbb{G}_m - \{\lambda\}$, by $\gamma := \Sigma\beta_j - \Sigma\alpha_i$.
(1) The group G is reductive. If γ, $\Sigma\alpha_i$, and $\Sigma\beta_j$ are all in $\mathbb{Q}$ (i.e., if $\det\mathcal{H}$ is of finite order), then $G^0 = G^{0,\mathrm{der}}$. Otherwise, $G^0 = \mathbb{G}_m G^{0,\mathrm{der}}$.
(2) The group $G^{0,\mathrm{der}}$ is either {1}, SL(n), SO(n), or (if n is even) Sp(n).
(3) if $\exp(2\pi i\gamma) \neq \pm 1$, $G^{0,\mathrm{der}} = \{1\}$ or SL(n).
(4) if $\exp(2\pi i\gamma) = -1$, $G^{0,\mathrm{der}} = \{1\}$ or SL(n) or SO(n).
(5) if $\exp(2\pi i\gamma) = +1$, $G^{0,\mathrm{der}} =$ SL(n) or (for n even) Sp(n).
(6) if γ is irrational, G = GL(n).

proof The local monodromy around λ is a pseudoreflection of determinant $\exp(2\pi i\gamma)$. So if $\mathcal{H} \mid \mathbb{G}_m - \{\lambda\}$ is Lie irreducible, the theorem is an immediate consequence of the Pseudoreflection Theorem 1.5. In view of the preceding Lemma, the only other case is when $\mathcal{H} \mid \mathbb{G}_m - \{\lambda\}$

is the tensor product $W \otimes K$ of a D.E. W of rank one with an irreducible D.E. K of rank n whose G_{gal} is finite. In this case G^0 is either {1} or $\mathbb{G}_m$, depending on whether or not W, or equivalently $\det\mathcal{H} \mid \mathbb{G}_m - \{\lambda\}$, is of finite order. So (1) through (4) hold (trivially) in this case. If γ is either in $\mathbb{Z}$ or is irrational, then we cannot be in this case, for then local monodromy around λ is a either a unipotent pseudoreflection or is $\mathrm{Diag}(\exp(2\pi i\gamma), 1, 1, \ldots, 1)$, no power of which is scalar. QED

We can be more precise about the distinguishing the the various Lie-irreducible cases. (We will discuss later, in 5.4-5.5 and then again in 8.17, how to detect the case when $G^{0,\mathrm{der}}$ is {1}.)

Corollary 3.5.8.1 Notations and hypotheses as above, suppose further that $G^{0,\mathrm{der}} \neq \{1\}$. Then $G^{0,\mathrm{der}}$ is SO(n) (respectively Sp(n)) if and only if there exists $\delta \in \mathbb{C}$ such that $\mathcal{H} \otimes x^\delta := \mathcal{H}_\lambda(\alpha_i + \delta\text{'s};\ \beta_j + \delta\text{'s})$ is self dual and its autoduality pairing is symmetric (resp. alternating).

proof Notice that $G^{0,\mathrm{der}}$ is the same for any twist $\mathcal{H} \otimes x^\delta$ as for $\mathcal{H}$. So if some twist $\mathcal{H} \otimes x^\delta$ is self dual, then $G^{0,\mathrm{der}}$ is **contained in** SO(n) or Sp(n), depending on the "sign" of the autoduality. In view of the paucity of choices for $G^{0,\mathrm{der}}$, $G^{0,\mathrm{der}}$ must **be** SO(n) or Sp(n).

Conversely, suppose that $G^{0,\mathrm{der}}$ is SO(n) (so $n \geq 3$ since SO(2) is not semisimple) or Sp(n). We must distinguish several cases.

If $G^{0,\mathrm{der}}$ is Sp(n), then G must be contained in the normalizer in GL(n) of Sp(n), which is $\mathbb{G}_m\mathrm{Sp}(n)$. Since the only scalars in Sp(n) are ± 1, we can construct a character χ of G by writing an element of G as tA (t in $\mathbb{G}_m$, A in Sp(n)) and defining $\chi(g) = \chi(tA) := t^2$. This character of G corresponds to a rank one D.E. which is in the tensor subcategory $\langle\mathcal{H}\rangle$. Now $\mathcal{H}$ has regular singularities at $0, \lambda$, and ∞, and its local monodromy around λ is a unipotent pseudoreflection (otherwise we can't have $G^{0,\mathrm{der}} = \mathrm{Sp}(n)$). Therefore every object in $\langle\mathcal{H}\rangle$ is regular singular at 0, λ, and ∞ and has unipotent local monodromy at λ. Therefore the rank one D.E. on $\mathbb{G}_m - \{\lambda\}$ corresponding to χ is regular singular at 0, λ, and ∞ and has **trivial** local monodromy at λ, so it must be the D.E. for x^δ for some δ. Then $\mathcal{H} \otimes x^{-\delta/2}$ has its differential galois group inside $\pm\mathrm{Sp}(n) = \mathrm{Sp}(n)$, as required.

If $G^{0,\mathrm{der}}$ is SO(n), $n \geq 3$, then G must be contained in the normalizer in GL(n) of SO(n), which is $\mathbb{G}_m O(n)$. As O(n) contains no scalars except ± 1, so we get a character χ of $\mathbb{G}_m O(n)$ by $\chi(tA):=t^2$. The corresponding rank one D.E. is in $\langle\mathcal{H}\rangle$. But $\mathcal{H}$ has regular singularities at $0, \lambda$, and ∞, and its local monodromy around λ is a true reflection (otherwise we can't have $G^{0,\mathrm{der}} = SO(n)$), so any object in $\langle\mathcal{H}\rangle$ is regular singular at 0, λ, and ∞. Let us denote by T_λ the local monodromy of $\mathcal{H}$ around λ. If we can show that $\chi(T_\lambda) = 1$, then just as above χ corresponds to the D.E. for some x^δ, and $\mathcal{H}\otimes x^{-\delta/2}$ has its differential galois group inside $\pm O(N) = O(N)$, as required. To show that $\chi(T_\lambda) = 1$, suppose not; then $\chi(T_\lambda) = \xi^2$, where $\xi^2 \neq 1$. This means that $T_\lambda = \xi^{-1}A$ with A in O(n). But in a suitable basis $e_1,\ldots, e_n$ of the representation space, the reflection T_λ is Diag(-1, 1,..., 1). So we find that the matrix $A:=\mathrm{Diag}(-\xi, \xi,\ldots, \xi) \in O(n)$ for some nondegenerate quadratic form $\langle\ ,\ \rangle$. To see that this is impossible for $n \geq 3$, we argue as follows. Denote by V the line $\mathbb{C}e_1$, and by W the $\mathbb{C}$-span of $e_2,\ldots, e_n$. Writing vectors in the form v+w, with $v\in V$ and $w\in W$, we have $A(v+w) = -\xi v + \xi w$. As $A\in O(N)$, we have

$$\langle v+w, v+w\rangle = \langle A(v+w), A(v+w)\rangle = \langle -\xi v + \xi w, -\xi v + \xi w\rangle.$$

Expanding out, we find

$$\langle v, v\rangle + 2\langle v, w\rangle + \langle w, w\rangle = \xi^2\langle v, v\rangle - 2\xi^2\langle v, w\rangle + \xi^2\langle w, w\rangle.$$

Taking v=0 (resp. w=0), we see that $\xi^2 \neq 1$ forces $\langle w, w\rangle = 0 = \langle v, v\rangle$. Therefore V and W are totally isotropic, and so $W\cap\perp(V) = 0$. But both W and $\perp(V)$ are codimension one subspaces, so $W\cap\perp(V) = 0$ is impossible if $n \geq 3$. QED

In the case $n\neq m$, we have

Theorem 3.6 Suppose that $\mathcal{H}:=\mathcal{H}_\lambda(\alpha_i\text{'s}; \beta_j\text{'s})$ is an irreducible hypergeometric $\mathcal{D}$-module on $\mathbb{G}_m$ of type (n,m), $n \neq m$, which is not Kummer induced. Let N:=max(n,m) be the rank of $\mathcal{H}$, and G its differential galois group. Then

(1) G is reductive. If $\det\mathcal{H}$ is of finite order (i.e., if $|n-m| > 1$ and if $\Sigma\alpha_i \in \mathbb{Q}$ when $n > m$ (resp. $\Sigma\beta_j \in \mathbb{Q}$ when $m > n$)) , then $G^0 = G^{0,\mathrm{der}}$; otherwise $G^0 = \mathbb{G}_m G^{0,\mathrm{der}}$.

(2) If $|n-m|$ is odd, $G^{0,\mathrm{der}}$ is SL(N). If $|n-m| = 1$ then G is GL(N).

(3) If $|n-m|$ is even, then $G^{0,\mathrm{der}}$ is SL(N) or SO(N) or (if N is even) SP(N), or $|n-m|=6$, N=7,8 or 9, and $G^{0,\mathrm{der}}$ is one of

N=7: the image of G_2 in its 7-dim'l irreducible representation
N=8: the image of Spin(7) in the 8-dim'l spin representation
the image of SL(3) in the adjoint representation
the image of SL(2)×SL(2)×SL(2) in std⊗std⊗std
the image of SL(2)×Sp(4) in std⊗std
the image of SL(2)×SL(4) in std⊗std
N=9: the image of SL(3)×SL(3) in std⊗std.

proof This theorem, "mise pour memoire", is just the special case d=1 of 2.11.10. QED

The discrimination among the various possible cases is aided by

Corollary 3.6.1 Notations and hypotheses as above, $G^{0,\mathrm{der}}$ is contained in SO(N) (resp. in Sp(N)) if and only if there exists $\delta \in \mathbb{C}$ such that $\mathcal{H}\otimes x^{\delta} := \mathcal{H}_{\lambda}(\alpha_i + \delta\text{'s}; \beta_j + \delta\text{'s})$ is self dual and its autoduality pairing is symmetric (resp. alternating). Moreover, if N is odd, then $G^{0,\mathrm{der}}$ is contained in SO(N) if and only if there exists $\delta \in \mathbb{C}$ such that $\mathcal{H}\otimes x^{\delta} := \mathcal{H}_{\lambda}(\alpha_i + \delta\text{'s}; \beta_j + \delta\text{'s})$ has its $G_{gal} \subset SO(N)$.

proof The proof is entirely analogous to that of 3.5.8.1. If some twist $\mathcal{H}\otimes x^{\delta}$ of $\mathcal{H}$ is self dual, then $G^{0,\mathrm{der}}$ is certainly contained in SO(N) or in Sp(N), depending on the sign of the autoduality.

If $G^{0,\mathrm{der}} \subset O(N)$ (resp. Sp(N)) , then $G \subset \mathbb{G}_m O(N)$ (resp. $\mathbb{G}_m Sp(N)$). Indeed, if Γ is any irreducible subgroup of GL(N) which respects a nonzero bilinear form $\langle\ ,\ \rangle$, then its normalizer in GL(N) lies in in the corresponding similitude group. [For if $A \in GL(N)$ normalizes Γ then the form $(x,y) := \langle Ax, Ay\rangle$ is also Γ-invariant, so a scalar multiple of $\langle\ ,\ \rangle$.] Now consider the character χ of G defined by $\chi(g) = \chi(tA) := t^2$. The corresponding rank one D.E. on $\mathbb{G}_m$ is in $\langle\mathcal{H}\rangle$. Now $|n-m|$ is nonzero (by assumption) and even (otherwise $G^{0,\mathrm{der}}$ is SL(N)), hence $|n-m| \geq 2$. Therefore all slopes of $\mathcal{H}$ at 0 or ∞ are $\leq 1/|n-m| < 1$, and hence every object of $\langle\mathcal{H}\rangle$ has all its slopes < 1 at 0 or ∞. So any rank one object in $\langle\mathcal{H}\rangle$ has slope zero at 0 and ∞, so is the D.E. for x^{δ} for some δ. Taking the δ corresponding to χ, $\mathcal{H}\otimes x^{-\delta/2}$ has its differential galois group in $\pm O(N) = O(N)$ (resp. in $\pm Sp(N) = Sp(N)$).

If N is odd, then O(N) is the product $\{\pm 1\}\times SO(N)$, so if $G_{gal} \subset O(N)$

but $G_{gal} \not\subset SO(N)$, its projection onto the $\{\pm 1\}$ factor is a character of order two corresponding to a rank one object of $\langle\mathcal{H}\rangle$, necessarily the D.E. for $x^{1/2}$. QED

Lemma 3.6.2 Let $\mathcal{H} := \mathcal{H}_\lambda(\alpha_i\text{'s}; \beta_j\text{'s})$ be a hypergeometric $\mathcal{D}$-module on $\mathbb{G}_m$ of type (n,m), $n > m$. Then $G_{gal} \subset SL(n)$ if and only if $n-m \geq 2$ and $\Sigma\alpha_i \in \mathbb{Z}$.

proof If $n-m=1$, then $\det\mathcal{H}$ has slope 1 at ∞, so nontrivial. If $n-m \geq 2$, then $\det(\mathcal{H})$ is necessarily the D.E. for x^δ, some δ; looking at zero we see that $\delta \equiv \Sigma\alpha_i \bmod \mathbb{Z}$. QED

3.7 Intrinsic characterization of hypergeometric equations

We now turn to the intrinsic characterization of irreducible hypergeometric $\mathcal{D}$-modules on $\mathbb{G}_m$ among all irreducible $\mathcal{D}$-modules on $\mathbb{G}_m$.

Theorem 3.7.1 Let $\mathcal{M}$ be an irreducible, nonpunctual holonomic $\mathcal{D}$-module on $\mathbb{G}_m$. Then $\mathcal{M}$ is hypergeometric if and only if its Euler characteristic $\chi(\mathbb{G}_m, \mathcal{M}) := \chi(H^*{}_{DR}(\mathbb{G}_m, \mathcal{M}))$ is -1.

proof It is obvious from the elementary Euler Poincare formula on $\mathbb{G}_m$ (2.9.13) that $\chi(\mathbb{G}_m, \mathcal{H}) = -1$ for any hypergeometric $\mathcal{H}$ on $\mathbb{G}_m$, irreducible or not.

Suppose now that $\mathcal{M}$ is an irreducible, nonpunctual $\mathcal{D}$-module on $\mathbb{G}_m$ with $\chi(\mathbb{G}_m, \mathcal{M}) = -1$. We will make essential use of

Lemma 3.7.2 (compare 3.5.5) If $\mathcal{M}$ is an irreducible, nonpunctual $\mathcal{D}$-module on $\mathbb{G}_m$ with $\chi(\mathbb{G}_m, \mathcal{M}) = -1$, then for any twist $\mathcal{M}\otimes x^\delta$,

$$\dim_{\mathbb{C}}\mathrm{Soln}_0(\mathcal{M}\otimes x^\delta) + \dim_{\mathbb{C}}\mathrm{Soln}_\infty(\mathcal{M}\otimes x^\delta) \leq 1.$$

proof The twist $\mathcal{M}\otimes x^\delta$ is also irreducible on $\mathbb{G}_m$, and its χ is the same as that of $\mathcal{M}$ (this is obvious from the Euler-Poincare formula), so it suffices to prove

$$\dim_{\mathbb{C}}\mathrm{Soln}_0(\mathcal{M}) + \dim_{\mathbb{C}}\mathrm{Soln}_\infty(\mathcal{M}) \leq 1.$$

Let $k: \mathbb{G}_m \to \mathbb{P}^1$ denote the inclusion. Then by 2.9.8 we have a short exact sequence on $\mathbb{P}^1$

$$0 \to k_{!*}\mathcal{M} \to k_*\mathcal{M} \to \delta_0\otimes\mathrm{Hom}_{\mathbb{C}}(\mathrm{Soln}_0, \mathbb{C}) \oplus \delta_\infty\otimes\mathrm{Hom}_{\mathbb{C}}(\mathrm{Soln}_\infty, \mathbb{C}) \to 0$$

Let us admit temporarily that

(*) $\dim_{\mathbb{C}} H^1{}_{DR}(\mathbb{P}^1, k_*\mathcal{M}) = 1$, $H^2{}_{DR}(\mathbb{P}^1, k_{!*}\mathcal{M}) = 0$.

Then as $H^1{}_{DR}(\mathbb{P}^1, \delta_0) \approx \mathbb{C} \approx H^1{}_{DR}(\mathbb{P}^1, \delta_\infty)$, the long exact cohomology sequence gives us a surjection

$$H^1{}_{DR}(\mathbb{P}^1, k_*\mathcal{M}) \twoheadrightarrow \mathrm{Hom}_{\mathbb{C}}(\mathrm{Soln}_0, \mathbb{C}) \oplus \mathrm{Hom}_{\mathbb{C}}(\mathrm{Soln}_\infty, \mathbb{C}),$$

whence the required inequality on dimensions.

To prove (*), we argue as follows. We have

$$H^i{}_{DR}(\mathbb{P}^1, k_*\mathcal{M}) \approx H^i{}_{DR}(\mathbb{G}_m, \mathcal{M}),$$

and only H^0 and H^1 are possibly nonzero. But $H^0 = \mathrm{Hom}_{\mathcal{D}}(\mathcal{O},\mathcal{M})$ vanishes, because if not then $\mathcal{M}$, being irreducible, would be isomorphic to $\mathcal{O}$, which is nonsense because $\chi(\mathbb{G}_m, \mathcal{O}) = 0$. This proves that $\dim_{\mathbb{C}} H^1{}_{DR}(\mathbb{P}^1, k_*\mathcal{M}) = 1$. To prove the vanishing of $H^2{}_{DR}(\mathbb{P}^1, k_{!*}\mathcal{M})$, it is equivalent by duality to prove the vanishing of $H^0{}_{DR}(\mathbb{P}^1, k_{!*}(\mathcal{M}^*))$. But $k_{!*}(\mathcal{M}^*) \subset k_*(\mathcal{M}^*)$, so $H^0{}_{DR}(\mathbb{P}^1, k_{!*}(\mathcal{M}^*)) \subset H^0{}_{DR}(\mathbb{P}^1, k_*(\mathcal{M}^*)) \approx H^0{}_{DR}(\mathbb{G}_m, \mathcal{M}^*)$ which vanishes (otherwise $\mathcal{M}^* \approx \mathcal{O}$ by irreducibility, whence $\mathcal{M} \approx \mathcal{O}$, nonsense). QED

Let $j: U \to \mathbb{G}_m$ be the inclusion of a dense open set such that $j^*\mathcal{M}$ is a D.E. on U. Because $\mathcal{M}$ is irreducible, $\mathcal{M} \approx j_{!*}j^*\mathcal{M}$. Since $\chi(\mathbb{G}_m)=0$, the Euler-Poincare formula gives

$$-1 = \chi(\mathbb{G}_m, j_{!*}j^*\mathcal{M}) = -\mathrm{Irr}_0(\mathcal{M}) - \mathrm{Irr}_\infty(\mathcal{M}) - \Sigma_{\alpha\in\mathbb{G}_m - U}\, \mathrm{totdrop}_\alpha,$$

i.e.,

$$\mathrm{Irr}_0(\mathcal{M}) + \mathrm{Irr}_\infty(\mathcal{M}) + \Sigma_{\alpha\in\mathbb{G}_m - U}\, \mathrm{totdrop}_\alpha = 1.$$

As all the lefthand terms are nonnegative integers, there are three possibilities.

(case 1: reg) $\mathrm{Irr}_0(\mathcal{M}) = 0 = \mathrm{Irr}_\infty(\mathcal{M})$, $\mathrm{totdrop}_\alpha = 0$ for all $\alpha \in \mathbb{G}_m - U$ save one, call it λ, and $\mathrm{totdrop}_\lambda = 1$.

In this case, $\mathrm{totdrop}_\lambda \neq 0$ forces $\mathrm{drop}_\lambda \neq 0$ (by 2.9.10), whence $\mathrm{drop}_\lambda = 1$ and $\mathrm{Irr}_\lambda(\mathcal{M}) = 0$. This means exactly that $\mathcal{M} \mid \mathbb{G}_m - \{\lambda\}$ is an irreducible D.E. with regular singularities at $0, \lambda, \infty$, and its local monodromy around λ is a pseudoreflection.

(case 2: 0-irreg) $\mathrm{Irr}_0(\mathcal{M})=1$, $\mathrm{Irr}_\infty(\mathcal{M})=0$, $\mathrm{totdrop}_\alpha = 0$ for all $\alpha \in \mathbb{G}_m - U$.

(case 3: ∞-irreg) $\mathrm{Irr}_0(\mathcal{M})=0$, $\mathrm{Irr}_\infty(\mathcal{M})=1$, $\mathrm{totdrop}_\alpha = 0$ for all $\alpha \in \mathbb{G}_m - U$.

In cases 2 and 3, which are interchanged by multiplicative inversion, $\mathcal{M}$ is an irreducible D.E. on $\mathbb{G}_m$, and the sum of its irregularities at 0 and ∞ is 1.

By multiplicative inversion, we may assume henceforth that we are in case 1 or case 3, i.e., that 0 is a regular singularity. By the lemma above, the formal decomposition of $\mathcal{M}\otimes\mathbb{C}((x))$ is of the form

$$\mathcal{M}\otimes\mathbb{C}((x)) \approx \oplus_{\alpha}\ \mathbb{C}((x))[D]/\mathbb{C}((x))[D](D-\alpha)^{n_\alpha},$$

and that of $\mathcal{M}\otimes\mathbb{C}((1/x))$ is of the form

$$\mathcal{M}\otimes\mathbb{C}((1/x)) \approx \oplus_{\beta}\ \mathbb{C}((1/x))[D]/\mathbb{C}((1/x))[D](D-\beta)^{m_\beta}\oplus\ (\mathcal{M}\otimes\mathbb{C}((1/x)))_{\text{slope}>0},$$

and no α with $n_\alpha\neq 0$ is congruent mod $\mathbb{Z}$ to any β with $m_\beta\neq 0$.

Let n be the generic rank of $\mathcal{M}$, and let m be the dimension of its slope=0 part at ∞. Let $\alpha_1, \ldots, \alpha_n$ be the α's with multiplicities $n_\alpha > 0$ occuring in $\mathcal{M}\otimes\mathbb{C}((x))$, $P(t) := \prod(t-\alpha_i)$, and let $\beta_1, \ldots, \beta_m$ be the β's with multiplicity $m_\beta > 0$ occurring in the slope zero part of $\mathcal{M}\otimes\mathbb{C}((1/x))$, $Q(t) := \prod(t-\beta_j)$. We choose all the α_i's and all the β_j's to lie in some fundamental domain for $\mathbb{C}/\mathbb{Z}$, say in $0 \le \mathrm{Re}(z) < 1$; by this choice of α's and β's we have

$$\gcd(P(t), P(t+k)) = 1 \text{ and } \gcd(Q(t), Q(t+k)) = 1 \text{ for any } k\neq 0 \text{ in } \mathbb{Z}.$$

By 2.11.4, we have

$$\mathcal{M}\otimes\mathbb{C}((x)) \approx \mathbb{C}((x))[D]/\mathbb{C}((x))[D]P(D),$$

$$\mathcal{M}\otimes\mathbb{C}((1/x)) \approx \mathbb{C}((1/x))[D]/\mathbb{C}((1/x))[D]Q(D)\ \oplus\ (\mathcal{M}\otimes\mathbb{C}((1/x)))_{\text{slope}>0}.$$

We will show that there exists $\lambda\in\mathbb{C}^\times$ such that $\mathcal{H}:=\mathcal{H}_\lambda(\alpha_i\text{'s};\ \beta_j\text{'s}) \approx \mathcal{M}$.

In case 1, one takes for λ the unique point in $\mathbb{G}_m$ where $\mathcal{M}$ is not a D.E. In this case the existence of the isomorphism $\mathcal{H} \approx \mathcal{M}$ is given by 3.5.4.

In case 3, we will give an algebraic version of the proof of 3.5.4 in the $\mathcal{D}$-module context. We must first explain how to choose λ in this case. We will compare the slope decompositions of $\mathcal{M}\otimes\mathbb{C}((1/x))$ and of $\mathcal{H}_1(\alpha_i\text{'s};\ \beta_j\text{'s})\otimes\mathbb{C}((1/x))$. Let us write

$$W := (\mathcal{M}\otimes\mathbb{C}((1/x)))_{\text{slope}>0}, \qquad V := (\ \mathcal{H}_1(\alpha_i\text{'s};\ \beta_j\text{'s})\otimes\mathbb{C}((1/x)))_{\text{slope}>0}.$$

W and V both have rank $d:=n-m$, and all slopes $1/d$.

Det$\mathcal{M}$ has slope zero at 0, and slope ≤ 1 (in fact $\le 1/d$) at ∞, so it is the D.E. for $x^\delta e^{\gamma x}$, for some δ, $\gamma \in\mathbb{C}$. Moreover, $\gamma\neq 0$ if and only if $d=1$ (if $d > 1$, then det$\mathcal{M}$ has slope zero at ∞, while for $d=1$ it has slope 1). Looking at the local expression for $\mathcal{M}$ near 0, we find $\delta \equiv \Sigma\alpha_i$ mod $\mathbb{Z}$. Looking near ∞, we see that $\det(W)\otimes x^{\Sigma\beta_j} \approx x^\delta e^{\gamma x}\mathbb{C}((1/x))$ as D.E.'s on

$\mathbb{C}((1/x))$, i.e.,

$$\det(W) \approx x^{\Sigma\alpha_i - \Sigma\beta_j} e^{\gamma x}\mathbb{C}((1/x)), \text{ with } \gamma \neq 0 \text{ iff } d=1.$$

Repeating this same argument for $\mathcal{H}_1(\alpha_i\text{'s}; \beta_j\text{'s})$ instead of $\mathcal{M}$, we find

$$\det(V) \approx x^{\Sigma\alpha_i - \Sigma\beta_j} e^{\delta' x}\mathbb{C}((1/x)), \text{ with } \delta' \neq 0 \text{ iff } d=1.$$

If $d=1$, these are expressions for W and V themselves, and they show that some multiplicative translate of V is isomorphic to W. If $d \geq 2$, the $\gamma=\delta'=0$, and $\det(V) \approx \det(W)$, so by 3.4.1 some multiplicative translate of V is isomorphic to W. So in either case, some multiplicative translate $\mathcal{H} := \mathcal{H}_\lambda(\alpha_i\text{'s}; \beta_j\text{'s})$ of $\mathcal{H}_1(\alpha_i\text{'s}; \beta_j\text{'s})$ has $\mathcal{H}\otimes\mathbb{C}((1/x)) \approx \mathcal{M}\otimes\mathbb{C}((1/x))$. This determines λ. Of course the choice of the α_i's was made in such a way that $\mathcal{H}\otimes\mathbb{C}((x)) \approx \mathcal{M}\otimes\mathbb{C}((x))$.

So it remains only to prove the following

Rigidity Theorem bis 3.7.3 Let $\mathcal{M}$ and $\mathcal{H}$ be irreducible D.E.'s on $\mathbb{G}_m$ such that

(1) $\chi(\mathbb{G}_m, \mathcal{M}) = \chi(\mathbb{G}_m, \mathcal{H}) = -1$.

(2) There exists an isomorphism $\mathcal{H}\otimes\mathbb{C}((x)) \approx \mathcal{M}\otimes\mathbb{C}((x))$, and $\mathcal{M}\otimes\mathbb{C}((x))$ has all slopes zero.

(3) There exists an isomorphism $\mathcal{H}\otimes\mathbb{C}((1/x)) \approx \mathcal{M}\otimes\mathbb{C}((1/x))$.

Denote by $\mathit{Hom}(\mathcal{H}, \mathcal{M})$ the internal hom D.E. $\mathcal{H}^*\otimes_{\mathcal{O}}\mathcal{M}$ on $\mathbb{G}_m$, and by $k: \mathbb{G}_m \to \mathbb{P}^1$ the inclusion. Then

(1) $\chi(\mathbb{P}^1, k_{!*}\mathit{Hom}(\mathcal{H}, \mathcal{M})) = 2 > 0$,

(2) There exists an isomorphism $\mathcal{H} \approx \mathcal{M}$.

proof To prove (1) $\Rightarrow$ (2), we argue exactly as in the proof of 3.5.4. The positivity of $\chi = h^0 + h^2 - h^1$ implies that at least one of H^0 or H^2 is nonzero. The $H^0{}_{DR}(\mathbb{P}^1, k_{!*}\mathit{Hom}(\mathcal{H}, \mathcal{M}))$ is $\mathrm{Hom}_{\mathcal{D}}(\mathcal{O}, k_{!*}\mathit{Hom}(\mathcal{H}, \mathcal{M})) = \mathrm{Hom}_{\mathcal{D}}(k_{!*}\mathcal{O}, k_{!*}\mathit{Hom}(\mathcal{H}, \mathcal{M}))$ = (by 2.9.1.3) $\mathrm{Hom}_{\mathcal{D}}(\mathcal{O}, \mathit{Hom}(\mathcal{H}, \mathcal{M})) = \mathrm{Hom}_{\mathcal{D}}(\mathcal{H}, \mathcal{M})$, any nonzero element of which is necessarily an isomorphism by irreducibility. Similarly, $H^2{}_{DR}(\mathbb{P}^1, k_{!*}\mathit{Hom}(\mathcal{H}, \mathcal{M}))$ is dual to $H^0{}_{DR}(\mathbb{P}^1, k_{!*}\mathit{Hom}(\mathcal{M}, \mathcal{H})) = \mathrm{Hom}_{\mathcal{D}}(\mathcal{M}, \mathcal{H})$.

So it remains to show that $\chi(\mathbb{P}^1, k_{!*}\mathit{Hom}(\mathcal{H}, \mathcal{M})) = 2$. By the Euler-Poicare formula, we have

$\chi(\mathbb{P}^1, k_{!*}\mathit{Hom}(\mathcal{H}, \mathcal{M})) = -\mathrm{Irr}_0 - \mathrm{Irr}_\infty + \dim_{\mathbb{C}}\mathrm{Soln}_0 + \dim_{\mathbb{C}}\mathrm{Soln}_\infty$.

We will prove that, denoting by n the rank of $\mathcal{M}$, by m the rank of its slope zero part at ∞, and by $d := n-m$, we have

(1) $\dim_{\mathbb{C}} \mathrm{Soln}_0 = n$.

(2) $\mathrm{Irr}_0 = 0$.

(3) $\dim_{\mathbb{C}} \mathrm{Soln}_\infty = m + 1$.

(4) $\mathrm{Irr}_\infty = d-1 + 2m$.

Since $n = d + m$, this will give $\chi(\mathbb{P}^1, k_{!*}\mathcal{H}om(\mathcal{H}, \mathcal{M})) = 2$.

First notice that solutions of any D.E. are the same as horizontal sections of its dual, so we have

$$\mathrm{Soln}_0 = \mathrm{Hom}_{\mathcal{D}}(\mathcal{M}\otimes\mathbb{C}((x)), \mathcal{H}\otimes\mathbb{C}((x))) \approx \mathrm{End}_{\mathcal{D}}(\mathcal{M}\otimes\mathbb{C}((x))),$$

$$\mathrm{Soln}_\infty = \mathrm{Hom}_{\mathcal{D}}(\mathcal{M}\otimes\mathbb{C}((1/x)), \mathcal{H}\otimes\mathbb{C}((1/x))) \approx \mathrm{End}_{\mathcal{D}}(\mathcal{M}\otimes\mathbb{C}((1/x))),$$

the final isomorphisms because by hypothesis we have $\mathcal{M}\otimes\mathbb{C}((x)) \approx \mathcal{H}\otimes\mathbb{C}((x))$ and $\mathcal{M}\otimes\mathbb{C}((1/x)) \approx \mathcal{H}\otimes\mathbb{C}((1/x))$. Similarly, the irregularities in question are those of $\mathcal{E}nd(\mathcal{M})\otimes\mathbb{C}((x))$ and of $\mathcal{E}nd(\mathcal{M})\otimes\mathbb{C}((1/x))$ respectively.

We have seen that there exist polynomials $P = P_n$ and $Q = Q_m$ such that

$$\mathcal{M}\otimes\mathbb{C}((x)) \approx \mathbb{C}((x))[D]/\mathbb{C}((x))[D]P(D),$$
$$\mathcal{M}\otimes\mathbb{C}((1/x)) \approx \mathbb{C}((1/x))[D]/\mathbb{C}((1/x))[D]Q(D) \oplus W,$$

and such that

$\gcd(P(t), P(t+k)) = 1$ and $\gcd(P(t), P(t+k)) = 1$ for any $k\neq 0$ in $\mathbb{Z}$ $\mathrm{End}_{\mathcal{D}}(\mathcal{M}\otimes\mathbb{C}((x)))$ is thus the kernel of $P(D)$ acting on the left $\mathcal{D}$-module $\mathbb{C}((x))[D]/\mathbb{C}((x))[D]P(D) \approx \mathbb{C}((x))\otimes_{\mathbb{C}}(\mathbb{C}[D]/\mathbb{C}[D]P(D))$, and the relative primality ($\gcd(P(t), P(t+k)) = 1$ for any $k\neq 0$ in $\mathbb{Z}$) shows that this kernel is the subspace $\mathbb{C}[D]/\mathbb{C}[D]P(D)$ of "constant terms". This proves (1). And (2) is obvious, because $\mathcal{M}$ and hence $\mathcal{E}nd\mathcal{M}$ have all 0-slopes zero.

To prove (3) and (4), notice that because W has all slopes > 0 and is irreducible, while $\mathbb{C}((1/x))[D]/\mathbb{C}((1/x))[D]Q(D)$ has all slopes zero, there are no nonzero homs between them in either direction, so we have $\mathrm{End}_{\mathcal{D}}(\mathcal{M}\otimes\mathbb{C}((1/x))) = \mathrm{End}_{\mathcal{D}}(\mathbb{C}((1/x))[D]/\mathbb{C}((1/x))[D]Q(D)) \oplus \mathrm{End}_{\mathcal{D}}(W)$. These two terms have dimensions m (proven just as above) and 1 (by the irreducibility of W) respectively, which proves (3). To prove (4), which only concerns slopes, it suffices to write $\mathcal{M}\otimes\mathbb{C}((1/x))$ as

(rank m, all slopes 0) $\oplus$ W, W of rank d, all slopes $1/d$.

Then $\mathcal{E}nd\mathcal{M}\otimes\mathbb{C}((1/x))$ is

(rank m^2, all slopes 0) $\oplus$ $\mathcal{E}nd W$ $\oplus$

$\oplus$ (rank m, all slopes 0)$\otimes W$ $\oplus$ $W^*\otimes$(rank m, all slopes 0).

The last two terms contribute $2md$ slopes $1/d$, so a total contribution of $2m$. So it remains to see that $\mathcal{E}nd W$ has d slopes 0 and $d^2 - d$ slopes

1/d. For this it suffices to check that $[d]^*(End W) \approx End([d]^*W)$ has d slopes 0 and $d^2 - d$ slopes 1. But this is obvious, since for any W of rank d with all slopes 1/d, by Levelt's structure theorem (cf [Ka-DGG],2.6.6]) there exist $\delta \in \mathbb{C}$ and $\gamma \in \mathbb{C}^\times$ such that

$$[d]^*W \approx \oplus_{\zeta \in \mu_d} (\text{the D.E. for } x^\delta e^{\zeta\gamma x}), \text{ and hence}$$

$$End([d]^*W) \approx \oplus_{\zeta,\eta \in \mu_d} (\text{the D.E. for } e^{(\zeta-\eta)\gamma x}).\ \text{QED}$$

Remarks 3.7.4
(1) The irreducible, **punctual** $\mathcal{D}$-modules on $\mathbb{G}_m$ are precisely the delta-modules $\delta_\lambda := \mathcal{D}/\mathcal{D}(\lambda - x)$, $\lambda \in \mathbb{C}^\times$; they also have $\chi(\mathbb{G}_m, \delta_\lambda) = -1$. The operators $\lambda - x$ are precisely the hypergeometric operators $\text{Hyp}_\lambda(\varnothing, \varnothing)$ of the excluded type (0,0).
(2) The proof given of 3.7.3 is just a $\mathcal{D}$-module translation of the topological proof of 3.5.4. Of course one can give a similar $\mathcal{D}$-module proof of the 3.5.4 itself.

As an application of this intrinsic characterization, we can also partially analyse the semisimplification of non-irreducible hypergeometrics.

Lemma 3.7.5 For any holonomic $\mathcal{D}$-module $\mathcal{M}$ on $\mathbb{G}_m$, $\chi(\mathbb{G}_m, \mathcal{M}) \le 0$, and $\chi(\mathbb{G}_m, \mathcal{M}) = 0$ if and only if $\mathcal{M}$ is a successive extension of the objects $x^\alpha \mathbb{C}[x, x^{-1}]$.
proof Any holonomic $\mathcal{D}$-module, having finite length, is a finite successive extension of irreducible holonomics. By the additivity of χ, we are reduced to the case where $\mathcal{M}$ is irreducible. If $\mathcal{M}$ is punctual, then it a delta-module with $\chi = -1$. If $\mathcal{M}$ is nonpunctual, then for any dense open set $j: U \to \mathbb{G}_m$ such that $j^*\mathcal{M}$ is a D.E. on U, $\mathcal{M} = j_{!*}j^*\mathcal{M}$, and the Euler-Poincare formula for $\chi(\mathbb{G}_m, \mathcal{M}) = \chi(\mathbb{G}_m, j_{!*}j^*\mathcal{M})$ is

$$\chi(\mathbb{G}_m, \mathcal{M}) =$$
$$= \text{rank}_{\mathcal{O}}(j^*\mathcal{M})\chi(\mathbb{G}_m) - \Sigma_{x \in \mathbb{P}^1 - \mathbb{G}_m} \text{Irr}_x(\mathcal{M}) - \Sigma_{\alpha \in \mathbb{G}_m - U} \text{totdrop}_\alpha.$$

Since $\chi(\mathbb{G}_m) = 0$, we see that $\chi(\mathbb{G}_m, \mathcal{M}) \le 0$. By 2.9.10 we have equality if and only if $\mathcal{M}$ is an irreducible D.E. on $\mathbb{G}_m$ which is regular singular at both 0 and ∞, in which case $\mathcal{M}$ is an $x^\alpha \mathbb{C}[x, x^{-1}]$ by 2.11.2.

QED

Corollary 3.7.5.1 Let $\mathcal{M}$ be a holonomic $\mathcal{D}$-module on $\mathbb{G}_m$ with

$\chi(\mathbb{G}_m, \mathcal{M}) = -1$.

If $\mathcal{M}$ is not irreducible, its semisimplification $\mathcal{M}^{ss}$ is the direct sum $\mathcal{N} \oplus \mathcal{T}$ where $\mathcal{N}$ is irreducible with $\chi(\mathbb{G}_m, \mathcal{N}) = -1$ and where $\mathcal{T}$ is a direct sum of $x^{\alpha}\mathbb{C}[x, x^{-1}]$'s.

Corollary 3.7.5.2 Let $\mathcal{H} := \mathcal{H}_\lambda(\alpha_i\text{'s}; \beta_j\text{'s})$ be a hypergeometric $\mathcal{D}$-module of type (n,m) which is not irreducible. Let $\gamma_1, \ldots, \gamma_r$ be the set with multiplicity of common exponents mod $\mathbb{Z}$ of the α's and the β's, i.e., the polynomial $\prod_{k=1,\ldots,r}(t - \exp(2\pi i\gamma_k))$ is the gcd of the two polynomials $\prod_{k=1,\ldots,n}(t - \exp(2\pi i\alpha_k))$ and $\prod_{k=1,\ldots,m}(t - \exp(2\pi i\beta_k))$. Remumber the α's and the β's so that $\alpha_i \equiv \beta_i \equiv \gamma_i$ mod $\mathbb{Z}$ for $i = 1, \ldots, r$.

Then for some $\mu \in \mathbb{C}^\times$, we have

$$\mathcal{H}^{ss} \approx \oplus x^{\gamma_i}\mathbb{C}[x, x^{-1}] \oplus \mathcal{H}_\mu(\alpha_{r+1}, \ldots, \alpha_n; \beta_{r+1}, \ldots, \beta_m).$$

Moreover, if $n = m$ then $\mu = \lambda$.

proof Apply the above corollary to $\mathcal{H}$. In view of the intrinsic characterization of irreducible hypergeometrics, the $\mathcal{N}$ is an irreducible hypergeometric. The exponents at 0 and ∞ of $\mathcal{N} \oplus \mathcal{T}$ must be those of $\mathcal{H}$, so we get the asserted forms for $\mathcal{N}$ and $\mathcal{T}$. If $n = m$, then $\mu = \lambda$ because it is the unique point of $\mathbb{G}_m$ where $\mathcal{H}$ or equivalently $\mathcal{H}^{ss}$ is not a D.E..QED

Another application is to analyse behavior under the d'th power map $[d]: \mathbb{G}_m \to \mathbb{G}_m$. The following lemma gives an intrinsic proof of the Kummer Induction Formula 3.5.6.1, but without specifying the multiplicative translates involved.

Lemma 3.7.6 Suppose $\mathcal{M}$ is an irreducible holonomic $\mathcal{D}$-module on $\mathbb{G}_m$, and suppose that $\chi(\mathbb{G}_m, \mathcal{M})$ is relatively prime to d. Then $[d]_*\mathcal{M}$ is irreducible on $\mathbb{G}_m$, and $\chi(\mathbb{G}_m, [d]_*\mathcal{M}) = \chi(\mathbb{G}_m, \mathcal{M})$.

proof Since [d] is finite etale, for any holonomic $\mathcal{D}$-module $\mathcal{M}$ on $\mathbb{G}_m$, we have

$$\chi(\mathbb{G}_m, [d]_*\mathcal{M}) = \chi(\mathbb{G}_m, \mathcal{M}).$$

We also have

$$\chi(\mathbb{G}_m, [d]^*\mathcal{M}) = d\chi(\mathbb{G}_m, \mathcal{M}),$$

say because $[d]_*[d]^*\mathcal{M} \approx \oplus_{i \bmod d} \mathcal{M}\otimes x^{i/d}$, and $\mathcal{M}\otimes x^{\alpha}$ has the same χ as $\mathcal{M}$ (2.11.1).

Now $[d]^*[d]_*\mathcal{M} \approx \oplus_{\zeta\in\mu_d} [x \mapsto \zeta x]^*\mathcal{M}$ is a sum of d pairwise nonisomorphic irreducibles. [For if $\mathcal{M} \approx [x \mapsto \zeta x]^*\mathcal{M}$ for some root of unity ζ of exact order $r \mid d$, $r > 1$, then $\mathcal{M}$, being irreducible, descends through the finite etale map $[r]$: $\mathbb{G}_m \to \mathbb{G}_m$, say $\mathcal{M} \approx [r]^*\mathcal{N}$, whence $\chi(\mathbb{G}_m, \mathcal{M}) = r\chi(\mathbb{G}_m, \mathcal{N})$ is divisible by r, contradiction.] Therefore the only subobjects $\mathcal{L}$ of $[d]^*[d]_*\mathcal{M}$ are the partial direct sums of the objects $[x \mapsto \zeta x]^*\mathcal{M}$, and of these only 0 and $[d]^*[d]_*\mathcal{M}$ are stable by $\mathcal{L} \mapsto [x \mapsto \zeta x]^*\mathcal{L}$ for every $\zeta \in \mu_d$. If $[d]_*\mathcal{M}$ has a proper nonzero subobject $\mathcal{K}$, then $[d]^*\mathcal{K}$ is a proper nonzero subobject $\mathcal{L}$ of $[d]^*[d]_*\mathcal{M}$ which is stable by $\mathcal{L} \mapsto [x \mapsto \zeta x]^*\mathcal{L}$ for every $\zeta \in \mu_d$, contradiction. QED

In the same vein, we have

Lemma 3.7.7 Suppose $\mathcal{M}$ is an irreducible holonomic $\mathcal{D}$-module on $\mathbb{G}_m$ with $\chi(\mathbb{G}_m, \mathcal{M}) \neq 0$, and that for some $\mu\in\mathbb{C}^\times$ there exists an isomorphism $\mathcal{M} \approx [x \mapsto \mu x]^*\mathcal{M}$. Then μ is a root of unity of order dividing $\chi(\mathbb{G}_m, \mathcal{M})$.

proof If μ is a root of unity, say of exact order r, then $\mathcal{M}$ descends through $[r]$ and hence r divides $\chi(\mathbb{G}_m, \mathcal{M})$. If μ is not a root of unity, then $\mathcal{M}$ has $\mathrm{Irr}_0 = 0 = \mathrm{Irr}_\infty$ (cf. [Ka-DGG, 2.3.8]), and consequently $\mathcal{M}$ must fail to be a D.E. at some points of $\mathbb{G}_m$ if it is to have $\chi(\mathbb{G}_m, \mathcal{M}) \neq 0$. So the set of its singularities is a finite nonempty subset of $\mathbb{G}_m$ which is stable by $x\mapsto \mu x$, whence μ must be a root of unity. QED

3.8 Direct Sums, Tensor Products, and Kummer Inductions

Lemma 3.8.1 Suppose that $\mathcal{H}$ and $\mathcal{H}'$ are irreducible nonpunctual hypergeometric $\mathcal{D}$-modules on $\mathbb{G}_m$, of types (n,m) and (n',m') respectively, whose generic ranks max(n,m) and max(n',m') are both ≥ 2. Suppose that there exists a dense open set $j: U \to \mathbb{G}_m$, a rank one

D.E. $\mathcal{L}$ on U, and an isomorphism of $j^*\mathcal{H} \approx j^*\mathcal{H}'\otimes\mathcal{L}$ in D.E.(U/$\mathbb{C}$). Then

(1) (n,m) = (n',m').
(2) If n = m, denoting by λ (resp. λ') the unique singularity of $\mathcal{H}$ (resp. $\mathcal{H}'$) in $\mathbb{G}_m$, we have $\lambda = \lambda'$.

(3) If (n,m) is not (2,1),or (1,2) or (2,2), then $\mathcal{L}$ is $x^\alpha\mathcal{O}_U$ for some $\alpha\in\mathbb{C}$, and $\mathcal{H} \approx \mathcal{H}'\otimes x^\alpha$ as $\mathcal{D}$-modules on $\mathbb{G}_m$.

proof Both $\mathcal{H}$ and $\mathcal{H}'$ have all their slopes at ∞ (resp. 0) ≤ 1. Since $\mathcal{L}$ is a direct factor of $\mathit{Hom}(j^*\mathcal{H}', j^*\mathcal{H})$, its slope at ∞ (resp. 0) is ≤ 1, so either 0 or 1.

Suppose that $\mathcal{L}$ has its ∞-slope 1. Either $\mathcal{H}'$ has all its ∞-slopes < 1, or n' - m' =1 and $\mathcal{H}'$ has one ∞-slope =1 and m' ∞-slopes =0. In the first case, $\mathcal{H}$ would have all its ∞-slopes =1, which is possible only if (n,m) is (1,0), a case excluded by hypothesis. In the second case, $\mathcal{H}$ has **at least** m' ∞-slopes =1. So m' $\le$ 1. But m' $\ge$ 1 because $\mathcal{H}'$ has generic rank $\ge$ 2, so m'=1, $\mathcal{H}$ has exactly m' ∞-slopes =1, and n' (=1+m')= 2. So $\mathcal{H}$ has type (2,1), since it has the same generic rank 2 as $\mathcal{H}'$ and has a single ∞-slope =1. Thus we conclude that (n,m)=(n',m')=(2,1) if $\mathcal{L}$ has its ∞-slope 1.

Similarly, if $\mathcal{L}$ has its 0-slope =1, then (n,m)=(n',m')=(1,2).

If $\mathcal{L}$ has slope =0 at both 0 and ∞, then $\mathcal{H}$ and $\mathcal{H}'$ have the same slopes at both 0 and ∞, so they have the same types (n,m)=(n',m'). So (1) is proven in all cases.

If n$\neq$m, then both $\mathcal{H}$ and $\mathcal{H}'$ are D.E.'s on $\mathbb{G}_m$, so $j_{!*}\mathcal{L}$ is a D.E. on $\mathbb{G}_m$, being a direct factor of the D.E. $j_{!*}\mathit{Hom}(j^*\mathcal{H}', j^*\mathcal{H}) = \mathit{Hom}(\mathcal{H}', \mathcal{H})$. Therefore $j_{!*}\mathcal{L} \approx x^\alpha\mathbb{C}[x, x^{-1}]$ for some α. This proves (3) if n $\neq$ m.

If n=m, then $\mathcal{H}$ (resp. $\mathcal{H}'$) has a finite singularity at a unique point λ (resp. λ') in $\mathbb{G}_m$, and the local monodromy there is a pseudoreflection. We first show that $\lambda = \lambda'$. For if $\lambda \neq \lambda'$, the isomorphism

$$j^*\mathcal{H} \approx j^*\mathcal{H}'\otimes\mathcal{L}$$

shows that $j^*\mathcal{H}'\otimes\mathcal{L}$ has trivial monodromy at λ', which implies that $j^*\mathcal{H}'$ has scalar monodromy at λ'. But as the rank is $\ge$ 2, no pseudoreflection is scalar. This proves (2).

If n=m $\ge$ 3, then $\mathcal{L}$ must have trivial local monodromy at λ' (since the product tA of a nonzero scalar t with a pseudoreflection A in GL(n $\ge$ 3) has $\ge$ n-1 $\ge$ 2 eigenvalues t, so cannot be either trivial or a pseudoreflection unless t=1). So again we find that $j_{!*}\mathcal{L} \approx x^\alpha\mathbb{C}[x, x^{-1}]$

for some α. Therefore our isomorphism is $j^*\mathcal{H} \approx j^*(\mathcal{H}'\otimes x^{\alpha})$. Applying $j_{!*}$ yields $\mathcal{H} \approx \mathcal{H}'\otimes x^{\alpha}$, as required. QED

Proposition 3.8.2 Suppose that $\mathcal{H}_1, \ldots, \mathcal{H}_n$ are $n \geq 2$ irreducible nonpunctual hypergeometric $\mathcal{D}$-modules on $\mathbb{G}_m$, with $\mathcal{H}_i$ of rank $N_i \geq 2$. Suppose that
(1) if $N_i = 2$, $\mathcal{H}_i$ is of type (2,0) or (0,2).
(2)for each i, the differential galois group $G_i \subset GL(N_i)$ of $\mathcal{H}_i$ (restricted to some dense open U where it is a D.E) has $G_i^{0,der}$ one of the groups

$SL(N_i)$, any $N_i \geq 2$,
$Sp(N_i)$, any even $N_i \geq 4$,
$SO(N_i)$, $N_i = 7$ or any $N_i \geq 9$,
$SO(3)$, if $N_i = 3$ and no $N_j = 2$,
$SO(5)$, if $N_i = 5$ and no $N_j = 4$,
$SO(6)$, if $N_i = 6$ and no $N_j = 4$,
$G_2 \subset SO(7)$, if $N_i = 7$,
$Spin(7) \subset SO(8)$ if $N_i = 8$, and no $N_j = 7$

Suppose that for all $i \neq j$, and all $\alpha \in \mathbb{C}^\times$, there exist no isomorphisms from $\mathcal{H}_i\otimes x^{\alpha}$ to either $\mathcal{H}_j$ or to its adjoint $(\mathcal{H}_j)^*$. Then the differential galois group G of $\oplus\mathcal{H}_i$ has $G^{0,der} = \prod G_i^{0,der}$, and that of $\otimes\mathcal{H}_i$ has $G^{0,der}$ = the image of $\prod G_i^{0,der}$ in $\otimes \mathrm{std}_{n_i}$.

proof In view of the above Lemma, this is just the spelling out of the Goursat-Kolchin-Ribet Proposition 1.8.2. QED

Corollary 3.8.2.1 Let $\mathcal{H} := \mathcal{H}_\lambda(\alpha\text{'s}; \beta\text{'s})$ be an irreducible nonpunctual hypergeometric $\mathcal{D}$-module on $\mathbb{G}_m$ of rank $N \geq 2$. If $N = 2$, suppose that $\mathcal{H}$ is of type (2,0) or (0,2). Suppose that $\mathcal{H}$ is self-dual, and that G_{gal} (resp. $(G_{gal})^0$) is one of the groups G:

$Sp(N)$, if N even,
$SO(N)$, if $N \neq 4, 8$,
$G_2 \subset SO(7)$, if $N = 7$,
$Spin(7) \subset SO(8)$ if $N = 8$.

Let $d \geq 2$, and let $\mu_1, \ldots, \mu_d$ be d distinct elements of $\mathbb{C}^\times$. Then the direct sum

$$\oplus_i \mathcal{H}_{\lambda/\mu_i}(\alpha\text{'s};\ \beta\text{'s}) \ =\ \oplus_i [x \mapsto \mu_i x]^*\mathcal{H}$$

has G_{gal} (resp. $(G_{gal})^0$) the d-fold product group G^d.

proof Because $\mathcal{H}$ is self-dual, each $[x \mapsto \mu_i x]^*\mathcal{H}$ is self-dual, it suffices by 3.8.2 to check that there exist no isomorphisms

$$[x \mapsto \mu x]^*\mathcal{H} \approx \mathcal{H}\otimes x^\alpha$$

for any $\mu \neq 1$, and any α in $\mathbb{C}$. But $\mathcal{H}$ and $[x \mapsto \mu x]^*\mathcal{H}$ have the same exponents at zero (resp. ∞), so the map $x \mapsto x + \alpha$ must map the exponents at zero (resp. ∞) to themselves mod $\mathbb{Z}$. But then $\mathcal{H}\otimes x^\alpha \approx \mathcal{H}$, and so we obtain an isomorphism

$$[x \mapsto \mu x]^*\mathcal{H} \approx \mathcal{H}.$$

But for $\mu \neq 1$, no such isomorphism exists, thanks to 3.7.7. QED

In the case of Kummer induction, 3.8.2 gives:

Theorem 3.8.3 Let $\mathcal{H} := \mathcal{H}_\lambda(\alpha\text{'s};\ \beta\text{'s})$ be an irreducible nonpunctual hypergeometric $\mathcal{D}$-module on $\mathbb{G}_m$ of rank $N \geq 2$. If $N = 2$, suppose that $\mathcal{H}$ is of type (2,0) or (0,2). Suppose that $\mathcal{H}$ has $(G_{gal})^{0,\mathrm{der}}$ one of the groups G:

- $SL(N)$,
- $Sp(N)$, if N even,
- $SO(N)$, if $N \neq 4, 8$,
- $G_2 \subset SO(7)$, if $N = 7$,
- $Spin(7) \subset SO(8)$ if $N = 8$.

Fix an integer $d \geq 2$. Let $S \subset \mu_d(\mathbb{C})$ be a nonempty subset of $\mu_d(\mathbb{C})$ which is maximal among all nonempty subsets of $\mu_d(\mathbb{C})$ which satisfy the following condition:

whenever ζ_1 and ζ_2 are distinct elements of S, and $\delta \in \mathbb{C}$, there exists no isomorphism from $\mathcal{H}_{\lambda\zeta_1}(\alpha\text{'s};\ \beta\text{'s})\otimes x^\delta$ to either $\mathcal{H}_{\lambda\zeta_2}(\alpha\text{'s};\ \beta\text{'s})$ or to its adjoint $(\mathcal{H}_{\lambda\zeta_2}(\alpha\text{'s};\ \beta\text{'s}))^*$.

Then $[d]_*\mathcal{H}_\lambda(\alpha\text{'s};\ \beta\text{'s})$ has $(G_{gal})^{0,\mathrm{der}} \approx G^S$.

construction-proof For any D.E., pullback to a finite etale connected covering does not change $(G_{gal})^0$, nor a fortiori $(G_{gal})^{0,der}$. Now for any $\mathcal{D}$-module $\mathcal{M}$ on $\mathbb{G}_m$, we have

$$[d]^*[d]_*\mathcal{M} \approx \oplus_{\zeta \in \mu_d} [x \mapsto \zeta x]^*\mathcal{M}.$$

Applying this to $\mathcal{H}$, we find that the $(G_{gal})^{0,der}$ for $[d]_*\mathcal{H}$ is equal to that for

$$\oplus_{\zeta \in \mu_d} \mathcal{H}_{\lambda\zeta}(\alpha\text{'s}; \beta\text{'s}).$$

Since $(G_{gal})^{0,der}$ is insensitive to twisting any of the factors by rank one D.E.'s, the $(G_{gal})^{0,der}$ for $[d]_*\mathcal{H}$ is equal to that for

$$\oplus_{\zeta \in S} \mathcal{H}_{\lambda\zeta}(\alpha\text{'s}; \beta\text{'s}).$$

Now apply 3.8.2. QED

CHAPTER 4
Detailed Analysis of the Exceptional Cases

Detailed Analysis of the Exceptional Cases

(4.0) We now turn to a detailed discussion of the exceptional possibilities for the differential galois group G of an irreducible hypergeometric $\mathcal{D}$-module on $\mathbb{G}_m$ of type (n,m), $n \neq m$, which is not Kummer induced. Let N:=max(n,m). Recall that the exceptional possibilities for $G^{0,\mathrm{der}}$ can occur only for $|n-m|=6$, N=7,8 or 9:

N=7: the image of G_2 in its 7-dim'l irreducible representation

N=8: the image of Spin(7) in the 8-dim'l spin representation
the image of SL(3) in the adjoint representation
the image of SL(2)×SL(2)×SL(2) in std⊗std⊗std
the image of SL(2)×Sp(4) in std⊗std
the image of SL(2)×SL(4) in std⊗std

N=9: the image of SL(3)×SL(3) in std⊗std.

By inversion, we may and will assume n > m.

Proposition 4.0.1 (Ofer Gabber) For N=8, neither of the two groups
the image of SL(2)×Sp(4) in std⊗std
the image of SL(2)×SL(4) in std⊗std
occurs as $G^{0,\mathrm{der}}$ for a hypergeometric of type (8,2).

proof Suppose that $G^{0,\mathrm{der}}$ for a hypergeometric $\mathcal{H}$ of type (8,2) is one of the groups
the image of SL(2)×Sp(4) in std⊗std
the image of SL(2)×SL(4) in std⊗std.

By 1.8.5, the normalizer in GL(std⊗std) of this $G^{0,\mathrm{der}}$ is $\mathbb{G}_m G^{0,\mathrm{der}}$, so we have $G^{0,\mathrm{der}} \subset G \subset \mathbb{G}_m G^{0,\mathrm{der}}$. Therefore the conjugation-induced action of G on Lie($G^{0,\mathrm{der}}$) respects each of the two factors , and its action on Lie(SL(2)) defines a surjection of G onto SO(3). View this surjection as an irreducible three-dimensional representation ρ of G, and then view ρ as an irreducible rank three object V in the subcategory $\langle\mathcal{H}\rangle$ of D.E.($\mathbb{G}_m/\mathbb{C}$). Because V is in $\langle\mathcal{H}\rangle$, all its slopes at both 0 and ∞ are $\leq 1/6$. Since V has rank three, both Irr_0 and Irr_∞ are $\leq 3/6 < 1$, and hence V is entirely of slope zero at both 0 and ∞. But on $\mathbb{G}_m$ the only such irreducible D.E.'s are of rank one, contradiction. QED

4.1 The G_2 case

This case can arise only for hypergeometrics of type (7,1) or (1,7). By inversion, it suffices to treat the case (7,1). Notice that an irreducible hypergeometric of type (n,1) cannot be Kummer induced. Now $G_2 \subset SO(7)$, so if $G^{0,der}$ for $\mathcal{H}$ is G_2, then by 3.6.1 there exists a twist $\mathcal{H}\otimes x^{\delta}$ with $G \subset SO(7)$.

Lemma 4.1.1 Suppose $G \subset SO(7)$ and $G^{0,der}$ is G_2. Then G is G_2.
proof Indeed, G_2 is its own normalizer in SO(7) [every automorphism of G_2 is inner, and SO(7) contains no nontrivial scalars]. QED

Lemma 4.1.2 $\mathcal{H}_\lambda(\alpha_1, ..., \alpha_7; \beta_1)$ is irreducible with $G_{gal} \subset SO(7)$ if and only if there exist x,y,z $\in \mathbb{C}$ such that after renumbering we have $(\alpha_1, ..., \alpha_7) \equiv (0, x, -x, y, -y, z, -z) \bmod \mathbb{Z}^7$, and $\beta_1 \equiv 1/2 \bmod \mathbb{Z}$, and none of x,y,or z is $\equiv 1/2 \bmod \mathbb{Z}$.
proof First of all, such an equation $\mathcal{H}_\lambda(0, x, -x, y, -y, z, -z; 1/2)$ is irreducible, self dual and of determinant one. If $\mathcal{H}_\lambda(\alpha_1, ..., \alpha_7; \beta_1)$ is self dual, then β_1 must be 0 or 1/2 mod $\mathbb{Z}$. If G_{gal} lies in SO(7), then its local monodromy at zero must lie in SO(7). But the eigenvalues of any element of SO(7) are of the form $(1, a, a^{-1}, b, b^{-1}, c, c^{-1})$ for some a,b,c in $\mathbb{C}^\times$. As the eigenvalues of local monodromy at zero are the $\exp(2\pi i\alpha_j)$'s, we get the existence of x,y,z $\in \mathbb{C}$ such that after renumbering we have $(\alpha_1, ..., \alpha_7) \equiv (0, x, -x, y, -y, z, -z) \bmod \mathbb{Z}^7$. Since $\mathcal{H}$ is irreducible, β_1 cannot be 0 mod $\mathbb{Z}$, so $\beta_1 \equiv 1/2 \bmod \mathbb{Z}$. Irreducibility now insures that none of x,y,z can be $\equiv 1/2 \bmod \mathbb{Z}$. QED

Lemma 4.1.3 If $\mathcal{H}_\lambda(\alpha_1, ..., \alpha_7; \beta_1)$ is irreducible with $G_{gal} \subset G_2 \subset SO(7)$, then there exist x,y $\in \mathbb{C}$ such that after renumbering we have $(\alpha_1, ..., \alpha_7) \equiv (0, x, -x, y, -y, x+y, -x-y) \bmod \mathbb{Z}^7$, and $\beta_1 \equiv 1/2 \bmod \mathbb{Z}$, and none of x,y,or x+y is $\equiv 1/2 \bmod \mathbb{Z}$.

proof The eigenvalues of any element of G_2 in its seven-dimensional representation are of the form $(1, a, a^{-1}, b, b^{-1}, ab, (ab)^{-1})$ for some a,b in $\mathbb{C}^\times$. Proceed as above. QED

In view of the above lemmas, the problem of recognizing which hypergeometrics have $G^{0,\mathrm{der}} = G_2$ is sompletely solved by

Theorem 4.1.4 Let $x,y \in \mathbb{C}$ such none of x,y,or x+y is $\equiv 1/2 \bmod \mathbb{Z}$. For any $\lambda \in \mathbb{C}^\times$, $\mathcal{H} := \mathcal{H}_\lambda(0, x, -x, y, -y, x+y, -x-y; 1/2)$ has $G_{\mathrm{gal}} = G_2$.

In view of the preceding Lemmas, this implies

G_2 Recognition Theorem 4.1.5 Let $x,y \in \mathbb{C}$ such none of x,y,or x+y is $\equiv 1/2 \bmod \mathbb{Z}$. Then for any $\lambda \in \mathbb{C}^\times$, $\mathcal{H} := \mathcal{H}_\lambda(0, x, -x, y, -y, x+y, -x-y; 1/2)$ has $G_{\mathrm{gal}} = G_2$. These are all the hypergeometric of type (7,1) with $G_{\mathrm{gal}} = G_2$. The hypergeometrics of type (7,1) with $G^{0,\mathrm{der}} = G_2$ are precisely the x^δ twists of these.

proof of 4.1.4.The only two possibilities for G_{gal} are SO(7) or its subgroup G_2. These two cases may be distinguished by the fact that for SO(7), $\Lambda^3(\mathrm{std}_7)$ is irreducible, while G_2 has a non-zero (in fact one-dimensional) space of invariants in $\Lambda^3(\mathrm{std}_7)$. Thus we must show that for $\mathcal{H}$ as above, $\Lambda^3(\mathcal{H})$ has $H^0{}_{DR}(\mathbb{G}_m, \Lambda^3(\mathcal{H}))$ nonzero. Here is a proof suggested by Ofer Gabber, analogous to the proof of 3.7.3.

Denote by $j: \mathbb{G}_m \to \mathbb{P}^1$ the inclusion. Since $\Lambda^3(\mathcal{H})$ is self dual (because $\mathcal{H}$ is), its middle extension $j_{!*}\Lambda^3(\mathcal{H})$ is also self dual. By global duality, the two cohomology groups $H^i{}_{DR}(\mathbb{P}^1, j_{!*}\Lambda^3(\mathcal{H}))$ for i=0 and i=2 are dual to each other. By 2.9.1.3,

$$H^0{}_{DR}(\mathbb{P}^1, j_{!*}\Lambda^3(\mathcal{H})) = \mathrm{Hom}_{\mathcal{D}}(\mathcal{O}_{\mathbb{P}^1}, j_{!*}\Lambda^3(\mathcal{H})) =$$

$$= \mathrm{Hom}_{\mathcal{D}}(j_{!*}\mathcal{O}_{\mathbb{G}_m}, j_{!*}\Lambda^3(\mathcal{H})) = \mathrm{Hom}_{\mathcal{D}}(\mathcal{O}_{\mathbb{G}_m}, \Lambda^3(\mathcal{H})) = H^0{}_{DR}(\mathbb{G}_m, \Lambda^3(\mathcal{H})).$$

Therefore the nonvanishing of $H^0{}_{DR}(\mathbb{G}_m, \Lambda^3(\mathcal{H}))$ will result from the estimate

$$\chi(\mathbb{P}^1, j_{!*}\Lambda^3(\mathcal{H})) \geq 2 > 0.$$

By the Euler-Poicare formula, we have

$\chi(\mathbb{P}^1, j_{!*}\Lambda^3(\mathcal{H})) = -\mathrm{Irr}_0 - \mathrm{Irr}_\infty + \dim_{\mathbb{C}}\mathrm{Soln}_0 + \dim_{\mathbb{C}}\mathrm{Soln}_\infty$.

We will show that

(1) $\dim_{\mathbb{C}}\mathrm{Soln}_0 \geq 5$.

(2) $\mathrm{Irr}_0 = 0$.

(3) $\dim_{\mathbb{C}}\mathrm{Soln}_\infty = 2$.

(4) $\mathrm{Irr}_\infty = 5$.

In order to prove (1), let us denote by T the local monodromy of $\mathcal{H}$ around zero, and by P(T) its characteristic polynomial. We know that as $\mathbb{C}[T]$-module, $\mathcal{H}$ is $\mathbb{C}[T]/(P(T))$. In terms of the quantities

$a := \exp(2\pi ix)$, $b := \exp(2\pi iy)$,

the roots of P(T) are $(1, a, 1/a, b, 1/b, ab, 1/ab)$.
Since $\mathcal{H}$ is regular singular at zero, we have

$\dim_{\mathbb{C}}\mathrm{Soln}_0(\Lambda^3(\mathcal{H})) = \dim \mathrm{Ker}(T-1 \text{ acting on } \Lambda^3(\mathbb{C}[T]/(P(T)))$.

We claim that this dimension is ≥ 5. To see this, we will resort to a specialization argument to reduce to the case in which P has all distinct roots.

We first treat the case where P has all distinct roots. Then T is diagonalizable, say $T \approx \mathrm{Diag}(a_1, \ldots, a_7)$, hence $\Lambda^3(T)$ is diagonalizable with eigenvalues exactly all triple products $a_ia_ja_k$ with $i < j < k$. If we number the a_i so that they are $(1, a, 1/a, b, 1/b, ab, 1/ab)$, then the five triple products indexed by (1,2,3), (1,4,5), (1,6,7), (2,4,7), (3,5,6) are all 1, so $\dim\mathrm{Ker}(\Lambda^3(T) - 1) \geq 5$, as required.

In the general case, we argue as follows. Let us define, for indeterminates A, B , the polynomial

$P_{A,B}(T) := (T-1)(T-A)(T-1/A)(T-B)(T-1/B)(T-AB)(T-1/AB)$.

Then over the ring $R := \mathbb{C}[A, B][1/AB]$, we can form the R[T]-module $M := R[T]/(P_{A,B}(T))$, which is free of rank seven over R. The general case results immediately from the following elementary lemma, applied to $S := \mathrm{Spec}(R)$, $\mathcal{M} := \Lambda^3(M)$, $\mathcal{T} := \Lambda^3(T) - 1$.

Specialization Lemma 4.1.6 Let S be a scheme, $\mathcal{M}$ an $\mathcal{O}_S$-module which is locally free of finite rank n, and $\mathcal{T} \in \mathrm{End}_{\mathcal{O}_S}(\mathcal{M})$. For each point s in S, consider the induced endomorphism $\mathcal{T}_s$ of the n-dimensional $\kappa(s)$-vector space $\mathcal{M}_s$. For any integer $i \geq 1$, the set

$\{s \text{ in } S \text{ where } \dim\mathrm{Ker}(\mathcal{T}_s) \geq i\}$

is Zariski closed in S.

proof Since $\dim\mathrm{Ker}(\mathcal{T}_s) + \dim\mathrm{Im}(\mathcal{T}_s) = n$, this is also the set where $\dim\mathrm{Im}(\mathcal{T}_s) \leq n-i$. But $\dim\mathrm{Im}(\mathcal{T}_s) \leq n-i \Leftrightarrow \Lambda^{1+n-i}(\mathcal{T}_s) = 0$. Thus our set is the locus of vanishing of all minors of $\mathcal{T}$ of a given size. QED

This concludes the proof of (1). Since $\mathcal{H}$ is regular singular at zero, (2) is obvious. We now turn to the proofs of (3) and (4), both of which are tedious but straightforward. Let us denote by W the six-dimensional wild part of $\mathcal{H}\otimes\mathbb{C}((1/x))$. Since $\rho_1 = 1/2$,

$$\mathcal{H}\otimes\mathbb{C}((1/x)) \approx W \oplus x^{1/2}\mathbb{C}((1/x)), \text{ whence}$$
$$\Lambda^3(\mathcal{H}\otimes\mathbb{C}((1/x))) \approx \Lambda^3(W) \oplus \Lambda^2(W)\otimes x^{1/2}.$$

To prove (3) and (4), it then suffices to prove (a) and (b) below:

(a) $\Lambda^3(W)$ has a 1-dim'l solution space, and irregularity 3.

(b) $\Lambda^2(W)\otimes x^{1/2}$ has a 1-dim'l solution space, and irregularity 2.

Since $\mathcal{H}$ has trivial determinant, we see that $\det(W) \approx x^{1/2}\mathbb{C}((1/x))$. Denoting by $\mathcal{L}$ the rank one D.E. for e^x, it follows from 3.4.1.1 that W is a multiplicative translate of $[6]_*\mathcal{L}$. Since the assertions (a) and (b) are invariant under multiplicative translation, we may assume that

$$W \approx [6]_*\mathcal{L}.$$

(4.1.7) We now explain how to analyse the exterior powers of such a Kummer-induced W. It will be clearer if we consider a slightly more general situation. Fix a $\mathbb{C}$-valued fibre functor ω on D.E. $(\mathbb{C}((1/x))/\mathbb{C})$. For any polynomial f(x) in $\mathbb{C}[x]$, define

$\mathcal{L}_{f(x)} := e^{f(x)}\mathbb{C}((1/x))$ = the rank one D.E. for $e^{f(x)}$ over $\mathbb{C}((1/x))$,

and denote by

$L_{f(x)}$:= the one-dimensional $\mathbb{C}$-space $\omega(\mathcal{L}_{f(x)})$.

$\psi_{f(x)}$:= the corresponding character of I_∞.

In order to describe $[d]_*(\mathcal{L}_{f(x)})$, it is equivalent via descent theory to describe $[d]^*[d]_*(\mathcal{L}_{f(x)})$ with its canonical action of the covering group μ_d. Using the canonical isomorphism of functors

$$[d]^*[d]_*(?) \approx \oplus_{\zeta\in\mu_d} [x \mapsto \zeta x]^*(?),$$

this amounts to making explicit the the natural action of the group μ_d on the d-dimensional representation space

$$W(d, f(x)) := \oplus_{\zeta\in\mu_d} L_{f(\zeta x)} \text{ of } I_\infty.$$

Clearly an element $\mu \in \mu_d$ maps $L_{f(\zeta x)}$ to $L_{f(\mu\zeta x)}$, and μ^d induces the identity. So there exists an eigenbasis $\{e_\zeta \in L_{f(\zeta x)}\}$ of this representation space W(d, f(x)),

$$\gamma(e_\zeta) = \psi_{f(\zeta x)}(\gamma)e_\zeta \text{ for } \gamma \in I_\infty,$$

on which $\mu \in \mu_d$ acts by

$$[\mu](e_\zeta) := e_{\mu\zeta}.$$

Thus $W(d, f(x))$ with its action of μ_d corresponds via descent theory to $[d]_*(L_{f(x)})$. For any "construction of linear algebra" Constr, $Constr(W(d, f(x)))$ carries an induced μ_d action, and it corresponds via descent to $Constr([d]_*(L_{f(x)}))$. In order to avoid confusion, we will denote by $I_\infty(d) \subset I_\infty$ the two inertia groups in question, and by $P_\infty(d) = P_\infty$ their (common) wild inertia subgroup.

Lemma 4.1.7.1 In terms of the descent dictionary, for any construction of linear algebra Constr we have

(1) $Constr([d]_*(L_{f(x)}))^{P_\infty} = Constr(W(d, f(x)))^{P_\infty(d)}$

(2) $Constr(W(d, f(x)))^{P_\infty(d)} = Constr(W(d, f(x)))^{I_\infty(d)}$, and the action of I_∞ on $Constr([d]_*(L_{f(x)}))^{P_\infty}$ factors through its μ_d quotient.

(3) If χ is a character of I_∞ which is trivial on P_∞ but which does not factor through the μ_d quotient, then

$$(Constr([d]_*(L_{f(x)}))\otimes\chi)^{I_\infty} = 0.$$

(4) If χ is a character of I_∞ which factors through the μ_d quotient, then

$$(Constr([d]_*(L_{f(x)}))\otimes\chi)^{I_\infty} = ((\, Constr(W(d, f(x)))^{I_\infty(d)}\,)\otimes\chi)^{\mu_d}.$$

proof Assertion (1) is a tautology, since $P_\infty = P_\infty(d)$ is a subgroup of $I_\infty(d)$. For (2), the point is that $W(d, f(x))$ is the direct sum of the characters $\psi_{f(\zeta x)}$. For these characters on this form one has

$$\psi_{f(x)}\psi_{g(x)} = \psi_{f(x)+g(x)}.$$

Therefore any $Constr(W(d, f(x)))$ is a direct sum of characters of the form $\psi_{g(x)}$ where $g(x)$ is of the form $\Sigma f(\zeta_i x)$. Since characters of the form $\psi_{g(x)}$ satisfy

$$\psi_{g(x)} \text{ is trivial on } P_\infty(d) \Leftrightarrow \mathcal{L}_{g(x)} \text{ has slope zero} \Leftrightarrow$$
$$\Leftrightarrow g(x) \text{ is constant} \Leftrightarrow \psi_{g(x)} \text{ is trivial on } I_\infty(d),$$

we obtain (2). Assertions (3) and (4) then follow immediately. QED

We now turn to the explicit analysis of our $[6]_*\mathcal{L}$, which corresponds to W(6, x) with its μ_6 action. Pick a primitive sixth root of unity ζ, and denote by $\{e_i\}_{i \in \mathbb{Z}/6\mathbb{Z}}$ an eigenbasis of W(6, x) with $e_i \in L_{\zeta^i x}$, and the action of ζ given by $\zeta(e_i) = e_{i+1}$. We will analyse the exterior powers of W(6, x).

We already know that $\det(W) \approx x^{1/2}\mathbb{C}((1/x))$, so it must be the case that det(W(6, x)) is the unique character of order two of μ_6. We can see this directly, since ζ cyclically permutes the e_i, so maps $e_1\wedge e_2\wedge e_3\wedge e_4\wedge e_5\wedge e_6$ to $e_2\wedge e_3\wedge e_4\wedge e_5\wedge e_6\wedge e_1$. Since W is self dual, so are its exterior powers. Therefore if $0 \le i \le 6$, the wedge product pairing

$$\Lambda^i(W) \times \Lambda^{6-i}(W) \to \Lambda^6(W)$$

induces an isomorphism

$$\Lambda^{6-i}(W) \approx (\Lambda^i(W))\otimes x^{1/2}.$$

This cuts our work in half. However, for increased reliability we will not use it.

We now systematically list the P_∞-invariants among the wedge products of the e_i's in each Λ^j, and give the action of ζ on these invariants. A given wedge expression $e_{i_1}\wedge\ldots\wedge e_{i_j}$ with $1 \le i_1 < i_2 \ldots < i_j \le 6$ transforms under P_∞ by the character ψ_{ax} for $a =(\zeta)^{i_1} + \ldots + (\zeta)^{i_j}$, so it is is P_∞-invariant if and only if $(\zeta)^{i_1} + \ldots + (\zeta)^{i_j} = 0$; otherwise its irregularity is 1. We write $[i_1, \ldots, i_j]$ for $e_{i_1}\wedge\ldots\wedge e_{i_j}$:

i	basis of $(\Lambda^i(W(6, x)))^{P_\infty}$	action of ζ here, and its eigenvalues
1	none.	none
2	[1,4], [2,5], [3,6],	$\alpha\mapsto\beta\mapsto\gamma\mapsto-\alpha$, eigenvalues $-\mu_3$
3	[1,3,5], [2,4,6],	$\alpha\mapsto\beta\mapsto\alpha$, eigenvalues ± 1
4	[1,2,4,5], [2,3,5,6], [1,3,4,6],	$\alpha\mapsto\beta\mapsto-\gamma\mapsto\alpha$, eigenvalues μ_3.
5	none.	none

Thus we obtain the following table
(4.1.7.2)

i	$\mathrm{Irr}(\Lambda^i(W))$	$\dim((\Lambda^i(W))^{I_\infty})$	$\dim((\Lambda^i(W)\otimes x^{1/2})^{I_\infty})$
1	1	0	0
2	2	0	1
3	3	1	1
4	2	1	0
5	1	0	0

In particular, we see that (a) and (b) in the proof of 4.1.4 hold. This concludes the proof of the G_2 theorem 4.1.4. QED

4.2 The Spin(7), PSL(3) and SL(2)×SL(2)×SL(2) Cases

We now turn to the remaining possible exceptional values of $G^{0,\mathrm{der}}$ for hypergeometrics of rank eight. These can occur only for type (6,2) (or (2,6), by inversion). Both Spin(7) and PSL(3) are subgroups of SO(8) ⊂ O(8), while (the image of) SL(2)×SL(2)×SL(2) is a subgroup of Sp(8), so in virtue of 3.6.1, a x^{δ} twist reduces us to computing $G^{0,\mathrm{der}}$ for those $\mathcal{H}$'s with G ⊂ O(8) or G ⊂ Sp(8).

Our first observation is that if $\mathcal{H}$ has G ⊂ O(8), then the question of whether or not G ⊂ SO(8) (i.e., whether or not det$\mathcal{H}$ is trivial) is invariant under twisting $\mathcal{H}$ in such a way that it stays self dual. This a general fact about even orthogonal groups.

Lemma 4.2.1 Suppose V is a symmetrically autodual Lie-irreducible D.E. (on any X/$\mathbb{C}$) of rank n: $G_{\mathrm{gal}}(V) \subset O(n)$. Let L be any rank one D.E. such that V⊗L is autodual. Then $L^{\otimes 2}$ is trivial, and $G_{\mathrm{gal}}(V\otimes L) \subset O(n)$. In particular, if n is even, det(V) ≈ det(V⊗L).

proof Since both V and V⊗L are self dual, their determinants have order 1 or 2, so L is of finite order. Denote by χ the character of π_1^{diff} given by L, and by ρ the representation given by V. Since V is Lie-irreducible, so is V⊗L, and they define the same (once we fix a basis of the line ω(L), so as to be able to identify ω(V) with ω(v⊗L)) representation of the open subgroup Ker(χ) of π_1^{diff}. By Lie-irreducibility, there is a single (up to a $\mathbb{C}^\times$ factor) nonzero bilinear form ⟨ , ⟩ on ω(V) = ω(v⊗L) which is invariant by this open subgroup. By unicity, $G_{\mathrm{gal}}(V) \subset SO(\omega(V), \langle\, ,\rangle)$, and $G_{\mathrm{gal}}(V\otimes L) \subset O(\omega(v\otimes L), \langle\, ,\rangle)$. So for any $\gamma \in \pi_1^{\mathrm{diff}}$, both ρ(γ) and χ(γ)ρ(γ) lie in O(ω(v⊗L), ⟨ , ⟩), whence χ(γ), being a scalar in O(ω(v⊗L), ⟨ , ⟩), is ±1. Therefore $L^{\otimes 2}$ is trivial. If n is even, $\det(V\otimes L)=\det(V)\otimes L^{\otimes n} \approx \det(V)$. QED

The next two lemmas show that if G ⊂ O(8), then the Spin(7) case (resp. the PSL(3) case) is possible only for G ⊂ SO(8) (resp. G ⊄ SO(8)).

Lemma 4.2.2 if G ⊂ O(8) and $G^{0,\mathrm{der}}$ = Spin(7), then G = Spin(7) (and consequently G ⊂ SO(8)).

proof Spin(7) is its own normalizer in O(8). Indeed, every automorphism of Spin(7) is inner, and the only scalars in O(8), ± 1, also lie in the subgroup Spin(7), for instance because Spin(7) has a nontrivial center. QED

Lemma 4.2.3 If a hypergeometric $\mathcal{H}$ has $G \subset SO(8)$, then $G^{0,\mathrm{der}} \neq PSL(3)$.

proof The normalizer N of PSL(3) in SO(8) is $\pm$PSL(3). [Indeed, up to inner automorphisms, the only nontrivial automorphism of $\mathcal{SL}(3)$ (viewed as 3×3 matrices of trace zero) is the Cartan involution

$$C : X \mapsto -X^t.$$

So if we view O(8) as the orthogonal group of the Killing form on $\mathcal{SL}(3)$, the Cartan involution of $\mathcal{SL}(3)$ (now viewd inside Lie(O(8)) by the adjoint representation) is Ad(C). But det(C) = -1, so any element of N inducing an outer automorphism must have det = -1. As $N \subset SO(8)$ by its definition, every element of N induces an inner automorphism. And the only scalars in SO(8) are ± 1. We remark for later use that this same argument shows that the normalizer of PSL(3) in O(8) is the semidirect product $PSL(3) \ltimes \{\pm 1, \pm C\}$.]

Therefore if $G \subset SO(8)$ and $G^{0,\mathrm{der}} = PSL(3)$, then $G \subset \pm PSL(3)$. Projection onto the ± 1 is a character of G, so a rank one object of $\langle\mathcal{H}\rangle$, so an x^δ. So after an x^δ twist, we find an $\mathcal{H}$ with G =PSL(3). In virtue of the fact that one can lift projective representations of π_1^{diff} of an open, there exists a rank three D.E. V on $\mathbb{G}_m$ whose G_{gal} is SL(3) such that $End^0(V)$ is $\mathcal{H}$. Now the highest ∞-slope of $\mathcal{H}$ is 1/6. Since the adjoint representation of SL(3) has a finite kernel, it follows from the next lemma that the highest ∞-slope of V is also 1/6. Since V has rank three, this is impossible [the multiplicity of a slope is always a multiple of its exact denominator, (cf. [Ka-DGG],2.2.7.3)]. QED

Highest Slope Lemma 4.2.4 Let ω be a $\mathbb{C}$-valued fibre functor on D.E.($\mathbb{C}((1/x))/\mathbb{C}$), V a D.E. on $\mathbb{C}((1/x))$, $\rho: I_\infty \to GL(\omega(V))$ the corresponding representation. Suppose that G is a Zariski closed subgroup of $GL(\omega(V_\infty))$ such that $\rho(I_\infty) \subset G$. Let $\Lambda: G \to GL(d)$ be any representation of G with a finite kernel, say Γ, and denote by V_Λ the D.E. corresponding to the composite representation $\Lambda\circ\rho$ of I_∞. Then V and V_Λ have the same highest slope.

proof For any $x \geq 0$, V has all slopes $\leq x$ if and only if $\rho((I_\infty)^{(x+)}) = \{e\}$, and V_Λ has all slopes $\leq x$ if and only if $\rho((I_\infty)^{(x+)}) \subset \Gamma$ ([Ka-DGG], 2.5.3.6). Since $\rho((I_\infty)^{(x+)})$ is connected ([Ka-DGG], 2.6.4.2), these two conditions are equivalent. QED

Lemma 4.2.5 If $\mathcal{H}_\lambda(\alpha_1, ..., \alpha_8; \beta_1, \beta_2)$ is irreducible with $G_{gal} \subset SO(8)$, then there exist $x,y,z,w \in \mathbb{C}$ such that after renumbering we have

$$(\alpha_1, ..., \alpha_8) \equiv (x, -x, y, -y, z, -z, w, -w) \bmod \mathbb{Z}^8,$$

$$(\beta_1, \beta_2) \equiv (0, 1/2) \bmod \mathbb{Z},$$

and none of x,y,z,w is $\equiv 0$ or $1/2 \bmod \mathbb{Z}$.

proof This is an exercise in the Duality Recognition Theorem 3.4. Since the autoduality is symmetric, $\Sigma\alpha_i - \Sigma\beta_j \equiv 1/2 \bmod \mathbb{Z}$. Since $\det\mathcal{H}$ is trivial, $\Sigma\alpha_i \in \mathbb{Z}$, and hence $\beta_1+\beta_2 \equiv 1/2 \bmod \mathbb{Z}$. Since $\mathcal{H}$ is selfdual, while $\beta_1+\beta_2$ is not in $\mathbb{Z}$, we must have that $2\beta_1$ and $2\beta_2$ are in $\mathbb{Z}$. Therefore after renumbering $(\beta_1, \beta_2) \equiv (0, 1/2) \bmod \mathbb{Z}$. The eigenvalues of any element of SO(8) can be grouped into four pairs of inverses; looking at local monodromy at zero thus gives the existence of the x,y,z,w as asserted. None can be 0 or $1/2 \bmod \mathbb{Z}$ because of the assumed irreducibility of $\mathcal{H}$. QED

Lemma 4.2.6 If $\mathcal{H}_\lambda(\alpha_1, ..., \alpha_8; \beta_1, \beta_2)$ is irreducible with $G_{gal} \subset \mathrm{Spin}(7)$, then there exist $x,y,z \in \mathbb{C}$ such that after renumbering we have

$$(\alpha_1, ..., \alpha_8) \equiv (x, -x, y, -y, z, -z, x+y+z, -x-y-z) \bmod \mathbb{Z}^8,$$

$$(\beta_1, \beta_2) \equiv (0, 1/2) \bmod \mathbb{Z},$$

and none of x,y,z, or $x+y+z$ is $\equiv 0$ or $1/2 \bmod \mathbb{Z}$.

proof In the subroup Spin(7) of SO(8), the eigenvalues of any element are of the form $(a, 1/a, b, 1/b, c, 1/c, abc, 1/abc)$. Proceed as above. QED

Lemma 4.2.7 If $\mathcal{H}_\lambda(\alpha_1, ..., \alpha_8; \beta_1, \beta_2)$ is irreducible with $G_{gal} \subset O(8)$ and $\det\mathcal{H}$ nontrivial, then there exist $x,y,z,w \in \mathbb{C}$ such that after renumbering we have

$(\alpha_1, \ldots, \alpha_8) \equiv (0, 1/2, x, -x, y, -y, z, -z) \bmod \mathbb{Z}^8$,

$(\beta_1, \beta_2) \equiv (w, -w) \bmod \mathbb{Z}$,

and $w \not\equiv 0$ or $1/2 \bmod \mathbb{Z}$.

proof Since $\det\mathcal{H}$ is nontrivial, but $\mathcal{H}$ is selfdual, $\det\mathcal{H}$ has order two, so $\Sigma\alpha_i \equiv 1/2 \bmod \mathbb{Z}$. Since the autoduality is symmetric, we must have $\Sigma\beta_i$ in $\mathbb{Z}$. Autoduality forces the α_i to break into pairs of additive inverses mod $\mathbb{Z}$, and possibly a single $(0, 1/2) \bmod \mathbb{Z}$. But this $(0, 1/2)$ nod $\mathbb{Z}$ must occur, since $\Sigma\alpha_i \equiv 1/2 \bmod \mathbb{Z}$. QED

4.3 The PSL(3) Case: Detailed Analysis

Lemma 4.3.1 If $\mathcal{H}$ has $G := G_{gal} \subset O(8)$ and $G^{0,der} = PSL(3)$, then the quotient group $G/PSL(3)$ is cyclic of order two.

proof The normalizer N of PSL(3) in O(8) is the semidirect product $PSL(3) \ltimes \{\pm 1, \pm C\}$ of PSL(3) with the abelian (2,2) group $\{\pm 1\} \times \{1, C\}$. We have $PSL(3) \subset G \subset N$. Because we are on $\mathbb{G}_m$, whose topological π_1 is cyclic, any finite quotient group of G must be cyclic. Therefore $G/PSL(3)$ is a cyclic subgroup of the (2,2) group $N/PSL(3)$. Since $G \not\subset SO(8)$, this quotient is nontrivial, so it must be of order two. QED

Lemma 4.3.2 If $\mathcal{H} := \mathcal{H}_\lambda(\alpha_1, \ldots, \alpha_8; \beta_1, \beta_2)$ has $G := G_{gal} \subset O(8)$ and $G^{0,der} = PSL(3)$, there exist x in $\mathbb{C}$, $x \not\equiv \pm 1/3 \bmod \mathbb{Z}$, μ in $\mathbb{C}^\times$, and an isomorphism

$$[2]^*\mathcal{H}_\lambda(\alpha_1, \ldots, \alpha_8; \beta_1, \beta_2) \approx End^0(\mathcal{H}_\mu(x, -x, 0; \emptyset)).$$

This x, unique mod $\mathbb{Z}$ up to $x \mapsto -x$, is characterized by

$$\{2\alpha_1, 2\alpha_2, \ldots, 2\alpha_8\} = \{0, 0, x, x, -x, -x, 2x, -2x\}.$$

proof Projection of G onto $G/PSL(3)$ is a nontrivial character of order two of G, corresponding to the two-fold Kummer covering. Therefore $[2]^*\mathcal{H}$ has its $G_{gal} = PSL(3)$. Lifting this projective representation to an SL(3) representation, we get a rank three D.E. V on $\mathbb{G}_m$ with det(V) trivial and $End^0(V) \approx [2]^*\mathcal{H}$. By the Highest Slope Lemma 4.2.4, V has all its 0-slopes $=0$, and its highest ∞-slope is 1/3. Therefore **all** the ∞-slopes of V are 1/3. Consequently, V is I_∞-irreducible, and hence irreducible. By the intrinsic characterization of irreducible hypergeometrics, we see that V is a hypergeometric of type (3,0) with trivial determinant. Therefore $V \approx \mathcal{H}_\mu(x, y, -x-y; \emptyset)$ for some $x, y \in \mathbb{C}$,

$\mu \in \mathbb{C}^\times$.

Since $End^0(V) \approx [2]^*\mathcal{H}$, the isomorphism class of $End^0(V)$ is invariant under multiplicative translation $[x \mapsto -x]^*$. Let us define

$$W := [x \mapsto -x]^*(V).$$

Then V and W are two rank three D.E.'s, both with $G_{gal} = SL(3)$, and there exists an isomorphism of D.E.'s

$$End^0(V) \approx End^0(W).$$

Therefore there exists a rank one D.E. L on $\mathbb{G}_m$ such that

$$\text{either } V \otimes L \approx W \text{ or } V \otimes L \approx W^\vee.$$

This results from the (contrapositive of the) Goursat-Kolchin-Ribet Proposition 1.8.2, applied to $G_{gal}(V \oplus W, \omega)$ and its two representations $\omega(V)$ and $\omega(W)$, since the possibility that $G_{gal}(V \oplus W, \omega) = SL(\omega(V)) \times SL(\omega(W))$ is incompatible with the representations $End^0(\omega(V))$ and $End^0(\omega(V))$ of $G_{gal}(V \oplus W, \omega)$ being isomorphic. Since both V and W have G_{gal} in SL(3), $L^{\otimes 3}$ must be trivial. So L is x^δ for $\delta = 0$ or $\pm 1/3$.

We now analyze the cases. Recall that

$$V = \mathcal{H}_\mu(x, y, -x-y; \varnothing), \text{ whence}$$

$$W := [x \mapsto -x]^*(V) \approx \mathcal{H}_{-\mu}(x, y, -x-y; \varnothing), \text{ and hence}$$

$$W^\vee \approx \mathcal{H}_\mu(-x, -y, x+y; \varnothing)$$

If $V \otimes x^\delta \approx W$, then $\delta \not\equiv 0 \bmod \mathbb{Z}$ (if δ is in $\mathbb{Z}$, then $V \approx W$, whence $\mu = -\mu$ by 3.3, and this is impossible since μ is in $\mathbb{C}^\times$). So δ is $\pm 1/3 \bmod \mathbb{Z}$. Therefore the set $\{x, y, z\}$ is stable mod $\mathbb{Z}$ by $\alpha \mapsto \alpha + \delta$. So mod $\mathbb{Z}$ it contains all the δ-translates of say x, and as there are three of these it must be $\{x, x+\delta, x+2\delta\} = \{x, x+1/3, x+2/3\}$.mod $\mathbb{Z}$. As these must sum to 0 mod $\mathbb{Z}$, $3x \in \mathbb{Z}$, whence $V \approx \mathcal{H}_\mu(0, 1/3, 2/3; \varnothing)$, which is Kummer-induced, so not Lie-irreducible, contradicting $G_{gal}(V) = SL(3)$. So this case cannot arise.

What about the case $V \otimes x^\delta \approx W^\vee$? We may rewrite this

$$V \otimes x^\delta \approx W^\vee := ([x \mapsto -x]^*(V))^\vee \approx [x \mapsto -x]^*(V^\vee),$$

or, tensoring both sides by the translation-invariant $x^\delta \approx (x^{2\delta})^\vee$,

$$V \otimes x^{2\delta} \approx [x \mapsto -x]^*((V \otimes x^{2\delta})^\vee).$$

Replacing V by $V \otimes x^{2\delta}$, it suffices to find V's such that $V \approx W^\vee$ [since V and $V \otimes x^{2\delta}$ have the same End^0].

The condition $V \approx W^\vee$ is equivalent to

$$\{x, y, -x-y\} = \{-x, -y, x+y\} \bmod \mathbb{Z},$$

or that $\{x, y, -x-y\}$ is stable mod $\mathbb{Z}$ by negation. The set $\{x, y, -x-y\}$ mod $\mathbb{Z}$ is thus of the form $(x, -x, 0)$ for some x. To insure that this V is not Kummer induced, we need $x \not\equiv \pm 1/3 \bmod \mathbb{Z}$. Because $x \not\equiv \pm 1/3 \bmod \mathbb{Z}$, $\pm x \bmod \mathbb{Z}$ is uniqely determined (since the set $\{2\alpha_i\text{'s}\}$ is the set $\{0,0,\pm x,\pm x,\pm 2x\} \bmod \mathbb{Z}$, as follows from comparing local monodromies at zero). QED

Lemma 4.3.3 Let $\mathcal{H} := \mathcal{H}_\mu(x, -x, 0; \emptyset)$, with x in $\mathbb{C}$, $x \not\equiv \pm 1/3 \bmod \mathbb{Z}$, μ in $\mathbb{C}^\times$. Then$End^0(\mathcal{H})$ has $G_{gal} = PSL(3)$, and there exists a hypergeometric $\mathcal{H}_\lambda(\alpha_1, \ldots, \alpha_8; \beta_1, \beta_2)$ and an isomorphism

$$[2]^*\mathcal{H}_\lambda(\alpha_1, \ldots, \alpha_8; \beta_1, \beta_2) \approx End^0(\mathcal{H}).$$

Moreover, any such $\mathcal{H}_\lambda(\alpha_1, \ldots, \alpha_8; \beta_1, \beta_2)$ has nontrivial determinant and its $G_{gal} \subset O(8)$, with $(G_{gal})^0 = PSL(3)$.

proof Such an $\mathcal{H}$ has $G_{gal} = SL(3)$, being non-Kummer induced of type (3,0) with trivial determinant, so $End^0(\mathcal{H})$ has $G_{gal} = PSL(3)$. Such an $\mathcal{H}$ has $\mathcal{H}^\vee \approx [x \mapsto -x]^*\mathcal{H}$, so $End(\mathcal{H}) \approx \mathcal{H}\otimes\mathcal{H}^\vee \approx \mathcal{H}\otimes([x \mapsto -x]^*\mathcal{H})$ is isomorphic to its $[x \mapsto -x]^*$ transform. Now $End(\mathcal{H}) \approx End^0(\mathcal{H})\oplus(\text{triv})$ is a sum of two irreducibles of different ranks, so by Jordan Holder theory each is isomorphic to its $[x \mapsto -x]^*$ transform. Therefore $End^0(\mathcal{H})$ descends through the two-fold Kummer covering, so is of the form $[2]^*(W)$ for some (necessarily irreducible) D.E. W on $\mathbb{G}_m$ of rank eight, whose $(G_{gal})^0$ is PSL(3), and whose 0-slopes are all 0.

We now show that this W is hypergeometric. For this, it suffices, by the intrinsic characterization of hypergeometrics, that $\chi(\mathbb{G}_m, W) = -1$, or equivalently that $\chi(\mathbb{G}_m, End^0(\mathcal{H})) = -2$. Since $End^0(\mathcal{H})$ has 0-slopes all 0, we need to see that $End^0(\mathcal{H})$ has $Irr_\infty = 2$. To do this, we will completely analyze the I_∞-representation of $End^0(\mathcal{H})$. In virtue of 3.4.1.1, $\mathcal{H}$ as I_∞-representation is a multiplicative translate of $[3]_*\mathcal{L}$, where $\mathcal{L}$ is the rank one D.E. for e^x.

Lemma 4.3.4 Let $V := [3]_*\mathcal{L}$. Then as I_∞-representation, we have

$End^0(V) \approx$ (rank 6, slopes 1/3) $\oplus$ $x^{1/3}\mathbb{C}((1/x))$ $\oplus$ $x^{2/3}\mathbb{C}((1/x))$.

proof This is an easy computation using W(3, x) with its μ_3 action as in the discussion 4.1.7. In that approach, once we fix a primitive cube root of unity ζ, W(3,x) has an I_∞-eigenbasis $\{e_i\}_{i \bmod 3}$, ($\gamma(e_i) = \psi_{\zeta^i x}(\gamma)e_i$ for $\gamma \in I_\infty$) on which ζ acts by $e_i \mapsto e_{i+1}$. Similarly, the dual W(3, -x) has an I_∞-eigenbasis $\{f_i\}_{i \bmod 3}$, ($\gamma(f_i) = \psi_{-\zeta^i x}(\gamma)f_i$ for $\gamma \in I_\infty$) on which ζ acts by $f_i \mapsto f_{i+1}$. Now End(W(3, x)) is the tensor product W(3, x)$\otimes$W(3, -x), on which ζ acts by $e_i \otimes f_j \mapsto e_{i+1} \otimes f_{j+1}$. Among the basis vectors $e_i \otimes f_j$, only the $\{e_i \otimes f_i\}_{i \bmod 3}$ are P_∞ invariant; the others each have slope 1 (so slope 1/3 in $End^0(V)$). This gives the (rank 6, slopes 1/3) factor.

The three tame vectors $\{e_i \otimes f_i\}_{i \bmod 3}$ are cyclically permuted by ζ, so after removing the trivial character to come down from End to End^0, what remains are the two nontrivial characters of order three. QED

Thus End^0 has Irr_∞ = 2, and hence that there exists a hypergeometric $\mathcal{H}_\lambda(\alpha_1, ..., \alpha_8; \beta_1, \beta_2)$ and an isomorphism

$$[2]^*\mathcal{H}_\lambda(\alpha_1, ..., \alpha_8; \beta_1, \beta_2) \approx End^0(\mathcal{H}).$$

Consider the group G := $G_{gal}(\mathcal{H}_\lambda(\alpha_1, ..., \alpha_8; \beta_1, \beta_2))$. It contains PSL(3) with index dividing two. We have already seen that G $\neq$ PSL(3) (cf 4.3.1). Therefore the index is two. The normalizer in GL(8) of PSL(3) is PSL(3)$\ltimes$\{ $\mathbb{G}_m$, C\}. As C^2 = 1 and PSL(3) contains no nontrivial scalars, G must be $\pm$PSL(3) or PSL(3)$\ltimes$\{1, λC\} with $\lambda = \pm 1$. We cannot have G = $\pm$PSL(3) by 4.2.3. Thus G is PSL(3)$\ltimes$\{1, λC\}, $\lambda = \pm 1$, and this group lies in O(8) but not in SO(8). QED

Theorem 4.3.5 Suppose $\mathcal{H} := \mathcal{H}_\lambda(\alpha_1, ..., \alpha_8; \beta_1, \beta_2)$ has $G := G_{gal} \subset O(8)$ and det$\mathcal{H}$ nontrivial. Let x in $\mathbb{C}$, $x \not\equiv \pm 1/3 \bmod \mathbb{Z}$, μ in $\mathbb{C}^\times$. There exists an isomorphism

$$[2]^*\mathcal{H}_\lambda(\alpha_1, ..., \alpha_8; \beta_1, \beta_2) \approx End^0(\mathcal{H}_\mu(x, -x, 0; \emptyset))$$

if and only if mod $\mathbb{Z}$ **either**

$\{\alpha_1, ..., \alpha_8\}$ = \{0, 1/2, x/2, (1 + x)/2, -x/2, -(1 + x)/2, x, -x\} := A1

$\{\beta_1, \beta_2\}$ = \{1/3, 2/3\} := B1

or

$\{\alpha_1, ..., \alpha_8\}$ = \{0, 1/2, x/2, (1 + x)/2, -x/2, -(1 + x)/2, x + 1/2, -x- 1/2\}

:= A2

$\{\beta_1, \beta_2\} = \{1/6, 5/6\}$:= B2.

proof Suppose first that there exists an isomorphism

$$[2]^*\mathcal{H}_\lambda(\alpha_1, ..., \alpha_8; \beta_1, \beta_2) \approx \mathit{End}^0(\mathcal{H}_\mu(x, -x, 0; \varnothing)).$$

We will analyze separately the possibilities for the β's, and then for the α's. After doing this, we will figure out which pairs of possibilities can "go together".

Looking at the slope =0 part of $\mathit{End}^0(\mathcal{H}_\mu(x, -x, 0; \varnothing))$ at ∞, we see that $\{2\beta_1, 2\beta_2\} = \{1/3, 2/3\}$ mod $\mathbb{Z}$. As $\beta_1 + \beta_2 \in \mathbb{Z}$, the only possibilities for $\{\beta_1, \beta_2\}$ are $\{1/3, 2/3\}$ or $\{1/6, 5/6\}$.

We now turn to the α_i's. Suppose first that x is neither 0 nor 1/2 mod $\mathbb{Z}$, so that the local monodromy at 0 of $\mathcal{H}_\mu(x, -x, 0; \varnothing)$ is semisimple. Then that of $\mathit{End}^0(\mathcal{H}_\mu(x, -x, 0; \varnothing))$, and hence that of $\mathcal{H}_\lambda(\alpha_1, ..., \alpha_8; \beta_1, \beta_2)$ is also semisimple. Therefore there can be no repeats mod $\mathbb{Z}$ among the α_i's. The $2\alpha_i$'s mod $\mathbb{Z}$ are the exponents at 0 of $\mathit{End}^0(\mathcal{H}_\mu(x, -x, 0; \varnothing))$, namely $\{0,0, x, x, -x, -x, 2x, -2x\}$. Since the α_i's do not repeat mod $\mathbb{Z}$, they must be

$$\{0, 1/2, x/2, (1 + x)/2, -x/2, -(1 + x)/2, ?, ??\}.$$

Since $\Sigma\alpha_i \equiv 1/2$ mod $\mathbb{Z}$, the last two entries sum to zero, so are either $\pm x$ or $\pm(x + 1/2)$, as asserted.

If x=0, then, writing unip_n for an n-dimensional unipotent Jordan block, the local monodromy of $\mathcal{H}_\mu(x, -x, 0; \varnothing)$ at zero is unip_3, so that of End^0 is $\mathrm{unip}_5 \oplus \mathrm{unip}_3$. Since $\mathcal{H}_\lambda(\alpha_1, ..., \alpha_8; \beta_1, \beta_2)$ has at most one Jordan block for each slope =0 character at zero, its local monodromy at zero is either

$$(\mathrm{unip}_5) \oplus x^{1/2}\otimes(\mathrm{unip}_3) \quad \text{or} \quad x^{1/2}\otimes(\mathrm{unip}_5) \oplus (\mathrm{unip}_3),$$

as asserted.

If x = 1/2, then the local monodromy of $\mathcal{H}_\mu(x, -x, 0; \varnothing)$ at zero is

$$(\mathrm{unip}_1) \oplus x^{1/2}\otimes(\mathrm{unip}_2),$$

so that of End^0 is

$$(\mathrm{unip}_1) \oplus (\mathrm{unip}_3) \oplus x^{1/2}\otimes(\mathrm{unip}_2) \oplus x^{1/2}\otimes(\mathrm{unip}_2).$$

So the local monodromy of $\mathcal{H}_\lambda(\alpha_1, ..., \alpha_8; \beta_1, \beta_2)$ at zero, being self dual of nontrivial determinant, is either

$$(\text{unip}_1) \oplus x^{1/2}\otimes(\text{unip}_3) \oplus x^{1/4}\otimes(\text{unip}_2) \oplus x^{3/4}\otimes(\text{unip}_2)$$

or

$$x^{1/2}\otimes(\text{unip}_1) \oplus (\text{unip}_3) \oplus x^{1/4}\otimes(\text{unip}_2) \oplus x^{3/4}\otimes(\text{unip}_2),$$

as asserted.

The situation now is this. Fix x in $\mathbb{C}$, $x \not\equiv \pm 1/3 \bmod \mathbb{C}$, μ in $\mathbb{C}^\times$. We have proven that $End^0(\mathcal{H}_\mu(x, -x, 0; \varnothing))$ descends through [2], and that any descent is, for some λ in $\mathbb{C}^\times$, on the following list of four possibilities

$$\mathcal{H}_\lambda(A1, B1), \mathcal{H}_\lambda(A1, B2), \mathcal{H}_\lambda(A2, B1), \mathcal{H}_\lambda(A2, B2).$$

Now from the definition of the exponent sets A1, A2, B1, B2, we see that

$$\mathcal{H}_\lambda(A1, B1)\otimes x^{1/2} \approx \mathcal{H}_\lambda(A2, B2), \qquad \mathcal{H}_\lambda(A1, B2)\otimes x^{1/2} \approx \mathcal{H}_\lambda(A2, B1).$$

On the other hand, the only indeterminacy in descending through [2] any irreducible D.E. on $\mathbb{G}_m$ is the possibility of twisting by $x^{1/2}$ (cf 2.7.1). So the two descents of $End^0(\mathcal{H}_\mu(x, -x, 0; \varnothing))$ through [2] are either $\mathcal{H}_\lambda(A1, B1)$ and $\mathcal{H}_\lambda(A2, B2)$, which is precisely what we assert to be the case, or they are $\mathcal{H}_\lambda(A1, B2)$ and $\mathcal{H}_\lambda(A2, B1)$.

If $\mathcal{H}_\lambda(A1, B1)$ is not a descent of $End^0(\mathcal{H}_\mu(x, -x, 0; \varnothing))$ for any μ, then its $G := G_{gal}$ cannot have $G^{0,der} = PSL(3)$, thanks to Lemma 4.3.2. Since $\mathcal{H}_\lambda(A1, B1)$ is irreducible and not Kummer induced, with $G \subset O(8)$ and nontrivial determinant, the only other possibility is that $G = O(8)$. Similarly, if $\mathcal{H}_\lambda(A1, B2)$ is not a descent of $End^0(\mathcal{H}_\mu(x, -x, 0; \varnothing))$ for any μ, then its G_{gal} is O(8). Thus exactly one of $\mathcal{H}_\lambda(A1, B1)$, $\mathcal{H}_\lambda(A1, B2)$ has its $G_{gal} = O(8)$, and the others' G_{gal} has $G^0 = PSL(3)$. So the two cases are distinguished by the **dimensions** of their differential galois groups (dimO(8) = 28, dimPSL(3) = 8).

We will decide which one is which by using the specialization theorem. Fix λ, and regard x as a variable. In other words, denote by K the field $\mathbb{Q}(\lambda)$, and by R the polynomial ring K[x] in one variable x over K. Then on the scheme $(\mathbb{G}_m)_R$, both of our candidates

$$\mathcal{H}_\lambda(A1, B1) := \mathcal{H}_\lambda(0, 1/2, \pm x/2, \pm(1 + x)/2, \pm x; 1/3, 2/3),$$

$$\mathcal{H}_\lambda(A1, B2) := \mathcal{H}_\lambda(0, 1/2, \pm x/2, \pm(1 + x)/2, \pm x; 1/6, 5/6)$$

make sense as free $\mathcal{O}$-modules of rank eight endowed with integrable connections relative to the ground ring R.

In order to prove that $\mathcal{H}_\lambda(A1, B1)$ has $G^0 = PSL(3)$ for **every**

$x \not\equiv \pm 1/3 \bmod \mathbb{Z}$, it suffices to prove that $\mathcal{H}_\lambda(A1, B1)$ has $\dim G_{gal} \le 8$ for **every** x; by the specialization theorem it suffices to prove that $\mathcal{H}_\lambda(A1, B1)$ has $\dim G_{gal} \le 8$ at the generic point. This is equivalent to showing that $\mathcal{H}_\lambda(A1, B2)$ has $\dim G_{gal} > 8$ at the generic point. By the specialization theorem it suffices for this to find a **particular** x where $\mathcal{H}_\lambda(A1, B2)$ has $\dim G_{gal} > 8$. For this, we take $x = 1/3$. Then $\mathcal{H}_\lambda(A1, B2)|_{x=1/3}$ is the **reducible** hypergeometric

$$\mathcal{H}_\lambda(0, 1/2, \pm 1/6, \pm 2/3, \pm 1/3;\ 1/6, 5/6)$$

whose semisimplification is

$$x^{1/6}\mathbb{C}[x, x^{-1}] \ \oplus\ x^{5/6}\mathbb{C}[x, x^{-1}] \ \oplus\ \mathcal{H}_\lambda(0, 1/2, \pm 1/3, \pm 1/3;\ \emptyset).$$

Therefore the G_{gal} of $\mathcal{H}_\lambda(0, 1/2, \pm 1/3, \pm 1/3;\ \emptyset)$ is a quotient of that of $\mathcal{H}_\lambda(0, 1/2, \pm 1/6, \pm 2/3, \pm 1/3;\ 1/6, 5/6)$, hence has lower dimension. But $\mathcal{H}_\lambda(0, 1/2, \pm 1/3, \pm 1/3;\ \emptyset)$ has $G_{gal} = O(6)$ [being of type (6,0),not Kummer induced, and orthonally self dual with nontrivial determinant] which has dimension $15 > 8$. QED

Thus we find

PSL(3) Theorem 4.3.6 A hypergeometric $\mathcal{H} := \mathcal{H}_\lambda(\alpha_1, \ldots, \alpha_8;\ \beta_1, \beta_2)$ of type (8,2) has $G^{0,der} = PSL(3)$ if and only if there exists $\delta \in \mathbb{C}$ and $x \in \mathbb{C}$, $x \not\equiv \pm 1/3 \bmod \mathbb{Z}$,such that, mod $\mathbb{Z}$,we have

$$\{\alpha_i + \delta\} = \{0, 1/2, \pm x/2, \pm(1 + x)/2, \pm x\}, \qquad \{\beta_i + \delta\} = \{\pm 1/3\}.$$

4.4 The Spin(7) Case: Detailed Analysis

We have already seen (4.2.6) that if a hypergeometric of type (8,2) has $G^{0,der} = Spin(7)$, then a twist of it is (isomorphic to one) of the form $\mathcal{H}_\lambda(\pm x, \pm y, \pm z, \pm(x+y+z);\ 0, 1/2)$ for some $x, y, z \in \mathbb{C}$.

Spin(7) Theorem 4.4.1 Let $\lambda \in \mathbb{C}^\times$, $x, y, z \in \mathbb{C}$. If

$$\mathcal{H} := \mathcal{H}_\lambda(\pm x, \pm y, \pm z, \pm(x+y+z);\ 0, 1/2)$$

is irreducible and not Kummer induced of degree 2, then $G_{gal} = Spin(7)$.

proof This is very similar to the G_2 case. Since $\mathcal{H}$ is irreducible, not Kummer induced, and has $G_{gal} \subset SO(8)$, the only two possibilities for G_{gal} are Spin(7) or SO(8). We may distinguish these by the fact that $\Lambda^4(std_8)$ is SO(8)-irreducible, but has a one-dimensional space of

Spin(7) invariants. So it suffices to show that, denoting by $j: \mathbb{G}_m \to \mathbb{P}^1$ the inclusion, we have $\chi(\mathbb{P}^1, j_{!*}\Lambda^4(\mathcal{H})) \geq 2$. Since

$\chi(\mathbb{P}^1, j_{!*}\Lambda^4(\mathcal{H})) = -\mathrm{Irr}_0 - \mathrm{Irr}_\infty + \dim_{\mathbb{C}}\mathrm{Soln}_0 + \dim_{\mathbb{C}}\mathrm{Soln}_\infty$.

it suffices to show that we have

(1) $\dim_{\mathbb{C}}\mathrm{Soln}_0 \geq 8$.

(2) $\mathrm{Irr}_0 = 0$.

(3) $\dim_{\mathbb{C}}\mathrm{Soln}_\infty = 4$.

(4) $\mathrm{Irr}_\infty = 10$.

In order to prove (1), let us denote by T the local monodromy of $\mathcal{H}$ around zero, and by P(T) its characteristic polynomial. We know that as $\mathbb{C}[T]$-module, $\mathcal{H}$ is $\mathbb{C}[T]/(P(T))$. In terms of the quantities

$a := \exp(2\pi i x)$, $b := \exp(2\pi i y)$, $c := \exp(2\pi i z)$,

the roots of P(T) are (a, 1/a, b, 1/b, c, 1/c, abc, 1/abc).
Since $\mathcal{H}$ is regular singular at zero, we have

$\dim_{\mathbb{C}}\mathrm{Soln}_0(\Lambda^4(\mathcal{H})) = \dim \mathrm{Ker}(T-1$ acting on $\Lambda^4(\mathbb{C}[T]/(P(T))))$.

To show that this dimension is ≥ 8, it suffices by the specialization lemma 4.1.6 to treat the case in which P has all distinct roots.Then T is diagonalizable, say $T \approx \mathrm{Diag}(a_1, \ldots, a_8)$, hence $\Lambda^4(T)$ is diagonalizable with eigenvalues exactly all quadruple products $a_i a_j a_k a_n$ with $i < j < k < n$. If we number the a_i so that they are (a, 1/a, b, 1/b, c, 1/c, abc, 1/abc), then the eight quadruple products indexed by (1,2,3,4), (1,2,5,6), (1,2,7,8), (3,4,5,6), (3,4,7,8), (5,6,7,8), (1,3,5,8) and (2,4,6,7,) are all 1, so $\dim\mathrm{Ker}(\Lambda^4(T) - 1) \geq 8$, as required.

Since $\mathcal{H}$ is regular singular at zero, (2) is obvious. We now turn to the proofs of (3) and (4). Let us denote by W the six-dimensional wild part of $\mathcal{H}\otimes\mathbb{C}((1/x))$. Since (β_1, β_2) is (0, 1/2),

$$\mathcal{H}\otimes\mathbb{C}((1/x)) \approx W \oplus \mathbb{C}((1/x)) \oplus x^{1/2}\mathbb{C}((1/x)), \text{ whence}$$

$$\Lambda^4(\mathcal{H}\otimes\mathbb{C}((1/x))) \approx \Lambda^4(W) \oplus \Lambda^3(W) \oplus \Lambda^3(W)\otimes x^{1/2} \oplus \Lambda^2(W)\otimes x^{1/2}.$$

Since $\mathcal{H}$ has trivial determinant, we see that $\det(W) \approx x^{1/2}\mathbb{C}((1/x))$. Denoting by $\mathcal{L}$ the rank one D.E. for e^x, it follows from 3.4.1.1 that W is a multiplicative translate of $[6]_*\mathcal{L}$. Since the assertions (3) and (4) are invariant under multiplicative translation, we may assume that

$$W \approx [6]_*\mathcal{L}.$$

From the table 4.1.7.2, we read off that $\Lambda^4(\mathcal{H}\otimes\mathbb{C}((1/x)))$ has $\mathrm{Irr}_\infty = 10$

and $\dim \mathrm{Soln}_\infty = 4$, as required. QED

4.5 The SL(2)×SL(2)×SL(2) Case

We have already noted that if a hypergeometric $\mathcal{H}$ of type (8,2) has $G^{0,\mathrm{der}}$ = (the image of) SL(2)×SL(2)×SL(2), then a twist $\mathcal{H}\otimes x^\delta$ has its $G_{\mathrm{gal}} \subset \mathrm{Sp}(8)$, so $\mathcal{H}\otimes x^\delta$ is (isomorphic to one) of the form

$$\mathcal{H}_\lambda(\pm x, \pm y, \pm z, \pm t; \pm w)$$

for some x, y, z, t, w in $\mathbb{C}$.

We will denote by Γ the image of SL(2)×SL(2)×SL(2) in Sp(8), where Sp(8) is the symplectic group for $\mathrm{std}_2\otimes\mathrm{std}_2\otimes\mathrm{std}_2$ with the form $(x\otimes y\otimes z, u\otimes v\otimes w) := \langle x,u\rangle\langle y,v\rangle\langle z,w\rangle$. The action of the symmetric group S_3 on $\mathrm{std}_2\otimes\mathrm{std}_2\otimes\mathrm{std}_2$ visibly respects the symplectic form, so we may view S_3 as a subgroup of Sp(8). This S_3 normalizes Γ; as every outer automorphism of Γ is induced by the action of this S_3, we see that the normalizer of Γ in Sp(8) is $\Gamma\ltimes S_3$. Therefore if V is any rank eight D.E. with $G := G_{\mathrm{gal}} \subset \mathrm{Sp}(8)$ and with $G^{0,\mathrm{der}} = \Gamma$, then $\Gamma \subset G \subset \Gamma\ltimes S_3$.

Lemma 4.5.1 If a hypergeometric $\mathcal{H}$ of type (8,2) has $G := G_{\mathrm{gal}} \subset \mathrm{Sp}(8)$ and $G^{0,\mathrm{der}} = \Gamma$, then G is the subgroup $\Gamma\ltimes A_3$ of $\Gamma\ltimes S_3$, and there exist three rank two D.E.'s V_1, V_2, and V_3 on $\mathbb{G}_m$, each hypergeometric of type (2,0) with G_{gal} = SL(2), and an isomorphism

$$[3]^*\mathcal{H} \approx V_1\otimes V_2\otimes V_3.$$

proof The quotient G/Γ is cyclic, because we are on $\mathbb{G}_m$; as G/Γ is a subgroup of S_3, this quotient is either A_3, or it has order 1 or 2. So if it were not A_3, then $[2]^*\mathcal{H}$ has $G_{\mathrm{gal}} = \Gamma$. Viewing Γ as the quotient of SL(2)×SL(2)×SL(2) by the subgroup of $\{\pm1\}\times\{\pm1\}\times\{\pm1\}$ of triples with product 1, we can (by 2.2.2.1) lift the homomorphism $\pi_1{}^{\mathrm{diff}} \to \Gamma$ corresponding to $[2]^*\mathcal{H}$ to a homomorphism $\pi_1{}^{\mathrm{diff}} \to \mathrm{SL}(2)\times\mathrm{SL}(2)\times\mathrm{SL}(2)$. Thus there exist three rank two D.E.'s V_1, V_2, and V_3 on $\mathbb{G}_m$, all with G_{gal} = SL(2), and an isomorphism

$$[2]^*\mathcal{H} \approx V_1\otimes V_2\otimes V_3.$$

By the Highest Slope Lemma 4.2.4, the highest ∞-slope of $V_1\oplus V_2\oplus V_3$ is the same as that of $V_1\otimes V_2\otimes V_3 \approx [2]^*\mathcal{H}$, namely 1/3. But as V_i is of rank two, it cannot have any ∞-slope 1/3. Therefore the quotient G/Γ must be A_3.

Once G/Γ has order 3, $[3]^*\mathcal{H}$ has $G_{gal} = \Gamma$, and we repeat the lifting argument to produce three rank two D.E.'s V_1, V_2, and V_3 on $\mathbb{G}_m$, all with $G_{gal} = SL(2)$, and an isomorphism

$$[3]^*\mathcal{H} \approx V_1 \otimes V_2 \otimes V_3.$$

. By the Highest Slope Lemma 4.2.4, each V_i has its 0-slopes 0, and its ∞-slopes $\leq 1/2$. So the ∞-slopes of a given V_i are either both 1/2 or they are both 0. They cannot both be zero, for then V_i would be regular singular at both 0 and ∞, so reducible, contradicting the fact that its G_{gal} is SL(2). By the intrinsic characterization of irreducible hypergeometrics, each V_i is hypergeometric of type (2, 0). QED

Lemma 4.5.2 Suppose V_1, V_2, and V_3 are three hypergeometrics of type (2,0) on $\mathbb{G}_m$, such that $V_1 \oplus V_2 \oplus V_3$ has $G_{gal} = SL(2)\times SL(2)\times SL(2)$, and such that for some D.E. V on $\mathbb{G}_m$ there exists an isomorphism

$$[3]^*V \approx V_1 \otimes V_2 \otimes V_3.$$

Then

(1) V is hypergeometric of type (8,2), and there exists a twist $V\otimes x^\delta$ of V with $\delta = 0$ or $\pm 1/3$ such that after replacing $V \mapsto V\otimes x^\delta$ (which doesn't change $[3]^*V$), $G_{gal(V)} = \Gamma \ltimes A_3$.

(2)Fix a primitive cube root of unity ζ. After possibly renumbering the V_i and replacing certain of them by their $x^{1/2}$ twists, there exist isomorphisms $[x \mapsto \zeta x]^* V_i \approx V_{i+1}$, where the index i is read mod 3.

proof The G_{gal} of any such V obviously has $G^0 = \Gamma$, and G/G^0 is cyclic of order 1 or 3. In particular, V is irreducible. By 4.2.4, such a V is entirely of slope 0 at 0, and its highest ∞-slope is 1/6. Therefore $\chi(\mathbb{G}_m, V) = -1$, and so V is hypergeometric by the intrinsic characterization; by its slopes, it must be of type (8,2). By the previous lemma, some twist of V has its $G_{gal} = \Gamma \ltimes A_3 \subset Sp(8)$, so $G := G_{gal}(V) \subset \mathbb{G}_m Sp(8)$. The subgroup G^0 is $\Gamma \subset Sp(8)$, so $G \subset \mu_3 Sp(8)$. Twisting V by an x^δ, $\delta = 0$ or $\pm 1/3$ puts $G \subset Sp(8)$, whence G is $\Gamma \ltimes A_3$ by the previous lemma.

Let us denote by $T := [x \mapsto \zeta x]$, the multiplicative translation by the chosen primitive cube root of unity ζ. Since $V_1 \otimes V_2 \otimes V_3$ descends through [3], its isomorphism class is invariant under T^*. Therefore

$$V_1 \otimes V_2 \otimes V_3 \otimes T^*(V_1) \otimes T^*(V_2) \otimes T^*(V_3)$$

has G_{gal} a quotient of Γ. Since both $V_1 \oplus V_2 \oplus V_3$ and $T^*(V_1) \oplus T^*(V_2) \oplus T^*(V_3)$ have $G_{gal} = SL(2) \times SL(2) \times SL(2)$, it follows by the contrapositive of Goursat-Kolchin-Ribet 1.8.2 that for $1 \le i \le 3$ there exists a rank one D.E. L on $\mathbb{G}_m$, an index $1 \le j \le 3$, and an isomorphism

$$T^*(V_i) \approx V_j \otimes L.$$

Since both sides have trivial determinant, we see that $L^{\otimes 2}$ is trivial (i.e., L is $x^\delta \mathbb{C}[x, x^{-1}]$ for $\delta = 0$ or $1/2$). We claim that $i \neq j$. For if $i = j$, then $T^*(V_i) \approx V_i \otimes L$, and applying T^* again we find $T^*T^*(V_i) \approx T^*(V_i \otimes L) \approx V_i \otimes L \otimes L \approx V_i$. But as V_i is irreducible with $\chi = -1$, this is impossible by 3.7.7.

Apply this with $i = 1$, and renumber so the corresponding $j = 2$. Then replace V_2 by $V_2 \otimes L$, and we find

$$T^*(V_1) \approx V_2$$

Now apply this to $i = 2$. We find

$$\text{either } T^*(V_2) \approx V_1 \otimes L \text{ or } T^*(V_2) \approx V_3 \otimes L.$$

In the first case, this leads to $T^*T^*(V_1) \approx T^*(V_2) \approx V_1 \otimes L$, which is impossible as above. Therefore $T^*(V_2) \approx V_3 \otimes L$, so replacing V_3 by $V_3 \otimes L$ we now find

$$T^*(V_1) \approx V_2,\ T^*(V_2) \approx V_3.$$

From this and the fact that T is of order 3 we see that $T^*(V_3) \approx T^*T^*(V_2) \approx T^*T^*T^*(V_1) \approx V_1$, as asserted. QED

Theorem 4.5.3 Let V be a hypergeometric of type (2,0) with $G_{gal} = SL(2)$, i.e., V is (isomorphic to) $\mathcal{H}_\lambda(x, -x; \varnothing)$ for some x in $\mathbb{C}$, $x \not\equiv \pm 1/4 \bmod \mathbb{Z}$. Let ζ be a primitive cube root of unity, $T := [x \mapsto \zeta x]$. Then

(1) $V \oplus T^*V \oplus T^*T^*V$ has $G_{gal} = SL(2) \times SL(2) \times SL(2)$.

(2) For some $\mu \in \mathbb{C}^\times$, there exists an isomorphism

$$V\otimes T^*V\otimes T^*T^*V \approx [3]^*\mathcal{H}_\mu(\pm x/3,\ \pm(1+x)/3,\ \pm(2+x)/3,\ \pm x;\ \pm 1/4),$$

and this $\mathcal{H}$ is the unique descent of $V\otimes T^*V\otimes T^*T^*V$ whose $G_{gal} \subset Sp(8)$.

proof To prove (1), by 1.8.2, and the fact that V is self dual, it suffices that for any rank one D.E. L on $\mathbb{G}_m$, there exist no isomorphism of $V\otimes L$ with either T^*V or T^*T^*V. Suppose this were not the case. Looking at determinants, we see that $L^{\otimes 2}$ is trivial. Replacing ζ by ζ^2, we may assume $V\otimes L \approx T^*V$. We must have L nontrivial, by 3.7.7. Thus $V\otimes x^{1/2} \approx T^*V$. Comparing exponents at zero, we find that they must be $\pm 1/4$, contradiction. This proves (1).

To prove (2), notice that $V\otimes T^*V\otimes T^*T^*V$ is irreducible since $G_{gal} = \Gamma$, and its isomorphism class is T-invariant, so it descends through [3], and the descent is unique up to twisting by x^δ with $\delta = 0$ or $\pm 1/3$ (cf. 2.7.1). By 4.5.2, the descended D.E. is a hypergeometric of type (8,2) which may be uniquely ($Sp(8)\cap\mu_3 = \{e\}$) specified by requiring $G_{gal} \subset Sp(8)$. Let us denote by $\mathcal{H}$ this choice of descent.

It remains to determine the exponents of this $\mathcal{H}$. The exponents of $V\otimes T^*V\otimes T^*T^*V \approx [3]^*\mathcal{H}$ at zero are $\{\pm 3x, \pm x, \pm x, \pm x\}$. Suppose first that $x \not\equiv 0$ or $1/2 \bmod \mathbb{Z}$, so that V has semisimple local monodromy at zero. Then also $V\otimes T^*V\otimes T^*T^*V \approx [3]^*\mathcal{H}$ and consequently $\mathcal{H}$ itself have semisimple local monodromy around zero. Therefore the exponents mod $\mathbb{Z}$ of $\mathcal{H}$ must be, mod $\mathbb{Z}$, $(\pm x/3, \pm(1+x)/3, \pm(2+x)/3,$ **?, ??**). Because the local monodromy of $\mathcal{H}$ lies is $Sp(8) \subset SL(8)$, the last two exponents at zero are ±?, with ? either x or x + 1/3 or x + 2/3. If x is 0 (resp. 1/2), then the local monodromy of $[3]^*\mathcal{H}$ (resp. of $([3]^*\mathcal{H})\otimes x^{1/2}$) at zero is the tensor product of three unipotent Jordan blocks of size two, so it is the direct sum of unipotent Jordan blocks of sizes 4, 2, and 2. This means that the eight exponents at zero of $\mathcal{H}$ are all among $\{0, \pm 1/3\}$ (resp. among $\{1/2, \pm 1/6\}$) and that their multiplicities are, in some order, 4,2,2. So we see by inspection that our description of the possbiilities is correct in this case also.

To partially analyse the I_∞-representation attached to $\mathcal{H}$, we will use the descent method 4.1.7. After a multiplicative translation, V as I_∞-representation is $([2]_*\mathcal{L})\otimes x^{1/4}$ (since $\det([2]_*\mathcal{L})$ is $x^{1/2}$, while det(V) is trivial). Therefore $[2]^*V$ as I_∞-representation is $W(2,x)\otimes x^{1/2}$, $[2]^*(T^*V)$ is $W(2, \zeta x)\otimes x^{1/2}$ and $[2]^*(T^*T^*V)$ is $W(2, \zeta^2 x)\otimes x^{1/2}$.

Therefore as I_∞-representation, we have

$$[6]^*\mathcal{H} \approx [2]^*[3]^*\mathcal{H} \approx [2]^*(V\otimes T^*V\otimes T^*T^*V)$$
$$\approx W(2,x)\otimes x^{1/2}\otimes W(2, \zeta x)\otimes x^{1/2} \otimes W(2, \zeta^2 x)\otimes x^{1/2}$$
$$\approx (W(2,x)\otimes W(2, \zeta x)\otimes W(2, \zeta^2 x))\otimes x^{3/2}$$
$$\approx (\oplus_{\text{indep. } \pm\text{'s}}\ \mathcal{L}_{(\pm 1 \pm\zeta \pm\zeta^2)x})\otimes x^{3/2}$$
$$\approx (\text{rank } 6, \text{ slope } = 1) \oplus (\text{trivial of rank } 2)\otimes x^{3/2}.$$

Because $\mathcal{H}$ is hypergeometric of type (8,2) with $G_{gal} \subset Sp(8)$, its two ∞-exponents mod $\mathbb{Z}$ must be $\pm y$ for some y. By the above description of $[6]^*\mathcal{H}$ as I_∞-representation, we must have $6y \equiv 3/2 \bmod \mathbb{Z}$. So the ∞-exponents are either $\pm 1/4$ or $\pm 1/12$ or $\pm 5/12$.

Let us summarize the situation so far. We began with

$$V := \mathcal{H}_\lambda(x, -x; \emptyset),\ x \not\equiv \pm 1/4 \bmod \mathbb{Z},$$

and showed that there is a unique symplectic $\mathcal{H}$ of type (8,2) for which $[3]^*\mathcal{H} \approx V\otimes T^*V\otimes T^*T^*V$, and that this $\mathcal{H}$ is, for some $\mu\in\mathbb{C}^\times$, isomorphic to one of the following nine possibilities:

$\text{Poss}(i,j) := \mathcal{H}_\mu(\pm x/3, \pm(1 + x)/3, \pm(2+ x)/3, \pm(x + i/3); \pm(1/4 + j/3))$,

or, what is the same,

$\text{Poss}(i,j) := \mathcal{H}_\mu(\text{the six roots z of } 3z \equiv \pm x \bmod \mathbb{Z}, \pm(x + i/3); \pm(1/4 + j/3))$

where i and j run independently over the set $\{0, \pm 1\}$.

The **correct** $\mathcal{H}$ has $G_{gal} = \Gamma\ltimes A_3$, $\dim G_{gal} = 9$. Moreover, for any $x \not\equiv \pm 1/4$ or $\pm 1/12$ or $\pm 5/12 \bmod \mathbb{Z}$, **each** of the nine possibilities Poss(i,j) is irreducible, not Kummer induced, and has $G_{gal} \subset Sp(8)$. In view of the general classification theorem, for $x \not\equiv \pm 1/4$ or $\pm 1/12$ or $\pm 5/12 \bmod \mathbb{Z}$, **each** of the nine possibilities Poss(i,j) has G_{gal} **either** Sp(8) **or** $\Gamma\ltimes A_3$. We claim that the correct $\mathcal{H}$ is the **only one** of the nine possibilities Poss(i,j) which has $G_{gal} = \Gamma\ltimes A_3$. Indeed, suppose that $\mathcal{G}$ were another. By 4.5.1 and 4.5.2, $[3]^*\mathcal{G} \approx U\otimes T^*U\otimes T^*T^*U$, for some U of the form $\mathcal{H}_\alpha(\pm w, \emptyset)$. Comparing exponents at zero, we see that $w = \pm x$, whence some multiplicative translate $[x \mapsto \xi x]^*\mathcal{G}$ of $\mathcal{G}$ is also a symplectic [3]-descent of $[3]^*\mathcal{H}$. By uniqueness, we infer that $[x \mapsto \xi x]^*\mathcal{G} \approx \mathcal{H}$. Comparing exponents at both 0 and ∞, we see that $\mathcal{G} = \mathcal{H}$.

This being the case, it suffices to show that for any $x \not\equiv \pm 1/4$ or $\pm 1/12$ or $\pm 5/12 \bmod \mathbb{Z}$, the first possibility

$$\text{Poss}(0,0) := \mathcal{H}_\mu(\text{the six roots of } 3z \equiv \pm x \bmod \mathbb{Z}, \pm x; \pm 1/4)$$

has $\dim G_{gal} \le 9$. By the specialization theorem 2.4.1, it suffices to show that Poss(0,0) has $\dim G_{gal} \le 9$ for **generic** x. For this, it suffices to show that each of the **other** eight possibilities Poss(i,j) has $\dim G_{gal} \ge 10$ for **generic** x. By 2.4.1, it suffices to exhibit, for each of the other eight possibilities, a **single** numerical value $x_{i,j}$ of x for which the corresponding specialized equation has $\dim G_{gal} \ge 10$.

Consider first what happens when we specialize x to 1/4. If $i \ne j$, write $\{0, \pm 1\} = \{i, j, k\}$. Then Poss(i,j) specializes to

$$\mathcal{H}_\mu(\pm 1/12, \pm(1/12 + 1/3), \pm(1/12 - 1/3), \pm(1/4 + i/3); \pm(1/4 + j/3))$$
$$= \mathcal{H}_\mu(\pm(1/4 + i/3), \pm(1/4 + j/3), \pm(1/4 + k/3), \pm(1/4 + i/3); \pm(1/4 + j/3))$$

whose semisimplification is of the form

$$\mathcal{H}_{\mu'}(\pm(1/4 + i/3), \pm(1/4 + i/3), \pm(1/4 + k/3); \varnothing) \oplus$$
$$\oplus\ x^\delta\mathbb{C}[x, x^{-1}] \ \oplus\ x^{-\delta}\mathbb{C}[x, x^{-1}]$$

for $\delta := 1/4 + j/3$. Therefore G_{gal} for this specialization admits as quotient the G_{gal} of $\mathcal{H}_{\mu'}(\pm(1/4 + i/3), \pm(1/4 + i/3), \pm(1/4 + k/3); \varnothing)$. The six cases of $i \ne j$ are (i,k) = (0,1), (0, −1), (1,0), (1, −1), (−1, 1), (−1, 0), and for these $\mathcal{H}_{\mu'}(\pm(1/4 + i/3), \pm(1/4 + i/3), \pm(1/4 + k/3); \varnothing)$ is respectively

$$\mathcal{H}_{\mu'}(\pm 1/4, \pm 1/4, \pm 5/12; \varnothing)$$
$$\mathcal{H}_{\mu'}(\pm 1/4, \pm 1/4, \pm 1/12; \varnothing)$$
$$\mathcal{H}_{\mu'}(\pm 5/12, \pm 5/12, \pm 1/4; \varnothing)$$
$$\mathcal{H}_{\mu'}(\pm 5/12, \pm 5/12, \pm 1/12; \varnothing)$$
$$\mathcal{H}_{\mu'}(\pm 1/12, \pm 1/12, \pm 5/12; \varnothing)$$
$$\mathcal{H}_{\mu'}(\pm 1/12, \pm 1/12, \pm 1/4; \varnothing).$$

Each of these is irreducible, not Kummer induced (the exponents are not stable by $\alpha \mapsto \alpha + 1/2$ or by $\alpha \mapsto \alpha + 1/3$), and symplectically autodaual, so has $G_{gal} = Sp(6)$, which has dimension dimension $21 \ge 10$.

Thus the correct possibility has $i = j$. Now let us consider the effect of putting $x_{i,i} = 3/4$. Then

$\text{Poss}(i,i) := \mathcal{H}_\mu(\text{the six roots } z \text{ of } 3z \equiv \pm x \bmod \mathbb{Z}, \pm(x + i/3); \pm(1/4 + i/3))$

specializes, for i = 1, −1, to

$$\mathcal{H}_\mu(\text{the six roots of } 3z \equiv \pm 1/4 \bmod \mathbb{Z}, \pm(3/4 + 1/3); \pm(1/4 + 1/3))$$
$$= \mathcal{H}_\mu(\pm 1/12, \pm 1/4, \pm 5/12, \pm 1/12; \pm 5/12)$$

and to

$$\mathcal{H}_\mu(\text{the six roots of } 3z \equiv \pm 1/4 \bmod \mathbb{Z}, \pm(3/4 - 1/3); \pm(1/4 - 1/3))$$
$$= \mathcal{H}_\mu(\pm 1/12, \pm 1/4, \pm 5/12, \pm 5/12; \pm 1/12)$$

Their semisimplifications contain

$$\mathcal{H}_{\mu'}(\pm 1/12,\ \pm 1/12,\ \pm 1/4;\ \emptyset)$$

and

$$\mathcal{H}_{\mu'}(\pm 1/4,\ \pm 5/12,\ \pm 5/12;\ \emptyset)$$

respectively, both of which have $G_{gal} = Sp(6)$. So these cases are ruled out. The only remaining possibility Poss(0,0) must be the correct one. QED

Combining the above theorem with the two lemmas preceding it, we obtain

SL(2)×SL(2)×SL(2) Theorem 4.5.4 The hypergeometrics of type (8,2) whose $G := G_{gal}$ has $G^{0,der} = \Gamma :=$ the image of $SL(2)\times SL(2)\times SL(2)$ are precisely the x^δ twists of those with $G_{gal} = \Gamma\ltimes A_3$, and those with $G_{gal} = \Gamma\ltimes A_3$ are precisely those (isomorphic to one) of the form

$$\mathcal{H}_\mu(\pm x/3,\ \pm(1+x)/3,\ \pm(2+x)/3,\ \pm x;\ \pm 1/4)$$

for any μ in $\mathbb{C}^\times$, and any $x \not\equiv \pm 1/4 \bmod \mathbb{Z}$.

4.6 The SL(3)×SL(3) Case

In this section, we will analyze those irreducible hypergeometrics $\mathcal{H}$ of type (9,3) or (3,9) whose $G := G_{gal}$ has $G^{0,der} =$ (the image of) $SL(3)\times SL(3)$ in $SL(9)$. By inversion, it suffices to treat the case (9,3). Throughout this section, we will denote by

$$\Gamma := \text{the image of } SL(3)\times SL(3) \text{ in } SL(9).$$

By 1.8.4, the normalizer of Γ in $GL(9)$ is $\mathbb{G}_m\Gamma\ltimes S_2$. Therefore its normalizer in $SL(9)$ is $\mu_9(\Gamma\ltimes\{1, -\sigma\})$, where $-\sigma$ is the involution of $std_3\otimes std_3$ given by $x\otimes y \mapsto -y\otimes x$ [the change of sign achieves determinant 1].

Lemma 4.6.1 If an $\mathcal{H}$ of type (9,3) has $G^{0,der} = \Gamma$, some x^δ twist of $\mathcal{H}$ has $\Gamma \subset G_{gal} \subset \Gamma\ltimes\{1, -\sigma\}$. If $\mathcal{H}$ has $G_{gal} \subset SL(9)$, we can take $\delta \in (1/9)\mathbb{Z}$.

proof Given any $\mathcal{H}$ of type (9,3), an x^δ twist of it has $G_{gal} \subset SL(9)$ (cf. 3.6.2), and the same $G^{0,der}$. For such an $\mathcal{H}$, we have $\Gamma \subset G_{gal} \subset \mu_9(\Gamma\ltimes\{1, -\sigma\})$. The only scalars in $\Gamma\ltimes\{1, -\sigma\}$ are μ_3, so "the cube of the μ_9 factor" is a character $\chi: G_{gal} \to \mu_3$ which is precisely the obstruction to having $G_{gal} \subset \Gamma\ltimes\{1, -\sigma\}$. This character of G_{gal} corresponds to the rank one D.E. $x^\delta\mathbb{C}[x, x^{-1}]$ for $\delta \equiv 0$ or $\pm 1/3 \bmod \mathbb{Z}$. So $\mathcal{H}\otimes x^{-\delta/3}$ has $\Gamma \subset G_{gal} \subset \Gamma\ltimes\{1, -\sigma\}$. QED

Lemma 4.6.2 No hypergeometric $\mathcal{H}$ of type (9,3) has $\Gamma \subset G_{gal} \subset \mathbb{G}_m\Gamma$.
proof If such an $\mathcal{H}$ exists, a twist of it has $G_{gal} = \Gamma$. The universal covering of Γ is its triple covering by $SL(3)\times SL(3)$. Lifting the surjective classifying homomorphism $\pi_1^{diff} \to \Gamma$ through this covering, we find two rank three equations V and W on $\mathbb{G}_m$, such that $V\oplus W$ has $SL(3)\times SL(3)$ as its G_{gal}, and such that $V\otimes W \approx \mathcal{H}$. Now $\mathcal{H}$ has highest ∞-slope 1/6, and highest 0-slope =0, so by the Highest Slope Lemma 4.2.4 $V\oplus W$ has highest ∞-slope 1/6, and highest 0-slope =0. Therefore at least one of V or W has highest ∞-slope 1/6, which is impossible as both V and W have rank three. QED

Lemma 4.6.3 If a hypergeometric $\mathcal{H}$ of type (9,3) has $\Gamma \subset G_{gal} \subset \Gamma\ltimes\{1, -\sigma\}$, then $G_{gal} = \Gamma\ltimes\{1, -\sigma\}$. There exist two rank three D.E.'s V_1 and V_2 on $\mathbb{G}_m$, such that $V_1\oplus V_2$ has $SL(3)\times SL(3)$ as its G_{gal}, each V_i is hypergeometric of type (3,0) with $G_{gal} = SL(3)$, and there exists an isomorphism

$$[2]^*\mathcal{H} \approx V_1\otimes V_2.$$

proof By the above lemma we must have $G = \Gamma\ltimes\{1, -\sigma\}$.The quotient G/Γ is thus cyclic of order two. Therefore $[2]^*\mathcal{H}$ has $G_{gal} = \Gamma$. Lifting its surjective classifying homomorphism $\pi_1^{diff} \to \Gamma$ through the universal covering, we obtain two rank three equations V_1 and V_2 on $\mathbb{G}_m$, each with $G_{gal} = SL(3)$, such that $V_1\oplus V_2$ has $SL(3)\times SL(3)$ as its G_{gal}, and such that $V_1\otimes V_2 \approx [2]^*\mathcal{H}$. Now $[2]^*\mathcal{H}$ has highest ∞-slope 1/3, and highest 0-slope =0, so by 4.2.4 $V_1\oplus V_2$ has highest ∞-slope 1/3, and highest 0-slope =0. Therefore each of V_1 and V_2 has highest ∞-slope $\le$ 1/3, and highest 0-slope =0. In fact, both of V_1 and V_2 must have highest ∞-slope =1/3 [for if V_1 had highest ∞-slope < 1/3, it would, being of rank three, have all ∞-slopes =0, so would be regular singular at both 0 and ∞, so reducible, so would not have SL(3) for its G_{gal}]. By the intrinsic characterization of hypergeometrics, both V_i are hypergeometric of type (3,0). QED

Corollary 4.6.4 If a hypergeometric $\mathcal{H}$ of type (9,3) has

$$G_{gal} = \Gamma \ltimes \{1, -\sigma\},$$

then $\mathcal{H} \otimes x^{1/2}$ has $G_{gal} = \Gamma \ltimes \{1, \sigma\} = \Gamma \ltimes S_2$, and conversely.

proof $\mathcal{H}$ and $\mathcal{H} \otimes x^{1/2}$ both have $G^0 = \Gamma$, and G/G^0 of order two, corresponding to the Kummer covering of degree two. So if we view $\mathcal{H}$ and $\mathcal{H} \otimes x^{1/2}$ as representations ρ_1 and ρ_2 of $\pi_1{}^{diff}$, and denote by χ the unique nontrivial character of order two of $\pi_1{}^{diff}$, then $\rho_2 = \chi \otimes \rho_1$, and $Ker(\chi) = (\rho_1)^{-1}(\Gamma) = (\rho_2)^{-1}(\Gamma)$. Let $g \in \pi_1{}^{diff}$ have $\rho_1(g) = -\sigma$. Then $\chi(g) = -1$, since $-\sigma \notin \Gamma$, and so $\rho_2(g) = \sigma$. Since $\pi_1{}^{diff} = Ker(\chi) \cup gKer(\chi)$, $Image(\rho_2) = \Gamma \cup \sigma\Gamma$, as asserted. The converse is proven the same way. QED

Corollary 4.6.5 If $\mathcal{H}$ of type (9,3) has $G^{0,der} = \Gamma$, there exists an x^δ twist of $\mathcal{H}$ which has $G_{gal} = \Gamma \ltimes \{1, -\sigma\}$, and another which has $G_{gal} = \Gamma \ltimes \{1, \sigma\}$.

proof. This is immediate from the previous four results. QED

Lemma 4.6.6 If a hypergeometric $\mathcal{H}$ of type (9,3) has $G_{gal} = \Gamma \ltimes \{1, \sigma\}$, there exist two rank three D.E.'s V_1 and V_2 on $\mathbb{G}_m$, such that $V_1 \oplus V_2$ has $SL(3) \times SL(3)$ as its G_{gal}, each V_i is hypergeometric of type (3,0) with $G_{gal} = SL(3)$, and there exists an isomorphism

$$[2]^*\mathcal{H} \approx V_1 \otimes V_2.$$

proof Simply apply 4.6.3 to $\mathcal{H} \otimes x^{1/2}$, which in virtue of 4.6.4 has $G_{gal} = \Gamma \ltimes \{1, -\sigma\}$, but the same $[2]^*$ as $\mathcal{H}$. QED

Lemma 4.6.7 If $\mathcal{H}$ of type (9,3) has $G_{gal} = \Gamma \ltimes \{1, -\sigma\}$ or $\Gamma \ltimes \{1, \sigma\}$, then for $\delta \equiv 0$ or $\pm 1/3 \bmod \mathbb{Z}$, $G_{gal}(\mathcal{H} \otimes x^\delta) = G_{gal}(\mathcal{H})$.

proof For any δ, $G_{gal}(\mathcal{H} \otimes x^\delta)^{0,der} = \Gamma$, a group which contains μ_3. So if $3\delta \equiv 0 \bmod \mathbb{Z}$, then $G_{gal}(\mathcal{H} \otimes x^\delta) \subset G_{gal}(\mathcal{H})\mu_3 = G_{gal}(\mathcal{H})$, and $G_{gal}(\mathcal{H}) \subset G_{gal}(\mathcal{H} \otimes x^\delta)\mu_3 = G_{gal}(\mathcal{H} \otimes x^\delta)$. QED

Lemma 4.6.8 Suppose V_1 and V_2, are two hypergeometric of type (3,0) on $\mathbb{G}_m$, such that $V_1 \oplus V_2$ has $G_{gal} = SL(3) \times SL(3)$, and such that for some D.E. V on $\mathbb{G}_m$ there exists an isomorphism

$$[2]^*V \approx V_1 \otimes V_2.$$

Denote by $T := [x \mapsto -x]$, the multiplicative translation by -1. Then

(1) V is Lie-irreducible, and unique up to $V \mapsto V \otimes x^{1/2}$.

(2) V is hypergeometric of type (9,3), and there exists a unique twist $V \otimes x^{\alpha}$ of V with $\alpha = 0$ or $1/2$ such that after replacing $V \mapsto V \otimes x^{\alpha}$ (which doesn't change $[2]^*V$), det(V) is trivial (resp. nontrivial). This V has $G_{gal}(V) = \Gamma \ltimes \{1, -\sigma\}$ (resp. has $G_{gal}(V) = \Gamma \ltimes \{1, \sigma\}$).

(2)There exists an isomorphism $T^*V_1 \approx V_2 \otimes x^{\delta}$ for $\delta \equiv 0$ or $\pm 1/3$ mod $\mathbb{Z}$.

(3) For this choice of δ, we have:

$(V_1 \otimes x^{\delta}) \oplus T^*(V_1 \otimes x^{\delta})$ has $G_{gal} = SL(3) \times SL(3)$,

$[2]^*V \approx (V_1 \otimes x^{\delta}) \otimes T^*(V_1 \otimes x^{\delta})$

$G_{gal}(V) = \Gamma \ltimes \{1, -\sigma\}$ if det(V) trivial, $G_{gal}(V) = \Gamma \ltimes \{1, \sigma\}$ if not.

proof The G_{gal} of such a V obviously has $G^0 = \Gamma$, and G/G^0 of order at most two. In particular, V is Lie-irreducible, and (1) follows from 2.7.1. By 4.2.4, V has highest ∞-slope 1/6, and highest 0-slope =0. Therefore $\chi(\mathbb{G}_m, V) = -1$, and so V is hypergeometric by the intrinsic characterization; by its slopes, it must be of type (9,3). Since $[2]^*V$ has trivial determinant, det(V) is either trivial or is $x^{1/2}\mathbb{C}[x, x^{-1}]$. Since V has odd rank 9, exactly one of V or $V \otimes x^{1/2}$ has $G := G_{gal} \subset SL(9)$. Since $G^0 = \Gamma$ and G/G^0 has order ≤ 2, while $G \neq \Gamma$ by 4.6.2, we must have G/Γ of order two. Since G is not contained in $\mathbb{G}_m\Gamma$, G must contain a scalar times $-\sigma$, say $\xi(-\sigma)$, and $G = \Gamma \cup \Gamma\xi(-\sigma)$. So if $G \subset SL(9)$, $\xi \in \mu_9$. Since the square of every element of G lies in Γ, $(\xi(-\sigma))^2 = \xi^2$ is a scalar in Γ, so $\xi^2 \in \mu_3$. Therefore $\xi \in \mu_3$. But $\mu_3 \subset \Gamma$, so from $G = \Gamma \cup \Gamma\xi(-\sigma)$ we see that $G = \Gamma \cup \Gamma(-\sigma) = \Gamma \ltimes \{1, -\sigma\}$ if det(V) is trivial. By the previous lemma, it follows that $G = \Gamma \ltimes \{1, \sigma\}$ if det(V) is nontrivial.

Since $V_1 \otimes V_2$ descends through [2], its isomorphism class is invariant under T^*: $V_1 \otimes V_2 \approx T^*(V_1) \otimes T^*(V_2)$. Now consider the two D.E.'s $V_1 \oplus V_2$ and $T^*(V_1) \oplus T^*(V_2)$. Both have $G_{gal} = SL(3) \times SL(3)$. As representations of π_1^{diff} their isomorphism classes arise from (possibly different) liftings of the classifying map for $[2]^*V$. From the exact sequence for $\mathrm{Hom}_{\text{alg. gp.}}(\pi_1^{\text{diff}}, ?)$ applied to the central extension

$$0 \to \mu_3 \to SL(3)\times SL(3) \to \Gamma \to 0,$$

we see that any lifting is isomorphic to one of

$$V_1 \oplus V_2,$$
$$(V_1)\otimes x^{1/3} \oplus (V_2)\otimes x^{-1/3},$$
$$(V_1)\otimes x^{-1/3} \oplus (V_2)\otimes x^{1/3}.$$

Therefore either

$$T^*(V_1)\oplus T^*(V_2) \approx V_1 \oplus V_2, \text{ or}$$
$$T^*(V_1)\oplus T^*(V_2) \approx (V_1)\otimes x^{1/3} \oplus (V_2)\otimes x^{-1/3}, \text{ or}$$
$$T^*(V_1)\oplus T^*(V_2) \approx (V_1)\otimes x^{-1/3} \oplus (V_2)\otimes x^{1/3}.$$

By Jordan Holder theory, the isomorphism classes of the irreducible constituents of $T^*(V_1)\oplus T^*(V_2)$ are well defined, and they occur in any decomposition of $T^*(V_1)\oplus T^*(V_2)$ into a sum of irreducibles. But $T^*(V_1)\approx V_1$ is impossible by 3.7.7, and $T^*(V_1) \approx (V_1)\otimes x^{\pm 1/3}$ would imply that the exponents of V at zero are stable by $\alpha \mapsto \alpha + 1/3$, which in turn implies that V_1 is Kummer induced of degree three, which is impossible since $G_{gal}(V_1) = SL(3)$. So either

$$T^*(V_1) \approx V_2, \text{ or}$$
$$T^*(V_1) \approx (V_2)\otimes x^{-1/3}, \text{ or}$$
$$T^*(V_1) \approx (V_2)\otimes x^{1/3}.$$

This proves (2): $T^*V_1 \approx V_2\otimes x^{\delta}$ for $\delta \equiv 0$ or $\pm 1/3 \bmod \mathbb{Z}$. Assertion (3) is immediate from (1) and (2), for by (2) we have

$$(V_1\otimes x^{\delta}) \oplus T^*(V_1\otimes x^{\delta}) \approx (V_1\otimes x^{\delta}) \oplus (V_2\otimes x^{2\delta}). \qquad \text{QED}$$

Corollary 4.6.8.1 Let V be hypergeometric of type (9,3) with $G_{gal} = \Gamma \ltimes \{1, \sigma\}$. Then there exists a hypergeometric $\mathcal{H}$ of type (3,0) such that $\mathcal{H}\oplus T^*\mathcal{H}$ has $G_{gal} = SL(3)\times SL(3)$, and such that $[2]^*V \approx \mathcal{H}\otimes T^*\mathcal{H}$.

proof Simply combine 4.6.6 and 4.6.8. QED

Lemma 4.6.9 Let $\mathcal{H} := \mathcal{H}_\lambda(x, y, z; \emptyset)$ be a hypergeometric of type

(3,0), and $T := [x \mapsto -x]$ the multiplicative translation by -1. Then $\mathcal{H}\oplus T^*\mathcal{H}$ has $G_{gal} = SL(3)\times SL(3)$ if and only if x, y, z satisfy the following conditions mod $\mathbb{Z}$:

(1) $x + y + z \equiv 0 \bmod \mathbb{Z}$.

(2) $\{x, y, z\} \not\equiv \{0, 1/3, 2/3\} \bmod \mathbb{Z}$.

(3) none of x, y, z is $\equiv 0$ or $\pm 1/3 \bmod \mathbb{Z}$.

proof Condition (1) is that $\det\mathcal{H}$ be trivial. Once $\det\mathcal{H}$ is trivial, (2) is the condition that $\mathcal{H}$ not be Kummer induced. By 3.6 (2), $G_{gal} = SL(3)$ (since $|n-m| = 3$ is odd). So (1) and (2) together are equivalent to $G_{gal} = SL(3)$.

By Goursat-Kolchin-Ribet via 3.8.2, $\mathcal{H}\oplus T^*\mathcal{H}$ will **fail** to have its $G_{gal} = SL(3)\times SL(3)$ if and only if for some $\alpha\in\mathbb{C}$, either $\mathcal{H}\otimes x^{\alpha}$ or its adjoint $\mathcal{H}^*\otimes x^{-\alpha}$ is isomorphic to $T^*\mathcal{H}$. We now analyze these cases.

Suppose first that $\mathcal{H}\otimes x^{\alpha} \approx T^*\mathcal{H}$. Comparing determinants, we see that $\alpha \equiv 0$ or $\pm 1/3 \bmod \mathbb{Z}$. If $\alpha \equiv 0 \bmod \mathbb{Z}$, then $\mathcal{H} \approx T^*\mathcal{H}$, which is impossible by 3.7.7. Therefore $\mathcal{H}\otimes x^{1/3} \approx T^*\mathcal{H}$ or $\mathcal{H}\otimes x^{-1/3} \approx T^*\mathcal{H}$. In either case, the 0-exponents mod $\mathbb{Z}$ of $\mathcal{H}$ are stable by $x \mapsto x + 1/3$, in which case $\mathcal{H}$ would be Kummer induced of degree three, contradiction. So this case cannot arise.

Suppose now that $\mathcal{H}^*\otimes x^{-\alpha} \approx T^*\mathcal{H}$. Again $\alpha \equiv 0$ or $\pm 1/3 \bmod \mathbb{Z}$. So $(\mathcal{H}\otimes x^{2\alpha})^* \approx \mathcal{H}^*\otimes x^{-2\alpha} \approx T^*\mathcal{H}\otimes x^{-\alpha} = T^*\mathcal{H}\otimes x^{2\alpha} \approx T^*(\mathcal{H}\otimes x^{2\alpha})$, and the 0-exponents mod $\mathbb{Z}$ of $\mathcal{H}$ are stable by $x \mapsto -x -\alpha$. The only possibility for $\mathcal{H}$ compatible with (1) is $\mathcal{H}_{\lambda}(x, -x -\alpha, \alpha;\varnothing)$, and this $\mathcal{H}$ has $\mathcal{H}^*\otimes x^{-\alpha} \approx T^*\mathcal{H}$. It is precisely this sort of $\mathcal{H}$ which is ruled out by condition (3). QED

Theorem 4.6.10 Let $\mathcal{H} := \mathcal{H}_{\lambda}(x, y, z; \varnothing)$ be a hypergeometric of type (3,0), and $T := [x \mapsto -x]$ the multiplicative translation by -1. Suppose that x, y, z in $\mathbb{C}$ satisfy

$x+y+z \equiv 0 \bmod \mathbb{Z}$, and none of x, y, z is $\equiv 0$ or $\pm 1/3 \bmod \mathbb{Z}$,

i.e., suppose that $\mathcal{H}\oplus T^*\mathcal{H}$ has $G_{gal} = SL(3)\times SL(3)$.

Then

(1) There is a unique hypergeometric V of type (9,3) with $\det(V)$ nontrivial for which there exists an isomorphism $[2]^*V \approx \mathcal{H}\otimes T^*\mathcal{H}$.

(2) This V has $G_{gal(V)} = \Gamma\ltimes\{1, \sigma\}$.

(3) For some $\mu\in\mathbb{C}^{\times}$, V is isomorphic to

$\mathcal{H}_\mu(x, y, z, -x/2, -y/2, -z/2, 1/2 - x/2, 1/2 - y/2, 1/2 - z/2; 0, \pm 1/3)$.
(4) Any hypergeometric V of type (9,3) with $G_{gal} = \Gamma \ltimes \{1, \sigma\}$ is obtained in this way.

proof Assertion (4), "mise pour memoire" has already been proven (4.6.8.1). Since $\mathcal{H} \otimes T^*\mathcal{H}$ is irreducible on $\mathbb{G}_m$ and isomorphic to its T^* pullback, it descends through [2], i.e., it is of the form $[2]^*V$ for some D.E. V on $\mathbb{G}_m$. Therefore assertions (1) and (2) result from 4.6.8. The only subtle point is to compute the exponents mod $\mathbb{Z}$ of the unique [2]-descent V of $\mathcal{H} \otimes T^*\mathcal{H}$ whose determinant is nontrivial. Let us denote by $\{\alpha_i\}_{i=1,\dots,9}$ and $\{\beta_i\}_{i=1,\dots,3}$ the exponents of V at zero and ∞ respectively. Thus for some μ in $\mathbb{C}^\times$, V is $\mathcal{H}_\mu(\alpha_i\text{'s}; \beta_j\text{'s})$. Exactly as in the PSL(3) and SL(2)×SL(2)×SL(2) cases, we will first determine the exponents up to a few possibilities, and then use the specialization theorem to eliminate all but one of the possibilities.

Suppose first that x, y, z are all distinct mod $\mathbb{Z}$. Then $\mathcal{H}$ and $T^*\mathcal{H}$ has semisimple local monodromy at zero, so also $[2]^*V \approx \mathcal{H} \otimes T^*\mathcal{H}$ and hence V has semisimple local monodromy at zero. Therefore the $\{\alpha_i\}_{i=1,\dots,9}$ are all distinct mod $\mathbb{Z}$, and their doubles are those of $\mathcal{H} \otimes T^*\mathcal{H}$:

$$\{2\alpha_i\}_{i=1,\dots,9} = \{2x, 2y, 2z, x+y, x+y, x+z, x+z, y+z, y+z\}.$$

Since $x+y+z \equiv 0 \bmod \mathbb{Z}$, we may rewite this

$$\{2\alpha_i\}_{i=1,\dots,9} = \{2x, 2y, 2z, -z, -z, -y, -y, -x, -x\}.$$

Because the $\{\alpha_i\}_{i=1,\dots,9}$ are all distinct mod $\mathbb{Z}$, the α_i must include both the halves of $-x$, $-y$, and $-z$. The remaining three of them are some choices

$$\tilde{x} = x \textbf{ or } 1/2 + x, \quad \tilde{y} = y \textbf{ or } 1/2 + y, \quad \tilde{z} = z \textbf{ or } 1/2 + z$$

of halves of 2x, 2y, and 2z. Therefore the $\{\alpha_i\}_{i=1,\dots,9}$ are

$$\{\tilde{x}, \tilde{y}, \tilde{z}, -x/2, 1/2 - x/2, -y/2, 1/2 - y/2, -z/2, 1/2 - z/2\}.$$

Since det(V) is to be nontrivial, det(V) is $x^{1/2}\mathbb{C}[x, x^{-1}]$. As $x + y + z \equiv 0 \bmod \mathbb{Z}$, we must have $\tilde{x} + \tilde{y} + \tilde{z} \equiv 0$. This can happen only if either

$$\tilde{x} = x, \ \tilde{y} = y, \ \tilde{z} = z$$

or if **precisely one** among x, y, z, has $\tilde{t} = t$, and the other two have $\tilde{t} = 1/2 + t$.

If we no longer assume that x, y, and z are distinct mod $\mathbb{Z}$, no more than two of them can coincide mod $\mathbb{Z}$, since their sum is 0 mod $\mathbb{Z}$ and none is 0 or $\pm 1/3$ mod $\mathbb{Z}$. Permuting x, y, and z, we may assume that $y=x$, $z = -2x$. Then the local monodromy transformation of both $\mathcal{H}$ and of $T^*\mathcal{H}$ around zero has Jordan normal form

$$e^{2\pi i x}\otimes(\text{unip}_2) \quad \oplus \quad e^{-4\pi i x}\otimes(\text{unip}_1),$$

where by unip_n we mean a unipotent Jordan block of size n. Therefore the local monodromy transformation of $\mathcal{H}\otimes T^*\mathcal{H} \approx [2]^*V$ around zero has Jordan normal form

$$e^{4\pi i x}\otimes(\text{unip}_3) \;\oplus\; e^{4\pi i x}\otimes(\text{unip}_1) \;\oplus\; e^{-2\pi i x}\otimes(\text{unip}_2) \;\oplus$$
$$\oplus\; e^{-2\pi i x}\otimes(\text{unip}_2) \;\oplus\; e^{-8\pi i x}\otimes(\text{unip}_1).$$

Therefore V has five distinct mod $\mathbb{Z}$ exponents at zero, occurring with multiplicities 3,1,2,2,1, say $\{\alpha_1, \alpha_1, \alpha_1, \alpha_2, \alpha_3, \alpha_3, \alpha_4, \alpha_4, \alpha_5\}$, and

$$2\alpha_1 \equiv 2x \bmod \mathbb{Z},$$
$$2\alpha_2 \equiv 2x \bmod \mathbb{Z},$$
$$2\alpha_3 \equiv -x \bmod \mathbb{Z},$$
$$2\alpha_4 \equiv -x \bmod \mathbb{Z},$$
$$2\alpha_5 \equiv -4x \bmod \mathbb{Z}.$$

Since the α_i are distinct mod $\mathbb{Z}$, we have equalities mod $\mathbb{Z}$

$$\{\alpha_3, \alpha_3, \alpha_4, \alpha_4\} = \{-x/2, -x/2, 1/2 - x/2, 1/2 - x/2\},$$
$$\{\alpha_1, \alpha_1, \alpha_1, \alpha_2\} = \{x, x, x, 1/2 + x\}$$
$$\textbf{or} = \{ 1/2 + x, 1/2 + x, 1/2 + x, x\},$$
$$\alpha_5 = -2x \;\textbf{or} = 1/2 - 2x.$$

All of these possibilities are obtained from the general case's possibilities by specializing $y \mapsto x$.

Consider now the I_∞-representation.

Lemma 4.6.11 Notations as in the theorem, $\mathcal{H}\otimes T^*\mathcal{H} \approx [2]^*V$ as I_∞-representation is

$$(\text{rank } 6, \text{slopes } 1/3) \;\oplus\; \mathbb{C}((1/x)) \;\oplus\; x^{1/3}\mathbb{C}((1/x)) \;\oplus\; x^{2/3}\mathbb{C}((1/x)).$$

proof Denote by W the I_∞-representation attached to $\mathcal{H}$. Since $\mathcal{H}$ is of type (3,0) with trivial determinant, it follows from 3.4.1.1 that W is a multiplicative translate of $[3]_*\mathcal{L}$, where $\mathcal{L}$ is the rank one D.E. for e^x, and that T^*W is the dual of W. Therefore

$$W\otimes T^*W \approx \mathit{End}(W) \approx \mathit{End}^0(W) \oplus (\text{triv}),$$

and so the result has already been proven in 4.3.4. QED

From this lemma, we see that the ∞-exponents β_i satisfy

$\{2\beta_i\}_{i=1,2,3} = \{0, 1/3, 2/3\}$.

So the unique [2]-descent V of $\mathcal{H}\otimes T^*\mathcal{H}$ whose determinant is nontrivial (resp. trivial) is $\mathcal{H}_\mu(\alpha_i\text{'s}; \beta_j\text{'s})$, where

$\{\alpha_i\text{'s}\} = \{\tilde{x}, \tilde{y}, \tilde{z}, -x/2, -y/2, -z/2, 1/2 - x/2, 1/2 - y/2, 1/2 - z/2\}$,

$\tilde{x} = x$ **or** $1/2 + x$, $\tilde{y} = y$ **or** $1/2 + y$, $\tilde{z} = z$ **or** $1/2 + z$,

$x+y+z \equiv 0 \bmod \mathbb{Z}$,

$\tilde{x}+ \tilde{y}+ \tilde{z} \equiv 0$ (resp. $\equiv 1/2$) mod $\mathbb{Z}$,

none of x, y, z is $\equiv 0$ or $\pm 1/3 \bmod \mathbb{Z}$,

$\{2\beta_i\}_{i=1,2,3} = \{0, 1/3, 2/3\}$,

$\beta_1 = 0$ **or** $1/2$, $\beta_2 = 1/6$ **or** $2/3$, $\beta_3 = 1/3$ **or** $5/6$.

In order to decide which choice is correct, we now embark on a series of lemmas.

Lemma 4.6.12 Suppose $x+y+z \equiv 0 \bmod \mathbb{Z}$, and none of x, y, z is $\equiv 0$ or $\pm 1/3 \bmod \mathbb{Z}$. The set with multiplicity $\{x, y, z\} \bmod \mathbb{Z}$ is uniquely determined by the set with multiplicity $S := \{2x, 2y, 2z, -x, -x, -y, -y, -z, -z\} \bmod \mathbb{Z}$.

proof Consider subsets with multiplicity $T := \{a, b, c\} \bmod \mathbb{Z}$ of S such that $a+b+c \equiv 0 \bmod \mathbb{Z}$, and such that $\{a, a, b, b, c, c\} \bmod \mathbb{Z}$ is a subset with multiplicity of S. Clearly $\{-x, -y, -z\}$ is such a T. We will show it is the only one.

Suppose first that T is drawn entirely from $\{-x, -y, -z\}$, but is not $\{-x, -y, -z\}$. Then T must (up to permutation of x, y, z,) be $\{-x, -x, -z\}$ or $\{-x, -x, -x\}$. The second case $\{-x, -x, -x\}$ is impossible, because none of x, y, z is $\equiv 0$ or $\pm 1/3 \bmod \mathbb{Z}$, while $a+b+c \equiv 0 \bmod \mathbb{Z}$ if $T = \{a, b, c\}$. If T is $\{-x, -x, -z\}$, with x and z distinct mod $\mathbb{Z}$, we claim that $y \equiv x \bmod \mathbb{Z}$. For if not, then either $y \equiv z \bmod \mathbb{Z}$ [in which case $-x$ and $-y$ each occur in S with multiplicity at lease four, so at least two out of $\{2x, 2y, 2z=2y\}$ are $-x \bmod \mathbb{Z}$. Therefore $2y \equiv -x \bmod \mathbb{Z}$, and our original $\{x, y, z\}$ is $\{-2y, y, y\}$. So our T, namely $\{-x, -x, -z\}$, is $\{2y, 2y, -y\}$; but this T fails to sum to 0 mod $\mathbb{Z}$, since none of x, y, z is $\equiv 0$ or $\pm 1/3 \bmod \mathbb{Z}$, contradiction] or x, y, and z are distinct mod $\mathbb{Z}$ [in which case at least two of $\{2x, 2y, 2z\}$ are $-x \bmod \mathbb{Z}$. By hypothesis, $2x \not\equiv -x \bmod \mathbb{Z}$, so we must have $2y \equiv 2z \equiv -x \bmod \mathbb{Z}$. Then our original $\{x, y, z\}$ is $\{-2y, y, z\}$, and as $x+y+z \equiv 0 \bmod \mathbb{Z}$ we see that $y \equiv z \bmod \mathbb{Z}$, contradiction].

Suppose now that T is not drawn entirely from $\{-x, -y, -z\}$. Then there is an element of S other than $-x, -y, -z$ which occurs in S with multiplicity ≥ 2, and it occurs in T. So at least two out of $\{2x, 2y, 2z\}$ coincide, and their common value is none of $-x, -y, -z \bmod \mathbb{Z}$.

If $2x \equiv 2y \equiv 2z \bmod \mathbb{Z}$, then at least two out of x, y, z coincide mod $\mathbb{Z}$; as not all of x, y, z coincide mod $\mathbb{Z}$, precisely two of x, y, z coincide. Permuting x, y, z, we may assume that $y \equiv x$, $z \equiv x + 1/2 \bmod \mathbb{Z}$. Since $x+y+z \equiv 0 \bmod \mathbb{Z}$, x is 1/2 or $\pm 1/6 \bmod \mathbb{Z}$, whence z is 0 or $\pm 1/3 \bmod \mathbb{Z}$, contradiction.

Permuting x, y, z if necessary, we may assume that $2x \equiv 2y \bmod \mathbb{Z}$, and $2x \not\equiv 2z \bmod \mathbb{Z}$. Since 2x is present in S exactly twice, T is {2x, ?, ??} where ? and ?? are drawn from {-x, -y, -z}.

If -x, -y, -z are pairwise distinct mod $\mathbb{Z}$, then none of them can occur in S with multiplicity > 3, so T is either {2x, -x, -y} or {2x, -x, -z} or {2x, -y, -z}. Since T sums to 0 mod $\mathbb{Z}$, we find either $y \equiv x \bmod \mathbb{Z}$ or $z \equiv x \bmod \mathbb{Z}$ or $y+z \equiv 2x \bmod \mathbb{Z}$. The first two contradict the pairwise distinctness, and the last forces $3x \equiv 0 \bmod \mathbb{Z}$, contradicting that none of x, y, z is $\equiv 0$ or $\pm 1/3 \bmod \mathbb{Z}$.

If two of -x, -y, -z coincide mod $\mathbb{Z}$, then exactly two coincide (since $x+y+z \equiv 0 \bmod \mathbb{Z}$ but none of x, y, z is $\equiv 0$ or $\pm 1/3 \bmod \mathbb{Z}$), so either $x \equiv y \bmod \mathbb{Z}$ or $x \equiv z \bmod \mathbb{Z}$ or $y \equiv z \bmod \mathbb{Z}$. Since $2x \equiv 2y \bmod \mathbb{Z}$, and $2x \not\equiv 2z \bmod \mathbb{Z}$, we must have $x \equiv y \bmod \mathbb{Z}$ and $x \not\equiv z \bmod \mathbb{Z}$. Then z is -2x mod $\mathbb{Z}$, S is { 2x with mult. 4, -x with mult. 4, -4x}, and the only possible T's are {-x, -x, 2x} or {-x, 2x, 2x}. The second is impossible since T sums to 0 mod $\mathbb{Z}$, and the first is {-x, -y, -z}. QED

Lemma 4.6.13 Suppose that $\mathcal{H}_\mu(\alpha_i\text{'s}; \beta_j\text{'s})$ is any hypergeometric of type (9,3) such that $\{2\beta_i\}_{i=1,2,3} = \{0, 1/3, 2/3\}$, and such that $\Sigma\alpha_i \equiv 1/2$ (resp. $\Sigma\alpha_i \equiv 0$) mod $\mathbb{Z}$. If $G := G_{gal}$ has $G^{0,der} = \Gamma$, then

$$G_{gal} = \Gamma \ltimes \{1, \sigma\} \text{ (resp. } G_{gal} = \Gamma \ltimes \{1, -\sigma\}).$$

proof By an $x^{1/2}$ twist the two cases are interchanged, so it suffices to treat the case in which $\Sigma\alpha_i \equiv 0 \bmod \mathbb{Z}$. Then $G_{gal} \subset SL(9)$, so if $G^{0,der} = \Gamma$, then by 4.6.1 there exists $\delta \in (1/9)\mathbb{Z}$ such that $\mathcal{H}_\mu(\alpha_i\text{'s}; \beta_j\text{'s}) \otimes x^\delta$ has $G_{gal} = \Gamma \ltimes \{1, -\sigma\}$. By 4.6.4 and 4.6.8.1, there exists a hypergeometric $\mathcal{H}$ of type (3,0) such that $\mathcal{H} \oplus T^*\mathcal{H}$ has $G_{gal} = SL(3) \times SL(3)$, and such that $[2]^*(\mathcal{H}_\mu(\alpha_i\text{'s}; \beta_j\text{'s}) \otimes x^\delta) \approx \mathcal{H} \otimes T^*\mathcal{H}$. In view of 4.6.11, the ∞-exponents of $\mathcal{H} \otimes T^*\mathcal{H}$ are {0, 1/3, 2/3}, whence $\{2\delta + 2\beta_i\}_{i=1,2,3} = \{0, 1/3, 2/3\}$. By hypothesis, the $\{2\beta_i\}_{i=1,2,3}$ are themselves {0, 1/3, 2/3}, whence 2δ mod $\mathbb{Z}$ is in {0, 1/3, 2/3}. Since δ has denominator dividing 9, we infer that $3\delta \equiv 0 \bmod \mathbb{Z}$. By 4.6.7, $\mathcal{H}_\mu(\alpha_i\text{'s}; \beta_j\text{'s}) = (\mathcal{H}_\mu(\alpha_i\text{'s}; \beta_j\text{'s}) \otimes x^\delta) \otimes x^{-\delta}$ has the same G_{gal} as $\mathcal{H}_\mu(\alpha_i\text{'s}; \beta_j\text{'s}) \otimes x^\delta$, namely $\Gamma \ltimes \{1, -\sigma\}$. QED

Lemma 4.6.14 Suppose that $x+y+z \equiv 0 \bmod \mathbb{Z}$, and that none of x, y, z is $\equiv 0$ or $\pm 1/3 \bmod \mathbb{Z}$. Suppose that $\mathcal{H}_\mu(\alpha_i\text{'s}; \beta_j\text{'s})$ is a hypergeometric of type (9,3) whose exponents satisfy

$\{\alpha_i\text{'s}\} = \{\tilde{x}, \tilde{y}, \tilde{z}, -x/2, -y/2, -z/2, 1/2 - x/2, 1/2 - y/2, 1/2 - z/2\}$,

$\tilde{x} = x$ **or** $1/2 + x$, $\tilde{y} = y$ **or** $1/2 + y$, $\tilde{z} = z$ **or** $1/2 + z$,

$\tilde{x}+ \tilde{y}+ \tilde{z} \equiv 0$ (resp. $\equiv 1/2$) mod $\mathbb{Z}$,

$\{2\beta_i\}_{i=1,2,3} = \{0, 1/3, 2/3\}$.

Suppose that $G^{0,\mathrm{der}} = \Gamma$. Then for some $\lambda \in \mathbb{C}^\times$, there exists an isomorphism

$$[2]^*(\mathcal{H}_\mu(\alpha_i\text{'s}; \beta_j\text{'s})) \approx \mathcal{H}_\lambda(x, y, z; \emptyset)\otimes T^*\mathcal{H}_\lambda(x, y, z; \emptyset).$$

proof twisting by $x^{1/2}$, it suffices to treat the case $\tilde{x}+ \tilde{y}+ \tilde{z} \equiv 0 \bmod \mathbb{Z}$. By the above Lemma, $\mathcal{H}_\mu(\alpha_i\text{'s}; \beta_j\text{'s})$ has $G_{\mathrm{gal}} = \Gamma\ltimes\{1, \sigma\}$. By 4.6.8.1, and 4.6.9, there exist X, Y, Z in $\mathbb{C}$ and λ in $\mathbb{C}^\times$ with $X+Y+Z \equiv 0 \bmod \mathbb{Z}$, none of X, Y, Z is $\equiv 0$ or $\pm 1/3 \bmod \mathbb{Z}$, and there exists an isomorphism

$$[2]^*(\mathcal{H}_\mu(\alpha_i\text{'s}; \beta_j\text{'s})) \approx \mathcal{H}_\lambda(X, Y, Z; \emptyset)\otimes T^*\mathcal{H}_\lambda(X, Y, Z; \emptyset).$$

Comparing exponents at zero, we find

$$\{2\alpha_i\} = \{2X, 2Y, 2Z, -X, -X, -Y, -Y, -Z, -Z\} \bmod \mathbb{Z}.$$

Now $\{2\alpha_i\}$ is itself $\{2x, 2y, 2z, -x, -x, -y, -y, -z, -z\}$, so by 4.6.12 it follows that $\{x, y, z\} = \{X, Y, Z\} \bmod \mathbb{Z}$. QED

Lemma 4.6.15 Suppose that $x+y+z \equiv 0 \bmod \mathbb{Z}$, and that none of x, y, z is $\equiv 0$ or $\pm 1/3 \bmod \mathbb{Z}$. Suppose that V_1 and V_2 are two hypergeometrics of type (9,3), of the form

$$V_1 = \mathcal{H}_{\mu_1}(A1; B1)$$

$$V_2 = \mathcal{H}_{\mu_2}(A2; B2)$$

where μ_1 and μ_2 are in $\mathbb{C}^\times$,where A1 and A2 are choices of $\{\alpha_i\text{'s}\}$ satisfying

$\{\alpha_i\text{'s}\} = \{\tilde{x}, \tilde{y}, \tilde{z}, -x/2, -y/2, -z/2, 1/2 - x/2, 1/2 - y/2, 1/2 - z/2\}$,

$\tilde{x} = x$ **or** $1/2 + x$, $\tilde{y} = y$ **or** $1/2 + y$, $\tilde{z} = z$ **or** $1/2 + z$,

$\tilde{x}+ \tilde{y}+ \tilde{z} \equiv 0 \bmod \mathbb{Z}$,

and where B1 and B2 are choices of $\{\beta_i\text{'s}\}$ satisfying

$\{2\beta_i\}_{i=1,2,3} = \{0, 1/3, 2/3\}$.

If both V_1 and V_2 have $G^{0,\mathrm{der}} = \Gamma$, then V_1 is a multiplicative translate of V_2.

proof By the previous lemma, for i=1,2 there exists $\lambda_i \in \mathbb{C}^\times$ and an isomorphism

$$[2]^*(V_i) \approx \mathcal{H}_{\lambda_i}(x, y, z; \varnothing)\otimes T^*\mathcal{H}_{\lambda_i}(x, y, z; \varnothing).$$

By a multiplicative translation, we may suppose $\lambda_1 = \lambda_2$. Then V_1 and V_2 are each [2]-descents of $\mathcal{H}_{\lambda_i}(x, y, z; \varnothing)\otimes T^*\mathcal{H}_{\lambda_i}(x, y, z; \varnothing)$ with nontrivial determinant, so by the unicity of such a descent V_1 and V_2 must be isomorphic. QED

Corollary 4.6.15.1 Suppose that $x+y+z \equiv 0 \bmod \mathbb{Z}$, and that none of x, y, z is $\equiv 0$ or $\pm 1/3 \bmod \mathbb{Z}$. Among all possible (A, B) where A is a choice of $\{\alpha_i\text{'s}\}$ satisfying

$\{\alpha_i\text{'s}\} = \{\tilde{x}, \tilde{y}, \tilde{z}, -x/2, -y/2, -z/2, 1/2 - x/2, 1/2 - y/2, 1/2 - z/2\}$,

$\tilde{x} = x$ **or** $1/2 + x$, $\tilde{y} = y$ **or** $1/2 + y$, $\tilde{z} = z$ **or** $1/2 + z$,

$\tilde{x} + \tilde{y} + \tilde{z} \equiv 0 \bmod \mathbb{Z}$,

and where B is a choice of $\{\beta_i\text{'s}\}$ satisfying

$\{2\beta_i\}_{i=1,2,3} = \{0, 1/3, 2/3\}$,

there is one and only one (A,B) which satisfies the following equivalent properties:

(1) There exists some $\mu \in \mathbb{C}^\times$ for which $\mathcal{H}_\mu(A, B)$ has $G^{0,\mathrm{der}} = \Gamma$.

(2) For every $\mu \in \mathbb{C}^\times$, $\mathcal{H}_\mu(A, B)$ has $G^{0,\mathrm{der}} = \Gamma$.

Lemma 4.6.16 Suppose that $x+y+z \equiv 0 \bmod \mathbb{Z}$, and that none of x, y, z is $\equiv 0$ or $\pm 1/3 \bmod \mathbb{Z}$. Let $\mu \in \mathbb{C}^\times$, and let

$A = \{x, y, z, -x/2, -y/2, -z/2, 1/2 - x/2, 1/2 - y/2, 1/2 - z/2\}$,

$B = \{0, 1/3, 2/3\}$,

Then

(1) $\mathcal{H}_\mu(A, B)$ is irreducible and not Kummer induced.

(2) If $G := G_{\mathrm{gal}}(\mathcal{H}_\mu(A, B))$ has $\dim G \le 16$, then $G = \Gamma \ltimes \{1, \sigma\}$, and for some $\lambda \in \mathbb{C}^\times$, there exists an isomorphism

$$[2]^*(\mathcal{H}_\mu(\alpha_i\text{'s}; \beta_j\text{'s})) \approx \mathcal{H}_\lambda(x, y, z; \varnothing)\otimes T^*\mathcal{H}_\lambda(x, y, z; \varnothing).$$

proof Since none of x, y, z is $\equiv 0$ or $\pm 1/3 \bmod \mathbb{Z}$, a fortiori none of their "halves" is either, whence the irreducibility. If $\mathcal{H}_\mu(A, B)$ is Kummer induced, it must be Kummer induced of degree 3 = gcd(9,3), in which case the set A is stable mod $\mathbb{Z}$ by $t \mapsto t + 1/3$. But then the set 2A of doubles is stable by $t \mapsto t + 2/3$. By the uniqueness 4.6.12, it follows that the set $\{x, y, z\}$ is stable by $t \mapsto t + 1/3$, in which case $\{x, y, z\}$ would be $\{x, x + 1/3, x + 2/3\}$, which is impossible since $x+y+z \equiv 0 \bmod \mathbb{Z}$, and none of x, y, z is $\equiv 0$ or $\pm 1/3 \bmod \mathbb{Z}$. Therefore $\mathcal{H}_\mu(A, B)$ is irreducible and not Kummer induced. From the general classification theorem 3.6, it now follows that $G^{0,der}$ is one of Γ, SO(9), or SL(9). So if $\dim G \le 16$ only the case $G^{0,der} = \Gamma$ is possible. The rest of assertion (2) now follows from 4.6.13 and 4.6.14. QED

Key Lemma 4.6.17 Suppose that $x+y+z \equiv 0 \bmod \mathbb{Z}$. Let $\mu \in \mathbb{C}^\times$. Let

$A = \{x, y, z, -x/2, -y/2, -z/2, 1/2 - x/2, 1/2 - y/2, 1/2 - z/2\}$,

$B = \{0, 1/3, 2/3\}$.

Then $\dim G_{gal}(\mathcal{H}_\mu(A, B)) \le 16$.

proof Since G_{gal} is invariant under multiplicative translation, it suffices to prove this when $\mu = 1$. By the specialization theorem, it suffices to treat the case when x and y are algebraically independent over $\mathbb{Q}$, and $z := -x -y$. More symmetrically, we work over the generic point of the parameter ring $\mathbb{Q}[x, y, z]/(x+y+z)\mathbb{Q}[x, y, z]$.

In this case, none of x, y, z is $\equiv 0$ or $\pm 1/3 \bmod \mathbb{Z}$. So by 4.6.15.1, among all possible $(\tilde{A}, \tilde{B})$ where $\tilde{A}$ is a choice of $\{\alpha_i\text{'s}\}$ satisfying

$\{\alpha_i\text{'s}\} = \{\tilde{x}, \tilde{y}, \tilde{z}, -x/2, -y/2, -z/2, 1/2 - x/2, 1/2 - y/2, 1/2 - z/2\}$,

$\tilde{x} = x$ **or** $1/2 + x$, $\tilde{y} = y$ **or** $1/2 + y$, $\tilde{z} = z$ **or** $1/2 + z$,

$\tilde{x} + \tilde{y} + \tilde{z} \equiv 0 \bmod \mathbb{Z}$,

and where $\tilde{B}$ is a choice of $\{\beta_i\text{'s}\}$ satisfying

$\{2\beta_i\}_{i=1,2,3} = \{0, 1/3, 2/3\}$,

there is one and only one $(\tilde{A}, \tilde{B})$ for which $\mathcal{H}_1(\tilde{A}, \tilde{B})$ has $G^{0,der} = \Gamma$.

Since x and y are independent variables, and $z= -x-y$, it is clear that for any of the $(\tilde{A}, \tilde{B})$ above, $\mathcal{H}_1(\tilde{A}, \tilde{B})$ is irreducible (indeed none of the $\tilde{A}$ exponents lies in $\mathbb{Q}$) and not Kummer induced (same argument as in 4.6.16 above). In view of the limited possibilities for $G := G_{gal}(\mathcal{H}_1(\tilde{A}, \tilde{B}))$, either $G^{0,der} = \Gamma$ or $\dim G > 16$.

Therefore among all possible $(\tilde{A}, \tilde{B})$ as above, there is one and only one $(\tilde{A}, \tilde{B})$ for which $\mathcal{H}_1(\tilde{A}, \tilde{B})$ has $\dim G_{gal} \le 16$.

We will first show by a symmetry argument that A is the correct $\tilde{A}$. Indeed, if σ is any permutation of the set $\{x, y, z\}$, then σ induces an automorphism, still denoted σ, of the parameter ring R := $\mathbb{Q}[x, y, z]/(x+y+z)\mathbb{Q}[x, y, z]$ over whose generic point we are working. This same permutation induces a permutation, also noted σ, of the four possibles $\tilde{A}$'s, and it is tautological that $\mathcal{H}_1(\sigma\tilde{A}, \tilde{B})$ is deduced from $\mathcal{H}_1(\tilde{A}, \tilde{B})$ by the extension of scalars σ: R → R. Since the formation of G_{gal} commutes with extensions of the ground field, it follows that for any σ, we have

$$\dim G_{gal}(\mathcal{H}_1(\sigma\tilde{A}, \tilde{B})) = \dim G_{gal}(\mathcal{H}_1(\tilde{A}, \tilde{B})).$$

Since there is a **unique** $(\tilde{A}, \tilde{B})$ where this dimension is ≤ 16, its $\tilde{A}$ must be a fixed under permutation of x, y, z. Among the the four possibles $\tilde{A}$'s, only A is fixed.

We next show by a symmetry argument that that correct $\tilde{B}$ is either {0, 1/3, 2/3} or {1/2, 5/6, 1/6}. Indeed, consider the automorphism τ of the parameter ring R := $\mathbb{Q}[x, y, z]/(x+y+z)\mathbb{Q}[x, y, z]$ given by $x \mapsto x + 1/3$, $y \mapsto y + 1/3$, $z \mapsto z - 2/3$. Let us denote by τA the image of A under τ. Then $\mathcal{H}_1(\tau A, \tilde{B})$ is obtained from $\mathcal{H}_1(A, \tilde{B})$ by the extension of scalars τ, so just as above we have

$$\dim G_{gal}(\mathcal{H}_1(\tau A, \tilde{B})) = \dim G_{gal}(\mathcal{H}_1(A, \tilde{B})).$$

On the other hand, given any $\tilde{B} := \{\beta_i\}$, let $\tilde{B} + 1/3 := \{\beta_i + 1/3\}$. Then $\mathcal{H}_1(A, \tilde{B})\otimes x^{1/3} \approx \mathcal{H}_1(\tau A, \tilde{B} + 1/3)$. So trivially

$$\dim G_{gal}(\mathcal{H}_1(A, \tilde{B})) = \dim G_{gal}(\mathcal{H}_1(\tau A, \tilde{B} + 1/3)),$$

while we have seen above that (replacing $\tilde{B}$ by $\tilde{B} + 1/3$)

$$\dim G_{gal}(\mathcal{H}_1(\tau A, \tilde{B} + 1/3)) = \dim G_{gal}(\mathcal{H}_1(A, \tilde{B} + 1/3)).$$

Therefore we have

$$\dim G_{gal}(\mathcal{H}_1(A, \tilde{B})) = \dim G_{gal}(\mathcal{H}_1(A, \tilde{B} + 1/3)).$$

Since there is a **unique** $(\tilde{A}, \tilde{B})$ where this dimension is ≤ 16, its $\tilde{B}$ must be a fixed under $\tilde{B} \mapsto \tilde{B} + 1/3$. Among the possible $\tilde{B}$'s, only {0, 1/3, 2/3} or {1/2, 5/6, 1/6} are so fixed.

It remains only to show that $\mathcal{H}_1(A; 1/2, 5/6, 1/6)$ has $\dim G_{gal} > 16$; for this is suffices, by the specialization theorem, to exhibit **particular** values of x and y where $\dim G_{gal} > 16$. For this, take $x = 1/6$, $y = -1/6$, $z = 0$. Then $\mathcal{H}_1(A; 1/2, 5/6, 1/6)$ becomes

$$\mathcal{H}_1(1/6, -1/6, 0, -1/12, 5/12, 1/12, 7/12, 0, 1/2; 1/2, 5/6, 1/6)$$

whose semisimplification contains, for some λ in $\mathbb{C}^\times$,

$$\mathcal{H}_\lambda(0, -1/12, 5/12, 1/12, 7/12, 0; \emptyset).$$

So it suffices to show that $\mathcal{H}_\lambda(0, -1/12, 5/12, 1/12, 7/12, 0; \emptyset)$ has its

G_{gal} of dimension > 16. This is a hypergeometric of type (6,0) which is (automatically) irreducible and which is not Kummer induced (its exponents are not stable by $\alpha \mapsto \alpha + 1/2$ or by $\alpha \mapsto \alpha + 1/3$). It is symplectic (by the Duality Recognition Theorem 3.4) So its G_{gal} is Sp(6), which has dimension 21 > 16. QED

Corollary 4.6.17.1 Suppose that $x+y+z \equiv 0 \bmod \mathbb{Z}$, and that none of x, y, z is $\equiv 0$ or $\pm 1/3 \bmod \mathbb{Z}$. Let $\mu \in \mathbb{C}^\times$, and let

$$A = \{x, y, z, -x/2, -y/2, -z/2, 1/2 - x/2, 1/2 - y/2, 1/2 - z/2\},$$
$$B = \{0, 1/3, 2/3\},$$

Then $G_{gal}(\mathcal{H}_\mu(A, B)) = \Gamma \ltimes \{1, \sigma\}$, and for some $\lambda \in \mathbb{C}^\times$, there exists an isomorphism

$$[2]^*(\mathcal{H}_\mu(\alpha_i\text{'s}; \beta_j\text{'s})) \approx \mathcal{H}_\lambda(x, y, z; \emptyset) \otimes T^*\mathcal{H}_\lambda(x, y, z; \emptyset).$$

proof This results formally from the preceding two lemmas. QED

This corollary establishes the truth of 4.6.10. Thus we obtain **SL(3)×SL(3) Theorem 4.6.18** Hypergeometrics V of type (9,3) with $G := G_{gal} = \Gamma \ltimes \{1, \sigma\}$ are precisely those (isomorphic to one) of the form

$$\mathcal{H}_\mu(x, y, z, -x/2, -y/2, -z/2, 1/2 - x/2, 1/2 - y/2, 1/2 - z/2; 0, \pm 1/3)$$

for some $\mu \in \mathbb{C}^\times$, and for some x, y, z in $\mathbb{C}$ which satisfy $x+y+z \equiv 0 \bmod \mathbb{Z}$, and none of x, y, z is $\equiv 0$ or $\pm 1/3 \bmod \mathbb{Z}$. Hypergeometrics V of type (9,3) with $G^{0,der} = \Gamma$ are precisely the x^δ twists of these.

CHAPTER 5
Convolution of $\mathcal{D}$-modules

5.1 Convolution of $\mathcal{D}$-modules; Generalities

(5.1.1) Given a smooth $\mathbb{C}$-scheme $X/\mathbb{C}$, we denote by $\mathcal{D}MOD(X)$ the abelian category of all sheaves of left $\mathcal{D}_X$-modules on X, by $D(X; \mathcal{D})$ its derived category, and by $D^{b,holo}(X)$ the full subcategory of $D(X; \mathcal{D})$ consisting of those objects K such that $\mathcal{H}^i(K)$ is holonomic for all i and such that $\mathcal{H}^i(K)$ vanishes for all but finitely many i. For morphisms $f: X \to Y$ between smooth separated $\mathbb{C}$-schemes of finite type, one knows (cf. [Ber], [Bor], [Ka-Lau], [Me-SO]) that these $D^{b,holo}$ support the full Grothendieck formalism of the "six operations". Of these, we will need only f_* and $f^!$, both of which have fairly concrete descriptions. (The operations $f_!$ and f^* are **defined** as the duals of these, and are consequently less amenable to direct inspection.)

(5.1.2) We will need f_* primarily when $f: X \to Y$ is **smooth** of relative dimension d; in this case one has $f_*K = \mathbf{R}f_*(K \otimes_{\mathcal{O}_X} \Omega^{\cdot}_{X/Y})[d]$, so except for the dimension shift we are "just" talking about relative De Rham cohomlogy:

$$\mathcal{H}^{i-d}(f_*K) = H^i_{DR}(X/Y, K), \text{ with its Gauss-Manin connection.}$$

The deep fact here is that for K a single holonomic left $\mathcal{D}_X$-module, each of the relative De Rham cohomology sheaves $H^i_{DR}(X/Y, K)$, with its Gauss-Manin connection, is holonomic on Y. The other case of f_* we will need is when $f: X \to Y$ is the inclusion of a $\mathbb{C}$-valued point $y \in Y(\mathbb{C})$. Then $f_*\mathcal{O}_X$ is the delta module δ_y.

(5.1.3) For a general $f : X \to Y$, and $\mathcal{M}$ on Y, $f^!\mathcal{M}$ is defined as

$$f^!\mathcal{M} = \mathcal{D}_{X\to Y} \overset{L}{\otimes}_{f^{-1}\mathcal{D}_Y} f^{-1}\mathcal{M}[\dim X - \dim Y],$$

where $\mathcal{D}_{X\to Y}$ is the $(\mathcal{D}_X, f^{-1}\mathcal{D}_Y)$-bimodule

$$\mathcal{D}_{X\to Y} := \mathit{DiffOps}(f^{-1}\mathcal{O}_Y, \mathcal{O}_X) = \mathcal{O}_X \otimes_{f^{-1}\mathcal{O}_Y} f^{-1}\mathcal{D}_Y.$$

(5.1.4) Here is a more concrete description in some important special cases. Denote by $f^+\mathcal{M}$ the naive pullback of $\mathcal{M}$ as module with integrable connection. If f is a flat morphism (e.g., if f is smooth), or if $\mathcal{M}$ is a flat $\mathcal{O}_Y$-module (e.g., if $\mathcal{M}$ is a D.E. on Y), then

$$f^!\mathcal{M} = f^+\mathcal{M}[\dim X - \dim Y].$$

For f etale, $f^! = f^+ = f^*$. For f smooth of relative dimension d, we have $f^* = f^![-2d] = f^+[-d]$.

If f is a closed immersion of codimension one, with X defined in Y by one equation x=0 in Y, then $f^!\mathcal{M}$ is the two term complex

$$\mathcal{M} \xrightarrow{m \mapsto xm} \mathcal{M}$$

placed in degrees zero and one.

(5.1.5) We will also use the following two elementary facts for an arbitrary f:

(1) (base change) given a cartesian diagram

$$\begin{array}{ccc} X' & \xrightarrow{G} & X : \mathcal{M} \\ \downarrow F & & f \downarrow \\ Y' & \xrightarrow{g} & Y \end{array}, \quad g^! f_* \mathcal{M} \approx F_* G^! \mathcal{M}.$$

(2) (projection formula) for any $\mathcal{O}$-coherent $\mathcal{D}$-module $\mathcal{L}$ on Y, denoting by $f^+\mathcal{L}$ its naive pullback as a D.E. on X, we have

$$(f_*K)\otimes_{\mathcal{O}_Y}\mathcal{L} \approx f_*(K\otimes_{\mathcal{O}_X} f^+\mathcal{L}).$$

(5.1.6) Given two smooth $\mathbb{C}$-schemes $X/\mathbb{C}$ and $Y/\mathbb{C}$, "external tensor product over $\mathbb{C}$" defines a bi-exact bilinear pairing,

$$\mathcal{D}MOD(X)\times\mathcal{D}MOD(Y) \to \mathcal{D}MOD(X\times_{\mathbb{C}}Y)$$

$$(\mathcal{M}, \mathcal{N}) \mapsto \mathcal{M}\times\mathcal{N},$$

which passes to $D^{b,holo}$.

If K and L are objects of $\mathcal{D}MOD(X)$, we define their "exotic" tensor product, denoted $K\otimes^! L$, in terms of the diagonal map $\Delta: X \to X\times_{\mathbb{C}}X$, by

$$K\otimes^! L := \Delta^!(K\times L).$$

(5.1.7) If $G/\mathbb{C}$ is a smooth separated $\mathbb{C}$-groupscheme, we denote by the group law by

$$\text{product}_G : G\times_{\mathbb{C}}G \to G.$$

We define the convolution of objects of $D^{b,holo}(G)$ by

(5.1.7.1) $$(K, L) \mapsto K*L := (\text{product}_G)_*(K\times L).$$

The operation of convolution is associative, and the δ-module δ_e supported at the identity of G is a two-sided identity object. [For if we denote by $\iota: e \to G$ the inclusion of the identity, then for any K in $D^{b,holo}(G)$, we have

$$K = \mathcal{O}_e \times K \text{ on } e \times_{\mathbb{C}} G \approx G,$$

$$(\iota \times id_G)_*(\mathcal{O}_e \times K) = (\iota_* \mathcal{O}_e) \times K = \delta_e \times K \text{ on } G \times_{\mathbb{C}} G.$$

Since the composite map $(\text{product}_G) \circ (\iota \times id_G)$ is id_G, the result follows.]

If G is commutative, then convolution is commutative as well.

(5.1.8) In general, even if we start with two holonomic $\mathcal{D}$-modules $\mathcal{M}$ and $\mathcal{N}$ on G, viewed as objects of $D^{b,holo}(G)$ which are concentrated in degree zero, their convolution $\mathcal{M} * \mathcal{N}$ is "really" an object of $D^{b,holo}(G)$, and not simply a single holonomic $\mathcal{D}$-module placed in degree zero. It is this "instability" of $\mathcal{D}$-modules themselves under convolution that makes $D^{b,holo}$ the natural setting.

(5.1.9) The following formal properties of convolution are quite useful.

(1a) If $\varphi: G \to H$ is a homomorphism of smooth separated $\mathbb{C}$-groupschemes of finite type, then

$$\varphi_*(K * L) \approx (\varphi_* K) * (\varphi_* L).$$

This results from the fact that $(\varphi \times \varphi)_*(K \times L) = (\varphi_* K) \times (\varphi_* L)$ (valid for any φ) and the fact that $\text{product}_H \circ (\varphi \times \varphi) = \varphi \circ \text{product}_G$ (φ being a homomorphism).

In the special case when H is the trivial group, this becomes: Denote by $\pi: G \to \text{Spec}(\mathbb{C})$ the structural map. Then for any two objects K, L in $D^{b,holo}(G)$, we have

$$\pi_*(K * L) \approx (\pi_* K) \otimes_{\mathbb{C}} (\pi_* L).$$

(1b) If $\varphi : G \to G$ is a homomorphism, then for any two objects K, L in $D^{b,holo}(G)$, we have

$$\varphi^!((\varphi_* K) * L) \approx K * (\varphi^! L).$$

This is base change for the following commutative diagram, whose outer square is cartesian (verification left to the reader):

$$\begin{array}{ccc}
G \times G & \xrightarrow{\quad id \times \varphi \quad} & G \times G \\
 & & \downarrow \varphi \times id \\
\text{prod}_G \downarrow & & G \times G \\
 & & \downarrow \text{prod}_G \\
G & \xrightarrow{\quad \varphi \quad} & G.
\end{array}$$

(1c) If $\varphi : G \to H$ is a homomorphism, then for K in $D^{b,holo}(G)$ and L in $D^{b,holo}(H)$, we have

$$\varphi^!((\varphi_* K)*L) \approx K*(\varphi^! L).$$

This is base change for the following commutative diagram, whose outer square is cartesian (verification left to the reader):

$$\begin{array}{cccc} & G\times G & \xrightarrow{\quad id\times\varphi \quad} & G\times H \\ & & & \downarrow \varphi\times id \\ prod_G & \downarrow & & H\times H \\ & & & \downarrow prodH \\ & G & \xrightarrow{\quad \varphi \quad} & H \end{array}$$

(2) For $g \in G(\mathbb{C})$ denote by $T_g : G \to G$ the map $x \mapsto gx$ "translation by g", and by δ_g the delta module supported at g. Then for $g \in G(\mathbb{C})$, we have

$$(T_g)_*(K*L) \approx ((T_g)_* K)*L,$$
$$(T_g)_*(L) \approx (\delta_g)*L.$$

The first results from $T_g \circ product_G = product_G \circ (T_g \times id_G)$. The second is the special case $K = \delta_e$ of the first, since $(T_g)_*(\delta_e) = \delta_g$, and δ_e is the convolutional identity.

(5.1.10) In discussing convolution, it is sometimes convenient to take a slightly assymmetric point of view. Denoting points of $G\times G$ as (x,y), we can factor the product map as the composition of pr_2 with the shearing involution shear: $(x,y) \mapsto (x^{-1}, xy)$ of $G\times G$. So we find, in an obvious notation,

$$K*L := (product_G)_*(K\times L) = (pr_2)_*(shear)_*(K\times L) =$$
$$= (pr_2)_*(K(x^{-1})\otimes L(xy)) := \int K(x^{-1})\otimes L(xy)dx$$
$$= (pr_2)_*((pr_1{}^+ inv_* K)\otimes(product_G)^+ L).$$

5.2 Convolution on $\mathbb{G}_m$ and Fourier transform on $\mathbb{A}^1$

We now turn to the case of particular interest to us, when G is $\mathbb{G}_m$.

Lemma 5.2.1 For α in $\mathbb{C}$, and K, L in $D^{b,holo}(\mathbb{G}_m)$, we have

$$(K\otimes x^\alpha)*(L\otimes x^\alpha) \approx (K*L)\otimes x^\alpha.$$

proof Let T_α denote the $\mathcal{D}$-module $x^\alpha\mathbb{C}[x, x^{-1}]$. The $\mathcal{D}$-module version of " $(xy)^\alpha = (x)^\alpha(y)^\alpha$ " is

$$\text{product}^+(T_\alpha) = T_\alpha \times T_\alpha,$$

so the assertion is immediate from the projection formula. QED

(5.2.2) Denote by inv: $\mathbb{G}_m \to \mathbb{G}_m$ the multiplicative inversion, by $j: \mathbb{G}_m \to \mathbb{A}^1$ the inclusion, $\partial := d/dx$ on $\mathbb{A}^1$, by $\mathcal{L}$ the $\mathcal{D}$-module $\mathcal{D}_{\mathbb{A}^1}/\mathcal{D}_{\mathbb{A}^1}(\partial - 1)$ on $\mathbb{A}^1$ which is the D.E. for e^x, and by $j^*\mathcal{L}$ its restriction to $\mathbb{G}_m$. Thus $j^*\mathcal{L} = \mathcal{D}_{\mathbb{G}_m}/\mathcal{D}_{\mathbb{G}_m}(x\partial - x) = \mathcal{H}_1(0; \varnothing) = \mathcal{H}(t, 1)$ is the basic hypergeometric of type (1,0) on $\mathbb{G}_m$.

Key Lemma 5.2.3 (Compare [Ka-GKM, 8.6.1]) Convolution with $j^*\mathcal{L} = \mathcal{H}_1(0; \varnothing) = \mathcal{H}(t, 1)$ on $\mathbb{G}_m$ and Fourier Transform on $\mathbb{A}^1$ are related as follows: for any holonomic $\mathcal{D}$-module $\mathcal{M}$ on $\mathbb{G}_m$, we have

$$j^*FT(j_*inv_*(\mathcal{M})) \approx \mathcal{M}*j^*\mathcal{L} = \mathcal{M}*\mathcal{H}_1(0; \varnothing) = \mathcal{M}*\mathcal{H}(t,1).$$

proof We have

$$FT(j_*inv_*(\mathcal{M})) = \int_{\mathbb{A}^1} (j_*inv_*(\mathcal{M}))(x)e^{xy}dx$$
$$= \int_{\mathbb{A}^1} (j_*inv_*(\mathcal{M}))(x)\mathcal{L}(xy)dx.$$

Denote by

$pr_1: \mathbb{G}_m\times\mathbb{G}_m \to \mathbb{G}_m$ the first projection,

$pr_2: \mathbb{G}_m\times\mathbb{G}_m \to \mathbb{G}_m$ the second projection,

$\tilde{j} := (j\times id_{\mathbb{G}_m}) : \mathbb{G}_m\times\mathbb{G}_m \to \mathbb{A}^1\times\mathbb{G}_m$ the inclusion,

$\tilde{pr}_2 : \mathbb{A}^1\times\mathbb{G}_m \to \mathbb{G}_m$ the second projection,

$\tilde{pr}_1 : \mathbb{A}^1\times\mathbb{G}_m \to \mathbb{A}^1$ the first projection

$\widetilde{\text{product}} : \mathbb{A}^1\times\mathbb{G}_m \to \mathbb{A}^1$ the multiplication $(x, \lambda) \mapsto x\lambda$.

So with these notations, we can rewite the above FT formula as

$$j^*FT(j_*inv_*\mathcal{M}) = (\tilde{pr}_2)_*((j_*inv_*\mathcal{M})(x)e^{xy}).$$

Using smooth base change and the projection formula, we find

$(\tilde{pr}_2)_*((j_*inv_*\mathcal{M})(x)e^{xy}) =$

$$= (\tilde{pr}_2)_*(\tilde{pr}_1{}^+(j_*inv_*\mathcal{M}).\otimes(\widetilde{\text{product}})^+\mathcal{L}) =$$

$= (\tilde{pr}_2)_*\tilde{j}_*((pr_1{}^+inv_*\mathcal{M})\otimes(\tilde{j})^+\tilde{product}^+\mathcal{L}) =$

$= (pr_2)_*((pr_1{}^+inv_*\mathcal{M})\otimes product^+j^*\mathcal{L}) =$

$= \mathcal{M}*j^*\mathcal{L}.$ QED

Corollary 5.2.3.1 For any holonomic $\mathcal{D}$-module $\mathcal{M}$ on $\mathbb{G}_m$, we have

$$inv_*j^*FT(j_*\mathcal{M}) \approx \mathcal{M}*(inv_*j^*\mathcal{L}).$$

proof The Key Lemma applied to $inv_*\mathcal{M}$ gives

$$j^*FT(j_*\mathcal{M}) = (inv_*\mathcal{M})*(j^*\mathcal{L}).$$

Because inv: $\mathbb{G}_m \to \mathbb{G}_m$ is a group homomorphism, we get

$$inv_*j^*FT(j_*\mathcal{M}) = (inv_*inv_*\mathcal{M})*(inv_*j^*\mathcal{L}) = \mathcal{M}*(inv_*j^*\mathcal{L}).$$ QED

5.3 Convolution of Hypergeometrics on $\mathbb{G}_m$

We begin by explaining the heuristic motivation. Let P, Q, R, and S be four nonzero polynomials in $\mathbb{C}[t]$. Recall the hypergeometric differential operators

$$Hyp(P, Q) := P(xd/dx) - xQ(xd/dx),$$
$$Hyp(R, S) := R(xd/dx) - xS(xd/dx),$$

and the associated $\mathcal{D}$-modules on $\mathbb{G}_m$

$$\mathcal{H}(P, Q) := \mathcal{D}/\mathcal{D}Hyp(P, Q),$$
$$\mathcal{H}(R, S) := \mathcal{D}/\mathcal{D}Hyp(R, S).$$

A formal series $f(x) := \Sigma a_n x^n$ is killed by Hyp(P, Q) if and only if its coefficients a_n satisfy the two-term recurrence relation

$$P(n)a_n = Q(n-1)a_{n-1}.$$

Similarly, a formal series $g(x) := \Sigma b_n x^n$ is killed by Hyp(R, S) if and only if its coefficients b_n satisfy the two-term recurrence relation

$$R(n)b_n = S(n-1)b_{n-1}.$$

Thus if $f(x) := \Sigma a_n x^n$ and $g(x) := \Sigma b_n x^n$ are formal series solutions of Hyp(P, Q) and of Hyp(R, S) respectively, then their "convolution"

$$(f*g)(x) := \Sigma a_n b_n x^n$$

is visibly a formal solution of Hyp(PR, QS). This suggests that, at least under reasonable hypotheses, one should have

$$\mathcal{H}(P, Q)*\mathcal{H}(R, S) \approx \mathcal{H}(PR, QS)$$

as $\mathcal{D}$-modules on $\mathbb{G}_m$.

Convolution Theorem 5.3.1 Suppose that P, Q, R, and S are four nonzero polynomials in $\mathbb{C}[t]$, such that the two polynomials PR and QS have no common zeroes mod $\mathbb{Z}$, i.e., whenever $(PR)(\alpha) = 0 = (QS)(\beta)$, $\alpha - \beta$ is not an integer. Then

$$\mathcal{H}(P, Q)*\mathcal{H}(R, S) \approx \mathcal{H}(PR, QS)$$

as $\mathcal{D}$-modules on $\mathbb{G}_m$.

proof We proceed by induction on deg(RS).

If deg(RS) = 0, then both R and S are nonzero constants, say r and s. Then Hyp(R, S) = r−sx, so $\mathcal{H}(R, S) = \delta_{r/s}$ is the delta module supported at r/s. Similarly, we see that $\mathcal{H}(PR, QS) = \mathcal{H}(rP, sQ) = (T_{r/s})_*\mathcal{H}(P, Q)$ (cf. 3.1). Since convolution is commutative on $\mathbb{G}_m$, the assertion to be proven is

$$(\delta_{r/s})*\mathcal{H}(P, Q) \approx (T_{r/s})_*\mathcal{H}(P, R),$$

which is the translation formula 5.1.9 (2) (with g = r/s).

Suppose now that R or S is nonconstant. By multiplicative inversion, it suffices to treat the case when R has degree ≥ 1. Twisting by x^α, we reduce by 5.2.1 to the case where the polynomial R(t) is divisible by t, say $R(t) = tR_0(t)$. By induction, we know that

$$\mathcal{H}(P, Q)*\mathcal{H}(R_0, S) \approx \mathcal{H}(PR_0, QS).$$

Since convolution is commutative,we know by induction that

$$\mathcal{H}(R_0, S)*\mathcal{H}(t, 1) \approx \mathcal{H}(t, 1)*\mathcal{H}(R_0, S) \approx \mathcal{H}(R, S).$$

Therefore by the associativity of convolution we obtain

$$\mathcal{H}(P, Q)*\mathcal{H}(R, S) \approx \mathcal{H}(P, Q)*\mathcal{H}(R_0, S)*\mathcal{H}(t, 1) \approx$$
$$\approx \mathcal{H}(PR_0, QS)*\mathcal{H}(t, 1).$$

So we are reduced to showing universally that

$$\mathcal{H}(P(t), Q(t))*\mathcal{H}(t, 1) \approx \mathcal{H}(tP(t), Q(t))$$

whenever tP(t) and Q(t) have no common zeroes mod $\mathbb{Z}$. By 5.2.3, we have, denoting by $j: \mathbb{G}_m \to \mathbb{A}^1$ the inclusion,

$$\mathcal{H}(P(t), Q(t))*\mathcal{H}(t, 1) \approx j^*FT(j_*\mathrm{inv}_*\mathcal{H}(P(t), Q(t)))$$
$$\approx j^*FT(j_*\mathcal{H}(Q(-t), P(-t))).$$

Since Q(−t) has no zeroes in $\mathbb{Z}$, it follows from 2.9.4, (3) ⇔ (5) that

$$j_*\mathcal{H}(Q(-t), P(-t)) := j_*j^*(\mathcal{D}_{\mathbb{A}^1}/\mathcal{D}_{\mathbb{A}^1}\mathrm{Hyp}(Q(-t), P(-t)))$$
$$\approx \mathcal{D}_{\mathbb{A}^1}/\mathcal{D}_{\mathbb{A}^1}\mathrm{Hyp}(Q(-t), P(-t)),$$

whence

$$j^*FT(j_*\mathcal{H}(Q(-t), P(-t))) \approx j^*FT(\mathcal{D}_{\mathbb{A}^1}/\mathcal{D}_{\mathbb{A}^1}\mathrm{Hyp}(Q(-t), P(-t)))$$
$$\approx j^*(\mathcal{D}_{\mathbb{A}^1}/\mathcal{D}_{\mathbb{A}^1}FT(\mathrm{Hyp}(Q(-t), P(-t))))$$

$$\approx \mathcal{D}_{\mathbb{G}_m}/\mathcal{D}_{\mathbb{G}_m}\mathrm{FT}(\mathrm{Hyp}(Q(-t), P(-t))).$$

It is a simple matter to compute FT, since $\mathrm{FT}(x) = \partial := d/dx$, $\mathrm{FT}(\partial) = -x$, and $\mathrm{FT}(-xd/dx) = \partial x = 1 + x\partial$. We find

$$\begin{aligned}\mathrm{FT}(\mathrm{Hyp}(Q(-t), P(-t))) &:= \mathrm{FT}(Q(-xd/dx) - xP(-xd/dx))\\ &= Q(\mathrm{FT}(-xd/dx)) - \partial P(\mathrm{FT}(-xd/dx))\\ &= Q(1 + xd/dx) -(1/x)(xd/dx)P(1 + xd/dx)\\ &= (-1/x)[(xd/dx)P(1 + xd/dx) - xQ(1 + xd/dx)]\\ &= (-1/x)\mathrm{Hyp}(tP(1 + t), Q(1 + t)).\end{aligned}$$

Therefore we have

$$\mathcal{D}_{\mathbb{G}_m}/\mathcal{D}_{\mathbb{G}_m}\mathrm{FT}(\mathrm{Hyp}(Q(-t), P(-t))) \approx \mathcal{H}(tP(1 + t), Q(1 + t)).$$

By 3.2 and 3.2.1, $\mathcal{H}(tP(1 + t), Q(1 + t)) \approx \mathcal{H}(tP(t), Q(t))$. QED

Making this explicit in terms of the exponents, we find

Explicit Variant 5.3.2 We have

$$\mathcal{H}_\lambda(\alpha\text{'s}; \beta\text{'s})*\mathcal{H}_\mu(\gamma\text{'s}, \delta\text{'s}) \approx \mathcal{H}_{\lambda\mu}(\alpha\text{'s}, \gamma\text{'s}; \beta\text{'s}, \delta\text{'s})$$

provided that $\mathcal{H}_{\lambda\mu}(\alpha\text{'s}, \gamma\text{'s}; \beta\text{'s}, \delta\text{'s})$ is irreducible, i.e., provided that no element of the set $\{\alpha\text{'s}, \gamma\text{'s}\}$ is congruent mod $\mathbb{Z}$ to any element of the set $\{\beta\text{'s}, \delta\text{'s}\}$.

Corollary 5.3.2.1 All irreducible hypergeometrics on $\mathbb{G}_m$ can be built out of δ_1 and $\mathcal{H}_1(0; \varnothing)$, using only the following operations on holonomic $\mathcal{D}$-modules on $\mathbb{G}_m$:

(1) convolution

(2) $\mathcal{M} \mapsto (T_\alpha)_*\mathcal{M}$

(3) $\mathcal{M} \mapsto \mathcal{M}\otimes x^\delta$

(4) $\mathcal{M} \mapsto \mathrm{inv}^*\mathcal{M}$.

Specializing the Key Lemma 5.2.3 and its Corollary 5.2.3.1 to the case of hypergeometrics, we find

Proposition 5.3.3 Let $\mathcal{H} := \mathcal{H}_\lambda(\alpha_1, \ldots, \alpha_n; \beta_1, \ldots, \beta_m)$, with no $\alpha_i - \beta_j$ in $\mathbb{Z}$.

If no β_j lies in $\mathbb{Z}$, then

$$j^*\mathrm{FT}(j_*\mathrm{inv}_*\mathcal{H}) \approx \mathcal{H}_\lambda(0, \alpha_1, \ldots, \alpha_n; \beta_1, \ldots, \beta_m).$$

If no α_i lies in $\mathbb{Z}$, then

$$\mathrm{inv}_* j^* \mathrm{FT}(j_* \mathcal{H}) \approx \mathcal{H}_{-\lambda}(\alpha_1, \ldots, \alpha_n; 0, \beta_1, \ldots, \beta_m).$$

Corollary 5.3.3.1 All irreducible hypergeometrics can be built out of the delta module δ_1 on $\mathbb{G}_m$ using only the the following operations on holonomic $\mathcal{D}$-modules on $\mathbb{G}_m$:

(1) $\mathcal{M} \mapsto j^*\mathrm{FT}(j_*\mathrm{inv}_*\mathcal{M})$

(2) $\mathcal{M} \mapsto \mathrm{inv}_*j^*\mathrm{FT}(j_*\mathcal{M})$

(3) $\mathcal{M} \mapsto (T_\alpha)_*\mathcal{M}$

(4) $\mathcal{M} \mapsto \mathcal{M}\otimes x^\delta$

(5) $\mathcal{M} \mapsto \mathrm{inv}^*\mathcal{M}$.

5.4 Motivic Interpretation of Hypergeometrics of type (n,n)

In our earlier discussion of the determination of $(G_{gal})^{0,der}$ for irreducible hypergeometrics of type (n,n),

$$\mathcal{H}_\lambda(\alpha_1, \ldots, \alpha_n; \beta_1, \ldots, \beta_n), \text{ no } \alpha_i - \beta_j \text{ lies in } \mathbb{Z},$$

we put aside the problem of recognizing when G_{gal} is a finite group. We now confront this problem. Since G_{gal} is invariant by multiplicative translation, it suffices to treat the case $\lambda = 1$. By 3.2.2(3), a trivial necessary condition for the finiteness of G_{gal} is that the α's and β's all lie in $\mathbb{Q}$. By the convolution theorem 5.3.1, $\mathcal{H}_1(\alpha_1, \ldots, \alpha_n; \beta_1, \ldots, \beta_n)$ is a multiple convolution:

$$\mathcal{H}_1(\alpha_1, \ldots, \alpha_n; \beta_1, \ldots, \beta_n) \approx \mathcal{H}_1(\alpha_1; \beta_1)*\mathcal{H}_1(\alpha_2; \beta_2)* \ldots *\mathcal{H}_1(\alpha_n; \beta_n).$$

As we will now explain, this expression for $\mathcal{H}_1(\alpha_1, \ldots, \alpha_n; \beta_1, \ldots, \beta_n)$ leads directly to its motivic interpretation.

(5.4.1) The first step, then, is to understand completely $\mathcal{H}_1(\alpha, \beta)$, when $\alpha, \beta \in \mathbb{Q}$ and $\alpha - \beta$ is not an integer. Let N be a common denominator for α and β, and define integers A, B, C by

$$A := N\alpha,\ B := N\beta,\ C := B - A.$$

Denote by U the open set $\mathbb{G}_m - \{1\}$ of $\mathbb{G}_m$, and by $j: U \to \mathbb{G}_m$ the inclusion. Denote by Z the subvariety of $U\times\mathbb{G}_m$ (coordinates x,y) of equation

$$y^N = x^A(1 - x)^C.$$

This Z is a smooth (y is always invertible) affine curve on which μ_N acts (on y). Via pr_1, Z is a finite etale μ_N-torsor over $\mathbb{G}_m - \{1\}$. We denote by π the etale map

$$\pi : Z \to \mathbb{G}_m, \ \pi(x, y) := x,$$

which factors through j as

$$\begin{array}{ccc} Z & & \\ \mathrm{pr}_1 \downarrow & \searrow \pi & \\ U \xrightarrow{\ j\ } & \mathbb{G}_m. & \end{array}$$

The map π is μ_N -equivariant, for the given action $(x,y) \mapsto (x, \zeta y)$ on Z, and the trivial action on $\mathbb{G}_m$. Therefore the group μ_N acts on the holonomic $\mathcal{D}$-module $\pi_*\mathcal{O}_Z$. This action gives a direct sum decomposition

$$\pi_*\mathcal{O}_Z \approx \oplus_{\text{char.'s } \chi \text{ of } \mu_N} (\pi_*\mathcal{O}_Z)_\chi$$

into isotypical components.

Lemma 5.4.2 For each **faithful** character χ_r of μ_N,

$$\chi_r(\zeta) := \zeta^r \text{ with } \gcd(r, N) = 1, \ 1 \le r < N,$$

the χ_r-isotypical component of $\pi_*\mathcal{O}_Z$ is given by

$$(\pi_*\mathcal{O}_Z)_{\chi_r} \approx \mathcal{H}_1(r\alpha; r\beta).$$

proof Since $\pi = j \circ \mathrm{pr}_1$, we have

$$\pi_*\mathcal{O}_Z \approx j_*\mathrm{pr}_{1*}\mathcal{O}_Z.$$

Because pr_1 is μ_N -equivariant, we have a direct sum decomposition

$$\mathrm{pr}_{1*}\mathcal{O}_Z \approx \oplus_{\text{char.'s } \chi \text{ of } \mu_N} (\mathrm{pr}_{1*}\mathcal{O}_Z)_\chi,$$

whose direct image by j is the decomposition

$$\pi_*\mathcal{O}_Z \approx \oplus_{\text{char.'s } \chi \text{ of } \mu_N} (\pi_*\mathcal{O}_Z)_\chi.$$

Thus for **any** character χ of μ_N,

$$(\pi_*\mathcal{O}_Z)_\chi \approx j_*((\mathrm{pr}_{1*}\mathcal{O}_Z)_\chi).$$

As an $\mathcal{O}_U$ -module, $\mathcal{O}_Z$ is free on $1, y, \ldots, y^{N-1}$. So for any r we have

$$(\mathrm{pr}_{1*}\mathcal{O}_Z)_{\chi_r} \approx y^r\mathcal{O}_U.$$

But $j^*\mathcal{H}_1(r\alpha, r\beta) := \mathcal{D}_U/\mathcal{D}_U\mathrm{Hyp}_1(r\alpha, r\beta) \approx y^r\mathcal{O}_U$ by the map $1 \mapsto y^r$, simply because $y^r = x^{r\alpha}(1 - x)^{r\beta - r\alpha}$, and $\mathrm{Hyp}_1(r\alpha, r\beta)$ is the monic first order operator on U which kills $x^{r\alpha}(1 - x)^{r\beta - r\alpha}$. So we have

$$(\mathrm{pr}_{1*}\mathcal{O}_Z)_{\chi_r} \approx j^*\mathcal{H}_1(r\alpha, r\beta).$$

If $\gcd(r, N) = 1$, then $r\alpha - r\beta$, the exponent at 1 of $\mathcal{H}_1(r\alpha, r\beta)$, is a noninteger, so by 2.9.4, $\mathcal{H}_1(r\alpha, r\beta) \approx j_*j^*\mathcal{H}_1(r\alpha, r\beta)$. Thus

$$(\pi_*\mathcal{O}_Z)_{\chi_r} \approx j_*((pr_{1*}\mathcal{O}_Z)_{\chi_r}) \approx j_*j^*\mathcal{H}_1(r\alpha, r\beta) \approx \mathcal{H}_1(r\alpha, r\beta). \quad \text{QED}$$

The next step is to interpret the convolution of two direct images as a direct image.

Lemma 5.4.3 Suppose that

$$f: X \to \mathbb{G}_m, \qquad g: Y \to \mathbb{G}_m,$$

are two smooth morphisms. Then the "product" morphism

$$fg : X\times_{\mathbb{C}}Y \to \mathbb{G}_m, \quad (fg)(x,y) := f(x)g(y),$$

is smooth, and there is a canonical isomorphism

$$(fg)_*\mathcal{O}_{X\times_{\mathbb{C}}Y} \approx (f_*\mathcal{O}_X)*(g_*\mathcal{O}_Y).$$

If in addition we are given a finite group G acting f-linearly on X and a finite group H acting g-linearly on Y, then for any irreducible representations ρ of G and χ of H, the corresponding isotypical components are related by

$$((fg)_*\mathcal{O}_{X\times_{\mathbb{C}}Y})_{\rho\otimes\chi} \approx ((f_*\mathcal{O}_X)_\rho)*((g_*\mathcal{O}_Y)_\chi).$$

proof The map fg is the composite of the smooth map $\text{product}_{\mathbb{G}_m}$ with the smooth map

$$f\times g : X\times_{\mathbb{C}}Y \to \mathbb{G}_m\times_{\mathbb{C}}\mathbb{G}_m,$$

so fg is smooth. Since the trivial $\mathcal{D}$-module $\mathcal{O}_{X\times_{\mathbb{C}}Y}$ is the external tensor product $(\mathcal{O}_X)\times(\mathcal{O}_Y)$, we have

$$\begin{aligned}(fg)_*\mathcal{O}_{X\times_{\mathbb{C}}Y} &= (\text{product}_{\mathbb{G}_m})_*(f\times g)_*((\mathcal{O}_X)\times(\mathcal{O}_Y))\\ &= (\text{product}_{\mathbb{G}_m})_*((f_*\mathcal{O}_X)\times(g_*\mathcal{O}_Y))\\ &:= (f_*\mathcal{O}_X)*(g_*\mathcal{O}_Y).\end{aligned}$$

In the presence of finite group actions, we have

$$f_*\mathcal{O}_X \approx \oplus_\rho (f_*\mathcal{O}_X)_\rho, \qquad g_*\mathcal{O}_Y \approx \oplus_\chi (g_*\mathcal{O}_Y)_\chi,$$

whence

$$(f_*\mathcal{O}_X)*(g_*\mathcal{O}_Y) \approx \oplus_{\rho,\chi} ((f_*\mathcal{O}_X)_\rho)*((g_*\mathcal{O}_Y)_\chi),$$

and this is visibly the G×H -isotypical decomposition of

$$(f_*\mathcal{O}_X)*(g_*\mathcal{O}_Y) \approx (fg)_*\mathcal{O}_{X\times_{\mathbb{C}}Y}. \qquad \text{QED}$$

Theorem 5.4.4 Let $\alpha_1, \dots, \alpha_n$ and $\beta_1, \dots, \beta_n$ be 2n rational numbers, such that for all (i, j), $\alpha_i - \beta_j$ is not an integer. Pick a common denominator N for all the α's and β's. For each i = 1, ... , n, define integers A(i), B(i), C(i) by

$$A(i) := N\alpha_i, \quad B(i) := N\beta_i, \quad C(i) := B(i) - A(i).$$

Denote by U the open set $\mathbb{G}_m - \{1\}$ of $\mathbb{G}_m$. Denote by Z(i) the subvariety of $U \times \mathbb{G}_m$ (coordinates x_i, y_i) of equation

$$y_i^N = x_i^{A(i)}(1 - x_i)^{C(i)}.$$

This Z(i) is a smooth (y_i is always invertible) affine curve on which μ_N acts (on y_i). We denote by $\pi(i)$ the etale, μ_N -equivariant map

$$\pi(i) : Z(i) \to \mathbb{G}_m, \ \pi(x_i, y_i) := x_i.$$

Denote by

$$Z := Z(1) \times_{\mathbb{C}} \dots \times_{\mathbb{C}} Z(n),$$

and by

$\pi : Z \to \mathbb{G}_m$ the "product" map $\pi(x_1, y_1, \dots, x_n, y_n) := \prod_i x_i$.

Then π is equivariant for the product action of $(\mu_N)^n$ on Z. For every n-tuple of **faithful** characters $(\chi_{r_1}, \dots, \chi_{r_n})$ of μ_N, i.e., $\gcd(r_i, N) = 1$ for each i, the $(\chi_{r_1}, \dots, \chi_{r_n})$-isotypical component of $\pi_*\mathcal{O}_Z$ is given by

$$(\pi_*\mathcal{O}_Z)(\chi_{r_1}, \dots, \chi_{r_n}) \approx \mathcal{H}_1(r_1\alpha_1, \dots, r_n\alpha_n; r_1\beta_1, \dots, r_n\beta_n).$$

Equivalently, for every n-tuple of **faithful** characters $(\chi_{r_1}, \dots, \chi_{r_n})$ of μ_N, the $(\chi_{r_1}, \dots, \chi_{r_n})$-isotypical component of the relative De Rham cohomology sheaf $H^i{}_{DR}(Z/\mathbb{G}_m) := R^i\pi_*\Omega^\bullet{}_{Z/\mathbb{G}_m}$ with its Gauss-Manin connection is given by

$$(H^{n-1}{}_{DR}(Z/\mathbb{G}_m))(\chi_{r_1}, \dots, \chi_{r_n}) \approx \mathcal{H}_1(r_1\alpha_1, \dots, r_n\alpha_n; r_1\beta_1, \dots, r_n\beta_n)$$

$$(H^i{}_{DR}(Z/\mathbb{G}_m))(\chi_{r_1}, \dots, \chi_{r_n}) = 0 \text{ if } i \neq n-1.$$

proof By 5.4.2 this is true for n=1. It then follows for general n by induction, thanks to 5.4.3 and the Convolution Theorem 5.3.1. QED

5.5 Application to Grothendieck's p-curvature conjecture

In 1969, Grothendieck pointed out that, for a general D.E. on a smooth variety Y over $\mathbb{C}$, "p-curvature zero for almost all primes p" of any arithmetic "thickening" was a necessary condition for the finiteness of G_{gal}, and he asked whether it was also a sufficient condition.

It was proven in [Ka-AS, 5.7] that the equivalence

(**) $\quad G_{gal}$ finite $\Leftrightarrow$ p-curvature zero for almost all primes p

holds for Picard-Fuchs equations, i.e. for (the restriction to a dense open set of Y of the) the relative De Rham cohomology sheaves $H^i_{DR}(X/Y)$ of smooth morphisms $f: X \to Y$, endowed with the Gauss-Manin connection. Moreover, if a finite group G acts f-linearly on X, and if we pick any irreducible $\mathbb{C}$-representation ρ of G and denote $\rho_1, \dots, \rho_d$ its distinct conjugates by $\mathrm{Aut}(\mathbb{C}/\mathbb{Q})$, this equivalence (**) was proven to hold for the direct factor $\oplus_j (H^i_{DR}(X/Y))_{\rho_j}$.

Let us apply this result to the smooth morphism $\pi: Z \to \mathbb{G}_m$, the finite group $G = (\mu_N(\mathbb{C}))^n$, and the one-dimensional representation ρ of G defined by $\rho(\zeta_1, \dots, \zeta_n) := \prod_i \zeta_i$. Thus in the notations of 5.4.4 ρ is $(\chi_1, \dots, \chi_1)$, d is $\varphi(N) := \mathrm{Card}((\mathbb{Z}/N\mathbb{Z})^\times)$, and the various ρ_i are the characters $(\chi_r, \dots, \chi_r)$, as r runs over $(\mathbb{Z}/N\mathbb{Z})^\times$. We find

p-Curvature Theorem 5.5.1 Let $\alpha_1, \dots, \alpha_n$ and $\beta_1, \dots, \beta_n$ be 2n rational numbers, such that for all (i, j), $\alpha_i - \beta_j$ is not an integer. Pick a common denominator N for all the α's and β's. The equivalence (**) holds for the (restriction to $\mathbb{G}_m - \{1\}$, where it is a D.E, of the) direct sum

$$\oplus_{r \bmod N,\ \gcd(r,N)=1} \quad \mathcal{H}_1(r\alpha_1, \dots, r\alpha_n; r\beta_1, \dots, r\beta_n).$$

The interpretation of "p-curvature zero for almost all primes p", due to Beukers-Heckman, is given by

Lemma 5.5.2 ([B-H, 4.9]) Let $\alpha_1, \dots, \alpha_n$ and $\beta_1, \dots, \beta_n$ be 2n rational numbers, such that for all (i, j), $\alpha_i - \beta_j$ is not an integer. Pick a common denominator N for all the α's and β's. Then

$$\mathcal{H}_1(\alpha_1, \dots, \alpha_n; \beta_1, \dots, \beta_n)$$

has p-curvature zero for almost all primes p if and only if the following two conditions hold:

(1) $\alpha_1, \dots, \alpha_n; \beta_1, \dots, \beta_n \bmod \mathbb{Z}$ are 2n distinct elements of $(1/N)\mathbb{Z}/\mathbb{Z}$.

(2) for each integer $1 \le r < N$ with $\gcd(r, N) = 1$, the two subsets

$$A_r = \{r\alpha_1, \dots, r\alpha_n\} \bmod \mathbb{Z}, \qquad B_r := \{r\beta_1, \dots, r\beta_n\} \bmod \mathbb{Z}$$

of $(1/N)\mathbb{Z}/\mathbb{Z}$ are **intertwined** in $(1/N)\mathbb{Z}/\mathbb{Z}$ in the sense that if we display their images under $x \mapsto \exp(2\pi i x)$ on the unit circle, then as we walk counterclockwise around the unit circle we **alternately** encounter one from each subset.

proof We may assume that all the α's and β's lie in the half open interval [0, 1). We begin by a direct analysis of the p-curvature. The operator

$$\mathrm{Hyp}_1(\alpha\text{'s}; \beta\text{'s}) := P(xd/dx\ ) - xQ(xd/dx)$$
$$P(t) := \prod(t - \alpha_i),\ Q(t) := \prod(t - \beta_i),$$

lies in $\mathbb{Z}[1/N][x, xd/dx]$, so it makes sense to reduce it mod p for any prime p > N. Notice for any γ and δ chosen from the set {α's, β's}, we have $\gamma = \delta$ iff $\gamma \equiv \delta$ mod p (indeed, $|N\gamma - N\delta| < N < p$, so $\gamma \equiv \delta$ mod p iff $N\gamma \equiv N\delta$ mod p iff $N\gamma = N\delta$ iff $\gamma = \delta$).

SubLemma 5.5.2.1 Let $\alpha_1, \ldots, \alpha_n$ and $\beta_1, \ldots, \beta_n$ be 2n rational numbers in [0, 1) such that for all (i, j), $\alpha_i - \beta_j$ is not an integer. Pick a common denominator N for all the α's and β's. Then for a fixed prime p > N,

$$\mathcal{H}_1(\alpha_1, \ldots, \alpha_n; \beta_1, \ldots, \beta_n) \bmod p$$

has p-curvature zero if and only if the following two conditions hold:

(1) the α's and β's have 2n distinct reductions mod p in $\mathbb{F}_p$.
(2) the reductions mod p of the α's are **intertwined** in $\mathbb{F}_p$ with those of the β's in the sense that as we walk through $\mathbb{F}_p$ in the standard order 0, 1, 2, 3... we **alternately** encounter α's and β's.

proof To say that the reduced equation has p-curvature zero is precisely to say that the reduced operator has n solutions in the rational function field $\mathbb{F}_p(x)$ which are linearly independent over the subfield $\mathbb{F}_p(x^p)$ ([Ka-AS, 6.0.5]). So if we view $\mathbb{F}_p(x)$ as a p-dimensional vector space V over the field $k := \mathbb{F}_p(x^p)$, then $\mathrm{Hyp}_1(\alpha\text{'s}; \beta\text{'s})$ mod p is a linear endomorphism L of V, and p-curvature zero means precisely that this linear endomorphism L has an n-dimensional kernel, or equivalently that L has rank p-n. Fix an integer d such that d+p-1 mod p is one of the β_i mod p. As basis of V over k we may take the elements

$$x^d, x^{d+1}, \ldots, x^{d+p-1},$$

and on this basis, the operator $L = P(xd/dx\ ) - xQ(xd/dx)$ mod p acts as

$$L(x^i) = P(i)x^i - Q(i)x^{i+1} \text{ for } i = d, d+1, \ldots, d+p-2$$
$$= P(i)x^i \text{ for } i = d+p-1.$$

Let us denote by a_i (resp. b_i) $\in \mathbb{Z}$ the unique lift of α_i mod p (resp. of β_i mod p) which lies in the interval [d, d+p-1]. Renumbering the b's, we may suppose that there are precisely $t \geq 1$ **distinct** elements among the b_i's, and that these are

$$d \leq b_1 < b_2 < \ldots < b_t = d+p-1.$$

Then

$$Q(b_1) = Q(b_2) = Q(b_t) = 0,$$

and these are all the zeroes mod p of Q on $\mathbb{F}_p$.
The operator L is visibly stable on each of the t subspaces $V_0, \ldots, V_{t-1}$ defined by

$$V_0 := \text{the span of those } x^j \text{ with } d \leq j \leq b_1,$$

$$V_i := \text{the span of those } x^j \text{ with } b_i < j \leq b_{i+1} \text{ for } i > 0.$$

Since V is the direct sum of the V_i, we have

$$\dim\mathrm{Ker}(L \mid V) = \Sigma_i \dim\mathrm{Ker}(L \mid V_i).$$

Now on each V_i, the matrix of L is of the form

$$\begin{pmatrix} ?000 & 0000 \\ *?00 & 0000 \\ 0*?0 & 0000 \\ 00*? & 0000 \\ 000* & ?000 \\ 0000 & *?00 \\ 0000 & 0*?0 \\ 0000 & 00*? \end{pmatrix}$$

where each subdiagonal * entry is **nonzero**. Now such a matrix, if of size d, has rank $\geq d-1$, since its lower left $d-1 \times d-1$ minor is nonzero. Being lower triangular, its rank is d $\Leftrightarrow$ all the diagonal entries are nonzero. Therefore we find that

$$\dim\mathrm{Ker}(L \mid V_i) = 1 \text{ if } P(a_j) = 0 \text{ for some } a_j \text{ with } b_{i-1} < a_j \leq b_i,$$
$$= 0 \text{ if not.}$$

Notice that we cannot have $a_j = b_i$, since by hypothesis $\alpha_j \neq \beta_i$, and we chose $p > N$ to avoid any coalescing mod p of α's and β's.
Therefore we have $\dim\mathrm{Ker}(L \mid V) = n$ if and only if there are precisely $t = n$ distinct b_i's, and when we walk from d to d+p-1 we alternately encounter a's and b's. QED

SubLemma 5.5.2.2 Let $\alpha_1, \ldots, \alpha_n, \beta_1, \ldots, \beta_n$ be 2n distinct rational numbers in [0, 1), with common denominator N. Let $p > N$ be a prime, and denote by r the unique integer r in $1 \le r < N$ with $\gcd(r, N) = 1$ for which

$$rp + 1 \equiv 0 \bmod N.$$

Then the following conditions are equivalent:

(1) the reductions mod p of the α's are **intertwined** in $\mathbb{F}_p$ with those of the β's in the sense that as we walk through $\mathbb{F}_p$ in the standard order 0, 1, 2, 3... we **alternately** encounter α's and β's.

(2)the two subsets

$$A_r = \{r\alpha_1, \dots, r\alpha_n\} \bmod \mathbb{Z}, \qquad B_r := \{r\beta_1, \dots, r\beta_n\} \bmod \mathbb{Z}$$

of $(1/N)\mathbb{Z}/\mathbb{Z}$ are **intertwined** in $(1/N)\mathbb{Z}/\mathbb{Z}$ in the sense that if we display their images under $x \mapsto \exp(2\pi i x)$ on the unit circle, then as we walk counterclockwise around the unit circle we **alternately** encounter one from each subset.

proof For a real number x, we denote by $\langle x \rangle$ its fractional part; by definition $\langle x \rangle$ lies in the half open interval [0, 1) and satisfies $\langle x \rangle \equiv x \bmod \mathbb{Z}$. Denote by $a_1, \dots, a_n, b_1, \dots, b_n$ arbitrary integers whose reductions mod p agree with those of $\alpha_1, \dots, \alpha_n, \beta_1, \dots, \beta_n$. Then condition (1) is that in the interval [0, 1), the fractional parts $\langle a_i/p \rangle$ are intertwined with the fractional parts $\langle b_i/p \rangle$. And condition (2) is that in the interval [0, 1), the fractional parts $\langle r\alpha_i \rangle$ are intertwined with the fractional parts $\langle r\beta_i \rangle$.

Notice that each of the $\langle r\alpha_i \rangle$ and each of the $\langle r\beta_i \rangle$ is one of the numbers $\{0, 1/N, 2/N, \dots, (N-1)/N\}$. So the minimal distance between any two of them is 1/N. So if we add to each of them non-negative real quantities which are each $< 1/N$, we will not alter their order in [0,1), and in particular we will not alter the question of whether or not they are intertwined. Since $p > N$, we have $1/p < 1/N$. So it remains only to observe that for each i we have the inequalities

$$1/p > \langle a_i/p \rangle - \langle r\alpha_i \rangle \ge 0,$$

$$1/p > \langle b_i/p \rangle - \langle r\beta_i \rangle \ge 0.$$

To see this, recall that by the definition of r we have

$$pr + 1 \equiv 0 \bmod N.$$

Let $\gamma := C/N$ be any fraction with C an integer, $0 \le C < N$. Then

$$\gamma \equiv \gamma(1 + pr) \bmod^{\times} p,$$

so γ mod p is also the reduction mod p of the integer

$$c := \gamma(1 + pr) = C((1 + pr)/N).$$

Then

$$\langle c/p\rangle = \langle\gamma(1 + pr)/p\rangle = \langle(\gamma/p) + r\gamma\rangle = \langle (\gamma/p) + \langle r\gamma\rangle\rangle.$$

Since $0 \le \gamma < 1$, we have $0 \le (\gamma/p) < 1/p$; since $0 \le \langle r\gamma\rangle \le (N-1)/N$ and $(N-1)/N < 1 - 1/p$, we have

$$\langle c/p\rangle = (\gamma/p) + \langle r\gamma\rangle, \text{ whence } 1/p > (\gamma/p) = \langle c/p\rangle - \langle r\gamma\rangle \ge 0.$$ QED

These two sublemmas together prove the lemma, since by Dirichlet's theorem, any r in $1 \le r < N$ with $\gcd(r, N) = 1$ occurs in 5.5.2.2 for an infinity of primes p. QED

Combining this lemma 5.5.2 with the p-curvature theorem, we obtain the complete description of irreducible hypergeometrics of type (n,n) with G_{gal} finite. This description was obtained independently by Beukers-Heckman by a different method.

Theorem 5.5.3 ([B-H], 4.8) Let $\alpha_1, \dots, \alpha_n$ and $\beta_1, \dots, \beta_n$ be 2n complex numbers, such that for all (i, j), $\alpha_i - \beta_j$ is not an integer. Let $\lambda \in \mathbb{C}^\times$. Then

$$\mathcal{H}_\lambda(\alpha_1, \dots, \alpha_n; \beta_1, \dots, \beta_n).$$

has G_{gal} finite if and only if the α's and β's are all rational numbers, say with common denominator N, such that the following two conditions hold:

(1) $\alpha_1, \dots, \alpha_n; \beta_1, \dots, \beta_n$ mod $\mathbb{Z}$ are 2n distinct elements of $(1/N)\mathbb{Z}/\mathbb{Z}$.

(2) for each integer $1 \le r < N$ with $\gcd(r, N) = 1$, the two subsets

$$A_r = \{r\alpha_1, \dots, r\alpha_n\} \bmod \mathbb{Z}, \qquad B_r := \{r\beta_1, \dots, r\beta_n\} \bmod \mathbb{Z}$$

of $(1/N)\mathbb{Z}/\mathbb{Z}$ are **intertwined** in $(1/N)\mathbb{Z}/\mathbb{Z}$

We refer the interested reader to their paper [B-H, 7.1, 8.3] for a detailed discussion of exactly **which** finite groups occur as G_{gal} for irreducible hypergeometrics of type (n,n).

CHAPTER 6

Fourier Transforms of Kummer Pullbacks of Hypergeometrics

6.1 Some D.E.'s on $\mathbb{A}^1$ as Kummer Pullbacks of Hypergeometrics

In 2.10.6, we proved that the rank seven D.E. on $\mathbb{A}^1$

$$\partial^7 - f\partial - f'/2,$$

has differential galois group G_{gal} the subgroup G_2 of $SO(7)$, for any polynomial $f(x)$ of degree $k \geq 1$ prime to six. In this section, we will show that this same result holds for $f(x) = x^k$ for **any** integer $k \geq 1$.

The idea is that equations on $\mathbb{A}^1$ of the form

$$\partial^n + \alpha x^k \partial + \beta x^{k-1}$$

are Kummer pullbacks of hypergeometrics of type either $(n, 1)$, if $\alpha \neq 0$, or type $(n, 0)$, for $\alpha = 0$. As a consequence, one can explicitly determine the group G_{gal} for all such equations.

More generally, for any non-negative integer $m < n$, and any polynomial $P_m(t) \in \mathbb{C}[t]$ of degree m, equations of the form

$$\partial^n + (x^{k-1})P_m(D); \quad D := x\partial,$$

are Kummer pullbacks of hypergeometrics of type (n, m); just as above (the case $m = 0$ or 1), this leads to an explicit determination of their G_{gal}'s. For example, we will find that for any integer $s \geq 1$,

$$\partial^8 + x^{2s-1}(D + s)(D - 7/2)$$

has G_{gal} the subgroup $Spin(7)$ of $SO(8)$.

Key Lemma 6.1.1 Let $q \geq n > m \geq 0$ be integers. Denote by

$$[q] : \mathbb{G}_m \rightarrow \mathbb{G}_m$$

the q'th power endomorphism, and by $j : \mathbb{G}_m \rightarrow \mathbb{A}^1$ the inclusion. Then for any $\beta_1, \ldots, \beta_m$, we have an isomorphism of $\mathcal{D}$-modules on $\mathbb{A}^1$

$$j_{!*}([q]^*\mathcal{H}_\lambda(-1/q, -2/q, \ldots, -n/q; \beta_1, \ldots, \beta_m)) \approx \mathcal{D}/\mathcal{D}L$$

for L the operator on $\mathbb{A}^1$

$$L := \partial^n - (q^{n-m}/\lambda)x^{q-n}\textstyle\prod_j(D - n - q\beta_j).$$

proof Since $\mathcal{H} := \mathcal{H}_\lambda(-1/q, -2/q, \ldots, -n/q; \beta_1, \ldots, \beta_m)$ is RS at the origin with local monodromy of **finite** order q, its pullback by $[q]^*\mathcal{H}$ is RS at zero with trivial monodromy, so it extends uniquely to a D.E. $\mathcal{M}$ on all of $\mathbb{A}^1$. By 2.9.1.1, we have $\mathcal{M} \approx j_{*!}(j^*\mathcal{M})$, so $\mathcal{M}$ is the unique DE on $\mathbb{A}^1$ with $j^*\mathcal{M} \approx [q]^*\mathcal{H}$. Therefore it suffices to construct on $\mathbb{G}_m$ an isomorphism

$$[q]^*\mathcal{H} \approx \mathcal{D}_{\mathbb{G}_m}/\mathcal{D}_{\mathbb{G}_m}L.$$

On $\mathbb{G}_m$, we have

$$[q]^*(x) = x^q, \quad [q]^*(dx/x) = qdx/x, \quad [q]^*(D) = D/q,$$

so direct calculation shows that, as operators, we have

$$[q]^*\mathcal{H} = \lambda\prod_{i=1,\ldots,n}((D/q) + i/q) - x^q\prod_{j=1,\ldots,m}((D/q) - \beta_j)$$

$$= (\lambda/q^n)\prod_{i=1,\ldots,n}(D + i) - (1/q^m)x^q\prod_{j=1,\ldots,m}(D - q\beta_j)$$

$$= (\lambda/q^n)(\partial^n x^n) - (1/q^m)x^{q-n}\Big(\prod_{j=1,\ldots,m}(D - n - q\beta_j)\Big)x^n$$

$$= L\circ(\lambda x^n/q^n).$$

So right multiplication by $\lambda x^n/q^n$ defines the required isomorphism $\mathcal{D}_{\mathbb{G}_m}/\mathcal{D}_{\mathbb{G}_m}L \approx [q]^*\mathcal{H}$ of $\mathcal{D}$-modules on $\mathbb{G}_m$. QED

Variant 6.1.2 Let $q \geq n > m \geq 0$ be integers. For any integer α we have

$$j_{!*}([q]^*\mathcal{H}_\lambda((\alpha - 1)/q, (\alpha - 2)/q, \ldots, (\alpha - n)/q; \beta_1, \ldots, \beta_m)) \approx \mathcal{D}/\mathcal{D}L$$

for L the operator on $\mathbb{A}^1$

$$L := \partial^n - (q^{n-m}/\lambda)x^{q-n}\prod_j(D + \alpha - n - q\beta_j).$$

proof The operator

$$\mathcal{H}(\alpha/q) := \mathcal{H}_\lambda((\alpha - 1)/q, (\alpha - 2)/q, \ldots, (\alpha - n)/q; \beta_1, \ldots, \beta_m))$$

is the $x^{\alpha/q}$ twist of

$$\mathcal{H} := \mathcal{H}_\lambda(-1/q, -2/q, \ldots, -n/q; \beta_1 - \alpha/q, \ldots, \beta_m - \alpha/q),$$

to which the above lemma applies. But $[q]^*\mathcal{H} \approx [q]^*(\mathcal{H}(\alpha/q))$. QED

Question 6.1.3 Let $q \geq n > m \geq 0$ be integers. If $\alpha_1, \ldots, \alpha_n$ are n elements in $(1/q)\mathbb{Z}$ which are all distinct mod $\mathbb{Z}$, then for any $\beta_1, \ldots, \beta_m$, $[q]^*\mathcal{H}_\lambda(\alpha_1, \alpha_2, \ldots, \alpha_n; \beta_1, \ldots, \beta_m))$ extends to a D.E. on $\mathbb{A}^1$ of rank n. The above variant gives an explicit formula for it when the α_i are **consecutive**. Are there similar formulae when the α_i are not consecutive?

Corollary 6.1.4 Let $q \geq n > m \geq 0$ be integers. Let α be an integer. If

$$\mathcal{H}_\lambda((\alpha -1)/q, (\alpha -2)/q, \ldots, (\alpha -n)/q; \beta_1, \ldots, \beta_m)$$

is irreducible and not Kummer induced, then for L the operator on $\mathbb{A}^1$

$$L := \partial^n - (q^{n-m}/\lambda)x^{q-n}\prod_j(D + \alpha - n - q\beta_j),$$

$\mathcal{D}/\mathcal{D}L$ is an irreducible $\mathcal{D}$-module on $\mathbb{A}^1$.

proof If $\mathcal{H}$ is irreducible and not Kummer induced, it is Lie-irreducible and hence $[q]^*\mathcal{H}$ is irreducible on $\mathbb{G}_m$, which implies that its middle extension $\mathcal{D}/\mathcal{D}L$ is irreducible on $\mathbb{A}^1$. QED

Corollary 6.1.5 Let $q \geq n > m \geq 0$ be integers. Let α be an integer. Then group G_{gal} for

$$\mathcal{H}_\lambda((\alpha -1)/q, (\alpha -2)/q, \ldots, (\alpha -n)/q; \beta_1, \ldots, \beta_m)$$

contains that for

$$L := \partial^n - (q^{n-m}/\lambda)x^{q-n}\prod_j(D + \alpha - n - q\beta_j)$$

as a subgroup of finite index d which is a divisor of q.

proof G_{gal} for $\mathcal{D}/\mathcal{D}L$ on $\mathbb{A}^1$ is the same as G_{gal} for its restriction to $\mathbb{G}_m$, i.e., the same as for $[q]^*\mathcal{H}$. Now apply [Ka-DGG, 1.4.5]. QED

Examples 6.1.6

We consider first the case of hypergeometrics of type (n, 0). Then we find

Type (n, 0) Let $q \geq n > 0$ be integers. For any integer α we have

$$j_{!*}([q]^*\mathcal{H}_\lambda((\alpha -1)/q, (\alpha -2)/q, \ldots, (\alpha -n)/q; \emptyset)) \approx \mathcal{D}/\mathcal{D}L$$

for L the operator of Airy type (cf. [Ka-DGG, 4.2]) on $\mathbb{A}^1$

$$L := \partial^n - (q^n/\lambda)x^{q-n}.$$

Type (n, 1) Let $q \geq n > 1$ be integers. For any integer α we have

$$j_{!*}([q]^*\mathcal{H}_\lambda((\alpha -1)/q, (\alpha -2)/q, \ldots, (\alpha -n)/q; \beta)) \approx \mathcal{D}/\mathcal{D}L$$

for L the operator on $\mathbb{A}^1$

$$L := \partial^n - (q^{n-1}/\lambda)x^{q-n}(D + \alpha - n - q\beta).$$

Special Case (7,1), $\alpha = 4$, $\beta = -1/2$: For any integer $q \geq 7$,

$$j_{!*}([q]^*\mathcal{H}_\lambda(3/q, 2/q, 1/q, 0, -1/q, -2/q, -3/q; -1/2)) \approx \mathcal{D}/\mathcal{D}L$$

for L the operator on $\mathbb{A}^1$

$$L := \partial^7 - (q^6/\lambda)x^{q-7}(D + (q-6)/2).$$

We have already proven (4.1.4) that

$$\mathcal{H}_\lambda(3/q, 2/q, 1/q, 0, -1/q, -2/q, -3/q; -1/2)$$

has $G_{gal} = G_2$. So the above corollary shows that

$$\partial^7 - (q^6/\lambda)x^{q-7}(D + (q-6)/2)$$

has $G_{gal} = G_2$ for every integer $q \geq 7$. If we define $k := q-6$, this gives

Corollary 6.1.7 For any integer $k \geq 1$, and any nonzero constant μ,

$$\partial^7 - \mu x^k\partial - \mu k x^{k-1}/2$$

has $G_{gal} = G_2$.

Type (n, 2) Let $q \geq n > 2$ be integers. For any integer α we have

$$j_{!*}([q]^*\mathcal{H}_\lambda((\alpha -1)/q, (\alpha -2)/q, \ldots, (\alpha -n)/q; \beta_1, \beta_2)) \approx \mathcal{D}/\mathcal{D}L$$

for L the operator on $\mathbb{A}^1$

$$L := \partial^n - (q^{n-2}/\lambda)x^{q-n}(D + \alpha - n - q\beta_1)(D + \alpha - n - q\beta_2).$$

Special Case (8, 2), $q = 2r+1$, $r \geq 4$, $\alpha = r+5$, $\beta_1 = 0$, $\beta_2 = 1/2$

$$j_{!*}([q]^*\mathcal{H}_\lambda((r+4)/(2r+1), (r+3)/(2r+1), \ldots, (r-3)/(2r+1); 0, 1/2)) \approx \mathcal{D}/\mathcal{D}L$$

for L the operator on $\mathbb{A}^1$

$$L := \partial^8 - ((2r+1)^6/\lambda)x^{2r-7}(D + r - 3)(D - 7/2).$$

This $\mathcal{H}$ has $G_{gal} = \mathrm{Spin}(7)$ in $SO(8)$, by 4.4.1. Writing $s := r-3$, we find

Corollary 6.1.8 For any integer $s \geq 1$, and any nonzero constant μ,

$$L := \partial^8 - \mu x^{2s-1}(D + s)(D - 7/2)$$

has $G_{gal} = \mathrm{Spin}(7)$ inside $SO(8)$.

6.2 Fourier Transforms of Kummer Pulllbacks of Hypergeometrics: A Remarkable Stability

The following result shows that we can obtain (a Kummer pullback of) any (sufficiently general) hypergeometric of type (n, m) with $n > m$ as the Fourier Transform of (a Kummer pullback of) a hypergeometric of type (n, n). We will see later the importance of this sort of "reduction to the RS case".

Theorem 6.2.1 Suppose that $n > m \geq 0$ are integers. Put $d := n - m$. Suppose we are given a hypergeometric of type (n, m),

$$\mathcal{H} := \mathcal{H}_\lambda(\alpha_1, \dots, \alpha_n; \beta_1, \dots, \beta_m) = \mathcal{H}(P, Q)$$

which satisfies the following three conditions:

(i) $\mathcal{H}$ is is irreducible (i.e., for all i, j, $\alpha_i - \beta_j$ is not in $\mathbb{Z}$),

(ii) $\mathcal{H}$ is not Kummer induced,

(iii) for all i, $d\alpha_i$ is not an integer.

Then we have isomorphisms of irreducible $\mathcal{D}$-modules on $\mathbb{A}^1$

(1) $j_![d]^*\mathcal{H}_\lambda(\alpha_i\text{'s}; \beta_j\text{'s}) \approx j_{!*}[d]^*\mathcal{H}_\lambda(\alpha_i\text{'s}; \beta_j\text{'s}) \approx j_*[d]^*\mathcal{H}_\lambda(\alpha_i\text{'s}; \beta_j\text{'s})$,

(2) $FT(j_*[d]^*\mathcal{H}_\lambda(\alpha_i\text{'s}; \beta_j\text{'s})) \approx$

$$\approx j_{!*}[d]^*(\mathcal{H}_{(-d)^d/\lambda}(1/d, 2/d, \dots, d/d, -\beta_j\text{'s}; -\alpha_i\text{'s})).$$

(3) $j_*[d]^*\mathcal{H}_\lambda(\alpha_i\text{'s}; \beta_j\text{'s}) \approx$

$$\approx FT(j_{!*}[d]^*(\mathcal{H}_{(d)^d/\lambda}(1/d, 2/d, \dots, d/d, -\beta_j\text{'s}; -\alpha_i\text{'s})))$$

proof On $\mathbb{G}_m$, $\mathcal{H}$ is Lie-irreducible by (i) and (ii), so $[d]^*\mathcal{H}$ is irreducible. It is RS at zero, and has all exponents at zero nonintegral, by (iii). Direct calculation gives

$$[d]^*\mathcal{H} \approx \mathcal{D}_{\mathbb{G}_m}/\mathcal{D}_{\mathbb{G}_m}L,$$

for L the operator

$$L := [d]^*\mathrm{Hyp}(P, Q) = P(D/d) - x^dQ(D/d).$$

By 2.9.4, on $\mathbb{A}^1$ we have

$$j_!(\mathcal{D}_{\mathbb{G}_m}/\mathcal{D}_{\mathbb{G}_m}L) \approx \mathcal{D}/\mathcal{D}L \approx j_*(\mathcal{D}_{\mathbb{G}_m}/\mathcal{D}_{\mathbb{G}_m}L),$$

$$\mathcal{D}/\mathcal{D}L \approx j_{!*}(\mathcal{D}_{\mathbb{G}_m}/\mathcal{D}_{\mathbb{G}_m}L).$$

Therefore $\mathcal{D}/\mathcal{D}L$ on $\mathbb{A}^1$ is irreducible, being the middle extension of an irreducible $\mathcal{D}$-module on $\mathbb{G}_m$. This proves (1). Moreover, $\mathcal{D}/\mathcal{D}L$ has some (n-m, to be precise) of its ∞-slopes $=1$, so $\mathcal{D}/\mathcal{D}L$ is not the trivial $\mathcal{D}$-module $\mathcal{O}_{\mathbb{A}^1}$.

Therefore $FT(\mathcal{D}/\mathcal{D}L)$ is irreducible on $\mathbb{A}^1$, and it is not δ_0, so it is the middle extension of its restriction to $\mathbb{G}_m$:

$$\mathrm{FT}(\mathcal{D}/\mathcal{D}L) \approx j_{!*}j^*(\mathrm{FT}(\mathcal{D}/\mathcal{D}L)).$$

Now $j^*(\mathrm{FT}(\mathcal{D}/\mathcal{D}L))$ is easy to calculate explicitly:

$$j^*(\mathrm{FT}(\mathcal{D}/\mathcal{D}L)) \approx j^*(\mathcal{D}/\mathcal{D}\mathrm{FT}(L)) = \mathcal{D}_{\mathbb{G}_m}/\mathcal{D}_{\mathbb{G}_m}\mathrm{FT}(L).$$

Since $\mathrm{FT}(D) = \mathrm{FT}(x\partial) = -\partial x = -x\partial - 1 = -D - 1$, we have

$$\mathrm{FT}(L) = \mathrm{FT}(P(D/d) - x^dQ(D/d))$$
$$= P((-1 -D)/d) - \partial^dQ((-1 -D)/d).$$

Since x^d is a unit in $\mathcal{D}_{\mathbb{G}_m}$, $\mathrm{FT}(L)$ and $x^d\mathrm{FT}(L)$ generate the same left ideal, so we have

$\mathcal{D}_{\mathbb{G}_m}/\mathcal{D}_{\mathbb{G}_m}\mathrm{FT}(L) = \mathcal{D}_{\mathbb{G}_m}/\mathcal{D}_{\mathbb{G}_m}K$, where K is is the operator

$$K := x^dP((-1 -D)/d) - x^d\partial^dQ((-1 -D)/d)$$

$= [d]^*M$, for M the operator

$$M := xP(-1/d - D) - [\textstyle\prod_{j=0,\dots,d-1}(dD -j)]Q(-1/d -D).$$
$$-M = d^d[\textstyle\prod_{j=0,\dots,d-1}(D -j/d)]Q(-1/d -D) - xP(-1/d - D).$$

So all in all we have

$$j^*(\mathrm{FT}(\mathcal{D}/\mathcal{D}L)) \approx [d]^*(\mathcal{D}_{\mathbb{G}_m}/\mathcal{D}_{\mathbb{G}_m}M).$$

If we twist $\mathcal{D}_{\mathbb{G}_m}/\mathcal{D}_{\mathbb{G}_m}M$ by $x^{1/d}$, we do not change its pullback by $[d]^*$, so we have

$j^*(\mathrm{FT}(\mathcal{D}/\mathcal{D}L)) \approx [d]^*(\mathcal{D}_{\mathbb{G}_m}/\mathcal{D}_{\mathbb{G}_m}M_1)$ for M_1 the operator

$$M_1 := d^d[\textstyle\prod_{j=1,\dots,d}(D -j/d)]Q(-D) - xP(-D).$$

If we return to α, β notation:

$$P(t) = \lambda\textstyle\prod(t - \alpha_i), \quad Q(t) = \textstyle\prod(t - \beta_j),$$

then

$$\lambda^{-1}(-1)^nM_1 = \mathrm{Hyp}_{(-d)^d/\lambda}(1/d, 2/d, \dots, d/d, -\beta_1, \dots, -\beta_m; -\alpha_1, \dots, -\alpha_n).$$

So all in all we have

$$j^*\mathrm{FT}(j_*[d]^*\mathcal{H}_\lambda(\alpha_i\text{'s}; \beta_j\text{'s}))\approx$$
$$\approx [d]^*(\mathcal{H}_{(-d)^d/\lambda}(1/d, 2/d, \dots, d/d, -\beta_1, \dots, -\beta_m; -\alpha_1, \dots, -\alpha_n)).$$

Now applying $j_{!*}$, to both sides yields

$$\mathrm{FT}(j_*[d]^*\mathcal{H}_\lambda(\alpha_i\text{'s}; \beta_j\text{'s})) \approx$$
$$\approx j_{!*}[d]^*(\mathcal{H}_{(-d)^d/\lambda}(1/d, 2/d, \dots, d/d, -\beta_1, \dots, -\beta_m; -\alpha_1, \dots, -\alpha_n)),$$

which is (2). By Fourier inversion, (2) gives

$$[x \mapsto -x]^*(j_*[d]^*\mathcal{H}_\lambda(\alpha_i\text{'s}; \beta_j\text{'s})) \approx$$
$$\approx \mathrm{FT}(j_{!*}[d]^*(\mathcal{H}_{(-d)^d/\lambda}(1/d, 2/d, \ldots, d/d, -\beta_j\text{'s}; -\alpha_i\text{'s}))),$$

which is nearly (3). Because $[x \mapsto -x]^*$ and j_* (trivially) commute, we have

$$[x \mapsto -x]^*(j_*[d]^*\mathcal{H}_\lambda(\alpha_i\text{'s}; \beta_j\text{'s})) = j_*[x \mapsto -x]^*([d]^*\mathcal{H}_\lambda(\alpha_i\text{'s}; \beta_j\text{'s}))$$
$$= j_*[x \mapsto (-x)^d]^*\mathcal{H}_\lambda(\alpha_i\text{'s}; \beta_j\text{'s}) = j_*[d]^*[x \mapsto (-1)^d x]^*\mathcal{H}_\lambda(\alpha_i\text{'s}; \beta_j\text{'s})$$
$$= j_*[d]^*\mathcal{H}_{(-1)^d\lambda}(\alpha_i\text{'s}; \beta_j\text{'s}),$$

so we have

$$j_*[d]^*\mathcal{H}_{(-1)^d\lambda}(\alpha_i\text{'s}; \beta_j\text{'s}) \approx$$
$$\approx \mathrm{FT}(j_{!*}[d]^*(\mathcal{H}_{(-d)^d/\lambda}(1/d, 2/d, \ldots, d/d, -\beta_j\text{'s}; -\alpha_i\text{'s}))).$$

Replacing λ by $(-1)^d\lambda$ gives (3). QED

6.3 Convolution of hypergeometrics with non-disjoint exponents, via a modified sort of hypergeometric

In this section, we will explore what happens to the convolution formula 5.3.2 when the exponents are not disjoint.

(6.3.1) Given a smooth connected $\mathbb{C}$-scheme X, with structural map

$$\pi : X \to \mathrm{Spec}(\mathbb{C}),$$

we say that an object K in $D^{b,\mathrm{holo}}(X)$ is PC (for perverse cohomology) if π_*K is a single $\mathcal{D}$-module, concentrated in degree zero. If X is n-dimensional, this is the same as requiring that

$$H^i{}_{DR}(X/\mathbb{C}, K) = 0 \quad \text{for } i \neq n.$$

Lemma 6.3.2 Let G be a smooth connected $\mathbb{C}$-groupscheme of finite type. The convolution $K*L$ of two PC objects K, L in $D^{b,\mathrm{holo}}(G)$ is again PC. Moreover, if we put $n := \dim_{\mathbb{C}}(G)$, then

$$H^n{}_{DR}(G/\mathbb{C}, K*L) \approx H^n{}_{DR}(G/\mathbb{C}, K)\otimes_{\mathbb{C}}H^n{}_{DR}(G/\mathbb{C}, L)$$

proof This is immediate from 5.1.9,(1a). QED

Lemma 6.3.3 Let $\mathcal{M}$ be a holonomic $\mathcal{D}$-module on $\mathbb{G}_m$, and $\mathcal{H}$ a hypergeometric of type (1, 0) or (0, 1).

(1) The convolution $\mathcal{M}*\mathcal{H}$ is a single (holonomic) $\mathcal{D}$-module on $\mathbb{G}_m$.

(2) If $\mathcal{M}$ is PC on $\mathbb{G}_m$, then $\mathcal{M}*\mathcal{H}$ is PC on $\mathbb{G}_m$.

proof By multiplicative inversion, we reduce to the case when $\mathcal{H}$ is of type (1, 0). Twisting by x^α, we reduce to the case when $\mathcal{H}$ is $\mathcal{H}_\lambda(0; \varnothing)$. A multiplicative translation reduces us to the case when $\mathcal{H}$ is $\mathcal{H}_1(0; \varnothing)$. In this case, assertion (1) results from the formula (5.2.3)

$$j^*FT(j_*inv_*(\mathcal{M})) \approx \mathcal{M}*\mathcal{H}_1(0; \varnothing).$$

Assertion (2) follows from the fact that $\mathcal{H}$ is PC on $\mathbb{G}_m$ (direct computation), and the above lemma. QED

Lemma 6.3.4 Let $\mathcal{M}$ be a holonomic $\mathcal{D}$-module on $\mathbb{G}_m$ which is PC. Then the convolution $\mathcal{M}*\mathcal{O}$ of $\mathcal{M}$ with the "constant" $\mathcal{D}$-module $\mathcal{O} := \mathcal{O}_{\mathbb{G}_m}$ is the constant $\mathcal{D}$-module $V\otimes_{\mathbb{C}}\mathcal{O}$, with V the finite-dimensional $\mathbb{C}$-space $H^1{}_{DR}(\mathbb{G}_m/\mathbb{C}, \mathcal{M}) = \pi_*\mathcal{M}$.

proof This is the base change for the cartesian diagram

$$\begin{array}{ccc} G\times G & \xrightarrow{(x,\,y)\,\mapsto\,(x,\,xy)} & G\times G : \mathcal{M}\times\mathcal{O} \\ \downarrow\text{product} & & \downarrow pr_2 \\ G & = & G. \end{array}$$

QED

Corollary 6.3.4.1 Let $\alpha \in \mathbb{C}$. Let $\mathcal{M}$ be a holonomic $\mathcal{D}$-module on $\mathbb{G}_m$ such that $\mathcal{M}\otimes x^{-\alpha}$ is PC. Then the convolution $\mathcal{M}*(x^\alpha\mathcal{O})$ is the $\mathcal{D}$-module $V\otimes_{\mathbb{C}}x^\alpha\mathcal{O}$, with $V := H^1{}_{DR}(\mathbb{G}_m/\mathbb{C}, \mathcal{M}\otimes x^{-\alpha}) = \pi_*\mathcal{M}$.

proof Indeed, $x^\alpha\otimes((\mathcal{M}\otimes x^{-\alpha})*\mathcal{O}) \approx \mathcal{M}*(x^\alpha\mathcal{O})$ by 5.2.1. QED

Key Lemma 6.3.5 Let $\alpha \in \mathbb{C}$. For any $\lambda, \mu \in \mathbb{C}^\times$, the convolution

$$\mathcal{H}_\lambda(\alpha; \varnothing)*\mathcal{H}_\mu(\varnothing; \alpha)$$

sits in a short exact sequence of $\mathcal{D}$-modules on $\mathbb{G}_m$

$$0 \to \delta_{\lambda\mu} \to \mathcal{H}_\lambda(\alpha; \varnothing)*\mathcal{H}_\mu(\varnothing; \alpha) \to x^\alpha\mathcal{O} \to 0.$$

proof By 5.2.1, twisting by $x^{-\alpha}$ reduces us to the case where $\alpha = 0$. By 5.1.9, (2), a multiplicative translation reduces us to the case where $\mu = \lambda = 1$. By 5.2.3, we have

$$j^*FT(j_*inv_*(\mathcal{M})) \approx \mathcal{M}*\mathcal{H}_1(0; \varnothing).$$

Applying this to $\mathcal{M} = \mathcal{H}_1(\varnothing; 0)$, we find

$$\mathcal{H}_1(0;\ \emptyset) * \mathcal{H}_1(\emptyset;\ 0) \approx j^*\mathrm{FT}(j_*\mathrm{inv}_*\mathcal{H}_1(\emptyset;\ 0))$$
$$\approx j^*\mathrm{FT}(j_*\mathcal{H}_{-1}(0;\ \emptyset)).$$

Now by inspection we have

$$\mathcal{H}_{-1}(0;\ \emptyset) := \mathcal{D}_{\mathbb{G}_m}/\mathcal{D}_{\mathbb{G}_m}(-x\partial - x)$$
$$= \mathcal{D}_{\mathbb{G}_m}/\mathcal{D}_{\mathbb{G}_m}(\partial + 1)$$
$$= j^*(\mathcal{D}_{\mathbb{A}^1}/\mathcal{D}_{\mathbb{A}^1}(\partial + 1))$$
$$= j^*(e^{-x}\mathbb{C}[x]),$$

so

$$j_*\mathcal{H}_{-1}(0;\ \emptyset) = j_*j^*(e^{-x}\mathbb{C}[x]) = e^{-x}\mathbb{C}[x,\ x^{-1}].$$

Consider the short exact sequence of $\mathcal{D}$-modules on $\mathbb{A}^1$

$$0 \to e^{-x}\mathbb{C}[x] \to e^{-x}\mathbb{C}[x,\ x^{-1}] \to e^{-x}\mathbb{C}[x,\ x^{-1}]/e^{-x}\mathbb{C}[x] \to 0.$$

The third term we may rewrite as

$$e^{-x}\mathbb{C}[x,\ x^{-1}]/e^{-x}\mathbb{C}[x] \approx e^{-x}\mathbb{C}((x))/e^{-x}\mathbb{C}[[x]] = \mathbb{C}((x))/\mathbb{C}[[x]]$$
$$\approx \mathcal{D}/\mathcal{D}x = \delta_0.$$

So the above exact sequence is

$$0 \to \mathcal{D}_{\mathbb{A}^1}/\mathcal{D}_{\mathbb{A}^1}(\partial + 1) \to j_*\mathcal{H}_{-1}(0;\ \emptyset) \to \delta_0 \to 0.$$

The restriction to $\mathbb{G}_m$ of its Fourier Transform is the required short exact sequence. QED

(6.3.6) We now introduce a modified notion of hypergeometric $\mathcal{D}$-module, which is by its very definition well-behaved with respect to convolution. Namely, we **define**

$$\mathrm{M}\mathcal{H}_1(\alpha_1, \ldots, \alpha_n;\ \emptyset) := \mathcal{H}_1(\alpha_1;\ \emptyset) * \ldots * \mathcal{H}_1(\alpha_n;\ \emptyset)$$
$$\mathrm{M}\mathcal{H}_1(\emptyset;\ \beta_1, \ldots, \beta_m) := \mathcal{H}_1(\emptyset;\ \beta_1) * \ldots * \mathcal{H}_1(\emptyset;\ \beta_m)$$
$$\mathrm{M}\mathcal{H}_1(\alpha_1, \ldots, \alpha_n;\ \beta_1, \ldots, \beta_m) :=$$
$$:= \mathcal{H}_1(\alpha_1;\ \emptyset) * \ldots * \mathcal{H}_1(\alpha_n;\ \emptyset) * \mathcal{H}_1(\emptyset;\ \beta_1) * \ldots * \mathcal{H}_1(\emptyset;\ \beta_m).$$

For $\lambda \in \mathbb{C}^\times$, we define

$$\mathrm{M}\mathcal{H}_\lambda(\alpha\text{'s};\ \beta\text{'s}) := [x \mapsto \lambda x]_*\mathrm{M}\mathcal{H}_1(\alpha\text{'s};\ \beta\text{'s}).$$

In view of the preceeding lemmas, we see that $\mathrm{M}\mathcal{H}_\lambda(\alpha\text{'s};\ \beta\text{'s})$ is a single holonomic $\mathcal{D}$-module, which is PC on $\mathbb{G}_m$ with $H^1{}_{DR}(\mathbb{G}_m/\mathbb{C},\ \mathrm{M}\mathcal{H}_\lambda)$ one-dimensional (by 3.7.1). The effect of x^γ twisting is given by

$$x^\gamma \otimes \mathrm{M}\mathcal{H}_\lambda(\alpha_i\text{'s};\ \beta_j\text{'s}) \approx \mathrm{M}\mathcal{H}_\lambda(\alpha_i + \gamma\text{'s};\ \beta_j + \gamma\text{'s}).$$

The effect of inversion is given by

$$\mathrm{inv}_*\mathcal{MH}_\lambda(\alpha_i\text{'s}; \beta_j\text{'s})) \approx \mathcal{MH}_{(-1)^{n+m}/\lambda}(-\beta_j\text{'s}; -\alpha_i\text{'s}).$$

By 5.3.2, we have

$$\mathcal{MH}_\lambda(\alpha\text{'s}; \beta\text{'s}) \approx \mathcal{H}_\lambda(\alpha\text{'s}; \beta\text{'s}) \text{ if } \mathcal{H}_\lambda(\alpha\text{'s}; \beta\text{'s}) \text{ is irreducible.}$$

By the associativity and commutativity of convolution, we have

$$\mathcal{MH}_\lambda(\alpha\text{'s}; \beta\text{'s})*\mathcal{MH}_\mu(\gamma\text{'s}; \delta\text{'s}) = \mathcal{MH}_{\lambda\mu}(\alpha\text{'s}, \gamma\text{'s}; \beta\text{'s}, \delta\text{'s})$$

whatever the exponents. In particular, we have

$$\mathcal{MH}_\lambda(\alpha\text{'s}; \beta\text{'s}) \approx \mathcal{H}_1(\alpha\text{'s}; \varnothing)*\mathcal{H}_\lambda(\varnothing; \beta\text{'s}),$$

expressing every $\mathcal{MH}$ as a convolution of irreducible hypergeometrics. Now the isomorphism class of an irreducible $\mathcal{H}_\lambda(\alpha\text{'s}; \beta\text{'s})$ depends only on λ and on the classes mod $\mathbb{Z}$ of its exponents. So by the functoriality of convolution, we find that

Scholie 6.3.7 The isomorphism class of $\mathcal{MH}_\lambda(\alpha\text{'s}; \beta\text{'s})$ depends only on λ and on the classes mod $\mathbb{Z}$ of the α's and the β's.

Open Question 6.3.8 If the α's and β's are not disjoint, is there a simple expression for $\mathcal{MH}_\lambda(\alpha\text{'s}; \beta\text{'s})$? Is it of the form $\mathcal{H}_\lambda(\tilde{\alpha}\text{'s}; \tilde{\beta}\text{'s})$ for some particular choice of modified exponents ($\tilde{\alpha}$'s; $\tilde{\beta}$'s) which are termwise congruent mod $\mathbb{Z}$ to (α's, β's)?

Cancelation Theorem 6.3.9 Suppose given a modified hypergeometric $\mathcal{MH}_\lambda(\alpha_i\text{'s}; \beta_j\text{'s})$ of type (n, m). Then for any $\gamma \in \mathbb{C}$, and any integer d, the modified hypergeometric $\mathcal{MH}_\lambda(\alpha_i\text{'s}, \gamma; \beta_j\text{'s}, \gamma+d)$ of type $(n+1, m+1)$ sits in a short exact sequence of $\mathcal{D}$-modules

$$0 \to \mathcal{MH}_\lambda(\alpha_i\text{'s}; \beta_j\text{'s}) \to \mathcal{MH}_\lambda(\alpha_i\text{'s}, \gamma; \beta_j\text{'s}, \gamma+d) \to V\otimes_{\mathbb{C}}(x^\gamma\mathcal{O}) \to 0,$$

where V is the 1-dimensional $\mathbb{C}$-space

$$V := H^1{}_{DR}(\mathbb{G}_m/\mathbb{C}, \mathcal{MH}_\lambda(\alpha_i - \gamma\text{'s}; \beta_j - \gamma\text{'s})).$$

proof We first reduce to the case $d=0$ by the Scholie above, then write

$$\mathcal{MH}_\lambda(\alpha_i\text{'s}, \gamma; \beta_j\text{'s}, \gamma) = \mathcal{MH}_\lambda(\alpha_i\text{'s}; \beta_j\text{'s})*\mathcal{MH}_1(\gamma, \gamma).$$

By definition, $\mathcal{MH}_1(\gamma, \gamma)$ is the convolution $\mathcal{H}_1(\gamma; \varnothing)*\mathcal{H}_1(\varnothing; \gamma)$, which by 6.3.5 sits in a short exact sequence of $\mathcal{D}$-modules on $\mathbb{G}_m$

$$0 \to \delta_1 \to \mathcal{H}_1(\gamma; \varnothing)*\mathcal{H}_1(\varnothing; \gamma) \to x^\gamma\mathcal{O} \to 0.$$

Convolving this exact sequence with $\mathcal{MH}_\lambda(\alpha_i\text{'s}; \beta_j\text{'s})$ yields, via 6.3.4.1, the required short exact sequence. QED

(6.3.10) In order to formulate the next result, it will be convenient to introduce the operator **Cancel** on both hypergeometrics and on modified hypergeometrics which "cancels" the exponents mod $\mathbb{Z}$ common to numerator and denominator. Given

$$\mathcal{H} := \mathcal{H}_\lambda(\alpha_1, \dots, \alpha_n; \beta_1, \dots, \beta_m)$$
$$\mathrm{M}\mathcal{H} := \mathrm{M}\mathcal{H}_\lambda(\alpha_1, \dots, \alpha_n; \beta_1, \dots, \beta_m)$$

of type (n, m), look to see how many of the α_i's are also β_j's mod $\mathbb{Z}$. If there are r such common exponents, renumber so that

$$\alpha_{n-k} \equiv \beta_{m-k} \bmod \mathbb{Z} \quad \text{for } k < r,$$
$$\alpha_i \not\equiv \beta_j \bmod \mathbb{Z} \quad \text{if } i \le n-r \text{ and } j \le m-r,$$

and define

$$\mathbf{Cancel}(\mathcal{H}_\lambda(\alpha_1, \dots, \alpha_n; \beta_1, \dots, \beta_m)) :=$$
$$:= \mathcal{H}_\lambda(\alpha_1, \dots, \alpha_{n-r}; \beta_1, \dots, \beta_{m-r}),$$
$$\mathbf{Cancel}(\mathrm{M}\mathcal{H}_\lambda(\alpha_1, \dots, \alpha_n; \beta_1, \dots, \beta_m)) :=$$
$$:= \mathrm{M}\mathcal{H}_\lambda(\alpha_1, \dots, \alpha_{n-r}; \beta_1, \dots, \beta_{m-r}).$$

Since the result of canceling is irreducible, we always have

$$\mathbf{Cancel}(\mathcal{H}) = \mathbf{Cancel}(\mathrm{M}\mathcal{H}),$$

whatever the exponents.

Semisimplification Theorem 6.3.11 The semisimplification of

$$\mathrm{M}\mathcal{H}_\lambda(\alpha_1, \dots, \alpha_n; \beta_1, \dots, \beta_m)$$

as holonomic $\mathcal{D}$-module on $\mathbb{G}_m$ is the direct sum

$$\mathbf{Cancel}(\mathrm{M}\mathcal{H}_\lambda(\alpha_1, \dots, \alpha_n; \beta_1, \dots, \beta_m)) \oplus \Big(\bigoplus_{\text{common exponents } \alpha} x^\alpha \mathcal{O}\Big).$$

proof This is immediate from the cancellation theorem 6.3.9. QED

6.4 Application to Fourier Transforms of Kummer Pullbacks of Hypergeometrics

Lemma 6.4.1 Let $d \ge 1$ be an integer, and $\mathcal{H}_\lambda(\alpha_i\text{'s}; \beta_j\text{'s})$ an irreducible hypergeometric. Then we have an isomorphism of $\mathcal{D}$-modules on $\mathbb{G}_m$

$$j^*\mathrm{FT}(j_*[d]^*\mathcal{H}_\lambda(\alpha_i\text{'s}; \beta_j\text{'s})) \approx$$
$$\approx [d]^*\mathrm{M}\mathcal{H}_{(-1)^{m-n}(d)^d/\lambda}(1/d, 2/d, \dots, d/d, -\beta_j\text{'s}; -\alpha_i\text{'s}).$$

proof By 5.2.3, for any $\mathcal{D}$-module $\mathcal{M}$ on $\mathbb{G}_m$,

$$j^*FT(j_*inv_*(\mathcal{M})) \approx \mathcal{M}*\mathcal{H}_1(0;\ \emptyset).$$

Applying this to $\mathcal{M} = inv_*[d]^*\mathcal{H}_\lambda(\alpha_i\text{'s};\ \beta_j\text{'s}) = [d]^*inv_*\mathcal{H}_\lambda(\alpha_i\text{'s};\ \beta_j\text{'s})$ gives

$$j^*FT(j_*[d]^*\mathcal{H}_\lambda(\alpha_i\text{'s};\ \beta_j\text{'s})) \approx ([d]^*inv_*\mathcal{H}_\lambda(\alpha_i\text{'s};\ \beta_j\text{'s}))*\mathcal{H}_1(0;\ \emptyset).$$

By 5.1.9, (1b), for any two $\mathcal{D}$-modules $\mathcal{M}$ and $\mathcal{N}$ on $\mathbb{G}_m$, and any integer $d \geq 1$, we have

$$([d]^*\mathcal{M})*\mathcal{N} \approx [d]^*(\mathcal{M}*([d]_*\mathcal{N})).$$

Thus we have

$$j^*FT(j_*[d]^*\mathcal{H}_\lambda(\alpha_i\text{'s};\ \beta_j\text{'s})) \approx$$
$$\approx [d]^*((inv_*\mathcal{H}_\lambda(\alpha_i\text{'s};\ \beta_j\text{'s}))*([d]_*(\mathcal{H}_1(0;\ \emptyset)))).$$

By the Kummer Induction Formula 3.5.6.1 we have

$$[d]_*(\mathcal{H}_1(0;\ \emptyset)) \approx \mathcal{H}_{d^d}(1/d, \ldots, d/d;\ \emptyset)$$
$$\approx M\mathcal{H}_{d^d}(1/d, \ldots, d/d;\ \emptyset).$$

By the inversion formula (3.1), we have

$$inv_*\mathcal{H}_\lambda(\alpha_i\text{'s};\ \beta_j\text{'s}) \approx \mathcal{H}_{(-1)^{n+m}/\lambda}(-\beta_j\text{'s};\ -\alpha_i\text{'s}).$$

Combining these, we find

$$j^*FT(j_*[d]^*\mathcal{H}_\lambda(\alpha_i\text{'s};\ \beta_j\text{'s})) \approx$$
$$\approx [d]^*(\mathcal{H}_{(-1)^{n+m}/\lambda}(-\beta_j\text{'s};\ -\alpha_i\text{'s})*\mathcal{H}_{d^d}(1/d, \ldots, d/d;\ \emptyset)),$$

and the result follows by the (tautological) convolution formula for $M\mathcal{H}$. QED

The following theorem encompasses both 6.2.1 and the irreducible case of Key Lemma 6.1.1 as special cases.

Theorem 6.4.2 Let $d \geq 1$ be an integer, and $\mathcal{H}$ an irreducible hypergeometric of type (n, m),

$$\mathcal{H} := \mathcal{H}_\lambda(\alpha_1, \ldots, \alpha_n;\ \beta_1, \ldots, \beta_m).$$

Then we have isomorphisms of $\mathcal{D}$-modules on $\mathbb{A}^1$

(1) $FT(j_{!*}[d]^*\mathcal{H}_\lambda(\alpha_i\text{'s};\ \beta_j\text{'s})) \approx$

$$\approx j_{!*}[d]^*(\mathbf{Cancel}\mathcal{H}_{(-1)^{n+m}(d)^d/\lambda}(1/d, 2/d, \ldots, d/d, -\beta_j\text{'s};\ -\alpha_i\text{'s})).$$

(2) $j_{!*}[d]^*\mathcal{H}_\lambda(\alpha_i\text{'s};\ \beta_j\text{'s}) \approx$

$$\approx FT(j_{!*}[d]^*(\mathbf{Cancel}\mathcal{H}_{(-1)^{n+m+d}(d)^d/\lambda}(1/d, 2/d, \ldots, d/d, -\beta_j\text{'s};\ -\alpha_i\text{'s}))).$$

proof The isomorphism (2) is obtained from (1) by Fourier inversion. It remains to prove (1).

We first claim that $FT(j_{!*}[d]^*\mathcal{H}_\lambda(\alpha_i\text{'s};\ \beta_j\text{'s}))$ is a direct sum of irreducibles, none of which is δ_0. By Fourier inversion, it is equivalent to show that $j_{!*}[d]^*\mathcal{H}_\lambda(\alpha_i\text{'s};\ \beta_j\text{'s})$ is a direct sum of irreducibles on $\mathbb{A}^1$, none of which is the "constant" $\mathcal{D}$-module $\mathcal{O}$. Since $j_{!*}$ carries irreducibles to irreducibles, it suffices to show that $[d]^*\mathcal{H}_\lambda(\alpha_i\text{'s};\ \beta_j\text{'s})$ is a direct sum of irreducibles on $\mathbb{G}_m$, none of which is $\mathcal{O}$. For this, we argue as follows.

Since $\mathcal{H}_\lambda(\alpha_i\text{'s};\ \beta_j\text{'s})$ is irreducible on $\mathbb{G}_m$, $[d]^*\mathcal{H}_\lambda(\alpha_i\text{'s};\ \beta_j\text{'s})$ is semisimple, a direct sum of irreducibles. These irreducible constituents are all μ_d-translates of each other (since $\mathcal{H}_\lambda(\alpha_i\text{'s};\ \beta_j\text{'s})$ is irreducible), and hence if any of them were constant then $[d]^*\mathcal{H}_\lambda(\alpha_i\text{'s};\ \beta_j\text{'s})$ would be constant. But then its Euler characteristic on $\mathbb{G}_m$ would be 0, rather than $-d$ (cf. 3.7.1, 3.7.5, proof of 3.7.6).

Therefore $FT(j_{!*}[d]^*\mathcal{H}_\lambda(\alpha_i\text{'s};\ \beta_j\text{'s}))$ is a sum of irreducibles on $\mathbb{A}^1$, none of which is the delta sheaf δ_0 at the origin. Since any irreducible $\mathcal{M}$ on $\mathbb{A}^1$ other than δ_0 satisfies $\mathcal{M} \approx j_{!*}j^*\mathcal{M}$, we have

$$FT(j_{!*}[d]^*\mathcal{H}_\lambda(\alpha_i\text{'s};\ \beta_j\text{'s})) \approx j_{!*}j^*FT(j_{!*}[d]^*\mathcal{H}_\lambda(\alpha_i\text{'s};\ \beta_j\text{'s})).$$

So to prove the theorem it suffices to prove that on $\mathbb{G}_m$ we have

$$j^*FT(j_{!*}[d]^*\mathcal{H}_\lambda(\alpha_i\text{'s};\ \beta_j\text{'s})) \approx$$
$$\approx [d]^*(\mathbf{Cancel}\mathcal{H}_{(-1)^{n+m}(d)^d/\lambda}(1/d,\ 2/d,\ \ldots,\ d/d,\ -\beta_j\text{'s};\ -\alpha_i\text{'s})).$$

Since both of these $\mathcal{D}$-modules are semisimple, it suffices to show that they have isomorphic semisimplifications. For this, we argue as follows.

We have a short exact sequence of $\mathcal{D}$-modules on $\mathbb{A}^1$

$$0 \to j_{!*}[d]^*\mathcal{H}_\lambda(\alpha_i\text{'s};\ \beta_j\text{'s}) \to j_*[d]^*\mathcal{H}_\lambda(\alpha_i\text{'s};\ \beta_j\text{'s}) \to V\otimes_{\mathbb{C}}\delta_0 \to 0,$$

for some punctual $\mathcal{D}$-module $V\otimes_{\mathbb{C}}\delta_0$ at zero. In view of the known structure of the local monodromy at zero of $\mathcal{H}_\lambda(\alpha_i\text{'s};\ \beta_j\text{'s})$, we see from 2.9.8 that V has dimension

$r := \mathrm{Card}(R)$,

$R := \{k \text{ in } \{1, \ldots, d\} \text{ such that } k/d \bmod \mathbb{Z} \text{ is among the } \alpha_i \bmod \mathbb{Z}\}$.

Taking the Fourier Transformed exact sequence, passing to semisimplifications, and restricting to $\mathbb{G}_m$, we find

$$j^*\mathrm{FT}(j_*[d]^*\mathcal{H}_\lambda(\alpha_i\text{'s};\ \beta_j\text{'s}))^{ss} \approx j^*\mathrm{FT}(j_{!*}[d]^*\mathcal{H}_\lambda(\alpha_i\text{'s};\ \beta_j\text{'s})) \oplus \mathcal{O}^r.$$

By the above lemma we have

$$j^*\mathrm{FT}(j_*[d]^*\mathcal{H}_\lambda(\alpha_i\text{'s};\ \beta_j\text{'s})) \approx$$

$$\approx\ [d]^*\mathrm{M}\mathcal{H}_{(-1)^{m-n}(d)^d/\lambda}(1/d,\ 2/d,\ \ldots,\ d/d,\ -\beta_j\text{'s};\ -\alpha_i\text{'s}).$$

By the semisimplification theorem 6.3.11, we have

$$\mathrm{M}\mathcal{H}_{(-1)^{m-n}(d)^d/\lambda}(1/d,\ 2/d,\ \ldots,\ d/d,\ -\beta_j\text{'s};\ -\alpha_i\text{'s})^{ss} \approx$$

$$\mathbf{Cancel}(\mathcal{H}_{(-1)^{m-n}(d)^d/\lambda}(1/d,\ 2/d,\ \ldots,\ d/d,\ -\beta_j\text{'s};\ -\alpha_i\text{'s}))\oplus$$

$$\oplus(\textstyle\bigoplus_{k \in R} x^{-k/d}\mathcal{O}).$$

Therefore after $[d]^*$ we have

$$[d]^*\mathrm{M}\mathcal{H}_{(-1)^{m-n}(d)^d/\lambda}(1/d,\ 2/d,\ \ldots,\ d/d,\ -\beta_j\text{'s};\ -\alpha_i\text{'s})^{ss} \approx$$

$$\approx\ [d]^*\mathbf{Cancel}(\mathcal{H}_{(-1)^{m-n}(d)^d/\lambda}(1/d,\ 2/d,\ \ldots,\ d/d,\ -\beta_j\text{'s};\ -\alpha_i\text{'s}))\oplus\ \mathcal{O}^r.$$

Comparing these two expressions for $j^*\mathrm{FT}(j_*[d]^*\mathcal{H}_\lambda(\alpha_i\text{'s};\ \beta_j\text{'s}))^{ss}$, and cancelling the common $\mathcal{O}^r$, we find the required isomorphism

$$j^*\mathrm{FT}(j_{!*}[d]^*\mathcal{H}_\lambda(\alpha_i\text{'s};\ \beta_j\text{'s})) \approx$$

$$\approx\ [d]^*\mathbf{Cancel}(\mathcal{H}_{(-1)^{m-n}(d)^d/\lambda}(1/d,\ 2/d,\ \ldots,\ d/d,\ -\beta_j\text{'s};\ -\alpha_i\text{'s})).\quad \text{QED}$$

Corollary 6.4.3 Let $d \geq 1$ be an integer, and $\mathcal{H}$ an irreducible hypergeometric of type (n, m),

$$\mathcal{H} := \mathcal{H}_\lambda(\alpha_1,\ \ldots,\ \alpha_n;\ \beta_1,\ \ldots,\ \beta_m).$$

Suppose in addition that for all i, $d\alpha_i$ is not in $\mathbb{Z}$. Then we have isomorphisms of $\mathcal{D}$-modules on $\mathbb{A}^1$

(0) $j_{!*}[d]^*\mathcal{H}_\lambda(\alpha_i\text{'s};\ \beta_j\text{'s}) \approx j_*[d]^*\mathcal{H}_\lambda(\alpha_i\text{'s};\ \beta_j\text{'s})$.

(1) $\mathrm{FT}(j_*[d]^*\mathcal{H}_\lambda(\alpha_i\text{'s};\ \beta_j\text{'s})) \approx$

$$\approx\ j_{!*}[d]^*(\mathcal{H}_{(-1)^{n+m}(d)^d/\lambda}(1/d,\ 2/d,\ \ldots,\ d/d,\ -\beta_j\text{'s};\ -\alpha_i\text{'s})).$$

(2) $j_*[d]^*\mathcal{H}_\lambda(\alpha_i\text{'s};\ \beta_j\text{'s}) \approx$

$$\approx\ \mathrm{FT}(j_{!*}[d]^*(\mathcal{H}_{(-1)^{n+m+d}(d)^d/\lambda}(1/d,\ 2/d,\ \ldots,\ d/d,\ -\beta_j\text{'s};\ -\alpha_i\text{'s}))).$$

proof If no $d\alpha_i$ is in $\mathbb{Z}$, then (0) holds by 2.9.8, and

$$\mathcal{H}_{(-1)^{n+m+d}(d)^d/\lambda}(1/d,\ 2/d,\ \ldots,\ d/d,\ -\beta_j\text{'s};\ -\alpha_i\text{'s})$$

is itself irreducible, so its own **Cancel**. QED

Corollary 6.4.4 Let $d \geq 1$ be an integer, and $\mathcal{H}$ an irreducible hypergeometric of type (n, m),

$$\mathcal{H} := \mathcal{H}_\lambda(\alpha_1, \dots, \alpha_n; \beta_1, \dots, \beta_m).$$

Suppose in addition that

$$\mathcal{H}_\lambda(\alpha_i\text{'s}; \beta_j\text{'s})$$

is not Kummer induced of any degree $d_1 > 1$ which divides d. Then

$$j_{!*}[d]^*\mathcal{H}_\lambda(\alpha_i\text{'s}; \beta_j\text{'s})$$

is irreducible on $\mathbb{A}^1$, and consequently the isomorphisms

$$FT(j_{!*}[d]^*\mathcal{H}_\lambda(\alpha_i\text{'s}; \beta_j\text{'s})) \approx$$

$$\approx j_{!*}[d]^*(\mathbf{Cancel}\mathcal{H}_{(-1)^{n+m}(d)^d/\lambda}(1/d, 2/d, \dots, d/d, -\beta_j\text{'s}; -\alpha_i\text{'s})),$$

$$j_{!*}[d]^*\mathcal{H}_\lambda(\alpha_i\text{'s}; \beta_j\text{'s}) \approx$$

$$\approx FT(j_{!*}[d]^*(\mathbf{Cancel}\mathcal{H}_{(-1)^{n+m+d}(d)^d/\lambda}(1/d, 2/d, \dots, d/d, -\beta_j\text{'s}; -\alpha_i\text{'s}))),$$

are isomorphisms of irreducibles on $\mathbb{A}^1$.

proof. Since $\mathcal{H} := \mathcal{H}_\lambda(\alpha_i\text{'s}; \beta_j\text{'s})$ is an irreducible $\mathcal{D}$-module, and $[d]$ is finite etale galois, either $[d]^*\mathcal{H}$ is isotypical or $\mathcal{H}$ is induced from an intermediate covering. So the hypothesis insures that $[d]^*\mathcal{H}$ is isotypical. We first show that if $[d]^*\mathcal{H}$ is isotypical, then it is irreducible.

If $[d]^*\mathcal{H}$ is isotypical, say $k \geq 1$ copies of an irreducible $\mathcal{K}$, then since the isomorphism class of $\mathcal{K}$ is μ_d-invariant, $\mathcal{K}$ itself descends through the cyclic covering $[d]$, to an irreducible $\mathcal{K}_0$. Therefore the natural map of $\mathcal{D}$-modules

$$\mathcal{K}_0 \otimes \mathit{Hom}_{\mathcal{D}}(\mathcal{K}_0, \mathcal{H}) \to \mathcal{H},$$

is an isomorphism. But $\mathit{Hom}_{\mathcal{D}}(\mathcal{K}_0, \mathcal{H})$ becomes constant of rank k after $[d]^*$, so it is a sum of k objects each of the form $x^\alpha\mathcal{O}$. with $d\alpha \in \mathbb{Z}$. Since $\mathcal{H} \approx \mathcal{K}_0 \otimes \mathit{Hom}_{\mathcal{D}}(\mathcal{K}_0, \mathcal{H})$ is irreducible, we have $k = 1$, and hence $[d]^*\mathcal{H}$ is irreducible on $\mathbb{G}_m$. Its middle extension $j_{!*}[d]^*\mathcal{H}$ is therefore irreducible on $\mathbb{A}^1$. QED

CHAPTER 7
The ℓ-adic Theory

7.1 Exceptional Sets of primes

Let $b \geq 1$ be an integer. Recall that in 2.8 we proved the following two statements:

Corollary 2.8.2.1 If α, β, γ, δ in $\mu_b(\mathbb{C})$ satisfy $\alpha - \beta = \gamma - \delta$, then either

	(1) $\alpha = \beta$ and $\gamma = \delta$
or	(2) $\alpha = \gamma$ and $\beta = \delta$
or b is even and	(3) $\alpha = -\delta$ and $\beta = -\gamma$.

Corollary 2.8.3.1 If α, β, γ in $\mu_b(\mathbb{C})$ satisfy $\alpha - \beta = \pm\gamma$, then 6 divides b, α/β is a primitive sixth root of unity, and $\pm\gamma/\beta$ is $(\alpha/\beta)^2$.

For each prime number p, consider the following three assertions:

$*$(p,b) If α, β, γ, δ in $\mu_b(\overline{\mathbb{F}}_p)$ satisfy $\alpha - \beta = \gamma - \delta$, then either

	(1) $\alpha = \beta$ and $\gamma = \delta$
or	(2) $\alpha = \gamma$ and $\beta = \delta$
or b is even and	(3) $\alpha = -\delta$ and $\beta = -\gamma$.

$**$(p,b) If α, β, γ in $\mu_b(\overline{\mathbb{F}}_p)$ satisfy $\alpha - \beta = \pm\gamma$, then 6 divides b, α/β is a primitive sixth root of unity, and $\pm\gamma/\beta$ is $(\alpha/\beta)^2$.

$***$(p,b) If α, β in $\mu_b(\overline{\mathbb{F}}_p)$ satisfy $\alpha = -\beta$, then b is even and $\alpha \neq \beta$.

Lemma 7.1.1 For each integer $b \geq 1$, there exist entirely explicit nonzero integers $N_1(b)$ and $N_2(b)$ such that **$*$(p,b)** holds for all primes p which do not divide $N_1(b)$, and **$**$(p,b)** holds for all primes p which do not divide $N_2(b)$. The assertion **$***$(p,b)** holds if $p \neq 2$.

proof Put $\zeta_b := \exp(2\pi i/b) \in \mathbb{C}$. Fix a prime p not dividing b, and fix a prime ideal π of the cyclotomic integer ring $\mathbb{Z}[\zeta_b]$ lying over p. If we view π as a ring homomorphism $\pi : \mathbb{Z}[\zeta_b] \to \overline{\mathbb{F}}_p$, then π induces a group isomorphism $\mu_b(\mathbb{C}) = \mu_b(\mathbb{Z}[\zeta_b]) \cong \mu_b(\overline{\mathbb{F}}_p)$. So **$*$(p,b)** is equivalent to the following condition:

if $(\alpha, \beta, \gamma, \delta) \in (\mu_b(\mathbb{Z}[\zeta_b]))^4$ is such that such $\pi((\alpha - \beta) - (\gamma - \delta)) = 0$ in $\overline{\mathbb{F}}_p$, then either

(1) $\alpha = \beta$ and $\gamma = \delta$

or (2) $\alpha = \gamma$ and $\beta = \delta$
or b is even and (3) $\alpha = -\delta$ and $\beta = -\gamma$.

This may be restated as follows: let $N_1(b) \in \mathbb{Z}[\zeta_b]$ denote the product $\prod((\alpha - \beta) - (\gamma - \delta))$, extended to all quadruples $(\alpha, \beta, \gamma, \delta)$ in $(\mu_b(\mathbb{C}))^4$ such that **none** of the three conditions

(1) $\alpha = \beta$ and $\gamma = \delta$
or (2) $\alpha = \gamma$ and $\beta = \delta$
or (3) b is even and $\alpha = -\delta$ and $\beta = -\gamma$

holds. This product is a **nonzero** element of $\mathbb{Z}[\zeta_b]$, in virtue of 2.8.2.1 recalled above. As the conditions (1), (2), (3) are Galois-invariant, $N_1(b)$ is in fact a nonzero element of $\mathbb{Z}$. Then for a prime p not dividing b, the condition ***(p,b)** holds if and only p is not a divisor of $N_1(b)$. In fact ***(p,b)** holds if and only p is not a divisor of $N_1(b)$, for if $\ell | b$ is any prime divisor of b, then

$$(\zeta_\ell - 1) = (\zeta_\ell - 1) - (1 - 1)$$

and hence ℓ itself is one of factors of $N_1(b)$.

The proof for ****(p,b)** is entirely analogous. One considers the element $N_2(b) := bN_+N_- \in \mathbb{Z}[\zeta_b]$ where $N_\pm := \prod(\pm \gamma - (\alpha - \beta))$, the product extended to all triples (α, β, γ) in $(\mu_b(\mathbb{C}))^3$ for which it is **not** the case that

$$\alpha/\beta \text{ is a primitive sixth root of unity, and } \pm\gamma/\beta \text{ is } (\alpha/\beta)^2.$$

Again $N_2(b)$ is nonzero in $\mathbb{Z}$, and for a prime p not dividing b, the condition ****(p,b)** holds if and only p is not a divisor of $N_2(b)$.

That the assertion *****(p,b)** holds if $p \neq 2$ is obvious. QED

Remark 7.1.2 If $b=1$, then $N_1(b) = -N_2(b) = 1$.

Remark 7.1.3 (Benji Fisher) The integer $N_2(b)$ can contain very large primes. The factors of N_+ include $(2 - \zeta) = (1 - (\zeta - 1))$ for every $\zeta \in \mu_b(\mathbb{C})$, so $N_2(b)$ is divisible by $2^b - 1$, which itself can have rather large prime factors!

7.2 ℓ-adic analogue of the main DE theorem 2.8.1

(7.2.1) In this section we fix a prime number p, an algebraically closed field k of characteristic p, a prime number $\ell \neq p$, and an algebraic closure $\overline{\mathbb{Q}}_\ell$ of $\mathbb{Q}_\ell$. We denote by ψ a nontrivial $\overline{\mathbb{Q}}_\ell$-valued additive character of a finite subfield $\mathbb{F}_q$ of k. We denote by $\mathcal{L}_\psi$ the lisse rank one $\overline{\mathbb{Q}}_\ell$-sheaf on $\mathbb{A}^1/\mathbb{F}_q$ (= $\mathbb{G}_a/\mathbb{F}_q$) obtained from the Artin-Schreier covering (Lang torsor)

$$\begin{array}{cccc} & \mathbb{G}_a & & x \\ 1 - F_q & \downarrow \ \} \ \mathbb{F}_q & & \downarrow \\ & \mathbb{G}_a & & x - x^q \end{array}$$

by extension of structural group via ψ. For any $\mathbb{F}_q$-scheme Y, and any function f on Y, we view f as a morphism f: $Y \to \mathbb{A}^1 = \mathbb{G}_a$, and we denote by $\mathcal{L}_{\psi(f)}$ the lisse rank one $\overline{\mathbb{Q}}_\ell$-sheaf $f^*\mathcal{L}_\psi$ on Y.

(7.2.2) For $\mathbb{F}_q$ any finite subfield of k, and χ a $\overline{\mathbb{Q}}_\ell$-valued character of $(\mathbb{F}_q)^\times$, we denote by $\mathcal{L}_\chi$ the lisse rank one $\overline{\mathbb{Q}}_\ell$-sheaf on $\mathbb{G}_m \otimes \mathbb{F}_q :=$ Spec($\mathbb{F}_q[x, x^{-1}]$) obtained from the Kummer covering (Lang torsor)

$$\begin{array}{cccc} & \mathbb{G}_m \otimes \mathbb{F}_q & & x \\ 1 - F_q & \downarrow \ \} \ (\mathbb{F}_q)^\times & & \downarrow \\ & \mathbb{G}_m \otimes \mathbb{F}_q & & x^{1-q} \end{array}$$

of degree 1 - q by extension of structural group via χ. For any $\mathbb{F}_q$-scheme Y, and any invertible function f on Y, we denote by $\mathcal{L}_{\chi(f)}$ the lisse rank one $\overline{\mathbb{Q}}_\ell$-sheaf $f^*\mathcal{L}_\chi$ on Y.

Over the algebraically closed field k, any connected finite etale covering of $\mathbb{G}_m$ which is tame at both 0 and ∞ is dominated by a suitable Kummer covering. This allows us to identify

$$\pi_1(\mathbb{G}_m \otimes k)^{\text{tame}} \approx \hat{\mathbb{Z}}(1)_{\text{not } p} \approx \varprojlim_{\text{finite subfields of } k} (\mathbb{F}_q)^\times,$$

with transition maps given by the norm. For χ any continuous $\overline{\mathbb{Q}}_\ell$-valued character of $\pi_1(\mathbb{G}_m \otimes k)^{\text{tame}} \approx \varprojlim_{\text{finite subfields of } k} (\mathbb{F}_q)^\times$, (we will often refer to such a χ simply as a "tame character") we denote by $\mathcal{L}_\chi$ the corresponding lisse rank one $\overline{\mathbb{Q}}_\ell$-sheaf on $\mathbb{G}_m \otimes k$. [For χ of finite order, this notion of $\mathcal{L}_\chi$ coincides with the one given above.] For Y any k-scheme, and f any invertible function on Y, we denote by $\mathcal{L}_{\chi(f)}$ the lisse rank one $\overline{\mathbb{Q}}_\ell$-sheaf $f^*\mathcal{L}_\chi$ on Y.

(7.2.3) Let X/k be a smooth connected affine curve over k, $x \in X$ a geometric point of X, $\bar{X}$ the complete nonsingular model of X. Let $\mathcal{F}$ be a lisse $\bar{\mathbb{Q}}_\ell$-sheaf on X of rank $n \geq 1$. We denote by π_1 the fundamental group $\pi_1(X, x)$, by ρ the n-dimensional $\bar{\mathbb{Q}}_\ell$-representation

$$\rho: \pi_1 \to GL(\mathcal{F}_x)$$

which $\mathcal{F}$ "is", and by

$$G_{geom} := \text{the Zariski closure of } \rho(\pi_1) \text{ in } GL(\mathcal{F}_x).$$

For each point at infinity $\infty \in \bar{X} - X$, we can speak of the upper-numbering "breaks" (or "slopes") of $\mathcal{F}$ as I_∞-representation (cf [Ka-GKM, Chap. 1]), and of their sum, the Swan conductor $\mathrm{Swan}_\infty(\mathcal{F}) \in \mathbb{Z}$ of $\mathcal{F}$ at ∞.

Given $\mathcal{F}$ as above, for any linear representation Λ of G_{geom}, say

$$\Lambda: G_{geom} \to GL(d),$$

the composite representation

$$\Lambda \circ \rho : \pi_1 \to GL(d)$$

gives rise to a lisse $\bar{\mathbb{Q}}_\ell$-sheaf of rank d on X, denoted $\mathcal{F}(\Lambda)$.

Highest Slope Lemma 7.2.4 (compare 4.2.4) Notations as above, suppose that the kernel Γ of $\Lambda: G_{geom} \to GL(d)$ is a finite subgroup of order prime to p. Then at every point $\infty \in \bar{X} - X$, $\mathcal{F}$ and $\mathcal{F}(\Lambda)$ have the same highest slope.

proof For any $x \geq 0$, $\mathcal{F}$ has all slopes $\leq x$ if and only if $\rho((I_\infty)^{(x+)}) = \{e\}$, and $\mathcal{F}(\Lambda)$ has all slopes $\leq x$ if and only if $\rho((I_\infty)^{(x+)}) \subset \Gamma$. Since $\rho((I_\infty)^{(x+)})$ is a p-group, and Γ is prime to p, these two conditions are equivalent. QED

Lifting Lemma 7.2.5 (compare 2.2.2.1) Notations as above, let

$$\rho : G \to H$$

be a surjective homomorphism of linear algebraic groups over $\bar{\mathbb{Q}}_\ell$, whose kernel Γ is a finite central subgroup of G. Then any continuous homomorphism $\varphi: \pi_1(X, x) \to H(\bar{\mathbb{Q}}_\ell)$ lifts to a homomorphism

$$\tilde{\varphi} : \pi_1(X, x) \to G(\bar{\mathbb{Q}}_\ell)$$

with $\varphi = \rho\tilde{\varphi}$.

proof The obstruction to lifting lies in $H^2(X, \Gamma)$, which vanishes because X is an affine curve over an algebraically closed field. QED

We say that $\mathcal{F}$ is irreducible if the corresponding representation ρ of π_1 on $\mathcal{F}_x$ is irreducible, or equivalently if the given representation of G_{geom} on $\mathcal{F}_x$ is irreducible. We say that $\mathcal{F}$ is Lie-irreducible if the given representation of G_{geom} on $\mathcal{F}_x$ is Lie-irreducible, or equivalently if the restriction of ρ to every open subgroup of π_1 remains irreducible.

The main results on and around Lie-irreducibility and its alternatives are summarized in the following theorem.

Theorem 7.2.6 (cf [Ka-MG], 2.7 and 3.5) Let X be a smooth connected affine curve over an algebraically closed field of characteristic $p > 0$, ℓ a prime number $\ell \neq p$, $\mathcal{F}$ a lisse $\overline{\mathbb{Q}}_\ell$-sheaf on X of rank $n \geq 1$ which is irreducible.

(1) $\mathcal{F}$ is either Lie-irreducible, or induced (i.e., the direct image $f_*\mathcal{G}$ of a lisse $\mathcal{G}$ on a finite etale connected covering $f : Y \to X$ of degree $d \geq 2$) or is, for some divisor $d \geq 2$ of n, a tensor product $\mathcal{G} \otimes \mathcal{H}$ where $\mathcal{G}$ is Lie-irreducible of rank n/d and where $\mathcal{H}$ is irreducible of rank d with corresponding representation $\rho_{\mathcal{H}}$ having finite image. Moreover, this tensor decomposition is unique up to twisting $(\mathcal{G}, \mathcal{H}) \mapsto (\mathcal{G} \otimes \mathcal{L}, \mathcal{H} \otimes \mathcal{L}^{-1})$ by some rank one $\mathcal{L}$ of finite order.

(2) If X is $\mathbb{A}^1$ and $p > 2n + 1$, then $\mathcal{F}$ is Lie-irreducible.

(3) If X is $\mathbb{A}^1$ and $p > n$, then $\mathcal{F}$ is not induced.

(4) If X is $\mathbb{G}_m$ and $p > 2n + 1$, then $\mathcal{F}$ is either Lie-irreducible or is Kummer-induced (i.e., of the form $[d]_*\mathcal{G}$ for some prime-to-p divisor $d \geq 2$ of n and some lisse $\mathcal{G}$ on $\mathbb{G}_m$ of rank n/d, where $[d]: \mathbb{G}_m \to \mathbb{G}_m$ denotes the d'th power map).

(5) If X is $\mathbb{G}_m$ and $p > n$, then if $\mathcal{F}$ is induced it is Kummer-induced.

(6)Supppose X is $\mathbb{G}_m - \{s\}$ for some $s \in k^\times$, and that $\mathcal{F}$ has pseudoreflection local monodromy at s (in the sense that under the action of the inertia group I(s) on $\mathcal{F}_x$ via ρ, the space of invariants has codimension one). If $\mathcal{F}$ is neither Lie-irreducible nor induced, then $\mathcal{F}$ is the tensor product $\mathcal{L} \otimes \mathcal{H}$ of a rank one $\mathcal{L}$ with a rank n $\mathcal{H}$ whose corresponding representation $\rho_{\mathcal{H}}$ has finite image. If in addition $\det(\mathcal{F})$ is of finite order, then $\mathcal{F}$ itself has ρ with finite image.

(7) Suppose X is $\mathbb{G}_m$, $p > 2n + 1$, and $\mathcal{F}$ has pseudoreflection local monodromy at 0. Then either $\mathcal{F}$ is Lie-irreducible or $\mathcal{F}$ has rank two and its local monodromy at 0 is a tame reflection.

(8) Suppose X is $\mathbb{G}_m$ - {s} for some s ∈ $k^\times$, p > n, and $\mathcal{F}$ has pseudoreflection local monodromy at s. Suppose $\mathcal{F}$ is induced, $\mathcal{F} = f_*\mathcal{G}$ for a lisse $\mathcal{G}$ on a finite etale connected covering f : Y → $\mathbb{G}_m$ - {s} of degree d ≥ 2. Then either the covering f is (the restriction to $\mathbb{G}_m$ - {s} of) the Kummer covering of degree d, or d = n and the covering is either a Belyi covering or an inverse Belyi covering of type (a,b) for some partition of n = a + b as the sum of two strictly positive integers. Moreover, in the case of a Belyi or inverse Belyi covering, local monodromy around s is a tame reflection.

proof Assertions (1), (2) and (4) are proven in [Ka-MG] as Prop.'s 1, 5, 6. Assertions (3) and (5) are proven in the first paragraphs of the proofs of [Ka-MG], Prop.'s 5 and 6 respectively. Assertion (6) follows from (1) by the argument of 3.5.7.

To prove assertion (7), we argue as follows. By (4), if $\mathcal{F}$ is not Lie-irreducible it is $[d]_*\mathcal{G}$, with $\mathcal{G}$ lisse on $\mathbb{G}_m$ and d ≥ 2 prime to p. We first observe that $\mathcal{G}$ is tame at zero. [If exactly M (counting with multiplicty) of the slopes of $\mathcal{G}$ at zero are > 0, then exactly dM of the slopes of $[d]_*\mathcal{G} = \mathcal{F}$ at zero are > 0. But $\mathcal{F}$ has at least n-1 of its slopes at zero =0, so M=0.] Once $\mathcal{G}$ is a tame I(0)-representation, the set with multiplicity of the characters Λ occuring in $[d]_*\mathcal{G} = \mathcal{F}$ at zero is stable under $\Lambda \mapsto \Lambda\otimes\chi_d$ for χ_d any character of order d. As all but at most one of the Λ are trivial, we see first that d = 2, then that n = 2.

To prove assertion (8), we first note that as p > n ≥ d, the galois closure of the covering f is necessarily tame, because prime to p. The proof is then entirely analogous to that of 3.5.2, making use of the fact that for p > n, 3.5.1 is equally valid with $\mathbb{C}$ replaced by an algebraically closed field of characteristic p. QED

Main ℓ-adic Theorem 7.2.7 (compare Main D.E. Theorem 2.8.1) Let X be a smooth connected affine curve over an algebraically closed field of characteristic p > 0, x ∈ X a geometric point of X, ℓ a prime number ℓ ≠ p, $\mathcal{F}$ a Lie-irreducible lisse $\overline{\mathbb{Q}}_\ell$-sheaf on X of rank n. Suppose that at some point $\infty \in \overline{X} - X$, the highest slope of $\mathcal{F}$, written a/b in lowest terms, is > 0 and occurs with multiplicity b. If b = n, suppose that p > n. If b < n, suppose that p does not divide the integer $2aN_1(b)N_2(b)$. Let $G := G_{geom} \subset GL(\mathcal{F}_x)$ be the Zariski closure of $\rho(\pi_1(X, x))$ in $GL(\mathcal{F}_x)$, G^0 its identity component, and $G^{0,der}$ the commutator subgroup of G^0.

Then G^0 is equal either to $G^{0,der}$ or to $\mathbb{G}_m G^{0,der}$, and the list of possible $G^{0,der}$ is given by:

(1) If b is odd, $G^{0,der}$ is $SL(\mathcal{F}_x)$.

(2) If b is even, then either $G^{0,der}$ is $SL(\mathcal{F}_x)$ or $SO(\mathcal{F}_x)$ or (if n is even) $SP(\mathcal{F}_x)$, or b=6, n=7,8 or 9, and $G^{0,der}$ is one of

n=7: the image of G_2 in its 7-dim'l irreducible representation

n=8: the image of Spin(7) in the 8-dim'l spin representation
the image of SL(3) in the adjoint representation
the image of $SL(2)\times SL(2)\times SL(2)$ in $std\otimes std\otimes std$
the image of $SL(2)\times Sp(4)$ in $std\otimes std$
the image of $SL(2)\times SL(4)$ in $std\otimes std$

n=9: the image of $SL(3)\times SL(3)$ in $std\otimes std$.

proof If $n = 1$, there is nothing to prove. If $b=n \geq 2$, this is proven in [Ka-MG, Thm.7] under the hypotheses that $p > 2n+1$ and that $\det(\mathcal{F})$ is of finite order prime to p. In fact, the proof given there works mutatis mutandis provided only that $p > n$ and that $\det(\mathcal{F})$ is of finite order prime to p. Let us explain how to reduce to this case. Denote by $\chi = \det(\rho)$ the character of π_1 given by $\det(\mathcal{F})$. The slope of χ at ∞ is an integer $s \leq a/b$, and as a/b has exact denominator $b \geq 2$, we must have $s < a/b$. We next claim that χ has an n'th root. Indeed, the obstruction lies in $H^2(X, \mu_n) = 0$ (cohomological dimension of open curves), so there exists a character Λ such that $\Lambda^n = \chi^{-1}$. We next claim that Λ has the same ∞-slope as χ. Indeed, for any real $x \geq 0$, a character ξ of I_∞ has slope $\leq x$ if and only if ξ kills the pro-p group $I^{(x+)}$. So raising characters to prime-to-p powers doesn't change their slopes. Therefore Λ has the same slope $s < a/b$ as χ. So $\mathcal{F}\otimes\Lambda$ has the the same highest slope a/b, the same rank n, and trivial determinant. Moreover, $\mathcal{F}\otimes\Lambda$ is still Lie-irreducible (= irreducible on all finite étale connected coverings) since this property is invariant under twisting by characters.

It remains only to remark that for any lisse $\mathcal{F}$, and any character Λ, the group $(G_{geom})^{0,der}$ is the same for $\mathcal{F}$ and for $\mathcal{F}\otimes\Lambda$. Indeed, we have trivial inclusions

$$G_{geom}(\mathcal{F}) \subset \mathbb{G}_m G_{geom}(\mathcal{F}\otimes\Lambda), \quad G_{geom}(\mathcal{F}\otimes\Lambda) \subset \mathbb{G}_m G_{geom}(\mathcal{F}).$$

Passing to connected components of the identity, we see that

$$\mathbb{G}_m(G_{geom}(\mathcal{F}))^0 = \mathbb{G}_m(G_{geom}(\mathcal{F}\otimes\Lambda))^0,$$

and passing to commutator subgroups we find

$$(G_{geom}(\mathcal{F}))^{0,der} = (G_{geom}(\mathcal{F}\otimes\Lambda))^{0,der},$$

as required. This concludes the proof in the case $b = n$.

We now treat the case $b < n$. The proof is very much analogous to that of the Main D.E. Theorem, and we will only indicate what changes need to be made. Exactly as in the proof of that theorem, we see that $\mathfrak{g} := \mathrm{Lie}(G^{0,der})$ is a semisimple Lie-subalgebra of $\mathrm{End}(\mathcal{F}_x)$ which acts irreducibly on $\mathcal{F}_x$. By its very construction, $\mathfrak{g}$ is normalized by **any** subgroup K of G.

We now use the slope hypothesis that at some point at infinity ∞ the highest slope is a/b in lowest terms and its multiplicity is b to construct a diagonal subgroup K of G, to which we will then apply Gabber's "torus trick" 1.0.

As a representation of I_∞, $\mathcal{F}$ is the direct sum

$$\mathcal{F} = \mathcal{F}_{a/b} \oplus \mathcal{F}_{< a/b} = (\text{slope } a/b, \text{ rank } b) \oplus (\text{all slopes} < a/b).$$

In order to describe the representation $\mathcal{F}_{a/b}$ of I_∞ explicitly, fix a uniformizing parameter $1/x$ at ∞. This identifies the ∞-adic completion of the function field of X with the Laurent series field $K:=k((1/x))$. Fix a b'th root t of x, and denote by K_b the Laurent series field $k((1/t))$. In this setting, $I := I_\infty$ is the local galois group $\mathrm{Gal}(K^{sep}/K)$, and $\mathrm{Gal}(K^{sep}/K_b)$ is its unique closed subgroup $I(b)$ of index b. The representation $\mathcal{F}_{a/b}$ of I is irreducible, and as p does not divide b, it is induced from a character χ of $I(b) = \mathrm{Gal}(K^{sep}/k((1/t)))$ of slope $=a$ (cf. [Ka-GKM,1.14]). Since p does not divide a, any such character χ is, by [Ka-GKM, 8.5.7.1], of the form

$$\chi = \mathcal{L}_{\psi(P_a(t))}\otimes(\text{a character of slope} \le a - 1),$$

where $P_a(t) \in k[t]$ is a polynomial of degree a. At the expense of scaling the parameter $1/x$, we may assume that $P_a(t)$ is monic, say $t^a + f_{<a}(t)$, with $f_{<a}(t)$ a polynomial of degree strictly less than a. As $\mathcal{L}_{\psi(f_{<a}(t))}$ has slope $\le a - 1$, we have

$$\chi = \mathcal{L}_{\psi(t^a)}\otimes(\text{a character of slope} \le a - 1).$$

Therefore the restriction of $\mathcal{F}_{a/b}$ to $I(b)$ is a direct sum

$$\oplus_{\zeta\in\mu_b(\overline{\mathbb{F}}_p)} \mathcal{L}_{\psi((\zeta t)^a)}\otimes(\text{a character of slope} \le a - 1).$$

Because $\gcd(a, b) = 1$, as ζ runs over $\mu_b(\overline{\mathbb{F}}_p)$ the ζ^a's are just a permutaion of the ζ's, so we may rewrite this as

$$\oplus_{\zeta \in \mu_b(\overline{\mathbb{F}}_p)} \mathcal{L}_{\psi(\zeta t^a)} \otimes (\text{a character of slope} \le a - 1).$$

Therefore as a representation of the **upper numbering subgroup** $I(b)^{(a)}$, we have

$$\mathcal{F} \mid I(b)^{(a)} \approx \oplus_{\zeta \in \mu_b(\overline{\mathbb{F}}_p)} \mathcal{L}_{\psi(\zeta t^a)} \oplus (\text{ trivial of rank } n - b).$$

For each ξ in k we denote by

$$\chi_\xi := \text{the character of } I(b)^{(a)} \text{ given by } \mathcal{L}_{\psi(\xi t^a)}$$

The key observation is that for ξ, ν in k we have

$$\chi_\xi \chi_\nu = \chi_{\xi+\nu},$$

$$\chi_\xi \text{ is trivial on } I(b)^{(a)} \text{ iff } \xi = 0.$$

Let

$$\Gamma := \text{the image of } I(b)^{(a)} \text{ in } G.$$

Then Γ is a diagonal subgroup of G, and the diagonal entries of Γ are the n characters

the b characters χ_ζ as ζ runs over $\mu_b(k) = \mu_b(\overline{\mathbb{F}}_p)$,

n-b repetitions of the trivial character χ_0.

We now apply the "torus trick" to Γ. The discussion from here on is exactly the same as in the proof of the Main D.E. Theorem 2.8.1, except that now one is analyzing all possible relations

Rel(b < n)($\overline{\mathbb{F}}_p$) $\alpha - \beta = \gamma - \delta$, where α, β, γ, δ lie in $\mu_b(\overline{\mathbb{F}}_p) \cup \{0\}$,

rather than in $\mu_b(\mathbb{C}) \cup \{0\}$. But our hypothesis on p insures that all three of $*$**(p,b)**, $**$**(p,b)**, and $***$**(p,b)** hold, in which case the analysis is exactly the same as it was over $\mathbb{C}$. QED

Remark 7.2.7.1 We could treat the case b=n by the above method directly, but doing so would require us to exclude all primes which divide $aN_1(n)$, rather than only those which are $\le$ n.

7.3 Construction of Irreducible Sheaves via Fourier Transform

In this section, we will explain the systematic use of the one-variable ℓ-adic Fourier Transform FT_ψ to construct irreducible lisse sheaves on open sets of $\mathbb{A}^1 := \mathrm{Spec}(k[x])$, k a perfect field of characteristic $p \ne \ell$. We will make free use of the basic facts about FT_ψ (cf. [Ka-TL], [Ka-GKM chpt.8], [Lau-TF]). Let us recall the basic set-up.

(7.3.1) On any smooth, geometrically connected curve C/k, a

constructible $\overline{\mathbb{Q}}_\ell$-sheaf $\mathcal{F}$ on C is called a **middle extension** if for some (or equivalently for every) nonempty open set $j: U \to C$ on which $j^*\mathcal{F}$ is lisse, we have $\mathcal{F} \approx j_*j^*\mathcal{F}$. Given a middle extension sheaf $\mathcal{F}$ on C, and a nonempty open set $j: U \to C$ on which $j^*\mathcal{F}$ is lisse, the sheaf $j_*(j^*(\mathcal{F})^\vee)$ (where $j^*(\mathcal{F})^\vee$ denotes the linear dual, i.e., the contragredient representation of $\pi_1(U, \bar{u})$) is again a middle extension, which is independent of the auxiliary choice of the open set U. This sheaf, denoted $D(\mathcal{F})$, is called the **dual** of the middle extension $\mathcal{F}$. A middle extension $\mathcal{F}$ on C is called **irreducible** if for some (or equivalently for every) nonempty open set $j: U \to C$ on which $j^*\mathcal{F}$ is lisse, $j^*\mathcal{F}$ is geometrically irreducible (i.e., irreducible as a representation of $\pi_1(U\otimes\bar{k}, u)$, for $u \in U\otimes\bar{k}$ any geometric point).

Lemma 7.3.2 Let $f: X \to Y$ be a dominating morphism of smooth, geometrically connected curves over k. If $\mathcal{F}$ is a middle extension on X, then $f_*\mathcal{F}$ is a middle extension on Y.

proof By [De-TF], $f_*\mathcal{F}$ is constructible. Let $j: U \to Y$ be the inclusion of a nonempty open set where $f_*\mathcal{F}$ is lisse. Because f is dominating, $f^{-1}(U)$ is a nonempty open set of X; we denote its inclusion by $h: f^{-1}(U) \to X$. Let $k: V \to f^{-1}(U)$ be a nonempty open set of $f^{-1}(U)$ where $\mathcal{F}$ is lisse. Then we have a commutative diagram

$$\begin{array}{ccccc} V & \xrightarrow{k} & f^{-1}(U) & \xrightarrow{h} & X \\ & & f_U\downarrow & \text{CART} & f\downarrow \\ & & U & \xrightarrow[j]{} & Y. \end{array}$$

Because $\mathcal{F}$ is a middle extension, $\mathcal{F} \approx h_*k_*(k^*h^*\mathcal{F}) = h_*h^*\mathcal{F}$, so taking f_* gives

$$f_*\mathcal{F} = f_*h_*h^*\mathcal{F} = j_*(f_U)_*h^*\mathcal{F} = j_*j^*f_*\mathcal{F}. \quad \text{QED}$$

(7.3.3) For any constructible $\overline{\mathbb{Q}}_\ell$-sheaf $\mathcal{F}$ on $\mathbb{A}^1$, its "naive Fourier Transform" $NFT_\psi(\mathcal{F})$ is the constructible $\overline{\mathbb{Q}}_\ell$-sheaf $\mathcal{F}$ on $\mathbb{A}^1$ defined (in terms of the two projections of $\mathbb{A}^2$ to $\mathbb{A}^1$ and the sheaf $\mathcal{L}_{\psi(xy)}$ on $\mathbb{A}^2$) as

$$\mathrm{NFT}_\psi(\mathcal{F}) := R^1\mathrm{pr}_{2!}(\mathrm{pr}_1{}^*\mathcal{F}\otimes\mathcal{L}_{\psi(xy)}).$$

By proper base change, the stalk of $\mathrm{NFT}_\psi(\mathcal{F})$ at any point a in $\mathbb{A}_1(\bar{k})$ is

$$(\mathrm{NFT}_\psi(\mathcal{F}))_a = H^1{}_c(\mathbb{A}^1\otimes\bar{k}, \mathcal{F}\otimes\mathcal{L}_{\psi(ax)}).$$

(7.3.4) A constructible $\bar{\mathbb{Q}}_\ell$-sheaf $\mathcal{F}$ on $\mathbb{A}^1$ is called **elementary** if it satisfies the following two conditions:

Elem(1) $\mathcal{F}$ has no (nonzero) punctual sections, i.e., $H_c{}^0(\mathbb{A}^1\otimes\bar{k}, \mathcal{F}) = 0$. [Equivalently, for every nonempty open set $j: U \to \mathbb{A}^1$ on which $j^*\mathcal{F}$ is lisse, $\mathcal{F} \hookrightarrow j_*j^*\mathcal{F}$.]

Elem(2) for every $t \in \bar{k}$, $H_c{}^2(\mathbb{A}^1\otimes\bar{k}, \mathcal{F}\otimes\mathcal{L}_{\psi(tx)}) = 0$.

(7.3.5) A constructible $\bar{\mathbb{Q}}_\ell$-sheaf $\mathcal{F}$ on $\mathbb{A}^1$ is called **Fourier** if it satisfies the following two conditions:

Fourier(1) for some (or equivalently for every) nonempty open set $j: U \to \mathbb{A}^1$ on which $j^*\mathcal{F}$ is lisse, we have $\mathcal{F} \approx j_*j^*\mathcal{F}$, i.e., $\mathcal{F}$ is a middle extension on $\mathbb{A}^1$.

Fourier(2) for every $t \in \bar{k}$, we have

$$H^0(\mathbb{A}^1\otimes\bar{k}, \mathcal{F}\otimes\mathcal{L}_{\psi(tx)}) = 0 = H_c{}^2(\mathbb{A}^1\otimes\bar{k}, \mathcal{F}\otimes\mathcal{L}_{\psi(tx)}).$$

Equivalently, for an $\mathcal{F}$ that satisfies Fourier(1), Fourier(2) is the condition that for some (or equivalently for every) nonempty open set $j: U \to \mathbb{A}^1$ on which $j^*\mathcal{F}$ is lisse, the geometric object $j^*\mathcal{F}| U\otimes_k\bar{k}$ has no subsheaf and no quotient sheaf of the form $j^*\mathcal{L}_{\psi(tx)}| U\otimes_k\bar{k}$ for any $t \in \bar{k}$. Notice that this condition is autodual.

Given a Fourier sheaf $\mathcal{F}$, its dual $D(\mathcal{F})$ as a middle extension is again a Fourier sheaf, called the **dual** of the Fourier sheaf $\mathcal{F}$.

(7.3.6) A Fourier sheaf $\mathcal{F}$ on $\mathbb{A}^1$ is called **irreducible** if $\mathcal{F}$ is irreducible as a middle extension, i.e., if for some (or equivalently for every) nonempty open set $j: U \to \mathbb{A}^1$ on which $j^*\mathcal{F}$ is lisse, $j^*\mathcal{F}$ is geometrically irreducible (i.e., irreducible as a representation of $\pi_1(U\otimes\bar{k}, u)$, for $u \in U\otimes\bar{k}$ any geometric point).

Thus a constructible $\overline{\mathbb{Q}}_\ell$-sheaf $\mathcal{F}$ on $\mathbb{A}^1$ is an irreducible Fourier sheaf if and only if it satisfies the following two conditions:
IrrFour(1) for some (or equivalently for every) nonempty open set $j: U \to \mathbb{A}^1$ on which $j^*\mathcal{F}$ is lisse, we have $\mathcal{F} \approx j_*j^*\mathcal{F}$, and $j^*\mathcal{F}$ is geometrically irreducible, i.e., $\mathcal{F}$ is an irreducible middle extension.
IrrFour(2) $\mathcal{F}$ is not geometrically isomorphic to $\mathcal{L}_{\psi(tx)}$ for any t in $\overline{k}$.

(7.3.7) If k is a finite field, X/k a smooth, geometrically connected curve, and $\mathcal{F}$ a constructible $\overline{\mathbb{Q}}_\ell$-sheaf on X, then for E a finite extension of k, a point $x \in X(E)$, and a geometric point $\overline{x} \in X(k^{sep})$ lying over x, the stalk $\mathcal{F}_{\overline{x}}$ of $\mathcal{F}$ at $\overline{x}$ is a finite-dimensional ℓ-adic representation of $\mathrm{Gal}(E^{sep}/E)$. We denote by F_E the geometric Frobenius element in this group (i.e., F_E is the inverse of $\alpha \mapsto \alpha^{\mathrm{Card}(E)}$). Thus we may speak of the trace, characteristic polynomial, eigenvalues, et cetera of F_E acting on $\mathcal{F}_{\overline{x}}$. We define the $\overline{\mathbb{Q}}_\ell$-valued trace function of $\mathcal{F}$ on X(E) by

$$x \in X(E) \mapsto \mathrm{Trace}(\, F_E \mid \mathcal{F}_{\overline{x}}\,) := \mathrm{Trace}(\mathcal{F})(E, x).$$

For a real number w, and an embedding $\iota: \overline{\mathbb{Q}}_\ell \to \mathbb{C}$, we say that $\mathcal{F}$ is punctually ι-pure of weight w if for every finite extension E of k, and for every point $x \in X(E)$, the eigenvalues α of F_E on $\mathcal{F}_{\overline{x}}$ all satisfy

$$|\iota(\alpha)| = \mathrm{Card}(E)^{w/2},$$

where |a| denotes the usual complex absolute value. We say that $\mathcal{F}$ is "pure of weight w" if for some (or equivalently for every) nonempty open set $j: U \to X$ on which $j^*\mathcal{F}$ is lisse, we have $\mathcal{F} \approx j_*j^*\mathcal{F}$, and $j^*\mathcal{F}$ is punctually ι-pure of weight w for every embedding $\iota: \overline{\mathbb{Q}}_\ell \to \mathbb{C}$.

We can now recall the first basic result (cf. [Ka-Lau, 2.1 and 2.2], [Ka-GKM, Chpt. 8], and [Ka-TL]) on Fourier Transform.
Theorem 7.3.8 (Brylinski, Deligne, Laumon) Let $\mathcal{F}$ be a constructible $\overline{\mathbb{Q}}_\ell$-sheaf $\mathcal{F}$ on $\mathbb{A}^1$.
(1) If $\mathcal{F}$ is elementary, then $\mathrm{NFT}_\psi(\mathcal{F})$ is elementary, and we have the inversion formula

$$[-1]^*\mathcal{F}(-1) \approx \mathrm{NFT}_\psi(\mathrm{NFT}_\psi(\mathcal{F})).$$

(2) If $\mathcal{F}$ is Fourier, then $\mathrm{NFT}_\psi(\mathcal{F})$ is Fourier, and

$$D(NFT_\psi(\mathcal{F})) \approx NFT_\psi([-1]^*(D(\mathcal{F})))(\mathbf{1}).$$

(3) If $\mathcal{F}$ is irreducible Fourier, then $NFT_\psi(\mathcal{F})$ is irreducible Fourier.

(4) If k is a finite field, ψ a nontrivial additive character of k, and if $\mathcal{F}$ is elementary, then for every finite extension E of k, the trace function of $NFT_\psi(\mathcal{F})$ on $\mathbb{A}^1(E) = E$ is (minus) the finite field Fourier Transform of that of $\mathcal{F}$:

$$\text{Trace}(NFT_\psi(\mathcal{F}))(E, y) = -\Sigma_{x \in E}\ \psi_E(yx)\text{Trace}(\mathcal{F})(E, x),$$

where $\psi_E(a) := \psi(\text{Trace}_{E/k}(a))$.

(5) If k is a finite field, and if $\mathcal{F}$ is elementary and pure of weight w, then $NFT_\psi(\mathcal{F})$ is elementary and pure of weight w+1.

Let us also recall (cf. [Ka-GKM], Chpt. 8) the numerology of the Fourier Transform.

Lemma 7.3.9 Suppose k is algebraically closed, and that $\mathcal{F}$ is elementary. Denote by rank($\mathcal{F}$) the generic rank of $\mathcal{F}$. Let $\mathcal{G} := NFT_\psi(\mathcal{F})$. For each $x \in \mathbb{A}^1(k) = k$, put $\text{drop}_x(\mathcal{F}) := \text{rank}(\mathcal{F}) - \dim(\mathcal{F}_x)$. Then

(1) For each t in $\mathbb{A}^1(k)$, $\dim(\mathcal{G}_t) =$

$= \text{Swan}_\infty(\mathcal{F}\otimes\mathcal{L}_{\psi(tx)}) - \text{rank}(\mathcal{F}) + \Sigma_{x \text{ in } k}(\text{Swan}_x(\mathcal{F}) + \text{drop}_x(\mathcal{F}))$.

(2) $\mathcal{G} := NFT_\psi(\mathcal{F})$ has generic rank

$= \Sigma_{\infty\text{-breaks } \lambda \text{ of } \mathcal{F}} \max(0, \lambda - 1) + \Sigma_{x \text{ in } k}(\text{Swan}_x(\mathcal{F}) + \text{drop}_x(\mathcal{F}))$.

(3) $\mathcal{G}$ is lisse at $t \in \mathbb{A}^1(k)$ if and only if all the ∞-slopes of $\mathcal{F}\otimes\mathcal{L}_{\psi(tx)}$ are ≥ 1.

7.4 Local monodromy of Fourier Transforms d'apres Laumon (cf. [Lau-TF], [Ka-TL])

We next review the analysis of the local monodromy of a Fourier Transform, via Laumon's "local Fourier Transform". For simplicity of exposition, we will throughout this section suppose that the field k is **algebraically closed**. Let us fix a geometric generic point $\overline{\eta}$ of $\mathbb{A}^1$, i.e., an algebraically closed overfield L of k(x). Denote by $k(x)^{sep}$ the separable closure of k(x) in L. For each point t of $\mathbb{P}^1(k) = \mathbb{A}^1(k) \cup \infty$, viewed as a discrete valuation of k(x)/k, pick a place $\tilde{t}$ of $k(x)^{sep}$ lying over it, and denote by $I(t) \subset \text{Gal}(k(x)^{sep}/k(x))$ the inertia group at $\tilde{t}$. Given a constructible $\overline{\mathbb{Q}}_\ell$-sheaf $\mathcal{F}$ on $\mathbb{A}^1$, its geometric generic fibre $\mathcal{F}_{\overline{\eta}}$ is an ℓ-adic representation of $\text{Gal}(k(x)^{sep}/k(x))$. We denote by $\mathcal{F}(t)$ the

I(t)- representation $\mathcal{F}_{\overline{\eta}}$ (cf [Ka-TL]). For s in $\mathbb{A}^1(k)$, we denote by $\mathcal{F}_s$ the stalk of $\mathcal{F}$ at s, viewed as a trivial I(s)-representation. To avoid confusion with Tate twists, we will denote the latter in boldface, e.g. $\mathcal{F}(-1)$ is the Tate twist and $\mathcal{F}(-1)$ is the representation of the inertia group at at he point s = -1.

Theorem of ℓ-adic Stationary Phase 7.4.1 (Laumon) For each point t in $\mathbb{A}^1(k) \cup \infty$, there is an exact functor

$$\mathrm{FT}_\psi\mathrm{loc}(t,\infty): (\ell\text{-adic } \mathrm{I}(t)\text{-rep's}) \to (\ell\text{-adic } \mathrm{I}(\infty)\text{-rep's})$$

such that if $\mathcal{F}$ is a constructible $\overline{\mathbb{Q}}_\ell$-sheaf on $\mathbb{A}^1$ which is the extension by zero of a lisse sheaf on a nonvoid open set $\mathbb{A}^1$ - S,there is a canonical direct sum decomposition of $\mathrm{NFT}_\psi(\mathcal{F})(\infty)$ as $\mathrm{I}(\infty)$-representation

$$\mathrm{NFT}_\psi(\mathcal{F})(\infty) = \bigoplus_{t \text{ in } S \cup \infty} \mathrm{FT}_\psi\mathrm{loc}(t,\infty)(\mathcal{F}(t)).$$

The functors $\mathrm{FT}_\psi\mathrm{loc}(t,\infty)$ have the following properties:

(1) For an $\mathrm{I}(\infty)$-representation N, $\mathrm{FT}_\psi\mathrm{loc}(\infty,\infty)(N) = 0$ if and only if all slopes of N are ≤ 1, and $\mathrm{FT}_\psi\mathrm{loc}(\infty,\infty)(N)$ has all slopes > 1. $\mathrm{FT}_\psi\mathrm{loc}(\infty,\infty)$ is an autoequivalence of the category of ℓ-adic $\mathrm{I}(\infty)$-representation with all slopes > 1; for N an $\mathrm{I}(\infty)$-representation with all slopes > 1, we have the inversion formula

$$\mathrm{FT}_\psi\mathrm{loc}(\infty,\infty)(\mathrm{FT}_\psi\mathrm{loc}(\infty,\infty)(N)) \approx [-1]^*N(-1).$$

If N has unique slope (a+b)/a with multiplicity a, then $\mathrm{FT}_\psi\mathrm{loc}(\infty,\infty)(N)$ has unique slope (a+b)/b with multiplicity b. [N.B.: We do **not** assume here that gcd(a,b) = 1.]

(2) For any I(0)-representation M, $\mathrm{FT}_\psi\mathrm{loc}(0,\infty)(M)$ has all slopes < 1.

(3) For s in $\mathbb{A}^1(k)$, denote by Add(s) : $x \mapsto x + s$ the additive translation by s. For L an I(s) representation, let $\mathrm{Add}(s)^*L$ denote the I(0)-representation obtained by identifying I(0) to I(s) by Add(s). Denote by y the Fourier Transform variable. Then

$$\mathrm{FT}_\psi\mathrm{loc}(s,\infty)(L) \approx (\mathrm{FT}_\psi\mathrm{loc}(0,\infty)(\mathrm{Add}(s)^*L)) \otimes \mathcal{L}_{\psi(sy)}.$$

In particular, for $s \in k^\times$, $\mathrm{FT}_\psi\mathrm{loc}(s,\infty)(L)$ has all slopes $=1$.

proof Everything except (3) is proven in [Ka-TL]. For the canonical extension $\mathcal{F}$ of L, we have $FT_\psi loc(\infty,\infty)(\mathcal{F}(\infty)) = 0$ by (1), since $\mathcal{F}(\infty)$ is tame, so stationary phase gives

$$NFT_\psi(\mathcal{F})(\infty) \approx FT_\psi loc(s,\infty)(L).$$

For this $\mathcal{F}$, $Add(s)^*\mathcal{F}$ is the canonical extension of $Add(s)^*L$, and so by the same argument we have

$$NFT_\psi(Add(s)^*\mathcal{F})(\infty) \approx FT_\psi loc(0,\infty)(Add(s)^*L).$$

Assertion (3) now follows from the fact that for any constructible $\overline{\mathbb{Q}}_\ell$-sheaf $\mathcal{F}$ on $\mathbb{A}^1$, we have the global formula

$$NFT_\psi(\mathcal{F}) \approx NFT_\psi(Add(s)^*\mathcal{F}) \otimes \mathcal{L}_{\psi(sy)}. \quad \text{QED}$$

Corollary 7.4.1.1 In the stationary phase decomposition, the individual pieces may be characterized as follows:

(1) $FT_\psi loc(\infty,\infty)(\mathcal{F}(\infty))$ has all slopes > 1

(2) $FT_\psi loc(0,\infty)(\mathcal{F}(0))$ has all slopes < 1.

(3) For $s \in k^\times$, $FT_\psi loc(s,\infty)(\mathcal{F}(s))$ has all slopes $=1$, and

$$(FT_\psi loc(s,\infty)(\mathcal{F}(s))) \otimes \mathcal{L}_{\psi(-sy)} \approx FT_\psi loc(0,\infty)((Add(s)^*\mathcal{F})(0))$$

has all slopes < 1.

Corollary 7.4.2 (Stationary Phase bis) Let $\mathcal{F}$ be a constructible $\overline{\mathbb{Q}}_\ell$-sheaf on $\mathbb{A}^1$ which has no punctual sections, and which is lisse on a nonvoid open set $\mathbb{A}^1 - S$. Then there is a canonical direct sum decomposition of $NFT_\psi(\mathcal{F})(\infty)$ as $I(\infty)$-representation

$$NFT_\psi(\mathcal{F})(\infty) = FT_\psi loc(\infty,\infty)(\mathcal{F}(\infty)) \oplus \bigoplus_{s \text{ in } S} FT_\psi loc(s,\infty)(\mathcal{F}(s)/\mathcal{F}_s).$$

proof Denote by $j: \mathbb{A}^1 - S \to \mathbb{A}^1$ the inclusion. Because $\mathcal{F}$ has no punctual sections, we have a short exact sequence of sheaves on $\mathbb{A}^1$

$$0 \to j_!j^*\mathcal{F} \to \mathcal{F} \to \oplus_{s \text{ in } S} (\mathcal{F}_s \text{ concentrated at } s) \to 0.$$

Taking Fourier Transform gives a short exact sequence of sheaves on $\mathbb{A}^1$,

$$0 \to \oplus_{s \text{ in } S} \mathcal{F}_s \otimes_{\overline{\mathbb{Q}}_\ell} \mathcal{L}_{\psi(sy)} \to NFT_\psi(j_!j^*\mathcal{F}) \to NFT_\psi(\mathcal{F}) \to 0.$$

Restricting to $I(\infty)$-representations, we get a short exact sequence

$$0 \to \oplus_{s \text{ in } S} \mathcal{F}_s \otimes_{\overline{\mathbb{Q}}_\ell} \mathcal{L}_{\psi(sy)} \to$$

$\to \mathrm{FT}_\psi\mathrm{loc}(\infty,\infty)(\mathcal{F}(\infty)) \oplus \bigoplus_{s \text{ in } S} \mathrm{FT}_\psi\mathrm{loc}(s,\infty)(\mathcal{F}(s)) \to \mathrm{NFT}_\psi(\mathcal{F})(\infty) \to 0.$
By 7.4.1.1, the term $\mathcal{F}_s \otimes_{\overline{\mathbb{Q}}_\ell} \mathcal{L}_{\psi(sy)}$ must land entirely inside in $\mathrm{FT}_\psi\mathrm{loc}(s,\infty)(\mathcal{F}(s))$. So it remains only to identify $\mathcal{F}_s \otimes_{\overline{\mathbb{Q}}_\ell} \mathcal{L}_{\psi(sy)}$ with $\mathrm{FT}_\psi\mathrm{loc}(s,\infty)(\mathcal{F}_s)$. But this is immediate from applying the above considerations to the constant sheaf with value $\mathcal{F}_s$, and $S = \{s\}$. QED

We next recall

Theorem 7.4.3 (Laumon) There is an exact functor

$$\mathrm{FT}_\psi\mathrm{loc}(\infty,0) : (\ell\text{-adic } I(\infty)\text{-rep's}) \quad \to \quad (\ell\text{-adic } I(0)\text{-rep's})$$

such that if $\mathcal{F}$ is a constructible $\overline{\mathbb{Q}}_\ell$-sheaf on $\mathbb{A}^1$ with no punctual sections, there is a four term exact sequence of $I(0)$-representations

$$0 \to H_c^1(\mathbb{A}^1, \mathcal{F}) \to \mathrm{NFT}_\psi(\mathcal{F})(0) \to \mathrm{FT}_\psi\mathrm{loc}(\infty,0)(\mathcal{F}(\infty)) \to H_c^2(\mathbb{A}^1, \mathcal{F}) \to 0.$$

If N is an $I(\infty)$-representation with all slopes ≥ 1, $\mathrm{FT}_\psi\mathrm{loc}(\infty,0)(N) = 0$.

Corollary 7.4.3.1 If $\mathcal{F}$ is elementary, and $\mathcal{G} := \mathrm{NFT}_\psi(\mathcal{F})$, then

$$\mathcal{G}(0)/\mathcal{G}_0 \approx \mathrm{FT}_\psi\mathrm{loc}(\infty,0)(\mathcal{F}(\infty)).$$

The fundamental interrelation of $\mathrm{FT}_\psi\mathrm{loc}(0,\infty)$ and $\mathrm{FTloc}_\psi(\infty,0)$ is given by

Theorem 7.4.4 (Laumon)

(1) $\mathrm{FT}_\psi\mathrm{loc}(0,\infty)$ and $[-1]^*\mathrm{FT}_\psi\mathrm{loc}(\infty,0)(\mathbf{1})$ are quasi-inverse equivalences of categories

$$(\ell\text{-adic } I(0)\text{-rep's}) \longleftrightarrow (\ell\text{-adic } I(\infty)\text{-rep's with all slopes} < 1);$$

For M an $I(0)$-representation, and N an $I(\infty)$-representation with all slopes < 1, we have the inversion formulas

$$\mathrm{FT}_\psi\mathrm{loc}(\infty,0)(\mathrm{FT}_\psi\mathrm{loc}(0,\infty)(M)) \approx [-1]^*M(-\mathbf{1}),$$

$$\mathrm{FT}_\psi\mathrm{loc}(0,\infty)(\mathrm{FT}_\psi\mathrm{loc}(\infty,0)(N)) \approx [-1]^*N(-\mathbf{1}).$$

(2) For χ any continuous $\overline{\mathbb{Q}}_\ell$-valued character of $\pi_1(\mathbb{G}_m \otimes k)^{\mathrm{tame}}$, with inverse character $\overline{\chi}$, we have

$$\mathrm{FT}_\psi\mathrm{loc}(0,\infty)(\mathcal{L}_\chi) \approx \mathcal{L}_{\overline{\chi}},\ \mathrm{FT}_\psi\mathrm{loc}(\infty,0)(\mathcal{L}_\chi) \approx \mathcal{L}_{\overline{\chi}}.$$

(3) If M is of the form $\mathcal{L}_\chi \otimes$(a unipotent Jordan block of size n), then $\mathrm{FT}_\psi\mathrm{loc}(0,\infty)(M)$ is of the form $\mathcal{L}_{\overline{\chi}} \otimes$(a unipotent Jordan block of size n).
If N is of the form $\mathcal{L}_\chi \otimes$(a unipotent Jordan block of size n), then $\mathrm{FT}_\psi\mathrm{loc}(\infty,0)(N)$ is of the form $\mathcal{L}_{\overline{\chi}} \otimes$(a unipotent Jordan block of size n).

(4) If M has unique slope $a/b > 0$ with multiplicity b, then $\mathrm{FT}_\psi\mathrm{loc}(0,\infty)(M)$ has unique slope $a/(a+b)$ with multiplicity a+b.
If N has unique slope $a/(a+b) < 1$ with multiplicity a+b, then $\mathrm{FT}_\psi\mathrm{loc}(\infty,0)(N)$ has unique slope a/b with multiplicity b.

proof Once (1) is proven, we argue as follows. One checks (2) by direct global calculation; (3) then follows because the functors carry indecomposables to indecomposables. They also carry irreducibles to irreducibles, and (4) then follows from a global calculation of their effects upon dimensions and Swan conductors.

A weaker version of (1) is proven somewhat clumsily in [Ka-TL, Prop. 12 and Thm. 13]. Here is a simple proof of it, based on the **bis** version 7.4.2 of stationary phase. Choose an auxiliary integer $k \geq 2$ which is prime to p (e.g., take k=2 unless p=2, in which case take k=3).

Let us begin with an I(0)-representation M, and denote by $\mathcal{M}$ its canonical extension to $\mathbb{G}_m$, extended by zero to $\mathbb{A}^1$. Then define $\mathcal{F} := \mathcal{M}\otimes\mathcal{L}_{\psi(x^k)}$. Since $\mathcal{L}_{\psi(x^k)}$ is lisse (in fact canonically trivial) at the origin, we have $\mathcal{F}(0) \approx M$ as I(0)-representation. The sheaf $\mathcal{F}$ is lisse on $\mathbb{G}_m$ and extended by zero across the origin, so it has no nonzero punctual sections; as all its ∞-slopes are $k \geq 2$, $\mathcal{F}$ is elementary. Let $\mathcal{G} := \mathrm{NFT}_\psi(\mathcal{F})$. By stationary phase applied to $\mathcal{F}$, we have

$$\mathcal{G}(\infty) \approx \mathrm{FT}_\psi\mathrm{loc}(\infty,\infty)(\mathcal{F}(\infty)) \oplus \mathrm{FT}_\psi\mathrm{loc}(0,\infty)(M).$$

Apply $\mathrm{FT}_\psi\mathrm{loc}(\infty,0)$; since $\mathrm{FT}_\psi\mathrm{loc}(\infty,0)$ kills (slopes ≥ 1), we get

$$\mathrm{FT}_\psi\mathrm{loc}(\infty,0)(\mathcal{G}(\infty)) \approx \mathrm{FT}_\psi\mathrm{loc}(\infty,0)(\mathrm{FT}_\psi\mathrm{loc}(0,\infty)(M)).$$

By 7.4.3.1 applied to $\mathcal{G}$, if we denote $\mathcal{H} := \mathrm{NFT}_\psi(\mathcal{G}) \approx [-1]^*\mathcal{F}(-\mathbf{1})$, we have

$$\mathcal{H}(0)/\mathcal{H}_0 \approx \mathrm{FT}_\psi\mathrm{loc}(\infty,0)(\mathcal{G}(\infty)).$$

Since $\mathcal{H} \approx [-1]^*\mathcal{F}(-\mathbf{1})$, we have $\mathcal{H}_0 = 0$, and

$$\mathcal{H}(0) \approx [-1]^*\mathcal{F}(0)(-\mathbf{1}) \approx [-1]^*M(-\mathbf{1}),$$

so we may rewrite this

$$[-1]^*M(-1) \approx FT_\psi loc(\infty,0)(\mathcal{G}(\infty)),$$

whence

$$[-1]^*M(-1) \approx FT_\psi loc(\infty,0)(FT_\psi loc(0,\infty)(M)).$$

Conversely, let us begin with an $I(\infty)$-representation N all of whose slopes are < 1. Let $\mathcal{N}$ denote its canonical extension to $\mathbb{G}_m$. Let $\mathcal{G} := \mathcal{N}\otimes\mathcal{L}_{\psi(1/x^k)}$, extended by zero across the origin. Then $\mathcal{G}$ has no nonzero punctual sections, and being totally wild at zero it is therefore elementary. Since $\mathcal{L}_{\psi(1/x^k)}$ extends across ∞ as a lisse sheaf (which is even canonically trivial at ∞), we have $\mathcal{G}(\infty) \approx N$ as $I(\infty)$-representations. Let $\mathcal{F} := NFT_\psi(\mathcal{G})$. By 7.4.3.1 we have

$$\mathcal{F}(0)/\mathcal{F}_0 \approx FT_\psi loc(\infty,0)(\mathcal{G}(\infty)) \approx FT_\psi loc(\infty,0)(N).$$

By stationary phase bis appled to $\mathcal{F}$, we have

$$FT_\psi loc(0,\infty)(\mathcal{F}(0)/\mathcal{F}_0) \approx \text{the slope} < 1 \text{ part of } NFT_\psi(\mathcal{F})(\infty).$$

But $NFT_\psi(\mathcal{F}) \approx [-1]^*\mathcal{G}(-1)$, so $NFT_\psi(\mathcal{F})(\infty) \approx [-1]^*N(-1)$ as $I(\infty)$-representation. As N is entirely of slope < 1, we obtain

$$FT_\psi loc(0,\infty)(\mathcal{F}(0)/\mathcal{F}_0) \approx [-1]^*N(-1), \text{ whence}$$

$$FT_\psi loc(0,\infty)(FT_\psi loc(\infty,0)(N)) \approx [-1]^*N(-1). \quad \text{QED}$$

Corollary 7.4.5 Let $\mathcal{F}$ be an elementary sheaf, $\mathcal{G} := NFT_\psi(\mathcal{F})$, and $s \in \mathbb{A}^1(k)$. Then
(1) $\mathcal{G}$ is lisse at s if and only if $\mathcal{F}(\infty)\otimes\mathcal{L}_{\psi(sx)}$ has all ∞-breaks ≥ 1.
(2) $\mathcal{G}$ is tame at s if and only if

$$\mathcal{F}(\infty)\otimes\mathcal{L}_{\psi(sx)} \approx (\text{all } \infty\text{-breaks} = 0)\oplus(\text{all } \infty\text{-breaks} \geq 1).$$

proof By translation, it suffices to treat the case $s=0$. Since $\mathcal{F}$ is elementary, $\mathcal{G}$ is elementary, $\mathcal{G}_0 \hookrightarrow \mathcal{G}(0)$, and we have

$$\mathcal{G}(0)/\mathcal{G}_0 \approx FT_\psi loc(\infty,0)(\mathcal{F}(\infty)).$$

But $FT_\psi loc(\infty,0)(\mathcal{F}(\infty))$ vanishes (resp. is tame) if and only if the part of $\mathcal{F}(\infty)$ of slope < 1 vanishes (reps. is tame). QED

Corollary 7.4.6 (Pseudoreflection Monodromy Criterion) Let $\mathcal{F}$ be a Fourier sheaf, $\mathcal{G} := NFT_\psi(\mathcal{F})$, and $s \in \mathbb{A}^1(k)$. Then $\mathcal{G}$ has pseudoreflection monodromy at s (in the sense that the subspace $\mathcal{G}(s)^{I(s)} = \mathcal{G}_s$ of $\mathcal{G}(s)$ is of codimension one in $\mathcal{G}(s)$) if and only if either

(1) $\mathcal{F}(\infty)\otimes\mathcal{L}_{\psi(sx)} \approx$ (1-dim'l, ∞-break $=0$)$\oplus$(all ∞-breaks ≥ 1)
$\approx \mathcal{L}_{\chi} \oplus$ (all ∞-breaks ≥ 1),

in which case I(s) acts on the line $\mathcal{G}(s)/\mathcal{G}_s$ by the tame character $\mathcal{L}_{\overline{\chi}(x-s)}$, or

(2) for some integer $n \geq 1$, we have

$$\mathcal{F}(\infty)\otimes\mathcal{L}_{\psi(sx)} \approx (n+1\text{ -dim'l, } \infty\text{-breaks} = n/(n+1))\oplus(\text{all } \infty\text{-breaks} \geq 1),$$

in which case I(s) acts on the line $\mathcal{G}(s)/\mathcal{G}_s$ by a character of slope n. In this latter case, there exists an element of the wild inertia group P(s) which acts on $\mathcal{G}(s)$ as a pseudreflection of determinant ζ_p.

proof By translation, it suffices to treat the case s=0. Since $\mathcal{F}$ is Fourier, $\mathcal{G}$ is Fourier. Therefore $\mathcal{G}_0 = \mathcal{G}(0)^{I(0)}$, and

$$\mathcal{G}(0)/\mathcal{G}_0 \approx \mathrm{FT}_{\psi}\mathrm{loc}(\infty,0)(\mathcal{F}(\infty)).$$

The cases listed are those where $\mathrm{FT}_{\psi}\mathrm{loc}(\infty,0)(\mathcal{F}(\infty))$ is of rank one. In the second case, the character is wild, so its restriction to the pro-p group P is nontrivial. QED

7.5 "Numerical" Explicitation of Laumon's Results

In this section we will make explicit the exact relation between the I(0) and I(∞) representations of a pair of Fourier sheaves $\mathcal{F}$ and $\mathcal{G}$ which are Fourier Transforms of each other. The only delicate part concerns the unipotent part of local monodromy.

We continue to suppose k algebraically closed throughout this section.

Lemma 7.5.1 Let $\mathcal{F}$ be a Fourier sheaf,

$$\mathcal{G} = \mathrm{NFT}_{\psi}(\mathcal{F}),\ [-1]^*\mathcal{F}(-1) = \mathrm{NFT}_{\psi}(\mathcal{G}).$$

Then

(1) $\dim(\mathcal{F}(0)^{I(0)}) = \dim(\mathcal{F}_0)$.

(2) $\dim(\mathcal{F}(\infty)^{I(\infty)}) \leq \dim(\mathcal{G}_0)$

(3) $\dim(\mathcal{G}(0)^{I(0)}) = \dim(\mathcal{G}_0)$.

(4) $\dim(\mathcal{G}(\infty)^{I(\infty)}) \leq \dim(\mathcal{F}_0)$

proof The first assertion holds because $\mathcal{F}$ is Fourier. For the second, denote by $j: \mathbb{A}^1 \to \mathbb{P}^1$ the inclusion, and consider the short exact sequence of sheaves on $\mathbb{P}^1$

$$0 \to j_!\mathcal{F} \to j_*\mathcal{F} \to (\text{the punctual sheaf } \mathcal{F}(\infty)^{I(\infty)} \text{ at } \infty) \to 0.$$

We have $H^0(\mathbb{P}^1, j_*\mathcal{F}) = H^0(\mathbb{A}^1, \mathcal{F}) = 0$ (since $\mathcal{F}$ is Fourier), so the coboundary of the long exact cohomology sequence given an injection

$$\mathcal{F}(\infty)^{I(\infty)} \hookrightarrow H^1(\mathbb{P}^1, j_!\mathcal{F}) := H^1{}_c(\mathbb{A}^1, \mathcal{F}) := \mathcal{G}_0,$$

which proves (2). Assertions (3) and (4) are simply (1) and (2) with the roles of $\mathcal{F}$ and $\mathcal{G}$ reversed. QED

(7.5.2) We will use this lemma in the following way. By part (3), the number of unipotent Jordan blocks in $\mathcal{G}(0)$ as $I(0)$-representation is precisely $\dim\mathcal{G}_0$. By part (2), the number of unipotent Jordan blocks in $\mathcal{F}(\infty)$ is **at most** $\dim\mathcal{G}_0$. We will adopt the convention (compare [Ka-GKM, 7.1.3, 7.5.1.3) that $\mathcal{F}(\infty)$ has **precisely** $\dim\mathcal{G}_0$ unipotent Jordan blocks, with the convention that some of these blocks are allowed to be of size zero. Similarly $\mathcal{G}(\infty)$ has precisely $\dim\mathcal{F}_0$ unipotent Jordan blocks, with the convention that some of these blocks are allowed to be of size zero. Using these conventions, we define polynomials in $\mathbb{Z}[T]$

$$P(\mathcal{F}, \infty, \mathbb{1}, T) := \Sigma_{\text{all } \dim\mathcal{G}_0 \text{ unip. blocks in } \mathcal{F}(\infty)} T^{(\text{dim. of block})},$$

$$P(\mathcal{G}, 0, \mathbb{1}, T) := \Sigma_{\text{all } \dim\mathcal{G}_0 \text{ unip. blocks in } \mathcal{G}(0)} T^{(\text{dim. of block})},$$

and similarly with the roles of $\mathcal{F}$ and $\mathcal{G}$ reversed. Notice that the polynomial $P(\mathcal{G}, 0, \mathbb{1}, T)$ has no constant term, while $P(\mathcal{F}, \infty, \mathbb{1}, T)$ may very well have a constant term (namely the number of "dummy" Jordan blocks, $\dim\mathcal{G}_0 - \dim(\mathcal{F}(\infty)^{I(\infty)})$). If $\mathcal{G}_0$ vanishes, these polynomials are identically zero. Clearly it is the same to know the isomorphism class of the unipotent parts of $\mathcal{F}(\infty)$ and $\mathcal{G}(0)$ as to know these two polynomials.

(7.5.3) For each nontrivial continuous $\overline{\mathbb{Q}}_\ell$-valued character χ of $\pi_1(\mathbb{G}_m \otimes k)^{\text{tame}}$, with inverse character denoted $\overline{\chi}$, we define polynomials in $\mathbb{Z}[T]$

$$P(\mathcal{F}, \infty, \overline{\chi}, T) := \Sigma_{\text{all unip. blocks in } \mathcal{L}_\chi \otimes \mathcal{F}(\infty)} T^{(\text{dim. of block})},$$

$$P(\mathcal{G}, 0, \overline{\chi}, T) := \Sigma_{\text{all unip. blocks in } \mathcal{L}_\chi \otimes \mathcal{G}(0)} T^{(\text{dim. of block})},$$

but this time with the "naive" convention that we only take the sum over as many Jordan blocks as there actually are. With this convention, these polynomials never have a constant term. They can be identically zero.

These conventions established, we can restate the "numerical" version of Laumon's results 7.4.1, 7.4.2, 7.4.4.

Theorem 7.5.4 (Laumon) Let $\mathcal{F}$ be a Fourier sheaf, $\mathcal{G} = \mathrm{NFT}_\psi(\mathcal{F})$, (and hence also $[-1]^*\mathcal{F}(-1) = \mathrm{NFT}_\psi(\mathcal{G})$). Then $\mathcal{G}(0)$ and $\mathcal{G}(\infty)$ are related to $\mathcal{F}(0)$ and $\mathcal{F}(\infty)$ by the following rules: Write

$$\mathcal{F}(\infty) = \mathcal{F}(\infty)_{\text{slope} > 1} \oplus \mathcal{F}(\infty)_{\text{slope} = 1} \oplus \mathcal{F}(\infty)_{0 < \text{slope} < 1} \oplus \mathcal{F}(\infty)_{\text{tame}}$$

$$\mathcal{F}(\infty)_{\text{slope} > 1} \approx \oplus(\text{slope } (a+b)/a, \text{ multiplicity } a)$$

$$\mathcal{F}(\infty)_{0 < \text{slope} < 1} \approx \oplus(\text{slope } c/(c+d), \text{ multiplicity } c+d),$$

$$\mathcal{F}(0) = \mathcal{F}(0)_{\text{slope} > 0} \oplus \mathcal{F}(0)_{\text{tame}}$$

$$\mathcal{F}(0)_{\text{slope} > 0} \approx \oplus(\text{slope } e/f, \text{ multiplicity } f)$$

Then we have

(1) $\mathcal{G}(\infty)_{\text{slope} > 1} \approx \oplus(\text{slope } (a+b)/b, \text{ multiplicity } b)$.

(2) $\mathcal{G}(\infty)_{0 < \text{slope} < 1} \approx \oplus(\text{slope } e/(e+f), \text{ multiplicity } e+f)$.

(3) $P(\mathcal{G}, \infty, \overline{\chi}, T) = P(\mathcal{F}, 0, \chi, T)$ for each nontrivial χ.

(4) $P(\mathcal{G}, \infty, \mathbb{1}, T) = P(\mathcal{F}, 0, \mathbb{1}, T)/T$.

(5) $\mathcal{G}(0)_{\text{slope} > 0} \approx \oplus(\text{slope } c/d, \text{ multiplicity } d)$.

(6) $P(\mathcal{G}, 0, \overline{\chi}, T) = P(\mathcal{F}, \infty, \chi, T)$ for each nontrivial χ.

(7) $P(\mathcal{G}, 0, \mathbb{1}, T) = T \times P(\mathcal{F}, \infty, \mathbb{1}, T)$.

proof Assertion (1) is 7.4.1 (1). Assertions (2) and (5) are 7.4.4 (4).

Assertions (3) and (4) are the same as (6) and (7) with the roles of $\mathcal{F}$ and $\mathcal{G}$ reversed, so it remains to prove (6) and (7). From 7.4.3.1 we have

$$\mathcal{G}(0)/\mathcal{G}_0 \approx \mathrm{FT}_\psi\mathrm{loc}(\infty,0)(\mathcal{F}(\infty));$$

taking $\overline{\chi}$-components, we see that (6) is just Thm 7.4.4 (3). As for (7), write

$$\mathcal{G}(0)^{\mathrm{unip}} \approx \oplus_{i=1 \text{ to } \dim\mathcal{G}_0} \mathrm{Unip}(n_i).$$

Then taking unipotent components gives

$$\mathrm{FT}_\psi\mathrm{loc}(\infty,0)((\mathcal{F}(\infty))^{\mathrm{unip}}) \approx (\mathcal{G}(0)/\mathcal{G}_0)^{\mathrm{unip}} = \mathcal{G}(0)^{\mathrm{unip}}/\mathcal{G}_0 =$$
$$= \mathcal{G}(0)^{\mathrm{unip}}/\mathcal{G}(0)^{I(0)} \approx \oplus_{i=1 \text{ to } \dim\mathcal{G}_0} \mathrm{Unip}(n_i - 1),$$

whence by 7.4.4 (3) we have

$$(\mathcal{F}(\infty))^{\mathrm{unip}} \approx \oplus_{i=1 \text{ to } \dim\mathcal{G}_0} \mathrm{Unip}(n_i - 1),$$

which is exactly (7). QED

7.6 Pseudoreflection Examples and Applications

In this section we discuss in detail examples of pseudoreflection local monodromy, and whenever possible apply to these examples the Pseudoreflection Thm. 1.5. We continue to suppose k algebraically closed throughout this section.

Example 7.6.1 $\mathcal{F}$ is a Fourier sheaf with

$$\mathcal{F}(\infty) \approx (1 \text{ dim'l tame}) \oplus (\text{all slopes} \geq 1)$$

Then $\mathcal{G} := \mathrm{NFT}_\psi(\mathcal{F})$ has pseudoreflection tame local monodromy at zero.

Example 7.6.2 $\mathcal{F}$ is a Fourier sheaf and

$$\mathcal{F}(\infty) \approx (1 \text{ dim'l, slope } =1) \oplus (\text{all slopes} < 1).$$

Then there is a unique s in $k^\times$ such $\mathcal{L}_{\psi(sx)}\otimes\mathcal{F}$ is of type (1), and $\mathcal{G} := \mathrm{NFT}_\psi(\mathcal{F})$ has pseudoreflection tame local monodromy at s. In this case $\mathcal{G}$ is lisse on $\mathbb{A}^1 - \{0, s\}$, and has all its ∞-slopes ≤ 1.

Conversely, if $\mathcal{G}$ is Fourier, lisse on $\mathbb{A}^1 - \{0, s\}$ with tame pseudoreflection local monodromy at s, and has all its ∞-slopes ≤ 1, then

$$\mathrm{NFT}_\psi(\mathcal{G})(\infty) \approx \mathrm{FT}_\psi\mathrm{loc}(s, \infty)(\mathcal{G}(s)/\mathcal{G}_s) \oplus \mathrm{FT}_\psi\mathrm{loc}(0, \infty)(\mathcal{G}(0)/\mathcal{G}_0)$$

is of this form. Thus these two classes of sheaves are interchanged by Fourier Transform. If we add the requirement that $\mathcal{F}$ be lisse on $\mathbb{G}_m$, this corresponds under Fourier Transform to the requirement that $\mathcal{G}$ have all its ∞-slopes < 1.

Theorem 7.6.2.1 Suppose $\mathcal{F}$ is an irreducible Fourier sheaf which is lisse on $\mathbb{G}_m$, with

$$\mathcal{F}(\infty) \approx (1 \text{ dim'l, slope} = 1) \oplus (\text{all slopes} < 1), \text{ say}$$
$$\approx (\mathcal{L}_{\psi(-sx)} \otimes \mathcal{L}_{\chi}) \oplus (\text{all slopes} < 1).$$

Then $\mathcal{G} := \mathrm{NFT}_{\psi}(\mathcal{F})$ is irreducible Fourier, lisse on $\mathbb{A}^1 - \{0, s\}$, has pseudoreflection tame local monodromy at s of determinant $\overline{\chi}(x-s)$, and all its ∞-slopes < 1. Moeover,
(1) If $p > 2\mathrm{rank}(\mathcal{F}) + 1$, then $\mathcal{F}$ is Lie-irreducible, the upper numbering subgroup $(I(\infty))^{(1)}$ acts on $\mathcal{F}(\infty)$ by pseudoreflections of determinant $\mathcal{L}_{\psi(-sx)}$, and $G := G_{\mathrm{geom}}$ for $\mathcal{F}$ has $G^{0,\mathrm{der}} = SL(\mathcal{F}_x)$.
(2) If $\mathcal{G}$ has $\det(\mathcal{G})$ of finite order, and if $\mathcal{G}$ is neither induced nor has finite geometric monodromy, then $\mathcal{G}$ is Lie-irreducible, and for it $G{:=}G_{\mathrm{geom}}$ has $G^{0,\mathrm{der}}$ one of the groups $SL(\mathcal{G}_x)$, $SO(\mathcal{G}_x)$, or (if $\mathrm{rank}(\mathcal{G})$ is even) $Sp(\mathcal{G}_x)$.

proof For (1), we get Lie-irreduciblity by observing that $\mathcal{F}$ cannot be Kummer induced because it has an ∞-slope (namely 1) occuring with multiplicity one. For (2), we have Lie irreducibility by 7.2.6(6). Now apply the pseudoreflection theorem 1.5. QED

Example 7.6.3 $\mathcal{F}$ is a Fourier sheaf of generic rank one, and the ∞-slope of $\mathcal{F}$ is ≤ 1. Then there is a unique s in k such that $\mathcal{L}_{\psi(sx)} \otimes \mathcal{F}$ is tame at ∞, say $\mathcal{L}_{\psi(sx)} \otimes \mathcal{F} \approx \mathcal{L}_{\chi(x)}$ as $I(\infty)$-representation, and $\mathcal{G}$ is lisse on $\mathbb{A}^1 - \{s\}$ with pseudoreflection tame local monodromy at s whose determinant is $\mathcal{L}_{\overline{\chi}(x-s)}$ as $I(s)$-representation. Moreover, at ∞ $\mathcal{G}$ has all slopes ≤ 1. For such a $\mathcal{G}$, its $\mathrm{NFT}_{\psi}(\mathcal{G})(\infty)$ is the single term $\mathrm{FT}_{\psi}\mathrm{loc}(s, \infty)(\mathcal{G}(s)/\mathcal{G}_s)$, which is 1-dim'l of slope ≤ 1. Thus these two classes of sheaves are interchanged by Fourier Transform.

Theorem 7.6.3.1 Suppose that $\mathcal{F}$ is a Fourier sheaf of generic rank one, with ∞-slope ≤ 1. Then there is a unique s in k such that

$$\mathcal{L}_{\psi(sx)} \otimes \mathcal{F} \approx \mathcal{L}_{\chi(x)} \text{ as } I(\infty)\text{-representation,}$$

and $\mathcal{G} := \mathrm{NFT}_{\psi}(\mathcal{F})$ is lisse on $\mathbb{A}^1 - \{s\}$ of rank

$$\mathrm{rank}(\mathcal{G}) = \Sigma_{x \text{ in } \mathbb{A}^1} (\mathrm{drop}_x(\mathcal{F}) + \mathrm{Swan}_x(\mathcal{F})).$$

with pseudoreflection local monodromy at s of determinant $\overline{\chi}(x-s)$. Suppose that $p > 2\mathrm{rank}(\mathcal{G}) + 1$, and that either $\mathrm{rank}(\mathcal{G}) \neq 2$ or that χ is not the character χ_2 of order two. Then $\mathcal{G}$ is Lie-irreducible, and for it $G:=G_{geom}$ has $G^{0,der}$ one of the groups $SL(\mathcal{G}_x)$, $SO(\mathcal{G}_x)$, or (if rank($\mathcal{G}$) is even) $Sp(\mathcal{G}_x)$. Moreover,

if $\chi \neq \mathbb{1}, \chi_2$, then $G^{0,der} = SL(\mathcal{G}_x)$;

if $\chi = \mathbb{1}$, then $G^{0,der} = SL(\mathcal{G}_x)$ or (for rank($\mathcal{G}$) even) $Sp(\mathcal{G}_x)$;

if $\chi = \chi_2$, then $G^{0,der} = SL(\mathcal{G}_x)$ or $SO(\mathcal{G}_x)$.

proof Notice that such an $\mathcal{F}$ is automatically irreducible Fourier (simply because it is Fourier of generic rank one). Therefore $\mathcal{G}$ is irreducible Fourier. By 7.2.6 (7), if $p > 2\mathrm{rank}(\mathcal{G}) + 1$, then $\mathcal{G}$ is Lie-irreducible unless $\mathcal{G}$'s local monodromy at s is a tame reflection (i.e., $\mathcal{L}_{\psi(sx)}\otimes\mathcal{F} \approx \mathcal{L}_{\chi(x)}$ as $I(\infty)$-representation with χ the character of order two) and $\mathcal{G}$ has rank two. If $\mathcal{G}$ is Lie-irreducible, then 1.5 gives the short list of possibilities for $G^{0,der}$. Let us calculate the rank of $\mathcal{G}$. By $\mathcal{F} \mapsto \mathcal{L}_{\psi(sx)}\otimes\mathcal{F}$, we may reduce to the case where $\mathcal{F}$ is tame at ∞, and $\mathcal{G}$ is lisse on $\mathbb{G}_m$. Since $\mathcal{G}$ then has pseudoreflection local monodromy at zero, we find

$\mathrm{rank}(\mathcal{G}) = 1 + \dim\mathcal{G}_0 = 1 + h^1{}_c(\mathbb{A}^1, \mathcal{F}) = 1 - \chi_c(\mathbb{A}^1, \mathcal{F})$

$= 1 - \chi(\mathbb{A}^1) + \mathrm{Swan}_\infty(\mathcal{F}) + \Sigma_{x \text{ in } \mathbb{A}^1} (\mathrm{drop}_x(\mathcal{F}) + \mathrm{Swan}_x(\mathcal{F}))$

$= \Sigma_{x \text{ in } \mathbb{A}^1} (\mathrm{drop}_x(\mathcal{F}) + \mathrm{Swan}_x(\mathcal{F}))$. QED

Example 7.6.4 Suppose that C is a connected smooth complete curve over k, with a marked point ∞. Fix an integer $n \geq 1$ such that n and n + 1 are both prime to p. Let $\mathcal{L}$ be an ℓ-adic sheaf on $C - \{\infty\}$ which is generically of rank one and is the direct image of its restriction to a nonempty open set where it is lisse. Suppose that $\mathrm{Swan}_\infty(\mathcal{L}) = n$. Let f be a rational function on C whose only pole is at ∞, of order n + 1. View f as a finite flat morphism $f: C - \{\infty\} \to \mathbb{A}^1$. (Because n+1 is prime to p, f is generically etale.) Then $\mathcal{F} := f_*\mathcal{L}$ has generic rank n+1, $\mathcal{F}$ is the direct image of its restriction to a nonempty open set where it is lisse, and all the ∞-slopes of $\mathcal{F}$ are n/(n+1). By its ∞-slopes, $\mathcal{F}$ is $I(\infty)$-irreducible. Therefore $\mathcal{F}$ is an irreducible Fourier sheaf. Thus $\mathcal{G} := NFT_\psi(\mathcal{F})$ is an irreducible Fourier sheaf, lisse on $\mathbb{G}_m$ with pseudoreflection local monodromy at zero whose determinant has slope n.

Theorem 7.6.4.1 Suppose that C is a connected smooth complete curve over k, with a marked point ∞. Fix an integer $n \geq 1$ such that n and n + 1 are both prime to p. Let $\mathcal{L}$ be an ℓ-adic sheaf on $C - \{\infty\}$ which is generically of rank one and is the direct image of its restriction to a nonempty open set where it is lisse. Suppose that $\mathrm{Swan}_\infty(\mathcal{L}) = n$. Let f be a rational function on C whose only pole is at ∞, of order n + 1. Define $\mathcal{F} := f_*\mathcal{L}$, $\mathcal{G} := \mathrm{NFT}_\psi(\mathcal{F})$. Then $\mathcal{G}$ is an irreducible Fourier sheaf, lisse on $\mathbb{G}_m$ of rank

$$\mathrm{rank}(\mathcal{G}) = n + 2\mathrm{genus}(C) + \Sigma_{x \text{ in } C - \{\infty\}} (\mathrm{drop}_x(\mathcal{L}) + \mathrm{Swan}_x(\mathcal{L})),$$

with pseudoreflection local monodromy at zero whose determinant has slope n. If If $p > 2\mathrm{rank}(\mathcal{G}) + 1$, then $G := G_{\mathrm{geom}}$ for $\mathcal{G}$ has $G^{0,\mathrm{der}} = SL(\mathcal{G}_x)$.

proof If $p > 2\mathrm{rank}(\mathcal{G}) + 1$, then $\mathcal{G}$ must be Lie-irreducible (by wildness of the pseudoreflection, cf. 7.2.6 (7)). In view of 1.5, we see that $\mathcal{G}$ must have $G^{0,\mathrm{der}} = SL(\mathcal{G}_x)$. Alternatively, once $\mathcal{G}$ is Lie-irreducible and $p \neq 2$ we can apply the Main ℓ-adic Theorem 7.2.7 to $\mathcal{G}$ at zero, with a/b = n/1. The conclusion we reach, namely $G^{0,\mathrm{der}} = SL(\mathcal{G}_x)$, is of course the same.

To see what "$p > 2\mathrm{rank}(\mathcal{G}) + 1$" means, let us calculate the rank of $\mathcal{G}$. Since $\mathcal{G}$ has pseudoreflection local monodromy at zero,

$$\begin{aligned}\mathrm{rank}(\mathcal{G}) &= 1 + \dim\mathcal{G}_0 = 1 + h^1{}_c(\mathbb{A}^1, f_*\mathcal{L}) = 1 - \chi_c(\mathbb{A}^1, f_*\mathcal{L})\\
&= 1 - \chi_c(C - \{\infty\}, \mathcal{L})\\
&= 1 - \chi(C - \{\infty\}) + \mathrm{Swan}_\infty(\mathcal{L}) + \Sigma_{x \text{ in } C - \{\infty\}} (\mathrm{drop}_x(\mathcal{L}) + \mathrm{Swan}_x(\mathcal{L}))\\
&= 2 - \chi(C) + \mathrm{Swan}_\infty(\mathcal{L}) + \Sigma_{x \text{ in } C - \{\infty\}} (\mathrm{drop}_x(\mathcal{L}) + \mathrm{Swan}_x(\mathcal{L})).\\
&= n + 2\mathrm{genus}(C) + \Sigma_{x \text{ in } C - \{\infty\}} (\mathrm{drop}_x(\mathcal{L}) + \mathrm{Swan}_x(\mathcal{L})).\end{aligned}$$

QED

7.7 A Highest Slope Application

We continue to suppose k algebraically closed throughout this section.

(7.7.1) Suppose that C is a connected smooth complete curve over k, with a marked point ∞. Fix integers $n \geq 1$ and $d \geq 1$ such that

$$\gcd(n, d) = 1, \ n \neq d, \text{ both n and d are prime to p.}$$

Let $\mathcal{L}$ be an ℓ-adic sheaf on $C - \{\infty\}$ which is generically of rank one and is the direct image of its restriction to a nonempty open set where it is lisse. Suppose that $\mathrm{Swan}_\infty(\mathcal{L}) = n$. Let f be a rational function on C whose only pole is at ∞, of order d. View f as a finite flat morphism $f: C - \{\infty\} \to \mathbb{A}^1$. (Because d is prime to p, f is generically etale.) Then $\mathcal{F} := f_*\mathcal{L}$ has generic rank d, $\mathcal{F}$ is the direct image of its restriction to a nonempty open set where it is lisse, and all the ∞-slopes of $\mathcal{F}$ are n/d. By its ∞-slopes, $\mathcal{F}$ is $I(\infty)$-irreducible. Therefore $\mathcal{F}$ is an irreducible Fourier sheaf. Therefore $\mathcal{G} := NFT_\psi(\mathcal{F})$ is an irreducible Fourier sheaf. To analyse $\mathcal{G}$ further, we must distinguish cases, according to whether $n > d$ or $d > n$.

(7.7.2) If $n > d$, then $\mathcal{G}$ is lisse on $\mathbb{A}^1$. Its rank is therefore equal to $\dim\mathcal{G}_0$, so we find

$$\begin{aligned}\mathrm{rank}(\mathcal{G}) &= \dim\mathcal{G}_0 = h^1{}_c(\mathbb{A}^1, f_*\mathcal{L}) = -\chi_c(\mathbb{A}^1, f_*\mathcal{L})\\ &= -\chi_c(C - \{\infty\}, \mathcal{L})\\ &= -\chi(C - \{\infty\}) + \mathrm{Swan}_\infty(\mathcal{L}) + \Sigma_{x \text{ in } C - \{\infty\}} (\mathrm{drop}_x(\mathcal{L}) + \mathrm{Swan}_x(\mathcal{L}))\\ &= n - 1 + 2\mathrm{genus}(C) + \Sigma_{x \text{ in } C - \{\infty\}} (\mathrm{drop}_x(\mathcal{L}) + \mathrm{Swan}_x(\mathcal{L})).\end{aligned}$$

If $p > 2\mathrm{rank}(\mathcal{G}) + 1$, then $\mathcal{G}$ is Lie-irreducible. As its highest ∞-slope is n/(n-d) with multiplicity n-d, we may apply the main ℓ-adic theorem 7.2.7 provided that p does not divide $2nN_1(n-d)N_2(n-d)$.

(7.7.3) If $d > n$, then $\mathcal{G}$ is lisse on $\mathbb{G}_m$. Since $\mathcal{G}(0)/\mathcal{G}_0 = FT_\psi loc(\infty,0)(\mathcal{F}(\infty))$ has rank d-n and all slopes n/(d-n), we find

$$\begin{aligned}\mathrm{rank}(\mathcal{G}) &= d - n + \dim\mathcal{G}_0 = d - n + h^1{}_c(\mathbb{A}^1, f_*\mathcal{L}) = d - n - \chi_c(\mathbb{A}^1, f_*\mathcal{L}) =\\ &= d - n - \chi_c(C - \{\infty\}, \mathcal{L}) =\\ &= d - n - \chi(C - \{\infty\}) + \mathrm{Swan}_\infty(\mathcal{L}) + \Sigma_{x \text{ in } C - \{\infty\}} (\mathrm{drop}_x(\mathcal{L}) + \mathrm{Swan}_x(\mathcal{L})) =\\ &= d - 1 + 2\mathrm{genus}(C) + \Sigma_{x \text{ in } C - \{\infty\}} (\mathrm{drop}_x(\mathcal{L}) + \mathrm{Swan}_x(\mathcal{L})).\end{aligned}$$

If $p > 2\mathrm{rank}(\mathcal{G}) + 1$, then $\mathcal{G}$ is either Lie-irreducible or Kummer-induced. In fact, $\mathcal{G}$ cannot be Kummer induced if $p > d$. [Notice that the condition $p > d$ is satisfied if $p > 2\mathrm{rank}(\mathcal{G}) + 1$, since $\mathrm{rank}(\mathcal{G}) \geq d-1$ and $d > n \geq 1$.] Once this point is established, we may apply the main ℓ-adic theorem 7.2.7 to $\mathcal{G}$ (its highest slope at zero is n/(d-n), multiplicity d-n) provided that $p > 2\mathrm{rank}(\mathcal{G}) + 1$ and p does not divide $2nN_1(d-n)N_2(d-n)$.

(7.7.4) We now explain why $\mathcal{G}$ cannot be Kummer induced. As I(0)-representation, we have

$$\mathcal{G}(0) \approx \mathcal{G}_0 \oplus (\text{rank d-n, all slopes n/(d-n)}),$$

which cannot be Kummer induced unless $\mathcal{G}_0$ vanishes. But

$$\dim\mathcal{G}_0 = n - 1 + 2\mathrm{genus}(C) + \Sigma_{x \text{ in } C - \{\infty\}} (\mathrm{drop}_x(\mathcal{L}) + \mathrm{Swan}_x(\mathcal{L})),$$

which can only vanish if all of the following conditions are satisfied:

(1) $n = 1$.

(2) genus(C) = 0, i.e., $C - \{\infty\}$ is $\mathbb{A}^1$.

(3) $\mathcal{L}$ is lisse on $\mathbb{A}^1$.

Therefore if $\mathcal{G}_0$ vanishes, $\mathcal{L}$ must be $\mathcal{L}_{\psi(tx)}$ for some $t \in k^\times$, f is a polynomial of degree d prime to p, and $\mathcal{F}$ is $f_*\mathcal{L}_{\psi(tx)}$. In this case, there is a simple sufficient condition which guarentees that $\mathcal{G}$ is not Kummer induced.

Lemma 7.7.5 Suppose that $f(x) \in k[x]$ is a polynomial of degree d, $1 < d < p$. Then $\mathcal{G} := \mathrm{NFT}_\psi(f_*\mathcal{L}_{\psi(tx)})$ is not Kummer-induced for any t in k.

proof We will analyze $\mathcal{G}(\infty)$. Let us denote by

$\beta_1, \ldots, \beta_r$ the critical values of f.

For each critical value β_i, denote by

$\alpha_{i,1}, \ldots, \alpha_{i,s(i)}$ the critical points of f (zeroes of f') in $f^{-1}(\beta)$.

Since $\mathcal{L}_{\psi(tx)}$ is lisse on $\mathbb{A}^1$, $\mathcal{F} := f_*\mathcal{L}_{\psi(tx)}$ is lisse outside the critical values $\beta_1, \ldots, \beta_r$ of f. For each critical point α, denote by $e(\alpha)$ its multiplicity as a zero of f'. Then

$$f(x) - f(\alpha) = (x - \alpha)^{1+e(\alpha)}(\text{a function invertible at } \alpha).$$

Since $1 + e(\alpha) \le 1 + d-1 < p$, we may rewrite this

$$f(x) - f(\alpha) = (\text{a uniformizing parameter at } \alpha)^{1+e(\alpha)}.$$

Therefore if we translate $\beta \mapsto 0$ the $I(\beta)$-representation $\mathcal{F}(\beta)/\mathcal{F}_\beta$ we get an isomorphism of $I(0)$-representations

$$\mathrm{Add}(\beta)^*(\mathcal{F}(\beta)/\mathcal{F}_\beta) \approx \oplus_{\alpha \mapsto \beta} ([1 + e(\alpha)]_*\overline{\mathbb{Q}}_\ell)/\overline{\mathbb{Q}}_\ell \approx$$
$$\approx \oplus_{\alpha \mapsto \beta} \oplus_{\text{all } \chi^{1+e(\alpha)} = \mathbb{1},\ \chi \text{ nontriv}} \mathcal{L}_{\chi(x)}.$$

Since $\mathcal{F}$ has all ∞-slopes 1/d (or 0, if t = 0) < 1, we have

$$\mathcal{G}(\infty) \approx \oplus_\beta \mathrm{FT}_\psi\mathrm{loc}(\beta, \infty)(\mathcal{F}(\beta)/\mathcal{F}_\beta) \approx$$
$$\approx \oplus_\beta \oplus_{\alpha \mapsto \beta} \oplus_{\text{all } \chi^{1+e(\alpha)} = \mathbb{1},\ \chi \text{ nontriv}} \mathcal{L}_{\psi(\beta y)} \otimes \mathcal{L}_{\overline{\chi}(y)}.$$

If $\mathcal{G}(\infty)$ is Kummer induced of degree $m \ge 2$ prime to p, say

$$\mathcal{G}(\infty) \approx [m]_*\mathcal{H} \text{ for some } I(\infty)\text{-representation } \mathcal{H},$$

then it is Kummer induced of degree q for any prime divisor q of m (since $[m]_*\mathcal{H} = [q]_*[m/q]_*\mathcal{H}$), so it suffices to show that $\mathcal{G}(\infty)$ is not Kummer induced of **prime** degree $m \neq p$. If it were so induced, then

$$[m]^*\mathcal{G}(\infty) \approx [m]^*[m]_*\mathcal{H} \approx \oplus_{\zeta \text{ in } \mu_m} [y \mapsto \zeta y]^*\mathcal{H},$$

and so in particular $\mathcal{H}$ itself is a direct factor of $[m]^*\mathcal{G}(\infty)$

$$\approx \oplus_\beta \oplus_{\alpha \mapsto \beta} \oplus_{\text{all } \chi^{1+e(\alpha)} = 1,\ \chi \text{ nontriv}} [m]^*(\mathcal{L}_{\psi(\beta y)} \otimes \mathcal{L}_{\overline{\chi}(y)}).$$

Therefore $\mathcal{H}$ is itself of the form

$$\oplus_\beta \oplus_{\alpha \mapsto \beta} \oplus_{\text{some } \chi^{1+e(\alpha)} = 1,\ \chi \text{ nontriv}} [m]^*(\mathcal{L}_{\psi(\beta y)} \otimes \mathcal{L}_{\overline{\chi}(y)}).$$

By the projection formula, $[m]_*\mathcal{H}$ is

$$\oplus_\beta \oplus_{\alpha \mapsto \beta} \oplus_{\text{some } \chi^{1+e(\alpha)} = 1,\ \chi \text{ nontriv}} (\mathcal{L}_{\psi(\beta y)} \otimes \mathcal{L}_{\overline{\chi}(y)}) \otimes [m]_*\overline{\mathbb{Q}}_\ell \approx$$

$$\approx \oplus_\beta \oplus_{\alpha \mapsto \beta} \oplus_{\substack{\text{some } \chi^{1+e(\alpha)} = 1,\ \chi \text{ nontriv} \\ \text{all } \rho^m = 1}} \mathcal{L}_{\psi(\beta y)} \otimes \mathcal{L}_{\overline{\chi}(y)} \otimes \mathcal{L}_{\rho(y)}.$$

If this is to be the expression of $\mathcal{G}(\infty)$

$$\approx \oplus_\beta \oplus_{\alpha \mapsto \beta} \oplus_{\text{all } \chi^{1+e(\alpha)} = 1,\ \chi \text{ nontriv}} \mathcal{L}_{\psi(\beta y)} \otimes \mathcal{L}_{\overline{\chi}(y)},$$

we see that for each critical value β of f, we have
(1) m divides the multiplicity with which $\mathcal{L}_{\psi(\beta y)}$ occurs in $\mathcal{G}(\infty)|P(\infty)$.
(2) the set of characters with multiplicity

$$\amalg_{\alpha \mapsto \beta} \amalg_{\text{all } \chi^{1+e(\alpha)} = 1,\ \chi \text{ nontriv}} \chi$$

is stable under $\chi \mapsto \chi\rho$, for any character ρ of order m.

Fix any critical value β. By (2), given ρ of order m, ρ is the ratio of two characters χ_1/χ_2 where each χ_i has order dividing $1 + e(\alpha_i)$ for some $\alpha_i \mapsto \beta$. Therefore as m is **prime**, at least one of the numbers $1 + e(\alpha_i)$ must be divisible by m, say $m \mid 1+e(\alpha_1)$. But then ρ^{-1} is itself a nontrivial character χ of order dividing $1+e(\alpha_1)$, and the stability under $\chi \mapsto \chi\rho$ implies that the trivial character is a member of the set in (2), contradiction. QED

Thus we obtain the following theorem.

Theorem 7.7.6 Suppose that C is a connected smooth complete curve over k, with a marked point ∞. Fix integers $n \geq 1$ and $d \geq 1$ such that $\gcd(n, d) = 1$, $n \neq d$, both n and d are prime to p.
Let $\mathcal{L}$ be an ℓ-adic sheaf on $C - \{\infty\}$ which is generically of rank one and is the direct image of its restriction to a nonempty open set where it is lisse. Suppose that $\mathrm{Swan}_\infty(\mathcal{L}) = n$. Let f be a rational function on C

whose only pole is at ∞, of order d. Define $\mathcal{F} := f_*\mathcal{L}$, $\mathcal{G} := \mathrm{NFT}_\psi(\mathcal{F})$, $G := G_{geom}$ for $\mathcal{G}$.

(case n > d) If $n > d$, then $\mathcal{G}$ is lisse on $\mathbb{A}^1$ of rank
$\mathrm{rank}(\mathcal{G}) = n - 1 + 2\mathrm{genus}(C) + \Sigma_{x \text{ in } C - \{\infty\}} (\mathrm{drop}_x(\mathcal{L}) + \mathrm{Swan}_x(\mathcal{L}))$.
$\mathrm{Det}(\mathcal{G})$ has finite p-power order q. If $p > 2\mathrm{rank}(\mathcal{G}) + 1$, then $\mathcal{G}$ is Lie-irreducible, $q = 1$ or p, and $G = \mu_q G^0$, with $G^0 = G^{0,der}$ semisimple.

(case d > n) If $d > n$, then $\mathcal{G}$ is lisse on $\mathbb{G}_m$ of rank
$\mathrm{rank}(\mathcal{G}) = d - 1 + 2\mathrm{genus}(C) + \Sigma_{x \text{ in } C - \{\infty\}} (\mathrm{drop}_x(\mathcal{L}) + \mathrm{Swan}_x(\mathcal{L}))$.
If $p > 2\mathrm{rank}(\mathcal{G}) + 1$, then $\mathcal{G}$ is Lie-irreducible.

In **either** of these two cases, if $p > 2\mathrm{rank}(\mathcal{G}) + 1$ and p does not divide $2nN_1(|n-d|)N_2(|n-d|)$, then $G^0 = G^{0,der}$ is one of

(1) If $|n-d|$ is odd, $G^{0,der}$ is $SL(\mathcal{G}_x)$.

(2) If $|n-d|$ is even, then either $G^{0,der}$ is $SL(\mathcal{G}_x)$ or $SO(\mathcal{G}_x)$ or (if $\mathrm{rank}(\mathcal{G})$ is even) $SP(\mathcal{G}_x)$, or $|n-d| = 6$, $\mathrm{rank}(\mathcal{G}) = 7, 8$ or 9, and $G^{0,der}$ is one of

$\mathrm{rank}(\mathcal{G})=7$: the image of G_2 in its 7-dim'l irred. representation
$\mathrm{rank}(\mathcal{G})=8$: the image of Spin(7) in the 8-dim'l spin representation
the image of SL(3) in the adjoint representation
the image of $SL(2)\times SL(2)\times SL(2)$ in $\mathrm{std}\otimes\mathrm{std}\otimes\mathrm{std}$
the image of $SL(2)\times Sp(4)$ in $\mathrm{std}\otimes\mathrm{std}$
the image of $SL(2)\times SL(4)$ in $\mathrm{std}\otimes\mathrm{std}$
$\mathrm{rank}(\mathcal{G})=9$: the image of $SL(3)\times SL(3)$ in $\mathrm{std}\otimes\mathrm{std}$.

proof This is the main ℓ-adic theorem 7.2.7. In the case $n > d$, we also use [Ka-MG, Prop. 5] to know that if $p > 2\mathrm{rank}(\mathcal{G}) + 1$, then $G = \mu_q G^0$, with $G^0 = G^{0,der}$ semisimple. QED

7.8 Fourier Transform-Stable Classes of Sheaves

In this section we will discuss in detail several classes of Fourier sheaves on $\mathbb{A}^1$ which are stable under NFT_ψ, with particular attention to "following" their highest slopes. Basically, all we are doing is spelling out Laumon's general results on the local monodromy of Fourier Transforms in some special cases where they provide input for the main ℓ-adic theorem 7.2.7. One of the principal applications of the material in this section will be to the definition and study of the characteristic p ℓ-adic sheaf analogue of the generalized hypergeometric $\mathcal{D}$-modules studied earlier.

In each example below of such a class $\mathcal{C}$, we denote by $\mathcal{F}$ and $\mathcal{G}$ members of the class $\mathcal{C}$. We write $\mathcal{F} \Leftrightarrow \mathcal{G}$ to indicate that $\mathcal{F}$ and $\mathcal{G}$ are Fourier Transforms of each other:

$$\mathcal{F} \Leftrightarrow \mathcal{G} \quad \text{means} \quad \mathcal{G} = \mathrm{NFT}_{\psi}(\mathcal{F}) \text{ and } [-1]^*\mathcal{F}(-1) = \mathrm{NFT}_{\psi}(\mathcal{G}).$$

We continue to suppose k algebraically closed throughout this section.

Class (1) The class of lisse sheaves on $\mathbb{A}^1$ with all ∞-breaks > 1. Indeed, any such sheaf is Fourier, and for Fourier sheaves, the two conditions "lisse on $\mathbb{A}^1$" and "all ∞-breaks > 1" are interchanged by Fourier Transform.

If $\mathcal{F} \Leftrightarrow \mathcal{G}$ in this class, then $\mathcal{F}$ and $\mathcal{G}$ have the same number of distinct ∞-breaks, and they correspond as follows:

	∞-break	multiplicity
$\mathcal{F}$	$(a+b)/a$	a
$\mathcal{G}$	$(a+b)/b$	b

$$\mathrm{rank}\mathcal{F} + \mathrm{rank}\mathcal{G} = \mathrm{Swan}_{\infty}(\mathcal{F}) = \mathrm{Swan}_{\infty}(\mathcal{G}).$$

On ∞-breaks themselves, the rule is $1+x \mapsto 1 + (1/x)$. So the biggest ∞-slope of $\mathcal{F}$ becomes **smallest** ∞-slope of $\mathcal{G}$, and vice versa. This class of sheaves is stable under additive translation and under $\otimes \mathcal{L}_{\psi(sx)}$.

Class (1 bis) The subclass of (1) which have a **unique** ∞-break.

Class (1 ter) The subclass of (1 bis) whose unique ∞-break has exact denominator the rank of the sheaf. Recall from [Ka-MG, Thm 9] that for such sheaves, if $p > 2\mathrm{rank} + 1$ we always have $G^0 = SL$ or (in even rank) Sp.

Class (2) Fourier sheaves which are lisse on $\mathbb{G}_m$, with all ∞-breaks $\neq 1$. Indeed for Fourier sheaves, the two conditions "lisse on $\mathbb{G}_m$" and "all ∞-breaks $\neq$ 1" are interchanged by Fourier Transform. If $\mathcal{F} \Leftrightarrow \mathcal{G}$ in this class, then visibly we have

$$\mathrm{Swan}_0(\mathcal{F}) + \mathrm{Swan}_{\infty}(\mathcal{F}) = \mathrm{Swan}_0(\mathcal{G}) + \mathrm{Swan}_{\infty}(\mathcal{G}).$$

In fact, if we calculate $\dim \mathcal{G}_0 = -\chi_c(\mathbb{A}^1, \mathcal{F})$ by the Euler-Poincare formula, we find

$$\begin{aligned}\dim \mathcal{F}_0 + \dim \mathcal{G}_0 &= \mathrm{Swan}_0(\mathcal{F}) + \mathrm{Swan}_\infty(\mathcal{F}) \\ &= \mathrm{Swan}_0(\mathcal{G}) + \mathrm{Swan}_\infty(\mathcal{G}).\end{aligned}$$

The polynomials (cf. 7.5.2, 7.5.3, 7.5.4) describing the tame parts of the local monodromies are interrelated by

$$P(\mathcal{G}, \infty, \overline{\chi}, T) = P(\mathcal{F}, 0, \chi, T) \text{ for each nontrivial } \chi.$$

$$P(\mathcal{G}, \infty, \mathbb{1}, T) = P(\mathcal{F}, 0, \mathbb{1}, T)/T.$$

$$P(\mathcal{G}, 0, \overline{\chi}, T) = P(\mathcal{F}, \infty, \chi, T) \text{ for each nontrivial } \chi.$$

$$P(\mathcal{G}, 0, \mathbb{1}, T) = T \times P(\mathcal{F}, \infty, \mathbb{1}, T).$$

(7.8.2.1) Within the class (2) of Fourier sheaves which are lisse on $\mathbb{G}_m$ and have all ∞-breaks $\neq 1$, there are various supplementary Fourier-stable conditions which define Fourier-stable subclasses. Here are two examples. The parameter k is a nonnegative integer.

(Σ**Card** = **k**)

Card{distinct $\neq 0$ ∞-breaks} + Card{distinct $\neq 0$ 0-breaks} = k,

(Σ**Swan** = **k**)

$\mathrm{Swan}_0(\mathcal{F}) + \mathrm{Swan}_\infty(\mathcal{F}) = k$.

(7.8.2.2) Between these conditions there are the obvious implications

(ΣSwan=k) $\Rightarrow$ (ΣCard $\leq$ k),

(ΣSwan $\neq$ 0) $\Leftrightarrow$ (ΣCard $\neq$ 0).

For any $k \geq 1$, and any tame character χ, the set of **irreducible** Fourier sheaves $\mathcal{F}$ of class(2) which satisfy (Σ**Card** = **k**) [resp. which satisfy (Σ**Swan** = **k**)] is stable by the operation

$$\mathcal{F} \mapsto j_*(\mathcal{L}_\chi \otimes j^*\mathcal{F}),$$

where $j: \mathbb{G}_m \to \mathbb{A}^1$ denotes the inclusion. (If k=0, the only such irreducibles are the $\mathcal{L}_\chi$ with χ nontrivial, and the condition "χ nontrivial" is clearly not stable under this operation.)

There is another stability property worth noting. For each integer

$n \geq 1$, denote by $[n]: x \mapsto x^n$ the n'th power map of $\mathbb{A}^1$ to itself.

Lemma 7.8.2.3 (compare 3.7.6) Suppose that $\mathcal{F}$ is a Fourier sheaf on $\mathbb{A}^1$ which is lisse on $\mathbb{G}_m$, with $k := \mathrm{Swan}_\infty(\mathcal{F}) + \mathrm{Swan}_0(\mathcal{F})$. Let $n \geq 1$ be an integer which is prime to p. Then
(a) If $n > k$, then $[n]_*\mathcal{F}$ is Fourier, lisse on $\mathbb{G}_m$, with all 0-breaks < 1 and and all ∞-break < 1, and $\mathrm{Swan}_\infty([n]_*\mathcal{F}) + \mathrm{Swan}_0([n]_*\mathcal{F}) = k$.
(b) If $\gcd(n, k) = 1$ and $\mathcal{F}$ is irreducible Fourier, then $[n]_*\mathcal{F}$ is irreducible Fourier.

proof (a) The condition **Fourier(1)** is stable by f_* for any finite map of $\mathbb{A}^1$ to itself, as is the condition that $H^0(\mathbb{A}^1, \mathcal{F}) = 0 = H^2_c(\mathbb{A}^1, \mathcal{F})$. That $[n]_*\mathcal{F}$ is lisse on $\mathbb{G}_m$, has all breaks $\leq k/n < 1$, and has $\mathrm{Swan}_\infty([n]_*\mathcal{F}) + \mathrm{Swan}_0([n]_*\mathcal{F}) = k$, is obvious. For $t \neq 0$, $([n]_*\mathcal{F})\otimes\mathcal{L}_{\psi(tx)}$ consequently has all ∞-breaks $=1$, so has vanishing H^0 and H^2_c. Therefore $[n]_*\mathcal{F}$ is Fourier.
(b) Suppose $\mathcal{F}$ irreducible Fourier. If $n=1$ or $\mathcal{F} = 0$, there is nothing to prove. If $n > 1$ and $\mathcal{F} \neq 0$, then $[n]_*\mathcal{F}$ has generic rank ≥ 2, and is the direct image of its lisse restriction to $\mathbb{G}_m$. So to show that $[n]_*\mathcal{F}$ is irreducible Fourier, it suffices to show that $[n]_*\mathcal{F} \mid \mathbb{G}_m$ is irreducible. By Frobenius reciprocity, this is the same as showing that $\mathcal{F} \mid \mathbb{G}_m$ has multiplicity one in $[n]^*[n]_*\mathcal{F} \mid \mathbb{G}_m \approx \oplus_{\zeta \in \mu_n(k)} [x \mapsto \zeta x]^*(\mathcal{F} \mid \mathbb{G}_m)$. If this were not the case, then there would exist $\zeta \neq 1$ in $\mu_n(k)$, say of exact order $d > 1$, $d|n$, and an isomorphism

$$\mathcal{F} \mid \mathbb{G}_m \approx [x \mapsto \zeta x]^*(\mathcal{F} \mid \mathbb{G}_m).$$

Since $\mathcal{F} \mid \mathbb{G}_m$ is irreducible, this isomorphism allows us to descend $\mathcal{F} \mid \mathbb{G}_m$ through the d-fold Kummer covering, whence $k := \mathrm{Swan}_\infty(\mathcal{F}) + \mathrm{Swan}_0(\mathcal{F})$ is divisible by d. But $d|n$, and $\gcd(n,k) = 1$. Therefore $d=1$. QED

Class (3) The subclass of (2) consisting of Fourier sheaves which are lisse on $\mathbb{G}_m$ and which have all ∞-breaks < 1. Indeed for Fourier sheaves, the condition "all ∞-breaks ≤ 1" is stable by Fourier transform. If $\mathcal{F} \Leftrightarrow \mathcal{G}$ in this class, then

$\mathcal{F}$ has ∞-break $a/(a+b) > 0$ with multiplicity $a+b \Leftrightarrow$
$\Leftrightarrow \mathcal{G}$ has 0-break $a/b > 0$ with multiplicity b.

On breaks themselves, the rule is

∞-break $x > 0$ for $\mathcal{F}$ $\Leftrightarrow$ 0-break $x/(1-x) > 0$ for $\mathcal{G}$.

0-break $y > 0$ for $\mathcal{F}$ $\Leftrightarrow$ ∞-break $y/(1+y) > 0$ for $\mathcal{F}$.

This rule is **order-preserving**., so we can read biggest 0-breaks in terms of biggest ∞-breaks.

In addition to

$$\dim\mathcal{F}_0 + \dim\mathcal{G}_0 = \mathrm{Swan}_0(\mathcal{F}) + \mathrm{Swan}_\infty(\mathcal{F}) = \mathrm{Swan}_0(\mathcal{G}) + \mathrm{Swan}_\infty(\mathcal{G}),$$

we have the additional symmetry

$$\mathrm{Swan}_\infty(\mathcal{F}) = \mathrm{Swan}_0(\mathcal{G}),\ \mathrm{Swan}_\infty(\mathcal{G}) = \mathrm{Swan}_0(\mathcal{F}).$$

(7.8.3.1) Within the class (3) of Fourier sheaves which are lisse on $\mathbb{G}_m$ and have all ∞-breaks < 1, there are (in addition to (**ΣCard = k**) and (**ΣSwan = k**)) a few more supplementary Fourier-stable conditions which define Fourier-stable subclasses. Again the parameter k is a nonnegative integer.

(**Card = k**)

Card{distinct $\neq 0$ ∞-breaks} = Card{distinct $\neq 0$ 0-breaks} = k.

(**ExDenom**): both the following conditions:

the highest ∞-break of $\mathcal{F}$, **if nonzero**, has multiplicity its exact denominator,

the highest 0-break of $\mathcal{F}$, **if nonzero**, has multiplicity its exact denominator.

(7.8.3.2) Between these we have the implications

(ΣSwan=k) $\Rightarrow$ (ΣCard $\le$ k),

(ΣSwan $\neq$ 0) $\Leftrightarrow$ (ΣCard $\neq$ 0)

(ΣSwan = 1) $\Rightarrow$ (ΣCard = 1) and (ExDenom).

(7.8.3.3) For any $k \ge 1$, and any tame character χ, the set of **irreducible** Fourier sheaves $\mathcal{F}$ of class(3) which satisfy (**ΣCard = k**) [resp. which satisfy (**ΣSwan = k**)] is stable by the operation

$$\mathcal{F} \mapsto j_*(\mathcal{L}_\chi \otimes j^*\mathcal{F}),$$

where $j: \mathbb{G}_m \to \mathbb{A}^1$ denotes the inclusion. The supplementary condition (**ExDenom**) is also stable by this operation.

7.9 Fourier Transforms of Tame Pseudoreflection Sheaves

(7.9.1) We continue to work over an algebraically closed field k of characteristic $p > 0$. We say that a constructible $\overline{\mathbb{Q}}_\ell$-sheaf $\mathcal{F}$ on $\mathbb{A}^1$ is a **tame pseudoreflection sheaf** if it satisfies the following three conditions:

(TPR1) $\mathcal{F}$ is everywhere tame, i.e., for every $t \in \mathbb{P}^1$, $\mathcal{F}(t)$ is a tame representation of $I(t)$.

(TPR2) for some (or equivalently for every) nonempty open set $j: U \to \mathbb{A}^1$ on which $j^*\mathcal{F}$ is lisse, we have $\mathcal{F} \approx j_*j^*\mathcal{F}$.

(TPR3) For every s in $\mathbb{A}^1(k)$ at which $\mathcal{F}$ is not lisse, $\mathcal{F}(s)/\mathcal{F}_s$ is one-dimensional (i.e., local monodromy at s is by tame pseudoreflections).

(7.9.2) We say that a tame pseudoreflection sheaf $\mathcal{F}$ is irreducible if in addition it satisfies

(IrrTPR) for some (or equivalently for every) nonempty open set $j: U \to \mathbb{A}^1$ on which $j^*\mathcal{F}$ is lisse, $j^*\mathcal{F}$ is irreducible (as rep'n of π_1).

(7.9.3) Any nonconstant irreducible tame pseudoreflection sheaf is an irreducible Fourier sheaf, and its set S of points of nonlissity in $\mathbb{A}^1$ is a finite nonempty subset of $\mathbb{A}^1(k)$.

Theorem 7.9.4 Let $\mathcal{F}$ be a nonconstant irreducible tame pseudoreflection sheaf on $\mathbb{A}^1$, with set $S = \{s_1, \dots, s_r\}$ of points of nonlissity in $\mathbb{A}^1(k) = k$. Then

(1) $\mathcal{G} := NFT_\psi(\mathcal{F})$ is an irreducible Fourier sheaf which is lisse on $\mathbb{G}_m$ of rank $r := \mathrm{Card}(S)$.

(2) The restriction of $\mathcal{G}(\infty)$ to the wild inertia group $P(\infty)$ is isomorphic to $\oplus_{i=1,\dots,r} \mathcal{L}_{\psi(s_i y)}$.

(3) $\mathcal{G}(\infty)$ is not Kummer induced.

(4) $\mathcal{G}$ is not Kummer induced.

(5) If $p > 2r + 1$, $\mathcal{G}$ is Lie-irreducible.

proof Since $\mathcal{F}$ is irreducible Fourier, so is $\mathcal{G}$. Since $\mathcal{F}(\infty)$ is tame, $\mathcal{G}$ is lisse on $\mathbb{G}_m$, and $\mathcal{G}(\infty) \approx \oplus_{i=1,\dots,r} FT_\psi loc(s_i, \infty)(\mathcal{F}(s_i)/\mathcal{F}_{s_i})$. By hypothesis each $\mathcal{F}(s_i)/\mathcal{F}_{s_i}$ is of the form $\mathcal{L}_{\chi_i(x-s_i)}$, so

$$\mathcal{G}(\infty) \approx \oplus_{i=1,\dots,r} \mathcal{L}_{\psi(s_i y)} \otimes \mathcal{L}_{\overline{\chi}_i(y)}.$$

This proves (1) and (2). Assertion (3) trivially implies (4), and (1) and (4) together imply (5). Assertion (3) holds because the distinct $\mathcal{L}_{\psi(s_i y)}$

each occur with multiplicity one, thanks to the following lemma.

Lemma 7.9.5 Let M be an ℓ-adic $I(\infty)$-representation whose restriction to $P(\infty)$ is a direct sum of r distinct characters $\mathcal{L}_{\psi(s_i y)}$ with multiplicities n_i. If M is Kummer induced of degree m, then m divides every n_i. In particular, if $\gcd(\text{all } n_i) = 1$, M is not Kummer induced.
proof As $I(\infty)$-representation, M is $\oplus_{i=1,\ldots,r} \mathcal{L}_{\psi(s_i y)} \otimes(\text{tame of dim. } n_i)$. If M is $[m]_* \mathcal{H}$ for some $I(\infty)$-representation $\mathcal{H}$, then (cf. the proof of 7.8.2.3) $\mathcal{H}$ is a direct factor of $[m]^* M$, so of the form

$$\mathcal{H} \approx \oplus_{i=1,\ldots,r} \mathcal{L}_{\psi(s_i y^m)} \otimes(\text{tame of dim. } d_i \le n_i),$$

whence $[m]_* \mathcal{H}$ is $\oplus_{i=1,\ldots,r} \mathcal{L}_{\psi(s_i y)} \otimes(\text{tame of dim. } md_i)$. QED

This lemma proves (3), and so concludes the proof of the theorem. QED

Theorem 7.9.6 Let $\mathcal{F}$ be a nonconstant irreducible tame pseudoreflection sheaf on $\mathbb{A}^1$, with set $S = \{s_1, \ldots, s_r\}$ of points of nonlissity in $\mathbb{A}^1(k) = k$. Supppose that
(1) among the numbers $s_i \in k$, there are no relations of the form

$$s_i - s_j = s_k - s_n$$

except for the trivial ones ($i=j$ and $k=n$) or ($i=k$ and $j=n$).
(2) $p > 2r + 1$.
Then $G := G_{geom}$ for $\mathcal{G} := NFT_\psi(\mathcal{F})$ has $G^{0,der} = SL(r)$.

proof Since $p > 2r + 1$, we know that $\mathcal{G}$ is Lie-irreducible. We apply Gabber's torus trick 1.0 to the diagonal subgroup which is the image of $P(\infty)$. The hypothesis (1) of nonrelations then forces $\mathrm{Lie}(G^{0,der})$ to contain the full maximal torus of $SL(r)$. If $r \le 2$, Lie-irreducibilty forces $G^{0,der} = SL(r)$. If $r \ge 3$, apply 1.2 and then eliminate the other possibilities SO or possibly Sp because their ranks are $< r-1$. QED

Here is a minor variant.
Theorem 7.9.7 Let $\mathcal{F}$ be a nonconstant irreducible tame pseudoreflection sheaf on $\mathbb{A}^1$, with set $S = \{s_1, \ldots, s_r\}$ of points of nonlissity in $\mathbb{A}^1(k) = k$. Supppose that

(1) the numbers s_i are all nonzero, the set S is stable under $s \mapsto -s$, and there are no relations of the form

$$s_i - s_j = s_k - s_n$$

except for the trivial ones (i=j and k=n) or (i=k and j=n) or ($s_i = -s_n$ and $-s_j = s_k$).

(2) $p > 2r + 1$.

Then $G := G_{geom}$ for $\mathcal{G} := NFT_\psi(\mathcal{F})$ has $G^{0,der} = SL(r)$ or $Sp(r)$ or $SO(r)$.

proof This time the hypothesis (1) of nonrelations and the torus trick 1.0 forces Lie($G^{0,der}$) to contain the full maximal torus of Sp(r), and we apply 1.2. QED

7.10 Examples

We continue to work over an algebraically closed field k of characteristic $p > 0$.

(7.10.1) (**Lefschetz pencils**) Start with a smooth connected projective variety $X \subset \mathbb{P}^N$ over k of dimension $n \geq 2$, and consider a Lefschetz pencil of hyperplane sections $X \cap H_t$ of X, with associated fibration $f: \tilde{X} \to \mathbb{P}^1$ (i.e., $f^{-1}(t) = X \cap H_t$). If $p \neq 2$ or if n-1 is odd, the quotient sheaf on $\mathbb{P}^1$

$$Ev^{n-1} := R^{n-1}f_*\overline{\mathbb{Q}}_\ell/(\text{the constant sheaf } H^{n-1}(X, \overline{\mathbb{Q}}_\ell))$$

is, when restricted to any $\mathbb{A}^1 \subset \mathbb{P}^1$, an irreducible tame pseudoreflection sheaf (cf [De-WI]).

(7.10.2) (**Supermorse functions**) This example is a slight generalization of a Lefschetz pencil of relative dimension $n-1 = 0$. Let C be a complete smooth connected curve over k, f a nonconstant rational function on C, D the divisor of poles of f. View f as a finite flat morphism

$$f: C - D \to \mathbb{A}^1,$$

whose degree we denote deg(f). We suppose that the differential df is not identically zero (i.e., f is not a p'th power), and denote by $Z \subset C - D$ the scheme of zeroes of df on C - D. We put $S := f(Z(k)) \subset \mathbb{A}^1(k)$. Then f makes C - D - Z a finite etale connected covering of $\mathbb{A}^1 - S$.

Consider the sheaf $f_*\overline{\mathbb{Q}}_\ell$ on $\mathbb{A}^1$. The trace morphism (for the finite flat morphism f) is a surjective map $\mathrm{Trace}_f : f_*\overline{\mathbb{Q}}_\ell \to \overline{\mathbb{Q}}_\ell$, whose restriction to the subsheaf $\overline{\mathbb{Q}}_\ell$ of $f_*\overline{\mathbb{Q}}_\ell$ is multiplication by deg(f). Thus

$$\mathcal{F} := \text{Kernel of } \mathrm{Trace}_f : f_*\overline{\mathbb{Q}}_\ell \to \overline{\mathbb{Q}}_\ell$$

is a direct factor of $f_*\overline{\mathbb{Q}}_\ell$ of generic rank deg(f) - 1.

Lemma 7.10.2.1 If deg(f) < p, $\mathcal{F}$ is a Fourier sheaf.

proof Since $\mathcal{F}$ is a direct factor of a sheaf (namely $f_*\overline{\mathbb{Q}}_\ell$) which is the direct image of its restriction to $\mathbb{A}^1 - S$, $\mathcal{F}$ shares this property. Since $\mathcal{F}(\infty)$ is tame (because deg(f) < p), we have

$$H^0(\mathbb{A}^1, \mathcal{F}\otimes\mathcal{L}_{\psi(tx)}) = 0 = H^2{}_c(\mathbb{A}^1, \mathcal{F}\otimes\mathcal{L}_{\psi(tx)})$$

for $t \neq 0$ by slopes. For $t = 0$ this vanishing persists. Because C - D is a smooth connected curve, $H^0(\mathbb{A}^1, f_*\overline{\mathbb{Q}}_\ell) = H^0(C - D, \overline{\mathbb{Q}}_\ell) = \overline{\mathbb{Q}}_\ell$, and $H^2{}_c(\mathbb{A}^1, f_*\overline{\mathbb{Q}}_\ell) = H^2{}_c(C - D, \overline{\mathbb{Q}}_\ell) = \overline{\mathbb{Q}}_\ell(-1)$, so the inclusion $\overline{\mathbb{Q}}_\ell \to f_*\overline{\mathbb{Q}}_\ell$ induces an isomorphism on both H^0 and $H^2{}_c$. QED

(7.10.2.2) We say that the function f on C - D is a supermorse function if it satisfies the following three conditions:

(SM1) deg(f) < p.

(SM2) all zeroes in C - D of the differential form df are simple, i.e., the scheme Z is finite etale over k.

(SM3) f separates the zeroes of df in C - D, i.e., Card(S) = Card(Z).

Lemma 7.10.2.3 If f is a supermorse function, then $\mathcal{F}$ is an irreducible tame reflection sheaf.

proof Since deg(f) < p, $f_*\overline{\mathbb{Q}}_\ell$ and hence its direct factor $\mathcal{F}$ is everywhere tame. By (SM2) and (SM3), $f_*\overline{\mathbb{Q}}_\ell$ is a pseudoreflection sheaf (indeed a reflection sheaf). To see that $\mathcal{F}$ is irreducible, it suffices to show that on the open set $\mathbb{A}^1 - S$ where $f_*\overline{\mathbb{Q}}_\ell$ is lisse, G_{geom} for $f_*\overline{\mathbb{Q}}_\ell$ is the full symmetric group $\mathfrak{S}_d$ in its standard d-dimensional representation. For then G_{geom} for $\mathcal{F}$ will be $\mathfrak{S}_d$ in its augmentation representation, which is irreducible.

The group G_{geom} for $f_*\overline{\mathbb{Q}}_\ell$ is intrinsically a subgroup Γ of $\mathfrak{S}_d$, well-defined up to conjugacy. In terms of a chosen geometric point ξ of $\mathbb{A}^1 - S$, Γ is the image of $\pi_1(\mathbb{A}^1 - S, \xi)$ acting on the finte set $f^{-1}(\xi)$, corresponding to the fact f makes C - D - Z a finite etale connected covering of $\mathbb{A}^1 - S$. Because C - D - Z is connected, the action of Γ is transitive. Because this covering is everywhere tame, Γ is generated by the conjugates of the images of the local inertia groups I(s) for each s in

S. By (SM2) and (SM3), each I(s) acts by a transposition. Thus Γ is a transitive subgroup of $\mathfrak{S}_d$ generated by transpositions, hence Γ is $\mathfrak{S}_d$ itself. QED

(7.10.3) We now consider in detail the special case of supermorse functions f on $\mathbb{A}^1$, i.e., supermorse polynomials. We maintain the notations

$$\mathcal{F} := \text{Kernel of Trace}_f : f_*\overline{\mathbb{Q}}_\ell \to \overline{\mathbb{Q}}_\ell, \quad \mathcal{G} := \mathrm{NFT}_\psi(\mathcal{F}),$$

$$Z := \text{the set of critical points of f in } \mathbb{A}^1(k),$$

$$S := f(Z) = \text{the set of critical values of f in } \mathbb{A}^1(k).$$

Lemma 7.10.4 Suppose that f is a supermorse polynomial in k[x] of degree n.

(1) $\mathcal{G}(0) \approx \bigoplus_{\text{all nontrivial } \chi \text{ with } \chi^n = \mathbb{1}} \mathcal{L}_\chi$.

(2) if $\Sigma_{z \text{ in } Z}\ f(z) = 0$, then $\det\mathcal{G}|\mathbb{G}_m \approx (\mathcal{L}_{\chi_2})^{\otimes(n-1)}$, where χ_2 denotes the tame character of order two.

(3) If n is odd and the function f is odd, then $\mathcal{G}|\mathbb{G}_m$ carries a symplectic autoduality.

proof Assertion (1) is obvious from 7.4.3.1 and 7.4.4, since

$$\mathcal{G}_0 = H^1{}_c(\mathbb{A}^1, \mathcal{F}) \subset H^1{}_c(\mathbb{A}^1, f_*\overline{\mathbb{Q}}_\ell) = H^1{}_c(\mathbb{A}^1, \overline{\mathbb{Q}}_\ell) = 0,$$

and

$$\mathcal{F}(\infty) = (f_*\overline{\mathbb{Q}}_\ell/\overline{\mathbb{Q}}_\ell)(\infty) \approx \bigoplus_{\text{all nontrivial } \chi \text{ with } \chi^n = 1} \mathcal{L}_\chi.$$

If $\Sigma_{z \text{ in } Z}\ f(z) = 0$, then by 7.9.4 (2) $\det\mathcal{G}$ is tame at ∞, and hence $\det\mathcal{G}$ is everywhere tame by (1). To evaluate it, we may proceed in two different fashions. It suffices to show that $(\det\mathcal{G})\otimes(\mathcal{L}_{\chi_2})^{\otimes(n-1)}$ is unramified **either** at ∞ or at zero. At zero, this is obvious from (1); if n is even, the nontrivial characters killed by n occur in inverse pairs except for χ_2, while for n odd they all occur in inverse pairs. At ∞, we can use the fact that $\mathcal{F}$ is a reflection sheaf to write

$$\mathcal{G}(\infty) \approx \bigoplus_{z \text{ in } Z} \mathcal{L}_{\chi_2(x)} \otimes \mathcal{L}_{\psi(f(z)x)},$$

which, as $\Sigma_{z \text{ in } Z}\ f(z) = 0$, gives $\det\mathcal{G}(\infty) \approx (\mathcal{L}_{\chi_2})^{\otimes(n-1)}$.

If n is odd and the function f is odd, then $\mathcal{G}$ is self-dual. In view of the behaviour of duality under Fourier Transform (cf 7.3.8), amounts to showing that

$$D(\mathcal{F}) \approx [-1]^*\mathcal{F}.$$

Since $f_*\overline{\mathbb{Q}}_\ell \approx \mathcal{F} \oplus \overline{\mathbb{Q}}_\ell$ with $\mathcal{F}$ irreducible Fourier, it suffices for this to show that $D(f_*\overline{\mathbb{Q}}_\ell) \approx [-1]^*(f_*\overline{\mathbb{Q}}_\ell)$. Now for any f, $f_*\overline{\mathbb{Q}}_\ell$ is self-dual, so it suffices to note that since f is odd,

$$[-1]^*(f_*\overline{\mathbb{Q}}_\ell) = [-1]_*(f_*\overline{\mathbb{Q}}_\ell) = f_*([-1]_*\overline{\mathbb{Q}}_\ell) \approx f_*\overline{\mathbb{Q}}_\ell.$$

Now let us make explicit exactly what this duality is fibre by fibre. For $t \neq 0$ in k, we have $H^1{}_c(\mathbb{A}^1, \mathcal{L}_{\psi(tx)}) = 0$, so

$$\mathcal{G}_t = H^1{}_c(\mathbb{A}^1, \mathcal{F}\otimes\mathcal{L}_{\psi(tx)}) = H^1{}_c(\mathbb{A}^1, (f_*\overline{\mathbb{Q}}_\ell)\otimes\mathcal{L}_{\psi(tx)}) = H^1{}_c(\mathbb{A}^1, \mathcal{L}_{\psi(tf(x))}).$$

The intrinsic dual to $\mathcal{G}_t$ is (up to a Tate twist)

$$H^1{}_c(\mathbb{A}^1, \mathcal{L}_{\psi(-tf(x))}) = H^1{}_c(\mathbb{A}^1, \mathcal{L}_{\psi(tf(-x))}) \overset{[-1]^*}{\approx} H^1{}_c(\mathbb{A}^1, \mathcal{L}_{\psi(tf(x))}).$$

To follow the signs, it is easiest if we view both $H^1{}_c(\mathbb{A}^1, \mathcal{L}_{\psi(\pm tf(x))})$ as direct factors of $H^1{}_c(W, \overline{\mathbb{Q}}_\ell)$, W the complete nonsingular model of the Artin-Schreier curve of equation $z^q - z = tf(x)$, where q is the cardinality of a finite subfield of k over which ψ is "defined". Because f(x) is odd, we can, define an involution A of W by $A(z,x) := (-z, -x)$. The autoduality of $\mathcal{G}_t$ occurs inside the autoduality of $H^1{}_c(W, \overline{\mathbb{Q}}_\ell)$ induced by the pairing

$$(\alpha, \beta) := \alpha\cdot A^*(\beta),$$

where $\alpha\cdot\beta$ is the cup-product pairing. Since cup-product is alternating, and A is an involution, we find

$$(\beta,\alpha) := \beta\cdot A^*(\alpha) = A^*(\beta\cdot A^*(\alpha)) = A^*(\beta)\cdot\alpha = -\alpha\cdot A^*(\beta) = -(\alpha,\beta).$$

QED

Theorem 7.10.5 Let n be an integer, $p > n \geq 3$. Then for any $a \neq 0$ in k, the polynomial $f(x) := x^n - nax$ is supermorse. Put

$$\mathcal{F} := \text{Kernel of Trace}_f : f_*\overline{\mathbb{Q}}_\ell \to \overline{\mathbb{Q}}_\ell, \quad \mathcal{G} := \mathrm{NFT}_\psi(\mathcal{F}).$$

If $p > 2n-1$ and if the condition $*(p, n-1)$ holds (cf. 7.1), then

(1) If n is even, G_{geom} for $\mathcal{G}$ is the group $\pm SL(n-1)$,

(2) If n is odd, G_{geom} for $\mathcal{G}$ is $Sp(n-1)$.

proof Since $a \neq 0$ and k is algebraically closed, $a = \alpha^{n-1}$ for some $\alpha \neq 0$. Then Z is $\alpha\mu_{n-1}$, and $S = f(Z)$ is $(1-n)a\alpha\mu_{n-1}$. Thus f is supermorse. Since $n \geq 3$, $\Sigma_{z \text{ in } Z}\ f(z) = 0$.

If If $p > 2n-1 = 2(n-1)+1$ and if the condition $*(p, n-1)$ holds, then 7.9.6 and 7.9.7 apply.

Consider first the case when n is even. Then n-1 is odd, and by $*(p, n-1)$ we are in the situation of 7.9.6. So G_{geom} contains SL(n-1). By 7.10.4, $\det\mathcal{G}$ is of order two.

If n is odd, then **7.40.4** shows that $G \subset Sp(n-1)$, so by the paucity of choice in 7.9.7, we must have $G = Sp(n-1)$. QED

Theorem 7.10.6 Let $n \geq 5$ be an integer. Let $g(x) \in \mathbb{Z}[x]$ be a monic polynomial of degree n, $\mathbb{Q}(g)/\mathbb{Q}$ the splitting field of g. Suppose that $\mathrm{Gal}(\mathbb{Q}(g)/\mathbb{Q})$ is $\mathfrak{S}_n$. Let $\alpha_1, \ldots, \alpha_n$ be the distinct roots of g, and let $f(x) \in \mathbb{Q}[x]$ be the unique primitive of g(x) such that $\Sigma_i f(\alpha_i) = 0$. Then
(1) $(n+1)f(x)$ is monic of degree n+1, with coefficients in $\mathbb{Z}[1/(n+1)!]$.
(2) there exists an explicitly computable nonzero element $N(f) \in \mathbb{Z}[1/(n+1)!]$ such that for any prime $p > 2n+1$ which does not divide N(f), the polynomial $f_p(x) := f(x) \bmod p$ in $\overline{\mathbb{F}}_p[x]$ is supermorse, and, putting

$$\mathcal{F}_p := \text{Kernel of } \mathrm{Trace}_{f_p} : (f_p)_*\overline{\mathbb{Q}}_\ell \to \overline{\mathbb{Q}}_\ell, \quad \mathcal{G}_p := \mathrm{NFT}_\psi(\mathcal{F}_p),$$

G_{geom} for $\mathcal{G}_p$ is

$$\begin{cases} SL(n) \text{ if n even} \\ \pm SL(n) \text{ if n odd.}\end{cases}$$

proof Let $F(x) := \int_0^x g(t)dt$. Then $(n+1)F$ is monic with coefficients in $\mathbb{Z}[1/(n+1)!]$. By galois theory and the monicity of g, $\sigma := \Sigma_i F(\alpha_i)$ lies in $\mathbb{Z}[1/(n+1)!]$. So $f(x) = F(x) - (1/n)\sigma$, and (1) is obvious.

We next claim that f separates the α_i. Indeed, if not then after renumbering the α's we have $f(\alpha_1) = f(\alpha_2)$. Applying elements of the galois group $\mathfrak{S}_n$, we deduce that $f(\alpha_1) = f(\alpha_i)$ for all $i = 1, \ldots, n$. Since $\Sigma f(\alpha_i) = 0$, we infer that $f(\alpha_i) = 0$ for all i, whence f is divisible by g, say $(n+1)f(x) = (x-a)g(x)$. Differentiating, we find

$$(n+1)g = g + (x-a)g',$$

whence ng(x) is divisible by g'(x). But g(x) has n distinct zeroes, so this is impossible.

Once f separates the zeroes of g(x), then for any nonempty subset S of $\{1, 2, \ldots, n\}$, the subfield $\mathbb{Q}(\text{all } f(\alpha_i), i \in S)$ of $\mathbb{Q}(\text{all } \alpha_i, i \in S)$ is in fact equal to it:

$$\mathbb{Q}(\text{all } f(\alpha_i),\ i \in S) = \mathbb{Q}(\text{all } \alpha_i,\ i \in S) \text{ inside } \mathbb{Q}(g).$$

For if τ is an element of $\mathrm{Gal}(\mathbb{Q}(g)/\mathbb{Q})$ which fixes $f(\alpha_i)$, then τ fixes α_i, simply because $\tau(f(\alpha_i)) = f(\tau(\alpha_i))$, f having coefficients in $\mathbb{Q}$, and, as f separates the α's, from $f(\alpha_i) = f(\tau(\alpha_i))$ we may infer $\alpha_i = \tau(\alpha_i)$.

Now suppose that for four indices i, j, k, m in $\{1, 2, \ldots, n\}$ we have a relation

$$f(\alpha_i) - f(\alpha_j) = f(\alpha_k) - f(\alpha_m)$$

in $\mathbb{Q}(g)$, but $i \neq j$, $i \neq k$, $k \neq m$, and $j \neq m$. Then either

$i = m$: $2f(\alpha_i) = f(\alpha_j) + f(\alpha_k)$, whence $\alpha_i \in \mathbb{Q}(\alpha_j, \alpha_k)$

or $i \neq m$: $f(\alpha_i) = f(\alpha_k) - f(\alpha_m) + f(\alpha_j)$, whence $\alpha_i \in \mathbb{Q}(\alpha_j, \alpha_k, \alpha_m)$.

Applying elements of $\mathrm{Gal}(\mathbb{Q}(g)/\mathbb{Q}) = \mathfrak{S}_n$, we conclude that all the α_i lie in $\mathbb{Q}(\alpha_1, \alpha_2, \alpha_3)$. Since $n \geq 5$ and $\mathrm{Gal}(\mathbb{Q}(g)/\mathbb{Q}) = \mathfrak{S}_n$, this is nonsense. Therefore if we define N(f) to be the product

$$N(f) := \prod_{i\neq j,\ i\neq k,\ k\neq m,\ j\neq m} ((f(\alpha_i) - f(\alpha_j)) - (f(\alpha_k) - f(\alpha_m))),$$

we see that N(f) is a **nonzero** element of $\mathbb{Z}[1/(n+1)!]$.

Now let A denote the integral closure of $\mathbb{Z}[1/N(f)(n+1)!]$ in $\mathbb{Q}(g)$. Then Spec(A) repesents the finite etale $\mathfrak{S}_n$-torsor over $\mathbb{Z}[1/N(f)(n+1)!]$ of all n-tuples of everywhere distinct critical points of f, with universal n-tuple $(\alpha_1, \ldots, \alpha_n)$. Spec(A) also represents the the finite etale $\mathfrak{S}_n$-torsor over $\mathbb{Z}[1/N(f)(n+1)!]$ of all n-tuples of everywhere distinct critical values of f, with universal n-tuple $(f(\alpha_1), \ldots, f(\alpha_n))$. By construction, for each quadruple of indices (i, j, k, m) with $i \neq j$, $i \neq k$, $k \neq m$, and $j \neq m$, the element $(f(\alpha_i) - f(\alpha_j)) - (f(\alpha_k) - f(\alpha_m))$ of A lies in $A^\times$.

It is clear that for any prime $p > n+1$ which does not divide N(f), f_p is supermorse and 7.9.6 applies to its $\mathcal{F}_p$. Therefore if in addition $p > 2n+1$, then G_{geom} for $\mathcal{G}_p$ contains SL(n). By 7.10.4 (2), we get $\det(\mathcal{G}_p) \approx (\mathcal{L}_{\chi_2})^{\otimes n}$. QED

7.11 Sato-Tate Laws for One-Variable Exponential Sums

In this chapter, most of the applications we have given of the main ℓ-adic theorem 7.2.7 have concerned the Fourier Transforms of sheaves which are essentially of rank one. This means that, over **finite** fields k, we are talking about one-variable exponential sums. Let us make this explicit. Fix a nontrivial additive character ψ of the finite field k. For E/k a finite extension, we denote by ψ_E the nontrivial additive character of E given by $x \mapsto \psi(\mathrm{Trace}_{E/k}(x))$. If we are given a multiplicative character χ of $k^\times$, we denote by χ_E the multiplicative character of $E^\times$ given by $x \mapsto \chi(\mathrm{Norm}_{E/k}(x))$.

In 7.6.3.1, we are concerned with the Fourier Transform of a Fourier sheaf $\mathcal{F}$ of generic rank one, whose ∞-break is ≤ 1. The archtypical example of this is the following. One takes a non-polynomial rational function $f(x)$ on $\mathbb{P}^1$ which has a pole of order ≤ 1 at ∞, say

$$f(x) = -sx + \text{holomorphic at } \infty,$$

all of whose poles have order prime to p. Let $S \subset \mathbb{A}^1$ denote the divisor of finite poles of f; viewing f as a morphism from $\mathbb{A}^1 - S$ to $\mathbb{A}^1$, it makes sense to speak of $\mathcal{L}_{\psi(f)}$ on $\mathbb{A}^1 - S$. The extension by zero of $\mathcal{L}_{\psi(f)}$ to all of $\mathbb{A}^1$ is then a Fourier sheaf $\mathcal{F}$ of generic rank one. For this $\mathcal{F}$, $\mathcal{G} := \mathrm{NFT}_\psi(\mathcal{F})$ has trace function

$$t \in E \mapsto -\Sigma_{x \text{ in } E - S(E)}\ \psi_E(f(x) + tx).$$

We can also add a multiplicative character to the story. Let χ be a nontrivial multiplicative character of $k^\times$, of order denoted order(χ), and $g(x)$ a nonzero rational function on $\mathbb{P}^1$ such that at any zero or pole of $g(x)$ in $\mathbb{A}^1 - S$, the order of zero or pole of g there is not divisible by order(χ). Let T denote the scheme of noninvertibility of g in $\mathbb{A}^1 - S$. We may view g as a morphism from $\mathbb{A}^1 - S - T$ to $\mathbb{G}_m$, and thus we may speak of the sheaf $\mathcal{L}_{\chi(g)}$ on $\mathbb{A}^1 - S - T$, and of the tensor product $\mathcal{L}_{\psi(f)} \otimes \mathcal{L}_{\chi(g)}$ on $\mathbb{A}^1 - S - T$. The extension by zero of $\mathcal{L}_{\psi(f)} \otimes \mathcal{L}_{\chi(g)}$ to all of $\mathbb{A}^1$ is then a Fourier sheaf $\mathcal{F}$ of generic rank one. For this $\mathcal{F}$, $\mathcal{G} := \mathrm{NFT}_\psi(\mathcal{F})$ has trace function

$$t \in E \mapsto -\Sigma_{x \text{ in } E - S(E) - T(E)}\ \psi_E(f(x) + tx)\chi_E(g(x)).$$

Theorems 7.6.4.1 and 7.7.6, and the results of the previous section on supermorse functions are all concerned with the following sort of situation: a complete smooth geometrically connected curve C over k, a rational function f on C with polar divisor D, and an $\mathcal{L}$ on C - D which is generically of rank one and is the direct image of its restriction to a nonempty open set where it is lisse. We are then concerned with the Fourier Transform either of $\mathcal{F} := f_*\mathcal{L}$ on $\mathbb{A}^1$ or, when $\mathcal{L}$ is the constant sheaf, with $\mathcal{F} :=$ Kernel of $\mathrm{Trace}_f : f_*\overline{\mathbb{Q}}_\ell \to \overline{\mathbb{Q}}_\ell$.

When $\mathcal{L}$ is the constant sheaf on C-D, then $\mathcal{G}$ has trace function

$$t \in E^\times \mapsto -\Sigma_{x \text{ in } C(E) - D(E)} \psi_E(tf(x))$$
$$0 \in E \mapsto \text{Card}(E) - \text{Card}(C(E) - D(E)).$$

In 7.7.6, $\mathcal{L}$ is nonconstant, D is a single rational point ∞ of C, the degree d of f is relatively prime to the ∞-break n of $\mathcal{L}$, and both n and d are prime to p. The archtypical example of such an $\mathcal{L}$ on C- $\{\infty\}$ is obtained from a rational function h on C with a pole of order n at ∞, and all poles of order prime to p. If we denote by C - S the open set where h is holomorphic, then the extension by zero from C - S of $\mathcal{L}_{\psi(h)}$ is an $\mathcal{L}$. For this $\mathcal{L}$, and $\mathcal{F} := f_*\mathcal{L}$, $\mathcal{G} := \text{NFT}_\psi(\mathcal{F})$ has trace function

$$t \in E \mapsto -\Sigma_{x \text{ in } C(E) - S(E)} \psi_E(tf(x) + h(x)).$$

Similarly, we can insert a multiplicative character. Let χ be a nontrivial multiplicative character of $k^\times$, of order denoted order(χ), and g(x) a nonzero rational function on C such that at any zero or pole of g(x) in C - S, the order of zero or pole of g there is not divisible by order(χ). Let T denote the scheme of noninvertibility of g in C - S. For $\mathcal{L}$ the extension by zero of $\mathcal{L}_{\psi(h)} \otimes \mathcal{L}_{\chi(g)}$ from C - S - T to C - $\{\infty\}$, and $\mathcal{F} := f_*\mathcal{L}$, $\mathcal{G} := \text{NFT}_\psi(\mathcal{F})$ has trace function

$$t \in E \mapsto -\Sigma_{x \text{ in } C(E) - S(E) - T(E)} \psi_E(tf(x) + h(x))\chi_E(g(x)).$$

In all of these examples, the sheaf $\mathcal{G}$ in question is pure of weight one. For any pure sheaf, we know from Deligne's fundamental results in **Weil II** that G_{geom} is semisimple. So we have determined, in these examples, $(G_{geom})^{0,der} = (G_{geom})^0$ up to a few possibilities. To the extent that one determines G_{geom} **precisely** (not just up to a few possibilities for its identity component) for a particular class of exponential sums, and works out exactly what if any twist will make all the Frobenii also lie in G_{geom}, one has, thanks to **Weil II**, proven an explicit equicharacteristic "Sato-Tate Law" for the distribution of the generalized "angles" of the exponential sums in question. Let us recall the precise statement (cf. [De-WII, 3.5]).

Theorem 7.11.1 (Deligne) Let C be a smooth geometrically connected curve over a finite field k of characteristic p, $\ell \neq p$, $\mathcal{F}$ a lisse $\overline{\mathbb{Q}}_\ell$ sheaf on C which is pure of weight zero, ρ the corresponding representation of $\pi_1(C, \xi)$, and G_{geom} := the Zariski closure of $\rho(\pi_1(C\otimes\overline{k}, \xi))$ its geometric monodromy group. Suppose that for every finite extension E/k and every $t \in C(E)$, each Frobenius $\text{Frob}_{E,t}$ for $\mathcal{F}$ has $\rho(\text{Frob}_{E,t}) \in G_{geom}$. Fix an embedding of $\overline{\mathbb{Q}}_\ell$ into $\mathbb{C}$, and a maximal compact subgroup K of the Lie group $G_{geom}(\mathbb{C})$. The conjugacy class of the semisimple part of each $\rho(\mathrm{Frob}_{E,t})$ meets K in a single conjugacy class, denoted $\vartheta(E, t)$. The conjugacy classes $\vartheta(E, t)$ are equidistributed in the space $K^\natural$ of conjugacy classes of K with respect to normalized Haar measure, in any of the three senses of of equidistrbiution of ([Ka-GKM, 3.5]).

Taking the direct image by the trace, we get the equidistribution of the traces.

Corollary 7.11.2 Hypotheses and notations as above, the traces

$$\mathrm{Trace}(\rho(\mathrm{Frob}_{E,t})) := \mathrm{Trace}(\mathrm{Frob}_{E,t} \mid \mathcal{G}_{\bar{t}})$$

are equidistributed in $\mathbb{C}$ with respect to the the direct image of normalized Haar measure on K by the trace map Trace: $K \to \mathbb{C}$.

7.12 Special Linear Examples

In this section, we give some examples where the Sato-Tate law is that given by a group containing the special linear group. Suppose we have a lisse pure sheaf $\mathcal{G}$ on C/k of rank N whose G_{geom} contains SL(N). If $t \in C(k)$ is any rational point, and if we denote by α any N'th root of $1/\det(\mathrm{Frob}_{k,t} \mid \mathcal{G}_{\bar{t}})$, then the arithmetically twisted sheaf $\alpha^{\deg}\otimes\mathcal{G}$ has all its Frobenii in G_{geom}, and we may apply Deligne's result to $\alpha^{\deg}\otimes\mathcal{G}$ and G_{geom}.

SL-Example(1) Let C be a complete smooth geometrically connected curve over k, ∞ a rational point on C. Take a rational function h(x) on C with a pole of order $n \geq 1$ at ∞ with n prime to p, and all other poles also of order prime to p. Take a second rational function g(x) on C which is nonzero. Take a rational function f(x) on C which is holomorphic on $C - \{\infty\}$ and which has a pole at ∞ of of order d prime to p with $d \neq n$, $\gcd(d, n)=1$. Let $j : C - \{\infty\} - S \to C - \{\infty\}$ be the inclusion of the open set of $C - \{\infty\}$ where h is holomorphic, and

$$k: C - \{\infty\} - S - T \to C - \{\infty\} - S$$

the inclusion of the open set of $C - \{\infty\} - S$ where g is invertible. Let χ be a nontrivial multiplicative character of $k^\times$, of order denoted $\mathrm{order}(\chi)$, such that at any zero or pole of g(x) in T, the order of zero or pole of g there is not divisible by order(χ). Take $\mathcal{L}$:= $j_*(\mathcal{L}_{\psi(h)} \otimes k_* \mathcal{L}_{\chi(g)})$ on C - {∞}, $\mathcal{F}$:= $f_*\mathcal{L}$ on $\mathbb{A}^1$, and $\mathcal{G}$:= $\mathrm{NFT}_\psi(\mathcal{F})$. Then $\mathcal{G}$ is lisse on $\mathbb{G}_m$ (indeed lisse on $\mathbb{A}^1$ if d < n), and its rank N is

$$N = \max(d,n) - 1 + 2\mathrm{genus}(C) + \mathrm{Card}(S(\bar{k})) + \mathrm{Card}(T(\bar{k})) + \\ + \Sigma_{\text{geom poles of h in } C - \{\infty\}} \text{(order of pole of h)}.$$

The trace function of $\mathcal{G}$ is

$$t \in E \mapsto -\Sigma_{x \text{ in } C(E) - S(E) - T(E)} \psi_E(tf(x) + h(x))\chi_E(g(x)).$$

If n-d is odd, p does not divide $2nN_1(|n-d|)N_2(|n-d|)$ and p > 2N + 1, then G_{geom} for $\mathcal{G}$ contains SL(N), by 7.7.6.

SL-Example(2) Let h(x) be a nonconstant rational function on $\mathbb{P}^1$ which is holomorphic at ∞, and has all its poles of order prime to p. Take a nonzero rational function g(x). Let j : $\mathbb{A}^1$ - S → $\mathbb{A}^1$ be the inclusion of the open set of $\mathbb{A}^1$ where h is holomorphic, and

$$k: \mathbb{A}^1 - S - T \to \mathbb{A}^1 - S$$

the inclusion of the open set of $\mathbb{A}^1$ - S where g is invertible. Let χ be a nontrivial multiplicative character of $k^\times$, of order denoted order(χ), such that at any zero or pole of g(x) in T, the order of zero or pole of g there is not divisible by order(χ). Take $\mathcal{F}$:= $j_*(\mathcal{L}_{\psi(h)} \otimes k_* \mathcal{L}_{\chi(g)})$, and $\mathcal{G}$:= $\mathrm{NFT}_\psi(\mathcal{F})$. Then $\mathcal{G}$ is lisse on $\mathbb{G}_m$ of rank N,

$$N = \mathrm{Card}(S(\bar{k})) + \mathrm{Card}(T(\bar{k})) + \Sigma_{\text{geom. poles of h in } \mathbb{A}^1} \text{(order of pole of h)}.$$

Denote by r := $\mathrm{ord}_\infty(g)$. **If χ^r has order ≥ 3**, and if p > 2N + 1, then G_{geom} for $\mathcal{G}$ contains SL(N). [Indeed, I(0) acts on $\mathcal{G}(0)$ by pseudoreflections of determinant $\mathcal{L}_{\bar{\chi}^r(x)}$, so apply 7.6.3.1.]

SL-Example(3) Let h(x) be a nonconstant rational function on $\mathbb{P}^1$ which is holomorphic at ∞, and has all its poles of order prime to p. Let j: $\mathbb{A}^1$ - S → $\mathbb{A}^1$ be the inclusion of the open set of $\mathbb{A}^1$ where h is holomorphic, $\mathcal{F}$:= $j_*\mathcal{L}_{\psi(h)}$, and $\mathcal{G}$:= $\mathrm{NFT}_\psi(\mathcal{F})$. Then $\mathcal{G}$ is lisse on $\mathbb{G}_m$ of rank

$$N := \Sigma_{\text{geom. poles } \alpha \text{ of h in } \mathbb{A}^1} (1 + n_\alpha),$$

where for each $\alpha \in S(\bar{k})$, n_α denotes the order of pole of h at α.
Its trace function is

$$t \in E \mapsto -\Sigma_{x \text{ in } E - S(E)} \psi_E(tx + h(x)).$$

The local monodromy of $\mathcal{G}$ at zero is a unipotent pseudoreflection, and as $I(\infty)$-representation $\mathcal{G}$ is

$$\oplus_{\text{geom. poles } \alpha \text{ of } h \text{ in } \mathbb{A}^1} \mathcal{L}_{\psi(\alpha y)} \otimes (\text{rank } 1 + n_\alpha, \text{ slope}= n_\alpha/(1+n_\alpha)).$$

Therefore $\det\mathcal{G}$, being lisse on $\mathbb{G}_m$, trivial at zero and of break ≤ 1 at ∞, must be geometrically isomorphic to $\mathcal{L}_{\psi(Ay)}$ for some $A \in k$. Looking at the above expression for $\mathcal{G}(\infty)$ we see that

$$\det\mathcal{G} \mid \mathbb{G}_m \otimes \bar{k} \approx \mathcal{L}_{\psi(Ay)} \text{ with}$$

$$A = \Sigma_{\text{geom. poles } \alpha \text{ of } h \text{ in } \mathbb{A}^1} (1 + n_\alpha)\alpha.$$

Suppose now that the rank N of $\mathcal{G}$ is prime to p. Then translating $h(x)$ by A/N, i.e., replacing $h(x)$ by $h(x + (A/N))$, we may and will suppose that $\det\mathcal{G}$ is geometrically trivial.

If in addition either $p > 2N + 1$, or $N=2$, then G_{geom} for $\mathcal{G} \mid \mathbb{G}_m$ is either $Sp(N)$ or $SL(N)$. To see this, we argue as follows. Since $\mathcal{G} \mid \mathbb{G}_m$ is pure, G_{geom} is semisimple. Since the local monodromy at zero is a unipotent pseudoreflection, we get the asserted possibilities for $(G_{geom})^0$. [If $p > 2N + 1$, by the paucity of choice in 7.6.3.1; if $N=2$ by the fact that G_{geom} is a semisimple ($\mathcal{G}$ being pure) subgroup of $SL(2)$, so is either $SL(2)$ or is finite: as the local monodromy at zero of $\mathcal{G}$ is a unipotent pseudoreflection, G_{geom} is not finite.] Since $\mathcal{G}$ has geometrically trivial determinant, if $(G_{geom})^0$ is $SL(N)$ we are done. If $(G_{geom})^0$ is $Sp(N)$, then $G_{geom} \subset \mu_N Sp(N)$, and the "square of the μ_N factor" is a character χ of G_{geom} of order dividing N. But as $\mathcal{G}$ is lisse on $\mathbb{G}_m$ and unipotent at zero, the character χ must be lisse on $\mathbb{A}^1$; as its order is prime to p, it must be trivial.

On the other hand, $\mathcal{G}$ is geometrically self dual if and only if the dual $\mathcal{L}_{\psi(-h(x))}$ of $\mathcal{L}_{\psi(h(x))}$ is geometrically isomorphic to $[-1]^*\mathcal{L}_{\psi(h(x))}$, i.e., if and only if $\mathcal{L}_{\psi(h(x)+h(-x))}$ is geometrically trivial. This function $h(x) + h(-x)$ has poles of order $\le \sup_\alpha(n_\alpha)$. So if $p > 2N + 1$, $\mathcal{G}$ is geometrically self dual if and only if $h(x) + h(-x)$ is constant. So all in all we get a complete determination of G_{geom} for $\mathcal{G}$, provided only that $p > 2N + 1$.

Theorem 7.12.3.1 Let $h(x)$ be a nonconstant rational function on $\mathbb{P}^1$ which is holomorphic at ∞ and has all poles of order prime to p,

$$j: \mathbb{A}^1 - S \to \mathbb{A}^1$$

the inclusion of the open set of $\mathbb{A}^1$ where h is holomorphic, $\mathcal{F} := j_*\mathcal{L}_{\psi(h)}$, and $\mathcal{G} := NFT_\psi(\mathcal{F})$. For each $\alpha \in S(\bar{k})$, let n_α denote the order of pole of h at α,

$$N := \Sigma_{\text{geom. poles } \alpha \text{ of h in } \mathbb{A}^1} (1 + n_\alpha) \in \mathbb{Z}$$

$$A := \Sigma_{\text{geom. poles } \alpha \text{ of h in } \mathbb{A}^1} (1 + n_\alpha)\alpha \in k.$$

Then $\mathcal{G}$ is lisse of rank N on $\mathbb{G}_m$.

Suppose that $p > 2N+1$, and define

$$H(x) := h(x + (A/N)).$$

Then G_{geom} for $\mathcal{G}$ is

$SL(N)$ if $A = 0$ and $H(x) + H(-x)$ is nonconstant,
$\mu_p SL(N)$ if $A \neq 0$ and $H(x) + H(-x)$ is nonconstant,
$Sp(N)$ if $A = 0$ and $H(x) + H(-x)$ is constant,
$\mu_p Sp(N)$ if $A \neq 0$ and $H(x) + H(-x)$ is constant.

SL-Example(4) Let h(x) be a nonconstant rational function on $\mathbb{P}^1$ which has a pole of order $n \geq 2$ at ∞, with n prime to p, and all other poles also of order prime to p. Take a nonzero rational function g(x). Let $j : \mathbb{A}^1 - S \to \mathbb{A}^1$ be the inclusion of the open set of $\mathbb{A}^1$ where h is holomorphic, and

$$k: \mathbb{A}^1 - S - T \to \mathbb{A}^1 - S$$

the inclusion of the open set of $\mathbb{A}^1 - S$ where g is invertible. Let χ be a nontrivial multiplicative character of $k^\times$, of order denoted order(χ), such that at any zero or pole of g(x) in T, the order of zero or pole of g there is not divisible by order(χ). Take $\mathcal{F} := j_*(\mathcal{L}_{\psi(h)} \otimes k_*\mathcal{L}_{\chi(g)})$, and $\mathcal{G} := NFT_\psi(\mathcal{F})$. Then $\mathcal{G}$ is lisse on $\mathbb{A}^1$ of rank N,

$$N = n - 1 + \mathrm{Card}(S(\bar{k})) + \mathrm{Card}(T(\bar{k})) + \Sigma_{\text{geom. poles of h in } \mathbb{A}^1} (\text{order of pole of h}).$$

We have already seen (SL-Example(1), d=1, $C - \{\infty\} = \mathbb{A}^1$) that **if n-1 is odd**, p does not divide $2nN_1(n-1)N_2(n-1)$ and $p > 2N + 1$, then G_{geom} for $\mathcal{G}$ contains SL(N).

In the special case when n=2, stationary phase shows that

$$\mathcal{G}(\infty) = (\text{break } 2, \text{rank}1) \oplus (\text{breaks} \leq 1),$$

so the upper numbering subgroup $I(\infty)^{(2)}$ acts as wild pseudoreflections on $\mathcal{G}(\infty)$. So if $p > 2N + 1$, we have $G_{geom} = \mu_p SL(N)$.

Suppose henceforth that $n \geq 3$, $p > 2N + 1$, either $n-1 \neq 6$ or

$N \notin \{7, 8, 9\}$, and also that p does not divide $2nN_1(n-1)N_2(n-1)$. Because $n \geq 3$, the largest ∞-break of $\mathcal{G}$ is $n/(n-1) \leq 3/2 < 2$, and consequently $\det\mathcal{G}$ has ∞-break ≤ 1. Since $\mathcal{G}$ is lisse on $\mathbb{A}^1$, there is a unique A in k such that $\det\mathcal{G}$ is geometrically isomorphic to $\mathcal{L}_{\psi(Ay)}$. Here is the "formula" for A:

Lemma 7.12.4.1 Write $h(x) = P(x)$ +(holomorphic at ∞), with P a polynomial of degree n, $P(x) = \Sigma_{i=1, \ldots, n} a_i x^i$. Define

$A_\infty := -(n-1)a_{n-1}/na_n \in k$,

$A_{\text{finite}} := \Sigma_{\text{geom. points } \alpha \text{ of } S} (1 + n_\alpha)\alpha + \Sigma_{\text{geom. points } \beta \text{ of } T} \beta \in k$.

Then

$$A = A_\infty + A_{\text{finite}}$$

proof of Lemma Since we know a priori that $\det\mathcal{G}$ is $\mathcal{L}_{\psi(Ay)}$, A is characterized by the property that $\mathcal{L}_{\psi(-Ay)} \otimes \det(\mathcal{G}(\infty))$ is tame at ∞. By stationary phase, $\mathcal{G}$ as $I(\infty)$-representation is the direct sum of three sorts of terms

$$FT_\psi \text{loc}(\infty,\infty)(\mathcal{F}(\infty)) \oplus$$

$$\oplus_{\text{geom. points } \alpha \text{ of } S} \mathcal{L}_{\psi(\alpha y)} \otimes (\text{rank } 1 + n_\alpha, \text{ slope}= n_\alpha/(1+n_\alpha))$$

$$\oplus_{\text{geom points } \beta \text{ of } T} \mathcal{L}_{\psi(\beta y)} \otimes (\text{rank } 1, \text{ slope}= 0).$$

Taking determinants, we find

$$\det\mathcal{G}(\infty) \approx \det(FT_\psi \text{loc}(\infty,\infty)(\mathcal{F}(\infty))) \otimes \mathcal{L}_{\psi(A_{\text{finite}} y)} \otimes (\text{tame}).$$

So we are reduced to computing $\det(FT_\psi \text{loc}(\infty,\infty)(\mathcal{F}(\infty)))$. Now $\mathcal{F}(\infty)$ as $I(\infty)$-representation is $\mathcal{L}_{\psi(P(x))} \otimes (\mathcal{L}_{\chi(x)})^{-\text{ord}_\infty(g)}$. So is enough to prove the lemma for all $\mathcal{F}$'s either of the form $\mathcal{L}_{\psi(P(x))}$ (if $\chi^{\text{ord}_\infty(g)}$ is trivial; strictly speaking we should write this $\mathcal{F}$ as $\mathcal{L}_{\psi(P(x))} \otimes \mathcal{L}_{\chi(g)}$ with g the constant function 1) or of the form $\mathcal{L}_{\psi(P(x))} \otimes \mathcal{L}_{\chi(x)}$ with nontrivial χ (namely $\chi^{-\text{ord}_\infty(g)}$).

For $\mathcal{F} := \mathcal{L}_{\psi(P(x))}$, the determinant formula is proven in [Ka-MG, Thm 17 (3)] under the sole hypothesis that P is a polynomial of degree $n \geq 3$ prime to p. In the case when $\mathcal{F}$ is $\mathcal{L}_{\psi(P(x))} \otimes \mathcal{L}_{\chi(x)}$ with nontrivial χ, a similar argument works also. Here is a sketch.

We know there exists a constant $A \in k$ such that the lisse sheaf $\mathcal{G} := NFT_\psi(\mathcal{L}_{\psi(P(x))} \otimes \mathcal{L}_{\chi(x)})$ on $\mathbb{A}^1/k$ has $\det\mathcal{G} \approx \mathcal{L}_{\psi(Ay)} \otimes \alpha^{\text{degree}}$. So A is determined by knowing $\det(\text{Frob}_{k,y} \mid \mathcal{G})$ for every rational point $y \in k$, since from the dependence rule

$$\det(\text{Frob}_{k,y} \mid \mathcal{G}) = \psi(Ay)\det(\text{Frob}_{k,0} \mid \mathcal{G})$$

we may calculate the additive character $y \mapsto \psi(Ay)$ of k, and this determines A itself. Now for $y \in k$,

$$\det(\mathrm{Frob}_{k,y} \mid \mathcal{G}) := \det(\mathrm{Frob}_k \mid H^1_c(\mathbb{G}_m \otimes \bar{k}, \mathcal{L}_{\psi(P(x)+yx)} \otimes \mathcal{L}_{\chi(x)}))$$

$$:= \det(\mathrm{Frob}_k \mid H^1_c(\mathbb{A}^1 \otimes \bar{k}, \mathcal{L}_{\psi(P(x)+yx)} \otimes j_!\mathcal{L}_{\chi(x)})),$$

where $j_!\mathcal{L}_{\chi(x)}$ is the extension by zero of $\mathcal{L}_{\chi(x)}$ from $\mathbb{G}_m$ to $\mathbb{A}^1$.

The **L-function** $L(T)$ of $\mathbb{A}^1/k$ with coefficients in $\mathcal{L}_{\psi(P(x)+yx)} \otimes j_!\mathcal{L}_{\chi(x)}$ is a polynomial of degree n, namely

$$L(T) = \det(1 - T\mathrm{Frob}_k \mid H^1_c(\mathbb{A}^1 \otimes \bar{k}, \mathcal{L}_{\psi(P(x)+yx)} \otimes j_!\mathcal{L}_{\chi(x)})).$$

Therefore $(-1)^n \times \det(\mathrm{Frob}_{k,y} \mid \mathcal{G})$ is the coefficient of T^n in $L(T)$. Let us write explicitly the additive expression of $L(T)$ as a sum over effective divisors in $\mathbb{A}^1$, i.e., over monic polynomials $f_d(x) := \Sigma(-1)^{d-i}S_{d-i}(f)x^i$ in $k[x]$, with Newton symmetric functions $N_j(f)$ of their roots, shows that for each integer $d \geq 1$, the coefficient of T^d in $L(T)$ is the sum

$$\Sigma_{\text{monic f of degree d}} \chi(S_d(f))\psi(\Sigma_{i=1,\dots,n} a_iN_i(f) + yS_1(f)),$$

where $\chi(S_d(f)) := 0$ if $S_d(f) = 0$.

Take $d=n$, and write the Newton symmetric functions N_i of n roots as isobaric polynomials in the elementary symmetric functions S_j of n roots. The top two are, for $n \geq 3$,

$$N_n = (-1)^{n+1}nS_n + (-1)^n nS_1S_{n-1} + Q$$

$$N_{n-1} = (-1)^n(n-1)S_{n-1} + + R,$$

where Q and R **do not** involve S_{n-1} or S_n.

So the coefficient of T^n in $L(T)$ is

$$\Sigma_{S_1,\dots,S_n} \chi(S_n)\psi(\Sigma_{i=1,\dots,n-2} a_iN_i + yS_1 + a_{n-1}N_{n-1} + a_nN_n),$$

and substituting for N_n and N_{n-1} this becomes

$$\Sigma_{S_n} \chi(S_n)\psi(a_n(-1)^{n+1}nS_n) \times$$

$$\times \Sigma_{S_1,\dots,S_{n-2}} \psi(\Sigma_{i=1,\dots,n-2} a_iN_i + yS_1 + a_{n-1}R + a_nQ) \times$$

$$\times \Sigma_{S_{n-1}} \psi(a_{n-1}(-1)^n(n-1)S_{n-1} + a_n(-1)^n nS_1S_{n-1}).$$

The final term vanishes unless $(n-1)a_{n-1} + na_nS_1$ vanishes, in which case it is $q :=$ Card(k). So only terms with $S_1 := -(n-1)a_{n-1}/na_n := A_\infty$ contribute to this expression, which is consequently

$$\psi(A_\infty y) \times q \times [\Sigma_{S_n} \chi(S_n)\psi(a_n(-1)^{n+1}nS_n)] \times$$

$\times[\Sigma_{S_2, \ldots, S_{n-2}} \psi(\Sigma_{i=1,\ldots,n-2} a_i N_i + a_{n-1}R + a_n Q) \,|S_1 := A_\infty]$.

The important thing is that it is of the form

$\psi(A_\infty y)\times$(a function of a_i's alone). QED for 7.12.4.1

Replacing (h(x), g(x)) by

$$(H(x), G(x)) := (h(x + (A/N)), g(x + (A/N))),$$

we reduce to the case where $\det\mathcal{G}$ is geometrically trivial. By [Ka-MG, Prop. 5] and 7.7.6, G_{geom} is then either SL(N) or SO(N) or, if N is even, Sp(N). But $\mathcal{G}$ is geometrically self dual if and only $D(\mathcal{F}) \approx [-1]^*\mathcal{F}$ geometrically, i.e., if and only if there exists a geometric isomorphism $\mathcal{L}_{\psi(-H(x))}\otimes\mathcal{L}_{\overline{\chi}(G(x))} \approx \mathcal{L}_{\psi(H(-x))}\otimes\mathcal{L}_{\chi(G(-x))}$, i.e., if and only if both of the following conditions are satisfied:

$H(x) + H(-x)$ is constant, say α,

$G(x)G(-x)$ is an order(χ)'th power in $\overline{k}(x)$.

So if either of these conditions fails to hold, then G_{geom} is SL(N).

Let us analyse the sign of the autoduality if both of these conditions are satisfied. Replacing H(x) by H(x) - (α/2) does not change $\mathcal{G}$ geometrically, and reduces us to the case when H is **odd**. Let r:= order(χ), and pick a rational function L(x) with

$$(L(x))^r = G(x)G(-x).$$

Since G(x)G(-x) is even, there is a unique **sign** $\varepsilon = \pm 1$, $\varepsilon^r = 1$, with

$$L(-x) = \varepsilon L(x).$$

Denote by $\boldsymbol{\chi} : \mu_r(k) \cong \mu_r(\overline{\mathbb{Q}}_\ell)$ the unique faithful character of $\mu_r(k)$ for which the character χ of $k^\times$ is

$$\chi(\alpha) := \boldsymbol{\chi}(\alpha^d),\ d := (\mathrm{Card}(k) - 1)/r.$$

The autoduality of $\mathcal{G}$ is symplectic if and only if $\varepsilon = 1$. To see this, view $\mathcal{G}_{\overline{t}}$ as the (ψ, $\boldsymbol{\chi}$)-eigenspace for the action of $(k,+)\times\mu_r$ on $H^1{}_c$ of the complete nonsingular model X of the curve in (x, z, w)-space of equation (q := Card(k))

$$z^q - z = H(x) + tx,\ w^r = G(x),$$

where (a,ζ) acts by $(x, z, w) \mapsto (x, z+a, \zeta w)$. Denote by A the automorphism of this curve defined by $A(x, z, w) := (-x, -z, L(x)/w)$. Notice that A^2 is $(x, z, w) \mapsto (x, z, \varepsilon w)$. The autoduality of $\mathcal{G}_{\overline{t}}$ is given in terms of the cup-product $\alpha\cdot\beta$ on $H^1{}_c(X\otimes\overline{k}, \overline{\mathbb{Q}}_\ell)$ as the pairing

$$(\alpha, \beta) := \alpha\cdot A^*(\beta)$$

on the (ψ, χ)-eigenspace. If $\varepsilon = 1$, then A is an involution, and this pairing is alternating. If $\varepsilon = -1$, then r must be even, $(A^2)^*(\alpha) = -\alpha$ for α in the (ψ, χ)-eigenspace, and the pairing is symmetric.

So all in all we get a fairly complete determination of G_{geom}.

Theorem 7.12.4.2 Let h(x) be a nonconstant rational function on $\mathbb{P}^1$ which has a pole of order $n \geq 2$ at ∞, with n prime to p, and all other poles also of order prime to p. Let g(x) be a nonzero rational function,

$$j : \mathbb{A}^1 - S \to \mathbb{A}^1$$

the inclusion of the open set of $\mathbb{A}^1$ where h is holomorphic, and

$$k: \mathbb{A}^1 - S - T \to \mathbb{A}^1 - S$$

the inclusion of the open set of $\mathbb{A}^1 - S$ where g is invertible. Let χ be a nontrivial multiplicative character of $k^\times$, of order denoted order(χ), such that at any zero or pole of g(x) in T, the order of zero or pole of g there is not divisible by order(χ). Take $\mathcal{F} := j_*(\mathcal{L}_{\psi(h)} \otimes k_* \mathcal{L}_{\chi(g)})$, and $\mathcal{G} := NFT_\psi(\mathcal{F})$. Then $\mathcal{G}$ is lisse on $\mathbb{A}^1$ of rank N,

$$N = n - 1 + \text{Card}(S(\bar{k})) + \text{Card}(T(\bar{k})) + \sum_{\text{geom. poles of h in } \mathbb{A}^1} (\text{order of pole of h}).$$

(1) If n=2 and $p > 2N + 1$, then G_{geom} for $\mathcal{G}$ is $\mu_p SL(N)$.

(2)Suppose $n \geq 3$, $p > 2N + 1$, and p does not divide $2nN_1(n-1)N_2(n-1)$. Write $h(x) = P(x)$ +(holomorphic at ∞), with P a polynomial of degree n, $P(x) = \Sigma_{i=1, \ldots ,n}\, a_i x^i$. Define $A \in k$ by

$A := -(n-1)a_{n-1}/na_n + \Sigma_{\text{geom. pts } \alpha \text{ of } S}\, (1 + n_\alpha)\alpha + \Sigma_{\text{geom. pts } \beta \text{ of } T}\, \beta$.

Define rational functions H(x), G(x) by

$$(H(x), G(x)) := (h(x + (A/N)), g(x + (A/N))).$$

Then we have

(2a) If n-1 is odd, G_{geom} for $\mathcal{G}$ is

$SL(N)$ if $A = 0$,
$\mu_p SL(N)$ if $A \neq 0$.

(2b)If either $n-1 \neq 6$ or $N \notin \{7, 8, 9\}$, then G_{geom} for $\mathcal{G}$ is

$SL(N)$ if $A = 0$,
$\mu_p SL(N)$ if $A \neq 0$,

unles both of the following conditions are satisfied:

$H(x) + H(-x)$ is constant, say α,
$G(x)G(-x)$ is an order(χ)'th power in $\bar{k}(x)$.

(2c) If either $n-1 \neq 6$ or $N \notin \{7, 8, 9\}$, and if both of the above conditions are satisfied, let $r := \mathrm{order}(\chi)$, and let $\varepsilon = \pm 1$ be the sign in $k^\times$ obtained by picking a rational function $L(x)$ with

$$(L(x))^r = G(x)G(-x), \text{ and writing } L(-x) = \varepsilon L(x).$$

Then G_{geom} for $\mathcal{G}$ is

$Sp(N)$ if $A = 0$ and $\varepsilon = 1$,

$\mu_p Sp(N)$ if $A \neq 0$ and $\varepsilon = 1$,

$SO(N)$ if $A = 0$ and $\varepsilon = -1$,

$\mu_p SO(N)$ if $A \neq 0$ and $\varepsilon = -1$.

7.13 Symplectic Examples

We will now give examples where the Sato-Tate law is that given by the symplectic group. Let us first explain why this case is particularly easy to handle. Choose a square root of p in $\overline{\mathbb{Q}}_\ell$, so for each $n \in \mathbb{Z}$ we can speak of the Tate twist sheaves $\overline{\mathbb{Q}}_\ell(n/2)$ on any $\mathbb{F}_p$-scheme, and of the twists $\mathcal{G}(n/2)$ of any given sheaf $\mathcal{G}$. In practice, when one shows that a lisse sheaf $\mathcal{G}$ of some even rank N which is pure of weight n has $G_{geom} = Sp(N)$, the proof shows that in fact $\mathcal{G}(n/2)$ is itself symplectically self-dual. If this is the case, then it is tautological that the Frobenii for $\mathcal{G}(n/2)$ land in $Sp(N)$, and so we can apply Deligne's general result directly to $\mathcal{G}(n/2)$, with $G_{geom} = Sp(N)$.

Sp-Example(1) Let $h(x) \in k(x)$ be an **odd** nonzero rational function which is holomorphic at ∞ and all of whose poles have order prime to p. Let g(x) be a nonzero rational function. Let $j : \mathbb{A}^1 - S \to \mathbb{A}^1$ be the inclusion of the open set of $\mathbb{A}^1$ where h is holomorphic, and

$$k: \mathbb{A}^1 - S - T \to \mathbb{A}^1 - S$$

the inclusion of the open set of $\mathbb{A}^1 - S$ where g is invertible. Let χ be a multiplicative character of $k^\times$, of order r, such that at any zero or pole of g(x) in T, the order of zero or pole of g there is not divisible by r. Suppose that there exists an **even** rational function L(x) such that $L(x)^r = g(x)g(-x)$. Take $\mathcal{F} := j_*(\mathcal{L}_{\psi(h)} \otimes k_*\mathcal{L}_{\chi(g)})$, $\mathcal{G} := NFT_\psi(\mathcal{F})$. Then $\mathcal{G}(1/2)$ is a lisse sheaf on $\mathbb{G}_m$ which is symplectically self-dual (by the same "embed in the Artin-Schreier covering" argument as in **SL-Example(4)** above) of (even) rank

$$N = \mathrm{Card}(S(\bar{k})) + \mathrm{Card}(T(\bar{k})) + \Sigma_{\text{geom. poles of h in } \mathbb{A}^1} (\text{order of pole of h}).$$

and pure of weight zero.The trace function of $\mathcal{G}(\mathbf{1/2})$ is

$$t \in E \mapsto -(\mathrm{Card}(E))^{-1/2}\sum_{x \text{ in } E - S(E) - T(E)} \psi_E(tx + h(x))\chi_E(g(x)).$$

If $p > 2N + 1$, or if $N=2$, then G_{geom} for $\mathcal{G}(\mathbf{1/2})|\mathbb{G}_m$ is $Sp(N)$. [If $N \neq 2$ and $p > 2N + 1$, by the paucity of choice in 7.6.3.1; if $N=2$ by the fact that G_{geom} is a semisimple ($\mathcal{G}$ being pure) subgroup of $SL(2)$, so is either $SL(2)$ or is finite: as the local monodromy at zero of $\mathcal{G}$ is a unipotent pseudoreflection, G_{geom} is not finite.]

Sp-Example(2) Take an **odd** rational function $h(x)$ with a pole of order $n \geq 1$ at ∞ with n prime to p, and all other poles also of order prime to p. Let $g(x)$ be a nonzero rational function. Let $j : \mathbb{A}^1 - S \to \mathbb{A}^1$ be the inclusion of the open set of $\mathbb{A}^1$ where h is holomorphic, and

$$k: \mathbb{A}^1 - S - T \to \mathbb{A}^1 - S$$

the inclusion of the open set of $\mathbb{A}^1 - S$ where g is invertible. Let χ be a multiplicative character of $k^\times$, of order r, such that at any zero or pole of $g(x)$ in T, the order of zero or pole of g there is not divisible by r. Suppose that there exists an **even** rational function $L(x)$ such that $L(x)^r = g(x)g(-x)$. Let $f(x)$ be an **odd** nonzero polynomial of degree d prime to p with $d \neq n$, $\gcd(d, n) = 1$. Take $\mathcal{L} := j_*(\mathcal{L}_{\psi(h)} \otimes k_*\mathcal{L}_{\chi(g)})$, $\mathcal{F} := f_*\mathcal{L}$, $\mathcal{G} := NFT_\psi(\mathcal{F})$. Then $\mathcal{G}(\mathbf{1/2})$ is lisse on $\mathbb{G}_m$ (indeed lisse on $\mathbb{A}^1$ if $d < n$) and just as above is symplectically self-dual. Its rank N is

$$N = \max(d,n) - 1 + \mathrm{Card}(S(\bar{k})) + \mathrm{Card}(T(\bar{k})) + \\ + \sum_{\text{geom. poles of } h \text{ in } \mathbb{A}^1} (\text{order of pole of } h).$$

The trace function of $\mathcal{G}(1/2)$ is

$$t \in E \mapsto -(\mathrm{Card}(E))^{-1/2}\sum_{x \text{ in } E - S(E) - T(E)} \psi_E(tf(x) + h(x))\chi_E(g(x)).$$

If $p > 2N+1$, p does not divide $2nN_1(|n-d|)N_2(|n-d|)$, and either $N \neq 8$ or $|n-d| \neq 6$, then G_{geom} for $\mathcal{G}(\mathbf{1/2})$ is $Sp(N)$, by the paucity of choice in 7.7.6.

Sp-Example(3) Fix $n \geq 3$ an odd integer, $a \in k^\times$, let $f(x) := x^n - nax$, $\mathcal{F} :=$ Kernel of $\mathrm{Trace}_f : f_*\bar{\mathbb{Q}}_\ell \to \bar{\mathbb{Q}}_\ell$, $\mathcal{G} := NFT_\psi(\mathcal{F})$. Here $\mathcal{G}(\mathbf{1/2}) \mid \mathbb{G}_m$ is symplectically self dual and lisse of rank $n-1$. Its trace function is

$$t \in E^\times \mapsto -(\mathrm{Card}(E))^{-1/2}\sum_{x \text{ in } E} \psi_E(tf(x))$$

$$0 \in E \mapsto 0.$$

If $p > 2n-1$ and if the condition $*(p, n-1)$ holds (cf. 7.1), then G_{geom} for $\mathcal{G}(1/2) \mid \mathbb{G}_m$ is $Sp(n-1)$.

7.14 Orthogonal Examples

We now give examples where the Sato-Tate law is that given by the orthogonal group. We work over a finite field k of characteristic $p \neq 2$, and denote by χ_2 the character of order two of $k^\times$.

O-Example(1) Let $h(x) \in k(x)$ be an **odd** nonzero rational function which is holomorphic at ∞ and all of whose poles have order prime to p. Let g(x) be a nonzero rational function. Let $j : \mathbb{A}^1 - S \to \mathbb{A}^1$ be the inclusion of the open set of $\mathbb{A}^1$ where h is holomorphic, and

$$k: \mathbb{A}^1 - S - T \to \mathbb{A}^1 - S$$

the inclusion of the open set of $\mathbb{A}^1 - S$ where g is invertible. Let χ be a multiplicative character of $k^\times$, of even order r, such that at any zero or pole of g(x) in T, the order of zero or pole of g there is not divisible by r. Suppose that there exists an **odd** rational function L(x) such that $L(x)^r = g(x)g(-x)$. Take $\mathcal{F} := j_*(\mathcal{L}_{\psi(h)} \otimes k_*\mathcal{L}_{\chi(g)})$, $\mathcal{G} := NFT_\psi(\mathcal{F})$. Then $\mathcal{G}(1/2)$ is a lisse sheaf on $\mathbb{G}_m$ which is orthogonally self-dual (by the argument in **SL-Example(4)** above) of rank

$$N = Card(S(\bar{k})) + Card(T(\bar{k})) + \Sigma_{\text{geom. poles of h in } \mathbb{A}^1} (\text{order of pole of h})$$

and pure of weight zero.The trace function of $\mathcal{G}(1/2)$ is

$$t \in E \mapsto -(Card(E))^{-1/2}\sum_{x \text{ in } E - S(E) - T(E)} \psi_E(tx + h(x))\chi_E(g(x)).$$

If $p > 2N+1$, and $N \neq 2$, then G_{geom} for $\mathcal{G}(1/2)$ is O(N). To see this, note that G_{geom} lies in O(N), so it must be either SO(N) or O(N), by the paucity of choice in 7.6.3.1. In fact G_{geom} is O(N), because the local monodromy of $\mathcal{G}$ around zero is, by 7.6.3.1, a reflection. (The odd rational function L(x) necessarily has odd ∞-valuation, so from the equation $g(x)g(-x) = L(x)^r$ we infer that $ord_\infty(g) = (r/2)\times$odd, so $\mathcal{F} \approx \mathcal{L}_{\chi_2}$ as $I(\infty)$-representation. Alternately, by 7.6.3.1 this local monodromy is a pseudoreflection which lies in O(N), and any orthogonal pseudoreflection is necessarily a reflection(cf. 1.5).) It is tautological that the Frobenii for $\mathcal{G}(1/2)$ land in O(N), and so we can apply Deligne's general result directly to $\mathcal{G}(1/2)$, with $G_{geom} = O(N)$.

Remark 7.14.1.1 If $N = 2$ in this example (e.g., $h(x)=1/x$, $g(x)=x$, $\chi = \chi_2$) then G_{geom} is **finite**, since it is a semisimple subgroup of $O(2)$.

O-Example(2) Take an **odd** rational function $h(x)$ with a pole of order $n \geq 1$ at ∞ with n prime to p, and all other poles also of order prime to p. Let $g(x)$ be a nonzero rational function. Let $j : \mathbb{A}^1 - S \to \mathbb{A}^1$ be the inclusion of the open set of $\mathbb{A}^1$ where h is holomorphic, and

$$k: \mathbb{A}^1 - S - T \to \mathbb{A}^1 - S$$

the inclusion of the open set of $\mathbb{A}^1 - S$ where g is invertible. Let χ be a multiplicative character of $k^\times$, of even order r, such that at any zero or pole of $g(x)$ in T, the order of zero or pole of g there is not divisible by r. Suppose that there exists an **odd** rational function $L(x)$ such that $L(x)^r = g(x)g(-x)$. Let $f(x)$ be an **odd** nonzero polynomial of degree d prime to p with $d \neq n$, $\gcd(d, n) = 1$. Take $\mathcal{L} := j_*(\mathcal{L}_{\psi(h)} \otimes k_*\mathcal{L}_{\chi(g)})$, $\mathcal{F} := f_*\mathcal{L}$, $\mathcal{G} := NFT_\psi(\mathcal{F})$. Then $\mathcal{G}(1/2)$ is lisse on $\mathbb{G}_m$ (indeed lisse on $\mathbb{A}^1$ if $d < n$) and just as above is orthogonally self-dual. Its rank N is

$$N = \max(d,n) - 1 + \mathrm{Card}(S(\bar{k})) + \mathrm{Card}(T(\bar{k})) + \\ + \Sigma_{\text{geom. poles of h in } \mathbb{A}^1} (\text{order of pole of h}).$$

The trace function of $\mathcal{G}(1/2)$ is

$$t \in E \mapsto -(\mathrm{Card}(E))^{-1/2} \sum_{x \text{ in } E - S(E) - T(E)} \psi_E(tf(x) + h(x))\chi_E(g(x)).$$

If $p > 2N+1$, p does not divide $2nN_1(|n-d|)N_2(|n-d|)$, and either $N \notin \{7,8\}$ or $|n - d| \neq 6$, then G_{geom} for $\mathcal{G}(1/2)$ is either $SO(N)$ or $O(N)$, by the paucity of choice in 7.7.6.

If in addition $n > d$, then $\mathcal{G}$ is lisse on $\mathbb{A}^1$, whence G_{geom} has no nontrivial prime-to-p quotients, so G_{geom} is $SO(N)$. Then $\det(\mathcal{G}(1/2))$ is a geometrically trivial character of order one or two, so it is either trivial or it is (the pullback to $\mathbb{A}^1$ of) "$(-1)^{\deg}$", the unique character of order two of $\mathrm{Gal}(k^{sep}/k)$. The question of **which one** is arithmetic, and in a given example can be decided by computing the determinant of Frobenius on $\mathcal{G}(1/2)_0$ and seeing whether it is 1 or -1. If we compute this sign ± 1 and choose an N'th root ε of it, then we can directly apply Deligne's general result to the slightly twisted sheaf $(\varepsilon)^{\deg} \otimes \mathcal{G}(1/2)$ with $G_{geom} = SO(N)$. [Another course of action: simply replace the given ground field k by its quadratic extension k_2, and directly apply Deligne's general result to $\mathcal{G}(1/2)$ on $\mathbb{A}^1/k_2$ with G_{geom} = SO(N).]

If, on the other hand, d > n, then G_{geom} is O(N). To show that G_{geom} is O(N), we make a series of reductions to a case where it is obvious. First of all, since G_{geom} is either SO(N) or O(N), and $\mathcal{G}$ is lisse on $\mathbb{G}_m$ (say with coordinate t), $\det(\mathcal{G}|\mathbb{G}_m)$ is geometrically either trivial or $\mathcal{L}_{\chi_2(t)}$. The key point is that these two possibilities on $\mathbb{G}_m$ are already distinguished by their I(0)-representations. By 7.4.3.1, we know that

$$\mathcal{G}(0)/\mathcal{G}_0 \approx FT_\psi loc(\infty,0)(\mathcal{F}(\infty))$$

as I(0)-representations. Taking determinants gives

$$\det(\mathcal{G})(0) \approx \det(FT_\psi loc(\infty,0)(\mathcal{F}(\infty)))$$

as I(0)-representations.

To exploit this, we look closely at $\mathcal{F}(\infty)$ as $I(\infty)$-representation. By definition, $\mathcal{F} := f_*j_*(\mathcal{L}_{\psi(h)}\otimes k_*\mathcal{L}_{\chi(g)})$. Write the odd rational function h(x) as the sum

h(x) = (an odd polynomial H(x) of degree n) + (a fct. holo. at ∞).

Then as $I(\infty)$-representation, $j_*(\mathcal{L}_{\psi(h)}\otimes k_*\mathcal{L}_{\chi(g)})$ is $\mathcal{L}_{\psi(H)}\otimes\mathcal{L}_{\chi(g)}$. From the equation $g(x)g(-x) = L(x)^r$ we infer that $\mathrm{ord}_\infty(g) = (r/2)\times\mathrm{odd}$, so $\mathcal{L}_{\chi(g)} \approx \mathcal{L}_{\chi_2}$ as $I(\infty)$-representation and hence

$$\mathcal{F}(\infty) \approx (f_*(\mathcal{L}_{\psi(H)}\otimes\mathcal{L}_{\chi_2(x)}))(\infty).$$

Therefore $\det(\mathcal{G})(0)$ as $I(\infty)$-representation is the same for the initial data (h, f, g, χ) as it is for the data (H, f, x, χ_2). We will now treat this case by a global argument.

For the data (H, f, x, χ_2), the lisse sheaf $\mathcal{G}$ on $\mathbb{G}_m\otimes\bar{k}$ (with parameter t) is

$$t \mapsto H^1_c(\mathbb{G}_m\otimes\bar{k}, \mathcal{L}_{\psi(H(x) + tf(x))}\otimes\mathcal{L}_{\chi_2(x)}).$$

Strictly speaking, we consider the product $\mathbb{G}_m\times\mathbb{G}_m$ with coordinates (x,t), the lisse sheaf $\mathcal{K} := \mathcal{L}_{\psi(H(x) + tf(x))}\otimes\mathcal{L}_{\chi_2(x)}$ on this product; then $\mathcal{G}$ is $R^1(pr_2)_!(\mathcal{K})$. Now $\mathcal{K}$ is tame at zero, and for t in $\mathbb{G}_m$ its Swan_∞ is d. Since the odd polynomial H has degree n < d, it follows from Deligne's semicontinuity theorem ([Lau-SCS]) that $\mathcal{G}$ makes sense as a lisse sheaf of rank d on the product space (the $\mathbb{G}_m$ of t's)×(the affine space of all odd polynomials H of degree $\le$ n), say $\mathbb{G}_m \times \mathbb{E}$, and $\mathcal{G}$ is orthogonally self-dual as a lisse sheaf on $\mathbb{G}_m \times \mathbb{E}$. Since we are not in characteristic 2, the Kunneth formula shows that for any point $e \in \mathbb{E}(\bar{k})$, the inclusion of $(\mathbb{G}_m)\otimes \bar{k}$ into $(\mathbb{G}_m \times \mathbb{E})\otimes \bar{k}$ by $t \mapsto (t,e)$ induces an isomorphism

$$H^1((\mathbb{G}_m \times \mathbb{E})\otimes \bar{k}, \mu_2) \approx H^1(\mathbb{G}_m \otimes \bar{k}, \mu_2),$$

whose inverse is induced by pullback along the projection of $(\mathbb{G}_m \times \mathbb{E})\otimes \bar{k}$ onto $(\mathbb{G}_m)\otimes \bar{k}$. Therefore $\det(\mathcal{G})$, viewed on $(\mathbb{G}_m \times \mathbb{E})\otimes \bar{k}$, is either trivial or it is $\mathcal{L}_{\chi_2(t)}$, and we can tell which by specializing H to be any particular odd polynomial of degree $\le$ n. We choose H = 0 for this purpose.

So now we are reduced to computing the determinant on $(\mathbb{G}_m)\otimes \bar{k}$ of the lisse sheaf

$$t \mapsto H^1{}_c(\mathbb{G}_m \otimes \bar{k}, \mathcal{L}_{\psi(tf(x))} \otimes \mathcal{L}_{\chi_2(x)}).$$

We know that its determinant is either trivial or is $\mathcal{L}_{\chi_2(t)}$. Since d is odd, the Kummer pullback $[d]^*(\mathcal{G} \mid \mathbb{G}_m)$ has the same determinant. This pullback sheaf is

$$t \mapsto H^1{}_c(\mathbb{G}_m \otimes \bar{k}, \mathcal{L}_{\psi(t^d f(x))} \otimes \mathcal{L}_{\chi_2(x)}).$$

On the fibre $\mathbb{G}_m$, we perform the automorphism $x \mapsto x/t$; this allows us to rewrite $[d]^*(\mathcal{G} \mid \mathbb{G}_m)$ as

$$t \mapsto H^1{}_c(\mathbb{G}_m \otimes \bar{k}, \mathcal{L}_{\psi(t^d f(x/t))} \otimes \mathcal{L}_{\chi_2(x/t)}) \approx$$

$$\approx \mathcal{L}_{\chi_2(t)} \otimes H^1{}_c(\mathbb{G}_m \otimes \bar{k}, \mathcal{L}_{\psi(t^d f(x/t))} \otimes \mathcal{L}_{\chi_2(x)}).$$

In other words, $[d]^*(\mathcal{G} \mid \mathbb{G}_m)$ is $\mathcal{L}_{\chi_2} \otimes \mathcal{H}$, where $\mathcal{H}$ is the lisse, orthogonally self-dual sheaf of rank d on $\mathbb{G}_m$

$$t \mapsto H^1{}_c(\mathbb{G}_m \otimes \bar{k}, \mathcal{L}_{\psi(t^d f(x/t))} \otimes \mathcal{L}_{\chi_2(x)}).$$

Because d is odd, $\det([d]^*(\mathcal{G} \mid \mathbb{G}_m)) = \det(\mathcal{L}_{\chi_2} \otimes \mathcal{H}) \approx \mathcal{L}_{\chi_2} \otimes \det\mathcal{H}$, so it suffices to show that $\det\mathcal{H}$ is geometrically constant.

Now write $f(x) = \Sigma a_i x^i$; then

$$t^d f(x/t) = \Sigma a_i t^{d-i} x^i = a_d x^d + (\text{terms of } x\text{-degree} < d \text{ in } k[t,x]).$$

By Deligne's semicontinuity theorem, the sheaf $\mathcal{H}$ extends to a lisse sheaf on $\mathbb{A}^1$, which is still orthogonally self-dual. Therefore $\det\mathcal{H}$ is lisse on $\mathbb{A}^1\otimes\bar{k}$ of order dividing two, and hence $\det\mathcal{H}$ is geometrically constant. This concludes the proof that $\mathcal{G}$ has $G_{geom} = O(N)$ if $d > n$, $p > 2N+1$, p does not divide $2nN_1(|n-d|)N_2(|n-d|)$, and either $N \notin \{7,8\}$ or $|n-d| \neq 6$.

We can then apply Deligne's general result to $\mathcal{G}(1/2)$ with Ggeom = O(N).

CHAPTER 8
ℓ-adic Hypergeometrics

8.1 Rapid Review of Perversity, Fourier Transform, and Convolution

(8.1.1) Let k be a perfect field of characteristic $p \neq \ell$. For variable separated k-schemes of finite type X/k, we can speak of $D^b{}_c(X, \overline{\mathbb{Q}}_\ell)$. For morphisms $f: X \to Y$ between separated k-schemes of finite type, one knows (cf. [De-WII] for the case when k is either algebraically closed or finite, [Ek], [Ka-Lau], [SGA 4, XVIII, 3]) that these $D^b{}_c$ support the full Grothendieck formalism of the "six operations". In this formalism, the (relative to k) dualizing complex K_X in $D^b{}_c(X, \overline{\mathbb{Q}}_\ell)$ is defined as $\pi^!\overline{\mathbb{Q}}_\ell$, where π denotes the structural morphism $\pi: X \to \mathrm{Spec}(k)$. In terms of K_X, the Verdier dual D(L) of an object L of $D^b{}_c(X, \overline{\mathbb{Q}}_\ell)$ is defined as $R\mathcal{H}om(L, K_X)$. One knows that $L \approx \mathbf{DD}(L)$ by the natural map. The duality theorem asserts that for $f : X \to Y$ a morphism of finite type between separated k-schemes of finite type, one has $\mathbf{D}(Rf_!L) \approx Rf_*\mathbf{D}(L)$, $\mathbf{D}(Rf_*L) \approx Rf_!\mathbf{D}(L)$. If X/k is a smooth separated k-scheme of finite type and everywhere of the same relative dimension, noted dimX, then K_X is $\overline{\mathbb{Q}}_\ell[2\dim X](\mathbf{dimX})$, and so $\mathbf{D}(L)$ is $R\mathcal{H}om(L, \overline{\mathbb{Q}}_\ell)[2\dim X](\mathbf{dimX})$.

(8.1.2) Given two separated k-schemes X/k and Y/k of finite type, "external tensor product over $\overline{\mathbb{Q}}_\ell$" defines a bi-exact bilinear pairing,

$$D^b{}_c(X, \overline{\mathbb{Q}}_\ell)\times D^b{}_c(Y, \overline{\mathbb{Q}}_\ell) \to D^b{}_c(X\times_k Y, \overline{\mathbb{Q}}_\ell)$$

$$(K, L) \mapsto K\times L := pr_1{}^*K\otimes pr_2{}^*L.$$

One knows that $\mathbf{D}(K\times L) = \mathbf{D}(K)\times\mathbf{D}(L)$.

An object K of $D^b{}_c(X, \overline{\mathbb{Q}}_\ell)$ is called **semiperverse** if its cohomology sheaves $\mathcal{H}^iK$ satisfy

$$\dim \mathrm{Supp}(\mathcal{H}^iK) \leq -i.$$

An object K of $D^b{}_c(X, \overline{\mathbb{Q}}_\ell)$ is called **perverse** if both K and its dual $\mathbf{D}(K)$ are semiperverse. If $f : X \to Y$ is an **affine** (respectively a **quasifinite**) morphism, then Rf_* (respectively $f_! = Rf_!$) preserves semiperversity. So if f is both affine and quasifinite (e.g., finite, or an affine immersion), then by duality both $f_! = Rf_!$ and Rf_* preserve perversity. If $f : X \to Y$ is a smooth morphism everywhere of relative dimension d, then $f^*[d]$ preserves perversity. In particular, if K is perverse on X, then its inverse image on $X\otimes_k\overline{k}$ is perverse on $X\otimes_k\overline{k}$. One knows that the full subcategory Perv(X) of $D^b{}_c(X, \overline{\mathbb{Q}}_\ell)$ consisting of perverse objects is an **abelian** category in which every object is of finite length. The objects of Perv(X) are sometimes called "perverse sheaves" on X. However, we will call them "perverse objects" to avoid confusion with "honest" sheaves.

(8.1.3) If X is smooth over k, everywhere of relative dimension dimX, the simplest example of a perverse object on X is provided by starting with a lisse sheaf $\mathcal{F}$ on X, and taking the object $\mathcal{F}[\dim X]$ of $D^b{}_c(X, \overline{\mathbb{Q}}_\ell)$ obtained by placing $\mathcal{F}$ in degree -dimX. The object $\mathcal{F}[\dim X]$ is trivially semiperverse, and its dual $D(\mathcal{F}[\dim X]) = (\mathcal{F}^\vee(\mathbf{dimX}))[\dim X]$, being of the same form, is also. If X is connected, and if $\mathcal{F}$ is irreducible as a lisse sheaf, i.e., as a representation of $\pi_1(X, x)$, then $\mathcal{F}[\dim X]$ is a simple object of Perv(X).

(8.1.4) Given a locally closed subscheme Y of X such that Y is affine, the inclusion $j: Y \to X$ is both affine and quasifinite (factor it as the open immersion of Y into its closure $\overline{Y}$, followed by the closed immersion of $\overline{Y}$ into X). So for a perverse object K on Y, both $j_!K$ and Rj_*K are perverse on X, and as functors from Perv(Y) to Perv(X) both $j_!$ and Rj_* are exact. There is a natural "forget supports" map from $j_!K$ to Rj_*K, and as Perv(X) is an abelian category it makes sense to form

$$j_{!*}(K) := \mathrm{Image}(j_!K \to Rj_*K) \in \mathrm{Perv}(X),$$

called the "middle extension" from Y to X of the perverse object K. The functor $j_{!*}$ is an exact functor from Perv(Y) to Perv(X), it carries simple objects to simple objects, and it commutes with duality. [The middle extension functor $j_{!*}$ can be defined for any open immersion, not just an affine one, but we will not have need of that more general case here.]

(8.1.5) One knows that for any simple object S of Perv(X) there exists an affine locally closed subscheme $j: Y \to X$ such that Y is smooth over k and irreducible, and an irreducible lisse sheaf $\mathcal{F}$ on Y such that S is $j_{!*}(\mathcal{F}[\dim Y])$. Given the simple object S, we construct Y and $\mathcal{F}$ as follows: the closure $\overline{Y}$ of Y is precisely the closure of the support of $\oplus_i \mathcal{H}^i S$, Y is any smooth affine open set of $\overline{Y}$ on which all the $\mathcal{H}^i S$ are lisse, and $\mathcal{F}$ is $\mathcal{H}^{-\dim Y}(S)|Y$.

An object S of Perv(X) is called geometrically simple if its inverse image on $X\otimes_k\overline{k}$ is simple. Of course "geometrically simple" $\Rightarrow$ "simple".

(8.1.6) Consider the special case when X/k is a smooth,

geometrically connected curve. Then an object K of $D^b{}_c(X, \overline{\mathbb{Q}}_\ell)$ is perverse if and only if

$\mathcal{H}^i K = 0$ for $i \neq -1, 0$,

$\mathcal{H}^{-1}K$ has no nonzero punctual sections,

$\mathcal{H}^0 K$ is punctual.

We call a perverse object K "nonpunctual" if $\mathcal{H}^0 K = 0$. If $\mathcal{F}$ is a lisse sheaf on an open nonempty open set $j: U \to X$, then the middle extension $j_{!*}(\mathcal{F}[1])$ is none other than $(j_*\mathcal{F})[1]$. It is for this reason that we adapted the terminology "middle extension" for sheaves of the type $j_*\mathcal{F}$ with $\mathcal{F}$ lisse on U. The dual $\mathbf{D}(j_{!*}(\mathcal{F}[1]))$ of such a middle extension is related to the naive dual $D(j_*\mathcal{F}) := j_*(\mathcal{F}^\vee)$ defined in 7.3.1 by

$$\mathbf{D}(j_{!*}(\mathcal{F}[1])) = j_{!*}(\mathbf{D}(\mathcal{F}[1])) = j_*(\mathcal{F}^\vee(\mathbf{1}))[1] = D(j_*\mathcal{F})(1)[1].$$

There are two types of simple perverse object on X:

(1) the punctual ones, whose Y is a single closed point x of X; the corresponding simple objects are $x_*\mathcal{F}$, where $\mathcal{F}$ is an irreducible representation of $\mathrm{Gal}(\overline{k}/k(x))$ [so if k is algebraically closed, only the delta sheaf $\delta_x := x_*\overline{\mathbb{Q}}_\ell$ supported at x].

(2) the nonpunctual ones, whose Y is a nonempty open set $j: U \to X$ of X; the corresponding simple objects are $(j_*\mathcal{F})[1]$, where $\mathcal{F}$ is an "arithmetically irreducible" lisse sheaf on U, i.e., one whose representation of $\pi_1(U, \overline{u})$ is irreducible [so the nonpunctual simples which are geometrically simple are precisely the $\mathcal{F}[1]$ where $\mathcal{F}$ is an "irreducible middle extension sheaf" in the terminology of 7.3.1].

(8.1.7) Consider now the particular case when X/k is $\mathbb{A}^1/k$. The derived category versions of Fourier Transform are defined by

$$FT_{\psi,!}(K) := R(pr_2)_!(pr_1{}^*K \otimes \mathcal{L}_{\psi(xy)})[1],$$

$$FT_{\psi,*}(K) := R(pr_2)_*(pr_1{}^*K \otimes \mathcal{L}_{\psi(xy)})[1].$$

Both are exact functors from $D^b{}_c(\mathbb{A}^1, \overline{\mathbb{Q}}_\ell)$ to itself, which are essentially interchanged by duality:

$$\mathbf{D}(FT_{\psi,!}K) = FT_{\psi,*}([-1]^* \cdot \mathbf{D}K)(\mathbf{1}).$$

It is easy to prove that $FT_{\psi,!}$ is essentially involutive:

$$FT_{\psi,!} \cdot FT_{\psi,!} \approx [-1]^*(-\mathbf{1});$$

by duality it follows that the same holds for $FT_{\psi,*}$.

The "miracle" of Fourier Transform is that there is really only one: the natural "forget supports" map $FT_{\psi,!} \to FT_{\psi,*}$ is an isomorphism. We denote it FT_ψ. As FT_ψ (viewed as $FT_{\psi,*}$) preserves semiperversity, it follows from the miracle that FT_ψ preserves perversity, and so defines an exact autoequivalence of $\mathrm{Perv}(\mathbb{A}^1)$. In particular, FT_ψ sends perverse simple objects to perverse simple objects.

The elementary sheaves $\mathcal{F}$ of 7.3.4 are precisely those for which both $K := \mathcal{F}[1]$ and $FT_\psi(K)$ are perverse and nonpunctual. For $\mathcal{F}$ elementary, we have

$$FT_\psi(\mathcal{F}[1]) = NFT_\psi(\mathcal{F})[1]$$

The Fourier sheaves $\mathcal{F}$ are those for which both $K := \mathcal{F}[1]$ and $FT_\psi(K)$ are perverse and are the middle extensions of their restrictions to all nonempty open sets. The irreducible Fourier sheaves $\mathcal{F}$ are those for which $K := \mathcal{F}[1]$ is a geometrically simple perverse object such that neither K nor $FT_\psi(K)$ is punctual.

(8.1.8) Suppose G is a smooth separated k-groupscheme of finite type of relative dimension noted dimG, $\pi: G\times_k G \to G$ the multiplication map, $e: \mathrm{Spec}(k) \to G$ the identity section. Given two objects K and L in $D^b_c(G, \overline{\mathbb{Q}}_\ell)$, we define their "compact" or "!" convolution, denoted $K *_! L$, by

$$K *_! L := R\pi_!(K\times L) \in D^b_c(G, \overline{\mathbb{Q}}_\ell).$$

We define their "$*$" convolution, denoted $K *_* L$, by

$$K *_* L := R\pi_*(K\times L) \in D^b_c(G, \overline{\mathbb{Q}}_\ell).$$

Duality interchanges the two sorts of convolution:

$$D(K *_! L) \approx D(K) *_* D(L),\ D(K *_* L) \approx D(K) *_! D(L).$$

By the Leray spectral sequence and the Kunneth formula, we have $\mathrm{Gal}(\overline{k}/k)$-equivariant isomorphisms of cohomology algebras

$$H_c{}^*(G\otimes \overline{k}, K *_! L) \approx H_c{}^*((G\times G)\otimes \overline{k}, K\times L) \approx H_c{}^*(G\otimes \overline{k}, K)\otimes H_c{}^*(G\otimes \overline{k}, L),$$

$$H^*(G\otimes \overline{k}, K *_* L) \approx H^*((G\times G)\otimes \overline{k}, K\times L) \approx H^*(G\otimes \overline{k}, K)\otimes H^*(G\otimes \overline{k}, L).$$

In general, even if we start with two constructible ℓ-adic sheaves $\mathcal{F}$ and $\mathcal{G}$ on G, and view them as objects of of $D^b_c(G, \overline{\mathbb{Q}}_\ell)$ which are concentrated in degree zero, their convolutions $\mathcal{F} *_! \mathcal{G}$ and $\mathcal{F} *_* \mathcal{G}$ are "really" objects of $D^b{}_c(G, \overline{\mathbb{Q}}_\ell)$, and **not** simply single sheaves placed in some degree. It is this "instability" of sheaves themselves under convolution that makes $D^b{}_c(G, \overline{\mathbb{Q}}_\ell)$ the natural setting for systematically discussing convolution.

(8.1.9) If K and L are semiperverse (resp. perverse) objects on G, then $K \times L$ is semiperverse (resp. perverse) on $G \times_k G$. Therefore if G is affine, and if K and L are both semiperverse on G, then $K *_* L$ is semiperverse on G. If K and L are both perverse on G and if moreover the natural "forget supports" map is an isomorphism $K *_! L \approx K *_* L$, then $K *_! L \approx K *_* L$ is perverse (its dual being $D(K) *_* D(L)$).

(8.1.10) The formal properties of the two sorts of convolution are easily established (cf. the analogous $\mathcal{D}$-module discussion in 5.1.8-9).

(1)Each sort of convolution is associative, and for each the δ-sheaf

$$\delta_e := e_* \overline{\mathbb{Q}}_\ell$$

supported at the identity of G is a two-sided identity object. If G is commutative, then each sort of convolution is commutative as well.

(2a) If φ: $G \to H$ is a homomorphism of smooth separated k-groupschemes of finite type, then for K and L on G we have

$$R\varphi_*(K *_* L) \approx (R\varphi_* K) *_* (R\varphi_* L),$$

$$R\varphi_!(K *_! L) \approx (R\varphi_! K) *_! (R\varphi_! L).$$

(2b) If $\varphi : G \to H$ is a homomorphism, then for K on G and L on H we have

$$\varphi^*((R\varphi_! K) *_! L) \approx K *_! (\varphi^* L),$$

$$\varphi^!((R\varphi_* K) *_* L) \approx K *_* (\varphi^! L).$$

These two relations are duals of each other. The first is proper base change for the following commutative diagram, whose outer square is cartesian (verification left to the reader):

$$\begin{array}{ccc} G\times G & \xrightarrow{\quad id\times\varphi \quad} & G\times H \\ & & \downarrow \varphi\times id \\ \text{mult} \downarrow & & H\times H \\ & & \downarrow \text{mult} \\ G & \xrightarrow{\quad \varphi \quad} & H \end{array}$$

(3) For $g \in G(k)$ denote by $T_g : G \to G$ the map $x \mapsto gx$ "left translation by g", and by $\delta_g := (T_g)_*(\delta_e)$ the delta sheaf supported at g. Then for $g \in G(k)$, we have

$$(T_g)_* = R(T_g)_* = (T_g)_! = R(T_g)_!$$

$$(T_g)_*(K*_*L) \approx ((T_g)_*K)*_*L,$$
$$(T_g)_*(K*_!L) \approx ((T_g)_*K)*_!L,$$
$$(T_g)_*(L) \approx (\delta_g)*L.$$

Moreover, if G is commutative, then for g, h in G(k), we have

$$(T_{gh})_*(K*_*L) \approx ((T_g)_*K)*_*((T_h)_*L),$$
$$(T_{gh})_*(K*_!L) \approx ((T_g)_*K)*_!((T_h)_*L).$$

(4) If G is commutative, geometrically connected, and defined over a finite subfield k_0 of k, then for every $\overline{\mathbb{Q}}_\ell$-valued character χ of $G(k_0)$, the associated lisse rank one $\mathcal{L}_\chi$ on G obtained from pushing out the Lang torsor by χ satisfies $\pi^*\mathcal{L}_\chi \approx \mathcal{L}_\chi \times \mathcal{L}_\chi$, whence by the projection formula

$$(K*_!L)\otimes\mathcal{L}_\chi \approx (K\otimes\mathcal{L}_\chi)*_!(L\otimes\mathcal{L}_\chi),$$
$$(K*_*L)\otimes\mathcal{L}_\chi \approx (K\otimes\mathcal{L}_\chi)*_*(L\otimes\mathcal{L}_\chi).$$

(8.1.11) We now recall (cf. [Ka-GKM, 8.6.1]) the relation between Fourier Transform on $\mathbb{A}^1$ and convolution on $\mathbb{G}_m$. Denote by

$j: \mathbb{G}_m \to \mathbb{A}^1$ the inclusion,

$\mathrm{inv}: \mathbb{G}_m \to \mathbb{G}_m$ the multiplicative inversion $x \mapsto x^{-1}$.

Proposition 8.1.12 (compare 5.2.3) For any object K in $D^b{}_c(\mathbb{G}_m, \overline{\mathbb{Q}}_\ell)$, we have canonical isomorphisms in $D^b{}_c(\mathbb{G}_m, \overline{\mathbb{Q}}_\ell)$:

$$(j^*\mathcal{L}_\psi)[1]*_!K \approx j^*FT_\psi(j_!\mathrm{inv}^*K),$$
$$(j^*\mathcal{L}_\psi)[1]*_*K \approx j^*FT_\psi(Rj_*\mathrm{inv}^*K),$$
$$(j^*\mathcal{L}_\psi)[1]*_!\mathrm{inv}^*K \approx j^*FT_\psi(j_!K),$$
$$(j^*\mathcal{L}_\psi)[1]*_*\mathrm{inv}^*K \approx j^*FT_\psi(Rj_*K),$$
$$(\mathrm{inv}^*j^*\mathcal{L}_\psi)[1]*_!K \approx \mathrm{inv}^*j^*FT_\psi(j_!K),$$
$$(\mathrm{inv}^*j^*\mathcal{L}_\psi)[1]*_*K \approx \mathrm{inv}^*j^*FT_\psi(Rj_*K),$$

proof The first is a formal consequence of the definitions of $*_!$ and of $FT_{\psi,!}$ (cf. 5.2.3).The second is the dual of the first, the third and fourth are the first two applied to inv^*K, and the last two are obtained from the third and fourth by applying $\mathrm{inv}^* = \mathrm{inv}_*$. QED

8.2 Definition of hypergeometric complexes and hypergeometric sums over finite fields

(8.2.1) We work over a finite field k of characteristic $p \neq \ell$. We denote by ψ a nontrivial $\overline{\mathbb{Q}}_\ell$-valued additive character of k. Let (n, m) be a pair of nonnegative integers. Let

$$(\chi\text{'s}) := (\chi_1, \dots, \chi_n)$$

be an (unordered) n-tuple of not necessarily distinct $\overline{\mathbb{Q}}_\ell$-valued multiplicative characters of $k^\times$, and let

$$(\rho\text{'s}) := (\rho_1, \dots, \rho_m)$$

be an (unordered) m-tuple of not necessarily distinct $\overline{\mathbb{Q}}_\ell$-valued multiplicative characters of $k^\times$.

(8.2.2) Given any such data, we define an object

$$\mathrm{Hyp}(!, \psi; \chi\text{'s}; \rho\text{'s}) = \mathrm{Hyp}(!, \psi; \chi_1, \dots, \chi_n; \rho_1, \dots, \rho_m)$$

in $D^b{}_c(\mathbb{G}_m, \overline{\mathbb{Q}}_\ell)$ as follows:

(1) if (n, m) = (0, 0), then $\mathrm{Hyp}(!, \psi; \emptyset; \emptyset) := \delta_1 := 1_*\overline{\mathbb{Q}}_\ell$ is the delta sheaf supported at 1.

(2) if (n, m) = (1, 0), then $\mathrm{Hyp}(!, \psi; \chi; \emptyset) := (j^*\mathcal{L}_\psi)\otimes\mathcal{L}_\chi[1]$.

(3) if (n, m) = (0, 1), then $\mathrm{Hyp}(!, \psi; \emptyset; \rho) := \mathrm{inv}^*((j^*\mathcal{L}_{\overline{\psi}})\otimes\mathcal{L}_{\overline{\rho}})[1]$.

(4) if (n, m) = (n, 0) with $n \geq 2$, then $\mathrm{Hyp}(!, \psi; \chi\text{'s}; \emptyset)$ is the n-fold mutiple convolution

$$\mathrm{Hyp}(!, \psi; \chi_1; \emptyset)*_!\mathrm{Hyp}(!, \psi; \chi_2; \emptyset)*_! \dots *_!\mathrm{Hyp}(!, \psi; \chi_n; \emptyset).$$

(5) if (n, m) = (0, m) with $m \geq 2$, then $\mathrm{Hyp}(!, \psi; \emptyset; \rho\text{'s})$ is the m-fold multiple convolution

$$\mathrm{Hyp}(!, \psi; \emptyset; \rho_1)*_!\mathrm{Hyp}(!, \psi; \emptyset; \rho_2)*_! \dots *_!\mathrm{Hyp}(!, \psi; \emptyset; \rho_m).$$

(6) in the general case, $\mathrm{Hyp}(!, \psi; \chi\text{'s}; \rho\text{'s})$ is defined to be

$$\mathrm{Hyp}(!, \psi; \chi\text{'s}; \emptyset)*_!\mathrm{Hyp}(!, \psi; \emptyset; \rho\text{'s}).$$

(8.2.3) Since ! convolution is associative and commutative, we have the general convolution formula

$$\mathrm{Hyp}(!, \psi; \chi\text{'s}; \rho\text{'s})*_!\mathrm{Hyp}(!, \psi; \Lambda\text{'s}; \Gamma\text{'s}) = \mathrm{Hyp}(!, \psi; \chi\text{'s} \cup \Lambda\text{'s}; \rho\text{'s} \cup \Gamma\text{'s})$$

for these objects. [This situation should be contrasted with the $\mathcal{D}$-module case, where we had an a priori definition of hypergeometric $\mathcal{D}$-modules "just" by writing down the corresponding DE, but where the convolution behaviour was a theorem. Here we lack a "simple" a priori definition of hypergeometrics, and we are essentially imposing their convolution behaviour as the definition.]

(8.2.4) Their behaviour under inversion is given by

$$\mathrm{inv}^*\mathrm{Hyp}(!, \psi; \chi\text{'s}; \rho\text{'s}) = \mathrm{Hyp}(!, \overline{\psi}; \overline{\rho}\text{'s}; \overline{\chi}\text{'s}).$$

(8.2.5) Tensoring with a Kummer sheaf $\mathcal{L}_\Lambda$ is also extremely simple:

$$\mathcal{L}_\Lambda \otimes \mathrm{Hyp}(!, \psi; \chi\text{'s}; \rho\text{'s}) = \mathrm{Hyp}(!, \psi; \Lambda\chi\text{'s}; \Lambda\rho\text{'s}).$$

(8.2.6) For E a finite extension of k, the pullback of Hyp(!, ψ; χ's; ρ's) to $\mathbb{G}_m \otimes_k E$ is Hyp(!, ψ_E; χ_E's; ρ_E's), where ψ_E (resp. χ_E, ρ_E) is the additive (resp. multiplicative) character of E (resp. $E^\times$) obtained from ψ (resp. χ, ρ) by composition with $\mathrm{Trace}_{E/k}$ (resp. $\mathrm{Norm}_{E/k}$). [Indeed the corresponding pullback of $\mathcal{L}_\psi$ (resp. $\mathcal{L}_\chi$, $\mathcal{L}_\rho$) is $\mathcal{L}_{\psi_E}$ (resp. $\mathcal{L}_{\chi_E}$, $\mathcal{L}_{\rho_E}$), cf [Ka-GKM, 4.3], and ! convolution commutes with arbitrary base change.]

(8.2.7) The trace function of Hyp(!, ψ; χ's; ρ's) is easily computed in terms of "hypergeometric sums", using the Lefschetz Trace Formula. For $(n, m) \neq (0, 0)$, the result is this. For each finite extension E of k, and each $t \in E^\times$, denote by

$$V(n, m; t) \subset (\mathbb{G}_m)^{n+m}$$

the hypersurface in $(\mathbb{G}_m)^{n+m}$, with coordinates $x_1, \ldots, x_n, y_1, \ldots, y_m$, defined by the equation

$$\Pi_i x_i = t(\Pi_j y_j).$$

Define the "hypergeometric sum" Hyp(ψ; χ's; ρ's)(E, t) $\in \mathbb{Q}(\psi, \chi\text{'s}, \rho\text{'s})$ to be the exponential sum

$$\mathrm{Hyp}(\psi; \chi\text{'s}; \rho\text{'s})(E, t) :=$$
$$\Sigma_{V(n, m; t)(E)}\, \psi_E(\Sigma_i x_i - \Sigma_j y_j)(\Pi_i \chi_{i,E}(x_i))(\Pi_j \overline{\rho}_{j,E}(y_j)).$$

Then

$$\Sigma(-1)^i \mathrm{Trace}(\mathrm{Frob}_{E,t} \mid \mathcal{H}^i(\mathrm{Hyp}(!; \psi; \chi\text{'s}; \rho\text{'s})))$$
$$= (-1)^{n+m}\mathrm{Hyp}(\psi; \chi\text{'s}; \rho\text{'s})(E, t).$$

(8.2.8) These hypergeometric sums, which include Kloosterman sums as the special case n=0 or m=0, when viewed as functions on $E^\times$, are related by **multiplicative** Fourier Transform to monomials in Gauss sums viewed as functions on the Pontrjagin dual of $E^\times$. The precise relation is this. For each finite extension E of k, and each multiplicative character Λ of $E^\times$, we have (by elementary calculation)

$\Sigma_{t \text{ in } E^\times} \Lambda(t)\mathrm{Hyp}(\psi;\ \chi\text{'s};\ \rho\text{'s})(E, t) = (\prod_i g(\psi_E, \Lambda\chi_{i,E}))(\prod_j g(\overline{\psi}_E, \overline{\Lambda}\overline{\rho}_{j,E}))$.

By multiplicative Fourier inversion, this gives ($q := \mathrm{Card}(E)$)

$$\mathrm{Hyp}(\psi;\ \chi\text{'s};\ \rho\text{'s})(E, t) = = (1/(q-1))\Sigma_{\Lambda \text{ on } E^\times} \overline{\Lambda}(t)(\prod_i g(\psi_E, \Lambda\chi_{i,E}))(\prod_j g(\overline{\psi}_E, \overline{\Lambda}\overline{\rho}_{j,E})).$$

The Plancherel formula gives, for any complex embedding of the field $\mathbb{Q}(\psi, \chi\text{'s}, \rho\text{'s})$,

$$\Sigma_{t \text{ in } E^\times} |\mathrm{Hyp}(\psi;\ \chi\text{'s};\ \rho\text{'s})(E, t)|^2 = = (1/(q-1))\Sigma_{\Lambda \text{ on } E^\times} |(\prod_i g(\psi_E, \Lambda\chi_{i,E}))(\prod_j g(\overline{\psi}_E, \overline{\Lambda}\overline{\rho}_{j,E}))|^2.$$

(8.2.9) From the Kunneth formula (cf. 8.1.8), and the Euler-Poincare formula for the case $n+m = 1$, we see that

$$H^i_c(\mathbb{G}_m\otimes_k\overline{k}, \mathrm{Hyp}(!, \psi;\ \chi\text{'s};\ \rho\text{'s})) = \text{1-dim'l if } i = 0$$
$$= 0 \text{ if } i \neq 0.$$

By the Lefschetz Trace Formula, it follows that for any finite extension E of k, the action of Frob_E on the one-dimensional space $H^0_c(\mathbb{G}_m\otimes_k\overline{k}, \mathrm{Hyp}(!, \psi;\ \chi\text{'s};\ \rho\text{'s}))$ is given by

(8.2.10) $$\mathrm{Trace}(\mathrm{Frob}_E \mid H^0_c(\mathbb{G}_m\otimes_k\overline{k}, \mathrm{Hyp}(!, \psi;\ \chi\text{'s};\ \rho\text{'s}))) = = (\prod_i(-g(\psi_E, \chi_{i,E})))(\prod_j(-g(\overline{\psi}_E, \overline{\rho}_{j,E}))).$$

Similarly, for any multiplicative character Λ of $k^\times$, we have

(8.2.11) $$H^i_c(\mathbb{G}_m\otimes_k\overline{k}, \mathcal{L}_\Lambda\otimes\mathrm{Hyp}(!, \psi;\ \chi\text{'s};\ \rho\text{'s})) = \text{1-dim'l if } i = 0,$$
$$= 0 \text{ if } i \neq 0,$$

and $$\mathrm{Trace}(\mathrm{Frob}_E \mid H^0_c(\mathbb{G}_m\otimes_k\overline{k}, \mathcal{L}_\Lambda\otimes\mathrm{Hyp}(!, \psi;\ \chi\text{'s};\ \rho\text{'s}))) = = (\prod_i(-g(\psi_E, \Lambda\chi_{i,E})))(\prod_j(-g(\overline{\psi}_E, \overline{\Lambda}\overline{\rho}_{j,E}))).$$

(8.2.12) In an entirely analogous way, we can use $*$ convolution to define objects $\mathrm{Hyp}(*, \psi;\ \chi\text{'s};\ \rho\text{'s})$ in $D^b_c(\mathbb{G}_m, \overline{\mathbb{Q}}_\ell)$ by replacing all occurrences of ! by $*$ in the above axioms 8.2.2 (1) through (6). There is a natural "forget supports" map

$$\mathrm{Hyp}(!, \psi;\ \chi\text{'s};\ \rho\text{'s}) \to \mathrm{Hyp}(*, \psi;\ \chi\text{'s};\ \rho\text{'s}),$$

which in general is not an isomorphism. Duality interchanges these two sorts of hypergeometrics:

$$D(\mathrm{Hyp}(!, \psi;\ \chi\text{'s};\ \rho\text{'s})) \approx \mathrm{Hyp}(*, \overline{\psi};\ \overline{\chi}\text{'s};\ \overline{\rho}\text{'s})(\mathbf{n+m}).$$

$$D(\mathrm{Hyp}(*,\ \psi;\ \chi\text{'s};\ \rho\text{'s})) \approx \mathrm{Hyp}(!,\ \overline{\psi};\ \overline{\chi}\text{'s};\ \overline{\rho}\text{'s})(\mathbf{n+m}).$$

(8.2.13) It will also be convenient to consider systematically the multiplicative translates of $\mathrm{Hyp}(!, \psi; \chi\text{'s}; \rho\text{'s})$ and of $\mathrm{Hyp}(*, \psi; \chi\text{'s}; \rho\text{'s})$. For each point $\lambda \in k^\times$, we define

$$\mathrm{Hyp}_\lambda(!,\ \psi;\ \chi\text{'s};\ \rho\text{'s}) := [x \mapsto \lambda x]_* \mathrm{Hyp}(!,\ \psi;\ \chi\text{'s};\ \rho\text{'s}),$$
$$\mathrm{Hyp}_\lambda(*,\ \psi;\ \chi\text{'s};\ \rho\text{'s}) := [x \mapsto \lambda x]_* \mathrm{Hyp}(*,\ \psi;\ \chi\text{'s};\ \rho\text{'s}).$$

(8.2.14) These objects enjoy the following basic properties:

$$\mathrm{inv}^*\mathrm{Hyp}_\lambda(!,\ \psi;\ \chi\text{'s};\ \rho\text{'s}) \approx \mathrm{Hyp}_{1/\lambda}(!,\ \overline{\psi};\ \overline{\rho}\text{'s};\ \overline{\chi}\text{'s}).$$
$$\mathrm{inv}^*\mathrm{Hyp}_\lambda(*,\ \psi;\ \chi\text{'s};\ \rho\text{'s}) \approx \mathrm{Hyp}_{1/\lambda}(*,\ \overline{\psi};\ \overline{\rho}\text{'s};\ \overline{\chi}\text{'s}).$$
$$\mathcal{L}_\Lambda \otimes \mathrm{Hyp}_\lambda(!,\ \psi;\ \chi\text{'s};\ \rho\text{'s}) \approx (\Lambda(\lambda))^{\deg} \otimes \mathrm{Hyp}_\lambda(!,\ \psi;\ \Lambda\chi\text{'s};\ \Lambda\rho\text{'s}).$$
$$\mathcal{L}_\Lambda \otimes \mathrm{Hyp}_\lambda(*,\ \psi;\ \chi\text{'s};\ \rho\text{'s}) \approx (\Lambda(\lambda))^{\deg} \otimes \mathrm{Hyp}_\lambda(*,\ \psi;\ \Lambda\chi\text{'s};\ \Lambda\rho\text{'s}).$$
$$D(\mathrm{Hyp}_\lambda(!,\ \psi;\ \chi\text{'s};\ \rho\text{'s})) \approx \mathrm{Hyp}_\lambda(*,\ \overline{\psi};\ \overline{\chi}\text{'s};\ \overline{\rho}\text{'s})(\mathbf{n+m}).$$
$$D(\mathrm{Hyp}_\lambda(*,\ \psi;\ \chi\text{'s};\ \rho\text{'s})) \approx \mathrm{Hyp}_\lambda(!,\ \overline{\psi};\ \overline{\chi}\text{'s};\ \overline{\rho}\text{'s})(\mathbf{n+m}).$$
$$\mathrm{Hyp}_\lambda(!,\ \psi;\ \chi\text{'s};\ \rho\text{'s}) *_! \mathrm{Hyp}_\mu(!,\ \psi;\ \Lambda\text{'s};\ \Gamma\text{'s}) =$$
$$= \mathrm{Hyp}_{\lambda\mu}(!,\ \psi;\ \chi\text{'s} \cup \Lambda\text{'s};\ \rho\text{'s} \cup \Gamma\text{'s}).$$
$$\mathrm{Hyp}_\lambda(*,\ \psi;\ \chi\text{'s};\ \rho\text{'s}) *_* \mathrm{Hyp}_\mu(*,\ \psi;\ \Lambda\text{'s};\ \Gamma\text{'s}) =$$
$$= \mathrm{Hyp}_{\lambda\mu}(*,\ \psi;\ \chi\text{'s} \cup \Lambda\text{'s};\ \rho\text{'s} \cup \Gamma\text{'s}).$$

8.3 Variant: Hypergeometric complexes over algebraically closed fields

(8.3.1) Suppose instead of working over a finite field we work over an algebraically closed field k. For ψ a nontrivial $\overline{\mathbb{Q}}_\ell$-valued additive character of a finite subfield k_0 of k, we can speak (cf 7.2.1) of the lisse rank one sheaf $\mathcal{L}_\psi$ on $\mathbb{A}^1 \otimes_{k_0} k$. For any tame $\overline{\mathbb{Q}}_\ell$-valued character χ of $\pi_1(\mathbb{G}_m \otimes_{k_0} k)$, with inverse character denoted $\overline{\chi}$, we can speak of the lisse, rank one $\overline{\mathbb{Q}}_\ell$-sheaves $\mathcal{L}_\chi$ and $\mathcal{L}_{\overline{\chi}}$ on $\mathbb{G}_m \otimes_{k_0} k$.

In terms of these objects, we define objects $\mathrm{Hyp}(!, \psi; \chi\text{'s}; \rho\text{'s})$ and $\mathrm{Hyp}(*, \psi; \chi\text{'s}; \rho\text{'s})$ in $D^b_c(\mathbb{G}_m \otimes_{k_0} k, \overline{\mathbb{Q}}_\ell)$ exactly as above.

(8.3.2) For each point $\lambda \in k^\times$, we define the translated objects

$$\mathrm{Hyp}_\lambda(!,\ \psi;\ \chi\text{'s};\ \rho\text{'s}) := [x \mapsto \lambda x]_* \mathrm{Hyp}(!,\ \psi;\ \chi\text{'s};\ \rho\text{'s}),$$

$$\mathrm{Hyp}_\lambda(*, \psi; \chi\text{'s}; \rho\text{'s}) := [x \mapsto \lambda x]_* \mathrm{Hyp}(*, \psi; \chi\text{'s}; \rho\text{'s}).$$

[When the χ's and ρ's are all of finite order, say all defined over k_0, and $\lambda \in (k_0)^\times$, these objects $\mathrm{Hyp}_\lambda(!, \psi; \chi\text{'s}; \rho\text{'s})$ and $\mathrm{Hyp}_\lambda(*, \psi; \chi\text{'s}; \rho\text{'s})$ are just the pullbacks to $\mathbb{G}_m \otimes_{k_0} k$ of the earlier defined objects on $\mathbb{G}_m/k_0$ with the same names.]

(8.3.3) The properties

$$\mathrm{inv}^* \mathrm{Hyp}_\lambda(!, \psi; \chi\text{'s}; \rho\text{'s}) \approx \mathrm{Hyp}_{1/\lambda}(!, \overline{\psi}; \overline{\rho}\text{'s}; \overline{\chi}\text{'s}),$$

$$\mathrm{inv}^* \mathrm{Hyp}_\lambda(*, \psi; \chi\text{'s}; \rho\text{'s}) \approx \mathrm{Hyp}_{1/\lambda}(*, \overline{\psi}; \overline{\rho}\text{'s}; \overline{\chi}\text{'s}),$$

$$\mathcal{L}_\Lambda \otimes \mathrm{Hyp}_\lambda(!, \psi; \chi\text{'s}; \rho\text{'s}) \approx \mathrm{Hyp}_\lambda(!, \psi; \Lambda\chi\text{'s}; \Lambda\rho\text{'s}),$$

$$\mathcal{L}_\Lambda \otimes \mathrm{Hyp}_\lambda(*, \psi; \chi\text{'s}; \rho\text{'s}) \approx \mathrm{Hyp}_\lambda(*, \psi; \Lambda\chi\text{'s}; \Lambda\rho\text{'s}),$$

$$D(\mathrm{Hyp}_\lambda(!, \psi; \chi\text{'s}; \rho\text{'s})) \approx \mathrm{Hyp}_\lambda(*, \overline{\psi}; \overline{\chi}\text{'s}; \overline{\rho}\text{'s})(\mathbf{n+m}),$$

$$D(\mathrm{Hyp}_\lambda(*, \psi; \chi\text{'s}; \rho\text{'s})) \approx \mathrm{Hyp}_\lambda(!, \overline{\psi}; \overline{\chi}\text{'s}; \overline{\rho}\text{'s})(\mathbf{n+m}),$$

$$\mathrm{Hyp}_\lambda(!, \psi; \chi\text{'s}; \rho\text{'s}) *_! \mathrm{Hyp}_\mu(!, \psi; \Lambda\text{'s}; \Gamma\text{'s}) = \\ = \mathrm{Hyp}_{\lambda\mu}(!, \psi; \chi\text{'s} \cup \Lambda\text{'s}; \rho\text{'s} \cup \Gamma\text{'s}),$$

$$\mathrm{Hyp}_\lambda(*, \psi; \chi\text{'s}; \rho\text{'s}) *_* \mathrm{Hyp}_\mu(*, \psi; \Lambda\text{'s}; \Gamma\text{'s}) = \\ = \mathrm{Hyp}_{\lambda\mu}(*, \psi; \chi\text{'s} \cup \Lambda\text{'s}; \rho\text{'s} \cup \Gamma\text{'s})$$

hold for these objects.

(8.3.4) From the general convolution formalism (cf. 8.1.8), and reduction to the case of hypergeometrics of type (1, 0) and (0, 1), we see that

$$H^i_c(\mathbb{G}_m, \mathrm{Hyp}_\lambda(!, \psi; \chi\text{'s}; \rho\text{'s})) = 0 \text{ for } i \neq 0,$$

$$H^0_c(\mathbb{G}_m, \mathrm{Hyp}_\lambda(!, \psi; \chi\text{'s}; \rho\text{'s})) \text{ is one-dimensional,}$$

and dually

$$H^i(\mathbb{G}_m, \mathrm{Hyp}_\lambda(*, \psi; \chi\text{'s}; \rho\text{'s})) = 0 \text{ for } i \neq 0,$$

$$H^0(\mathbb{G}_m, \mathrm{Hyp}_\lambda(*, \psi; \chi\text{'s}; \rho\text{'s})) \text{ is one-dimensional.}$$

Remark 8.3.5 This situation should be compared to the situation over $\mathbb{C}$ for the hypergeometric $\mathcal{D}$-modules $\mathcal{H}_\lambda(\alpha\text{'s}; \beta\text{'s})$. The tame characters χ's (resp. the ρ's) of $\pi_1(\mathbb{G}_m \otimes_{k_0} k)$ can be seen as playing the roles of the characters $x \mapsto \exp(2\pi i \alpha x)$ (resp. $x \mapsto \exp(2\pi i \beta x)$) of $\mathbb{Z} \approx \pi_1((\mathbb{G}_m)^{an})$. Having all the χ's and ρ's of finite order is analogous to having all the α's and b's rational. The choice of a ψ is required to define the objects $\mathrm{Hyp}(!, \psi; \chi\text{'s}; \rho\text{'s})$ and $\mathrm{Hyp}(*, \psi; \chi\text{'s}; \rho\text{'s})$ which are analogous to the $\mathcal{D}$-

module $\mathcal{H}_1(\alpha\text{'s}; \beta\text{'s})$. [One might "explain" the fact that in the $\mathcal{D}$-module case we define $\mathcal{H}_1(\alpha\text{'s}; \beta\text{'s})$ without having to make an analogous choice by the catch-phrase "there are many $\mathcal{L}_\psi$, but only one exp(x)".]

8.4 Basic Properties of Hypergeometric Complexes; Definition and Basic Properties of the Hypergeometric Sheaves $\mathcal{H}_\lambda(!, \psi; \chi\text{'s}; \rho\text{'s})$

(8.4.1) In this section we will establish the basic goemetric properties of the objects $\mathrm{Hyp}_\lambda(!, \psi; \chi\text{'s}; \rho\text{'s})$ and $\mathrm{Hyp}_\lambda(*, \psi; \chi\text{'s}; \rho\text{'s})$ of $D^b{}_c(\mathbb{G}_m, \overline{\mathbb{Q}}_\ell)$. We work over an algebraically closed field k of characteristic $p > 0$, $p \neq \ell$.

We say that $\mathrm{Hyp}_\lambda(!, \psi; \chi\text{'s}; \rho\text{'s})$ and $\mathrm{Hyp}_\lambda(*, \psi; \chi\text{'s}; \rho\text{'s})$ are defined over a finite subfield k_0 of k if $\lambda \in (k_0)^\times$, ψ is an additive character of k_0, and each of the χ's and each of the ρ's is a tame character of finite order defined over k_0 (i.e., each of the χ's and each of the ρ's has finite order dividing $\mathrm{Card}(k_0) - 1$).

We say that the χ's and the ρ's are **identical** if $n = m$ and if after possible renumbering we have $\chi_i = \rho_i$ for every $i = 1, \ldots, n$.

We say that the χ's and the ρ's are **disjoint** if $(n, m) \neq (0, 0)$ and if we have $\chi_i \neq \rho_j$ for any $i = 1, \ldots, n$ and for any $j = 1, \ldots, m$. [Thus if either $n = 0$ or if $m = 0$, but $n+m \neq 0$, then disjointness holds automatically.]

We denote by $\mathrm{mult}_0(\chi)$ (resp. $\mathrm{mult}_\infty(\rho)$) the multiplicity with which a particular character χ (resp. ρ) occurs among the χ's (resp. among the ρ's). In discussing ℓ-adic representations of inertia groups, for any integer $n \geq 1$, we denote by Unip(n) a unipotent Jordan block of size n (i.e., an indecomposable unipotent [and hence tame] n-dimensional ℓ-adic representation of the inertia group in question).

Theorem 8.4.2 Suppose that the χ's and ρ's are disjoint. Then
(1) $\mathrm{Hyp}_\lambda(!, \psi; \chi\text{'s}; \rho\text{'s})$ is simple perverse and nonpunctual, i.e., there exists an irreducible middle extension sheaf $\mathcal{H}_\lambda(!, \psi; \chi\text{'s}; \rho\text{'s})$ on $\mathbb{G}_m$ such that $\mathrm{Hyp}_\lambda(!, \psi; \chi\text{'s}; \rho\text{'s}) = \mathcal{H}_\lambda(!, \psi; \chi\text{'s}; \rho\text{'s})[1]$. The Euler characteristic $\chi(\mathbb{G}_m, \mathcal{H}_\lambda(!, \psi; \chi\text{'s}; \rho\text{'s})) = -1$.
(2) Denote by $j: \mathbb{G}_m \to \mathbb{A}^1$ the inclusion. Then $j_*\mathcal{H}_\lambda(!, \psi; \chi\text{'s}; \rho\text{'s})$ is an

irreducible Fourier sheaf on $\mathbb{A}^1$ except if (n, m) = (1, 0) and χ is trivial (in which case the sheaf $j_*\mathcal{H}_\lambda(!, \psi; \chi\text{'s}; \rho\text{'s})$ in question is $\mathcal{L}_{\psi(x/\lambda)}$).

(3) If no χ is trivial, then

$$j_!\mathcal{H}_\lambda(!, \psi; \chi\text{'s}; \rho\text{'s}) \approx j_*\mathcal{H}_\lambda(!, \psi; \chi\text{'s}; \rho\text{'s}) \approx Rj_*\mathcal{H}_\lambda(!, \psi; \chi\text{'s}; \rho\text{'s}),$$

whence $j_!\mathcal{H}_\lambda(!, \psi; \chi\text{'s}; \rho\text{'s})$ is an irreducible Fourier sheaf on $\mathbb{A}^1$.

(4) If $\mathrm{Hyp}_\lambda(!, \psi; \chi\text{'s}; \rho\text{'s})$ is defined over a finite subfield k_0 of k, then $\mathcal{H}_\lambda(!, \psi; \chi\text{'s}; \rho\text{'s})$ is pure of weight n + m - 1.

(5) The natural map $\mathrm{Hyp}_\lambda(!, \psi; \chi\text{'s}; \rho\text{'s}) \to \mathrm{Hyp}_\lambda(*, \psi; \chi\text{'s}; \rho\text{'s})$ is an isomorphism.

(6) If n > m, the sheaf $\mathcal{H}_\lambda(!, \psi; \chi\text{'s}; \rho\text{'s})$ is lisse of rank n on $\mathbb{G}_m$. As I(0)-representation it is tame, isomorphic to

$$\oplus_{\text{distinct } \chi\text{'s}} \mathcal{L}_\chi \otimes \mathrm{Unip}(\mathrm{mult}_0(\chi)).$$

As I(∞)-representation it has Swan conductor =1, and is isomorphic to the direct sum

$$(\text{dim. } n-m, \text{ brk. } 1/(n-m)) \bigoplus \oplus_{\text{distinct } \rho\text{'s}} \mathcal{L}_\rho \otimes \mathrm{Unip}(\mathrm{mult}_\infty(\rho)).$$

(7) If n < m, the sheaf $\mathcal{H}_\lambda(!, \psi; \chi\text{'s}; \rho\text{'s})$ is lisse of rank m on $\mathbb{G}_m$. As I(0)-representation it has Swan conductor =1, and is isomorphic to the direct sum

$$(\text{dim. } m-n, \text{ brk. } 1/(m-n)) \bigoplus \oplus_{\text{distinct } \chi\text{'s}} \mathcal{L}_\chi \otimes \mathrm{Unip}(\mathrm{mult}_0(\chi)).$$

As I(∞)-representation it is tame, isomorphic to

$$\oplus_{\text{distinct } \rho\text{'s}} \mathcal{L}_\rho \otimes \mathrm{Unip}(\mathrm{mult}_\infty(\rho)).$$

(8) If n = m, the sheaf $\mathcal{H}_\lambda(!, \psi; \chi\text{'s}; \rho\text{'s})$ is lisse of rank n on $\mathbb{G}_m - \{\lambda\}$, from which it is extended by direct image. I(λ) acts by tame pseudoreflections of determinant $\mathcal{L}_{\Lambda(x-\lambda)}$, for $\Lambda := \prod_i \rho_i / \prod_i \chi_i$.

As I(0)-representation it is tame, isomorphic to

$$\oplus_{\text{distinct } \chi\text{'s}} \mathcal{L}_\chi \otimes \mathrm{Unip}(\mathrm{mult}_0(\chi)).$$

As I(∞)-representation it is tame, isomorphic to

$$\oplus_{\text{distinct } \rho\text{'s}} \mathcal{L}_\rho \otimes \mathrm{Unip}(\mathrm{mult}_\infty(\rho)).$$

proof We proceed by induction on n+m. If n+m = 1, the theorem is obvious by inspection. Suppose the theorem has already been proven universally for all (n_0, m_0) with $n_0 + m_0 < n+m$; we must prove it universally for (n, m). Notice that assertions (2) and (3) follow from assertions (1), (6), (7) and (8).

We are thus "reduced" to proving assertions (1) and (4) - (8)..By multiplicative translation, we may assume $\lambda = 1$. By multiplicative inversion, we may suppose that $n \le m$. For any tame character Λ, (1) and (5) - (8) hold for $\mathrm{Hyp}_1(!, \psi; \chi\text{'s}; \rho\text{'s})$ if and only if they hold for $\mathcal{L}_\Lambda \otimes \mathrm{Hyp}_1(!, \psi; \chi\text{'s}; \rho\text{'s}) \approx \mathrm{Hyp}_1(!, \psi; \Lambda\chi\text{'s}; \Lambda\rho\text{'s})$. For (4), we have this same equivalence for any tame character Λ of finite order.

Suppose first that $0 < n \le m$. Then by picking Λ to be the inverse of one of the χ's we may assume that one of the χ's is trivial, say $\chi_n = \mathbb{1}$. To emphasize this, we will write our object of type (n, m) in the form

$$\mathrm{Hyp}_1(!, \psi; \mathbb{1}, \chi\text{'s}; \rho\text{'s}),$$

where now there are $n-1$ listed χ's in addition to $\mathbb{1}$.

Since $\{\mathbb{1}$, the χ's$\}$ and the ρ's are disjoint by hypothesis, none of the ρ's is trivial. Now apply the general formula relating convolution with Fourier Transform

$$(j^*\mathcal{L}_\psi)[1] *_! K \approx j^*\mathrm{FT}_\psi(j_!\mathrm{inv}^*K)$$

to the hypergeometric object K of type $(n-1, m)$

$$K := \mathrm{Hyp}_1(!, \psi; \chi\text{'s}; \rho\text{'s}).$$

We find

$$\mathrm{Hyp}_1(!, \psi; \mathbb{1}, \chi\text{'s}; \rho\text{'s}) \approx j^*\mathrm{FT}_\psi(j_!\mathrm{inv}^*K).$$
$$\approx j^*\mathrm{FT}_\psi(j_!\mathrm{inv}^*\mathrm{Hyp}_1(!, \psi; \chi\text{'s}; \rho\text{'s}))$$
$$\approx j^*\mathrm{FT}_\psi(j_!\mathrm{Hyp}_1(!, \overline{\psi}; \overline{\rho}\text{'s}; \overline{\chi}\text{'s})).$$

Since none of the $\overline{\rho}$'s is trivial, we have by induction that

$$j_!\mathrm{Hyp}_1(!, \overline{\psi}; \overline{\rho}\text{'s}; \overline{\chi}\text{'s}) = j_!\mathcal{H}_1(!, \overline{\psi}; \overline{\rho}\text{'s}; \overline{\chi}\text{'s})[1]$$
$$\approx j_*\mathcal{H}_1(!, \overline{\psi}; \overline{\rho}\text{'s}; \overline{\chi}\text{'s})[1]$$

is (an irreducible Fourier sheaf)[1]. Thus we find

$$\mathrm{Hyp}_1(!, \psi; \mathbb{1}, \chi\text{'s}; \rho\text{'s})[-1] \approx$$
$$\approx j^*\mathrm{FT}_\psi(j_*\mathcal{H}_1(!, \overline{\psi}; \overline{\rho}\text{'s}; \overline{\chi}\text{'s}))$$
$$= j^*\mathrm{NFT}_\psi(j_*\mathcal{H}_1(!, \overline{\psi}; \overline{\rho}\text{'s}; \overline{\chi}\text{'s})).$$

The key point is that $\mathrm{NFT}_\psi(j_*\mathcal{H}_1(!, \overline{\psi}; \overline{\rho}\text{'s}; \overline{\chi}\text{'s}))$ is irreducible Fourier. This shows first that $\mathrm{Hyp}_1(!, \psi; \mathbb{1}, \chi\text{'s}; \rho\text{'s})[-1]$ is an irreducible middle extension sheaf on $\mathbb{G}_m$, thus proving (1) and providing the required $\mathcal{H}_1(!, \psi; \mathbb{1}, \chi\text{'s}; \rho\text{'s})$. [Because the Euler characteristic of a convolution is the product of the Euler characteristics of the convolvees (cf 8.1.8), always $\chi(\mathbb{G}_m, \mathrm{Hyp}_1(!, \psi; \mathbb{1}, \chi\text{'s}; \rho\text{'s})) = 1$.]

If $\mathcal{H}_1(!, \psi; \mathbb{1}, \chi\text{'s}; \rho\text{'s})$ is defined over a finite subfield k_0 of k, then by induction it is pure of weight n+m-1 (cf. 7.3.8).

That the local monodromy at zero and at ∞ of $\mathcal{H}_1(!, \psi; \mathbb{1}, \chi\text{'s}; \rho\text{'s})$ is as asserted in the theorem follows (by induction) from the known effect (cf. 7.5.4) of NFT_ψ on the local monodromies at zero and ∞ of irreducible Fourier sheaves which are lisse on $\mathbb{G}_m$.

If $n < m$, $\mathcal{H}_1(!, \psi; \mathbb{1}, \chi\text{'s}; \rho\text{'s})$ is lisse on $\mathbb{G}_m$ because by induction it is the NFT of an irreducible Fourier which is lisse on $\mathbb{G}_m$ and all of whose ∞-slopes are $1/(m - (n-1)) < 1$. This gives (7).

If $n = m$, then there exists a unique $s \in k^\times$ such that on $\mathbb{G}_m - \{s\}$, $\mathcal{H}_1(!, \psi; \mathbb{1}, \chi\text{'s}; \rho\text{'s})$ is lisse of rank n,with tame pseudoreflection local monodromy at s. [The s in question is the unique element of $k^\times$ for which $\mathcal{H}_1(!, \overline{\psi}; \overline{\rho}\text{'s}; \overline{\chi}\text{'s})$ as $P(\infty)$-representation is $\mathcal{L}_{\psi(-sx)}\oplus(\text{trivial})$.] We must show that s = 1. If n=m=1, then by definition

$$\mathcal{H}_1(!, \overline{\psi}; \overline{\rho}\text{'s}; \overline{\chi}\text{'s}) := j^*(\mathcal{L}_{\overline{\psi}(x)})\otimes\mathcal{L}_{\overline{\rho}(x)} = j^*(\mathcal{L}_{\psi(-x)})\otimes\mathcal{L}_{\overline{\rho}(x)},$$

as required.

Suppose now n=m is ≥ 2. Since $\mathcal{H}_1(!, \psi; \mathbb{1}, \chi\text{'s}; \rho\text{'s})$ is a middle extension sheaf, s is the unique point in $k^\times$ where the rank of its stalk at s is n-1. Pick a partition n= A + B of n as the sum of two strictly positive integers, separate the $\{\mathbb{1}, \chi\text{'s}\}$ into a collection of A characters α's and B characters β's, and then separate the ρ's into a collection of A characters γ's and B characters δ's. Let

$$\mathcal{A} := \mathcal{H}_1(!, \psi, \alpha\text{'s}; \gamma\text{'s}),$$

$$\mathcal{B} := \mathcal{H}_1(!, \psi, \beta\text{'s}; \delta\text{'s}).$$

Then by induction $\mathcal{A}$ (resp. $\mathcal{B}$) is an irreducible middle extension sheaf on $\mathbb{G}_m$, lisse of rank A (resp. B) on $\mathbb{G}_m - \{1\}$, with tame pseudoreflection local monodromy at 1 and tame local monodromy at both zero and ∞. Moreover, we have(by part (1) and the definition of hypergeometrics)

$$\mathcal{H}_1(!, \psi; \mathbb{1}, \chi\text{'s}; \rho\text{'s})[-1] \approx \mathcal{A} *_! \mathcal{B}.$$

This implies that for any s in $k^\times$,

$$\dim(\mathcal{H}_1(!, \psi; \mathbb{1}, \chi\text{'s}; \rho\text{'s}))_s = -\chi(\mathbb{G}_m, \mathcal{A}\otimes[x \mapsto s/x]^*\mathcal{B}).$$

The sheaf $\mathcal{A}\otimes[x \mapsto s/x]^*\mathcal{B}$ is tame on $\mathbb{G}_m$, and its only nonlisseness is at the point 1 (where $\mathcal{A}$ drops by 1) and at s (where $\mathcal{B}$ drops by 1). So at any point $s \neq 1$, $\mathcal{A}\otimes[x \mapsto s/x]^*\mathcal{B}$ has two distinct drops, at 1 of size B and at s of size A; thus at $s \neq 1$, the Euler-Poincare formula gives

$$-\chi(\mathbb{G}_m, \mathcal{G}\otimes[x \mapsto s/x]^*\mathcal{B}) = B + A.$$

But at s = 1, the only drop is at 1, of size AB - (A-1)(B-1) = A + B - 1, so $\mathcal{H}_1(!, \psi; \mathbb{1}, \chi\text{'s}; \rho\text{'s})$ is nonlisse at 1, as asserted. This proves (8); the determinant of the tame local monodromy at 1 is forced by what it is at zero and ∞.

To prove (5), it is equivalent to show that under the natural pairing, $\mathcal{H}_1(!, \psi; \mathbb{1}, \chi\text{'s}; \rho\text{'s})$ and $\mathcal{H}_1(!, \overline{\psi}; \mathbb{1}, \overline{\chi}\text{'s}; \overline{\rho}\text{'s})$ are (up to a twist) duals. The first is $j^*\mathrm{NFT}_\psi(j_*\mathcal{H}_1(!, \overline{\psi}; \overline{\rho}\text{'s}; \overline{\chi}\text{'s}))$, and the second is $j^*\mathrm{NFT}_{\overline{\psi}}(j_*\mathcal{H}_1(!, \psi; \rho\text{'s}; \chi\text{'s}))$. In view of the duality behaviour of NFT (and the trivial remark that $\mathrm{NFT}_\psi\circ[-1]^* = \mathrm{NFT}_{\overline{\psi}}$), the result again follows by induction.

It remains to treat the case in which n=0. In this case we reduce by twisting by a suitable $\mathcal{L}_\Lambda$ to treating $\mathrm{Hyp}_1(!, \psi; \varnothing; \mathbb{1}, \rho\text{'s})$. We treat this case by applying the general formula

$$(\mathrm{inv}^*j^*\mathcal{L}_\psi)[1]*_!K \approx \mathrm{inv}^*j^*\mathrm{FT}_\psi(j_!K),$$

with ψ replaced by $\overline{\psi}$, to the object $K := \mathrm{Hyp}_1(!, \psi; \varnothing; \rho\text{'s})$. We find

$$\mathrm{Hyp}_1(!, \psi; \varnothing; \mathbb{1}, \rho\text{'s}) \approx \mathrm{inv}^*j^*\mathrm{FT}_{\overline{\psi}}(j_!\mathrm{Hyp}_1(!, \psi; \varnothing; \rho\text{'s})).$$

The proof by induction now proceeds as in the previous case. QED

Remark 8.4.3 The Kloosterman sheaves denoted

$$\mathrm{Kl}(\psi; \chi_1, \dots, \chi_n; 1, \dots, 1)$$

in [Ka-GKM] are precisely the sheaves $\mathcal{H}_1(!, \psi; \chi\text{'s}; \varnothing)$ of type (n, 0) above. The systematic use here of Fourier Transform to develop their basic properties is only hinted at there (cf. [Ka-GKM, Chapter 8]), and is independent of the method employed there.

(8.4.4) In terms of these Kloosterman sheaves

$$\mathrm{Kl}(\psi; \chi\text{'s}) := \mathcal{H}_1(!, \psi; \chi\text{'s}; \varnothing),$$

we can rewrite the definition of the hypergeometric complexes:

$$\mathrm{Hyp}(!, \psi; \chi\text{'s}; \varnothing) := \mathrm{Kl}(\psi, \chi\text{'s})[1]$$

$$\mathrm{Hyp}(!, \psi; \varnothing; \rho\text{'s}) := \mathrm{inv}^*\mathrm{Kl}(\overline{\psi}, \overline{\rho}\text{'s})[1]$$

$$\mathrm{Hyp}(!, \psi; \chi\text{'s}; \rho\text{'s}) := \mathrm{Kl}(\psi, \chi\text{'s})[1]*_!\mathrm{inv}^*\mathrm{Kl}(\overline{\psi}, \overline{\rho}\text{'s})[1].$$

$$\mathrm{Hyp}(*, \psi; \chi\text{'s}; \rho\text{'s}) := \mathrm{Kl}(\psi, \chi\text{'s})[1]*_*\mathrm{inv}^*\mathrm{Kl}(\overline{\psi}, \overline{\rho}\text{'s})[1].$$

This point of view will be useful now in establishing the perversity of the objects $\mathrm{Hyp}(!, \psi; \chi\text{'s}; \rho\text{'s})$.

Theorem 8.4.5 The objects $\mathrm{Hyp}_\lambda(!, \psi; \chi\text{'s}; \rho\text{'s})$ and $\mathrm{Hyp}_\lambda(*, \psi; \chi\text{'s}; \rho\text{'s})$ of $D^b{}_c(\mathbb{G}_m, \overline{\mathbb{Q}}_\ell)$ are perverse.

proof By multiplicative translation, we may suppose that $\lambda = 1$. If $(n, m) = (0, 0)$, then by definition

$$\mathrm{Hyp}(!, \psi; \varnothing; \varnothing) = \mathrm{Hyp}(*, \psi; \varnothing; \varnothing) = \delta_1$$

is perverse. If $(n, m) \neq (0, 0)$ but either $n=0$ or $m=0$, then the χ's and ρ's are automatically disjoint, and we may apply the previous theorem.

Suppose now that both n and m are strictly positive. Then

$$\mathrm{Hyp}(*, \psi; \chi\text{'s}; \rho\text{'s}) := \mathrm{Kl}(\psi, \chi\text{'s})[1] *_* \mathrm{inv}^*\mathrm{Kl}(\overline{\psi}, \overline{\rho}\text{'s})[1]$$

is the $*$ convolution of two perverse objects, and hence (cf. 8.1.9) is **semiperverse**. By the duality formulas (8.3.3)

$$D(\mathrm{Hyp}_\lambda(!, \psi; \chi\text{'s}; \rho\text{'s})) \approx \mathrm{Hyp}_\lambda(*, \overline{\psi}; \overline{\chi}\text{'s}; \overline{\rho}\text{'s})(\mathbf{n+m}),$$
$$D(\mathrm{Hyp}_\lambda(*, \psi; \chi\text{'s}; \rho\text{'s})) \approx \mathrm{Hyp}_\lambda(!, \overline{\psi}; \overline{\chi}\text{'s}; \overline{\rho}\text{'s})(\mathbf{n+m}),$$

it suffices to establish universally that

$$\mathrm{Hyp}(!, \psi; \chi\text{'s}; \rho\text{'s}) := \mathrm{Kl}(\psi, \chi\text{'s})[1] *_! \mathrm{inv}^*\mathrm{Kl}(\overline{\psi}, \overline{\rho}\text{'s})[1]$$

is semiperverse. For this we argue as follows. The sheaves $\mathrm{Kl}(\psi, \chi\text{'s})$ and $\mathrm{inv}^*\mathrm{Kl}(\overline{\psi}, \overline{\rho}\text{'s})$ are each lisse sheaves on $\mathbb{G}_m$ which are irreducible and which have Euler characteristic $= -1$. Now apply the following theorem.

Theorem 8.4.6 Suppose that $\mathcal{F}$ and $\mathcal{G}$ are lisse $\overline{\mathbb{Q}}_\ell$-sheaves on $\mathbb{G}_m$ over an algebraically closed field k of characteristic $p \neq \ell$. Suppose further that $\mathcal{G}$ is irreducible and that $\chi(\mathbb{G}_m, \mathcal{G}) \neq 0$. Then both $\mathcal{F}[1] *_! \mathcal{G}[1]$ and $\mathcal{F}[1] *_* \mathcal{G}[1]$ are perverse.

proof For any two lisse sheaves $\mathcal{F}$ and $\mathcal{G}$, the objects $\mathcal{F}[1]$ and $\mathcal{G}[1]$ are perverse, and hence $\mathcal{F}[1] *_* \mathcal{G}[1]$ is semiperverse. So by duality, it suffices to prove the semiperversity of $\mathcal{F}[1] *_! \mathcal{G}[1]$. Concretely, we must prove that the object $\mathcal{F} *_! \mathcal{G}$ has $\mathcal{H}^2$ punctual and $\mathcal{H}^i = 0$ for $i > 2$.

For $s \in k^\times$, we have, by proper base change,

$$\mathcal{H}^i(\mathcal{F} *_! \mathcal{G})_s = H^i{}_c(\mathbb{G}_m, \mathcal{F} \otimes [x \mapsto s/x]^*\mathcal{G}).$$

So the vanishing $\mathcal{H}^i = 0$ for $i > 2$ is obvious. We must show the vanishing of $H^2{}_c(\mathbb{G}_m, \mathcal{F} \otimes [x \mapsto s/x]^*\mathcal{G})$ for all but finitely many values of s. Writing $\mathcal{F}$ as a successive extension of irreducbles, we reduce to the case when $\mathcal{F}$ is irreducible. Then both $\mathcal{F}$ and $[x \mapsto s/x]^*\mathcal{G}$ are irreducible, so $H^2_c(\mathbb{G}_m, \mathcal{F}\otimes[x \mapsto s/x]^*\mathcal{G})$ is nonzero if and only if there exists an isomorphism $\mathcal{F}^\vee \approx [x \mapsto s/x]^*\mathcal{G}$. Now $\mathcal{H}^2(\mathcal{F} *_! \mathcal{G})$ is either punctual, or its stalk at s is nonzero for all s outside some finite subset T of $k^\times$.

So if $\mathcal{H}^2(\mathcal{F} *_! \mathcal{G})$ were nonpunctual, the isomorphism class of $[x \mapsto s/x]^*\mathcal{G}$ would be independent of s in $k^\times$ - T. Replacing $\mathcal{G}$ by a multiplicative translate of itself, we may assume that 1 is not in T, i.e., that $\mathcal{F}^\vee \approx \mathrm{inv}^*\mathcal{G}$. Since k is algebraically closed, $k^\times$ - T contains roots of unity of arbitrarily high order. If we take for s a root of unity ζ_N of order N, then the isomorphism class of $\mathrm{inv}^*\mathcal{G}$ is invariant under multiplicative translation by ζ_N. Since $\mathrm{inv}^*\mathcal{G}$ is irreducible, it descends through the N-fold Kummer covering of $\mathbb{G}_m$ by itself, and consequently $\chi(\mathbb{G}_m, \mathrm{inv}^*\mathcal{G}) \equiv 0 \bmod N$. Since we may choose N arbitrarily large, we must have $\chi(\mathbb{G}_m, \mathrm{inv}^*\mathcal{G}) = 0$, contradiction. This proves the theorem, and with it Theorem 8.4.5 above. QED

Corollary 8.4.6.1 Hypotheses and notations as in the theorem above, suppose that no multiplicative translate of $\mathcal{F}^\vee$ is isomorphic to $\mathrm{inv}^*\mathcal{G}$. Then $\mathcal{F}[1] *_! \mathcal{G}[1]$ is of the form $\mathcal{H}[1]$, for some sheaf $\mathcal{H}$ on $\mathbb{G}_m$ which has no nonzero punctual sections.
proof Indeed, in this case $H^2_c(\mathbb{G}_m, \mathcal{F}\otimes[x \mapsto s/x]^*\mathcal{G}) = 0$ for all s, so the perverse object $\mathcal{F}[1] *_! \mathcal{G}[1]$ is of the form $\mathcal{H}[1]$ for some sheaf $\mathcal{H}$ on $\mathbb{G}_m$. Because $\mathcal{H}[1]$ is perverse, $\mathcal{H}$ has no punctual sections. QED

Returning to $\mathrm{Hyp}_\lambda(!, \psi; \chi\text{'s}; \rho\text{'s})$ and $\mathrm{Hyp}_\lambda(*, \psi; \chi\text{'s}; \rho\text{'s})$, we can be much more precise about their structure.

Corollary 8.4.6.2 Suppose that the χ's and ρ's are not identical. Then the perverse object $\mathrm{Hyp}_\lambda(!, \psi; \chi\text{'s}; \rho\text{'s})$ is nonpunctual, i.e.,

$$\mathrm{Hyp}_\lambda(!, \psi; \chi\text{'s}; \rho\text{'s}) = \mathcal{H}_\lambda(!, \psi; \chi\text{'s}; \rho\text{'s})[1]$$

for a sheaf $\mathcal{H}_\lambda(!, \psi; \chi\text{'s}; \rho\text{'s})$ on $\mathbb{G}_m$ with no nonzero punctual sections.
proof If n=0 or m=0, this is proven in 8.4.2. By multiplicative translation, we may assume $\lambda = 1$. The Kloosterman expression

$$\mathrm{Hyp}(!,\ \psi;\ \chi\text{'s};\ \rho\text{'s}) := \mathrm{Kl}(\psi,\ \chi\text{'s})[1]*_!\mathrm{inv}^*\mathrm{Kl}(\overline{\psi},\ \overline{\rho}\text{'s})[1],$$

and 8.4.6.1 then give the existence of a sheaf $\mathcal{H}_1(!,\ \psi;\ \chi\text{'s};\ \rho\text{'s})$ on $\mathbb{G}_m$ with no punctual sections such that

$$\mathrm{Hyp}(!,\ \psi;\ \chi\text{'s};\ \rho\text{'s}) = \mathcal{H}_1(!,\ \psi;\ \chi\text{'s};\ \rho\text{'s})[1].\quad \text{QED}$$

Corollary 8.4.6.3 Suppose that the χ's and ρ's are not identical. Then

$$H^i{}_c(\mathbb{G}_m,\ \mathcal{H}_\lambda(!,\ \psi;\ \chi\text{'s};\ \rho\text{'s})) = 0 \text{ for } i \neq 1,$$

$$H^1{}_c(\mathbb{G}_m,\ \mathcal{H}_\lambda(!,\ \psi;\ \chi\text{'s};\ \rho\text{'s})) \text{ is one-dimensional.}$$

proof This is just the spelling out of 8.3.4 in the nonpunctual case. QED

Cancellation Theorem 8.4.7 Given arbitrary χ's and ρ's, and a tame character Λ, denote by V the one-dimensional $\overline{\mathbb{Q}}_\ell$-vector space

$$V := H^0{}_c(\mathbb{G}_m,\ \mathrm{Hyp}_\lambda(!,\ \psi;\ \Lambda^{-1}\chi\text{'s};\ \Lambda^{-1}\rho\text{'s})).$$

In the category $\mathrm{Perv}(\mathbb{G}_m)$, $\mathrm{Hyp}_\lambda(!,\ \psi;\ \Lambda,\ \chi\text{'s};\ \Lambda,\ \rho\text{'s})$ sits in a short exact sequence

$$0 \to V\otimes\mathcal{L}_\Lambda[1] \to \mathrm{Hyp}_\lambda(!,\ \psi;\ \Lambda,\ \chi\text{'s};\ \Lambda,\ \rho\text{'s}) \to \mathrm{Hyp}_\lambda(!,\ \psi;\ \chi\text{'s};\ \rho\text{'s})(-1)\to 0.$$

proof Twisting by $\mathcal{L}_{\overline{\Lambda}}$, we reduce to the case when Λ is $\mathbb{1}$. Then by definition we have

$$\mathrm{Hyp}_\lambda(!,\ \psi;\ \mathbb{1},\ \chi\text{'s};\ \mathbb{1},\ \rho\text{'s}) := \mathrm{Hyp}_\lambda(!,\ \psi;\ \chi\text{'s};\ \rho\text{'s})*_!\mathrm{Hyp}_1(!,\ \psi;\ \mathbb{1};\ \mathbb{1}).$$

We have the following lemma:

Lemma 8.4.8 (compare 6.3.5) Denote by $k : \mathbb{G}_m - \{1\} \to \mathbb{G}_m$ the inclusion. Then we have a canonical isomorphism $\mathrm{Hyp}_1(!,\ \psi;\ \mathbb{1};\ \mathbb{1}) \approx Rk_*\overline{\mathbb{Q}}_\ell[1]$. Writing $Rk_*\overline{\mathbb{Q}}_\ell[1]$ as an extension of its (shifted) cohomology sheaves gives a short exact sequence of perverse sheaves on $\mathbb{G}_m$

$$0 \to \overline{\mathbb{Q}}_\ell[1] \to \mathrm{Hyp}_1(!,\ \psi;\ \mathbb{1};\ \mathbb{1}) \to \delta_1(-1) \to 0.$$

proof By the fundamental relation 8.1.12 between Fourier Transform and convolution, denoting by $j: \mathbb{G}_m \to \mathbb{A}^1$ the inclusion, we have

$$(j^*\mathcal{L}_\psi)[1]*_!K \approx j^*\mathrm{FT}_\psi(j_!\mathrm{inv}^*K).$$

Applying this to $K := \mathrm{inv}^*j^*\mathcal{L}_{\overline{\psi}}[1]$, we find

$$\mathrm{Hyp}_1(!, \psi; \mathbb{1}; \mathbb{1}) \approx j^*\mathrm{FT}_\psi(j_!j^*\mathcal{L}_{\overline{\psi}}[1]) = j^*\mathrm{FT}_\psi(\mathcal{L}_{\overline{\psi}}\otimes j_!\overline{\mathbb{Q}}_\ell)[1].$$

Now $\mathrm{FT}_\psi(\mathcal{L}_{\overline{\psi}}\otimes K) = [x \mapsto x+1]_*\mathrm{FT}_\psi(K)$ for any K on $\mathbb{A}^1$, and it is proven in [Ka-PES, Prop. A2] (cf. 2.10.1(1) for the $\mathcal{D}$-module analogue) that $\mathrm{FT}_\psi(j_!\overline{\mathbb{Q}}_\ell) = Rj_*\overline{\mathbb{Q}}_\ell$. Thus we find

$$j^*\mathrm{FT}_\psi(\mathcal{L}_{\overline{\psi}}\otimes j_!\overline{\mathbb{Q}}_\ell)[1] \approx j^*[x \mapsto x+1]_*Rj_*\overline{\mathbb{Q}}_\ell[1] = Rk_*\overline{\mathbb{Q}}_\ell[1].$$ QED

Corollary 8.4.8.1 of Lemma 8.4.8 For any χ, the perverse object $\mathrm{Hyp}_1(!, \psi; \chi; \chi)$ sits in a short exact sequence of perverse sheaves on $\mathbb{G}_m$

$$0 \to \mathcal{L}_\chi[1] \to \mathrm{Hyp}_1(!, \psi; \chi; \chi) \to \delta_1(-1) \to 0.$$

proof Simply tensor with $\mathcal{L}_\chi$. QED

Apply 8.4.8 to calculate $\mathrm{Hyp}_\lambda(!, \psi; \chi\text{'s}; \rho\text{'s})*_!\mathrm{Hyp}_1(!, \psi; \mathbb{1}; \mathbb{1})$. The convolution $\mathrm{Hyp}_\lambda(!, \psi; \chi\text{'s}; \rho\text{'s})*_!\overline{\mathbb{Q}}_\ell$ is the constant sheaf with value

$$V := H^0{}_c(\mathbb{G}_m, \mathrm{Hyp}_\lambda(!, \psi; \chi\text{'s}; \rho\text{'s})).$$

The convolution $\mathrm{Hyp}_\lambda(!, \psi; \chi\text{'s}; \rho\text{'s})*_!\delta_1(-1)$ is $\mathrm{Hyp}_\lambda(!, \psi; \chi\text{'s}; \rho\text{'s})(-1)$, so we have the asserted short exact sequence

$$0 \to V[1] \to \mathrm{Hyp}_\lambda(!, \psi; \mathbb{1}, \chi\text{'s}; \mathbb{1}, \rho\text{'s}) \to \mathrm{Hyp}_\lambda(!, \psi; \chi\text{'s}; \rho\text{'s})(-1) \to 0.$$

QED

We now develop some of the immediate consequences of the cancellation theorem.

Cancellation Theorem bis 8.4.9 Given arbitrary χ's and ρ's, and a tame character Λ, denote by W the one-dimensional $\overline{\mathbb{Q}}_\ell$-vector space

$$W := H^0(\mathbb{G}_m, \mathrm{Hyp}_\lambda(!, \psi; \Lambda^{-1}\chi\text{'s}; \Lambda^{-1}\rho\text{'s})).$$

In the category $\mathrm{Perv}(\mathbb{G}_m)$, $\mathrm{Hyp}_\lambda(*, \psi; \Lambda, \chi\text{'s}; \Lambda, \rho\text{'s})$ sits in a short exact sequence

$$0 \to \mathrm{Hyp}_\lambda(*, \psi; \chi\text{'s}; \rho\text{'s})(-1) \to \mathrm{Hyp}_\lambda(*, \psi; \Lambda, \chi\text{'s}; \Lambda, \rho\text{'s}) \to W\otimes\mathcal{L}_\Lambda[1] \to 0.$$

proof This is the dual statement, with the dual proof. QED

Semisimplification Theorem 8.4.10 Suppose that the χ's and ρ's are disjoint. Let $r \geq 1$, and let $\Lambda_1, \ldots, \Lambda_r$ be r not necessarily distinct tame characters. In the category $\mathrm{Perv}(\mathbb{G}_m)$ over the algebraically closed field k, the semisimplifications of $\mathrm{Hyp}_\lambda(!, \psi; \Lambda_1, \ldots, \Lambda_r, \chi\text{'s}; \Lambda_1, \ldots, \Lambda_r, \rho\text{'s})$ and of $\mathrm{Hyp}_\lambda(*, \psi; \Lambda_1, \ldots, \Lambda_r, \chi\text{'s}; \Lambda_1, \ldots, \Lambda_r, \rho\text{'s})$ are each isomorphic to the direct sum

$$\mathrm{Hyp}_\lambda(!, \psi; \chi\text{'s}; \rho\text{'s}) \bigoplus \oplus_{i=1, \ldots, r} \mathcal{L}_{\Lambda_i}[1].$$

proof This is obvious by the cancellation theorem and the fact (8.4.2) that if the χ's and ρ's are disjoint, then

$$\mathrm{Hyp}_\lambda(!, \psi; \chi\text{'s}; \rho\text{'s}) \approx \mathrm{Hyp}_\lambda(*, \psi; \chi\text{'s}; \rho\text{'s})$$

is simple. QED

Corollary 8.4.10.1 The perverse object $\mathrm{Hyp}_\lambda(!, \psi; \chi\text{'s}; \rho\text{'s})$ is simple if and only if the χ's and ρ's are disjoint.

Theorem 8.4.11 Suppose that the χ's and ρ's are not identical. Write

$$\mathrm{Hyp}_\lambda(!, \psi; \chi\text{'s}; \rho\text{'s}) = \mathcal{H}_\lambda(!, \psi; \chi\text{'s}; \rho\text{'s})[1]$$

for a sheaf $\mathcal{H}_\lambda(!, \psi; \chi\text{'s}; \rho\text{'s})$ on $\mathbb{G}_m$ with no nonzero punctual sections. The local monodromy of $\mathcal{H}_\lambda(!, \psi; \chi\text{'s}; \rho\text{'s})$ is the following:

(1) If $n > m$, the sheaf $\mathcal{H}_\lambda(!, \psi; \chi\text{'s}; \rho\text{'s})$ is lisse of rank n on $\mathbb{G}_m$. As I(0)-representation it is tame, isomorphic to

$$\oplus_{\text{distinct } \chi\text{'s}} \mathcal{L}_\chi \otimes \mathrm{Unip}(\mathrm{mult}_0(\chi)).$$

As $I(\infty)$-representation it has Swan conductor $= 1$, and is isomorphic to the direct sum

$$(\text{dim. } n-m, \text{ brk. } 1/(n-m)) \bigoplus \oplus_{\text{distinct } \rho\text{'s}} \mathcal{L}_\rho \otimes \mathrm{Unip}(\mathrm{mult}_\infty(\rho)).$$

(2) If $n < m$, the sheaf $\mathcal{H}_\lambda(!, \psi; \chi\text{'s}; \rho\text{'s})$ is lisse of rank m on $\mathbb{G}_m$. As I(0)-representation it has Swan conductor $= 1$, and is isomorphic to the direct sum

$$(\text{dim. } m-n, \text{ brk. } 1/(m-n)) \bigoplus \oplus_{\text{distinct } \chi\text{'s}} \mathcal{L}_\chi \otimes \mathrm{Unip}(\mathrm{mult}_0(\chi)).$$

As $I(\infty)$-representation it is tame, isomorphic to

$$\oplus_{\text{distinct } \rho\text{'s}} \mathcal{L}_\rho \otimes \mathrm{Unip}(\mathrm{mult}_\infty(\rho)).$$

(3) If $n = m$, the sheaf $\mathcal{H}_\lambda(!, \psi; \chi\text{'s}; \rho\text{'s})$ is lisse of rank n on $\mathbb{G}_m - \{\lambda\}$, from which it is extended by direct image. $I(\lambda)$ acts by tame pseudoreflections of determinant $\mathcal{L}_{\Lambda(x-\lambda)}$, for $\Lambda := \prod_i \chi_i / \prod_i \rho_i$.

As I(0)-representation it is tame, isomorphic to

$$\oplus_{\text{distinct } \chi\text{'s}} \mathcal{L}_\chi \otimes \mathrm{Unip}(\mathrm{mult}_0(\chi)).$$

As $I(\infty)$-representation it is tame, isomorphic to

$$\oplus_{\text{distinct } \rho\text{'s}} \mathcal{L}_\rho \otimes \mathrm{Unip}(\mathrm{mult}_\infty(\rho)).$$

proof Suppose first $n \neq m$. Then it follows immediately from the cancellation theorem 8.4.7 (and 8.4.2 in the case when the χ's and ρ's are disjoint) that $\mathcal{H}_\lambda(!, \psi; \chi\text{'s}; \rho\text{'s})$ is lisse of rank $\max(n, m)$ on $\mathbb{G}_m$, and that the description claimed for its local monodromy at zero and ∞ is correct up to semisimplification. Similarly, if $n=m$, we see that the sheaf $\mathcal{H}_\lambda(!, \psi; \chi\text{'s}; \rho\text{'s})$ is lisse of rank n on $\mathbb{G}_m - \{\lambda\}$, and that the description claimed for its local monodromy at zero and ∞ is correct up to semisimplification.

To see that the description claimed for its local monodromy at zero and ∞ is absolutely correct, we must see that (universally, so after any $\mathcal{L}_\Lambda$ twist) the local monodromy at zero or at ∞ has at most a single unipotent Jordan block. This is a consequence of the fact that $H^1_c(\mathbb{G}_m, \mathcal{H}_\lambda(!, \psi; \chi\text{'s}; \rho\text{'s}))$ is one-dimensional. Indeed, if we denote by

$$\mathcal{F} := \mathcal{H}_\lambda(!, \psi; \chi\text{'s}; \rho\text{'s}),$$

and denote by $j: \mathbb{G}_m \to \mathbb{P}^1$ the inclusion, the coboundary of the short exact sequence of sheaves on $\mathbb{P}^1$

$$0 \to j_!\mathcal{F} \to j_*\mathcal{F} \to \mathcal{F}^{I(0)}\otimes\delta_0 \oplus \mathcal{F}^{I(\infty)}\otimes\delta_\infty \to 0,$$

gives an injective map

$$(\mathcal{F}^{I(0)}\oplus\mathcal{F}^{I(\infty)})/H^0(\mathbb{G}_m, \mathcal{F}) \to H^1_c(\mathbb{G}_m, \mathcal{F}).$$

Since $H^0(\mathbb{G}_m, \mathcal{F})$ injects into either $\mathcal{F}^{I(0)}$ or $\mathcal{F}^{I(\infty)}$, it follows that $\mathcal{F}^{I(0)}$ and $\mathcal{F}^{I(\infty)}$ each have dimension at most one.

If $n = m$, it remains to examine the local monodromy at λ of the sheaf $\mathcal{F} := \mathcal{H}_\lambda(!, \psi; \chi\text{'s}; \rho\text{'s})$. Denote by $k : \mathbb{G}_m - \{\lambda\} \to \mathbb{G}_m$ the inclusion. By the Cancellation Theorem 8.4.7 and 8.4.2, $\mathcal{F}$ is lisse on $\mathbb{G}_m - \{\lambda\}$ of rank n, and it has a one-dimensional drop at λ. Since $\mathcal{F}$ has no nonzero punctual sections, either $\mathcal{F}$ has pseudoreflection local monodromy at λ and $\mathcal{F} \approx k_*k^*\mathcal{F}$, or $k_*k^*\mathcal{F}$ is lisse on $\mathbb{G}_m$. In this latter case, $k_*k^*\mathcal{F}$ must be a successive extension of $\mathcal{L}_\Lambda$'s, becaue we already know that $\mathcal{F}$ is tame at both zero and ∞. But comparing the local monodromies of $\mathcal{F}$ at zero and ∞, we see that $\mathcal{F}$ cannot be a successive extension of $\mathcal{L}_\Lambda$'s. Therefore $\mathcal{F}$ has pseudoreflection local monodromy at λ, and $\mathcal{F} \approx k_*k^*\mathcal{F}$. Because $\chi(\mathbb{G}_m, \mathcal{F}) = -1$, $\mathcal{F}$ must be tame at λ. The determinant of the tame local monodromy at λ is determined by what it is at 0 and ∞, as indicated. QED

Corollary 8.4.11.1 Suppose that the χ's and ρ's are not identical. Then there exists a middle extension sheaf $\mathcal{H}_\lambda(*, \psi; \chi\text{'s}; \rho\text{'s})$ on $\mathbb{G}_m$ such that

$$\mathrm{Hyp}_\lambda(*, \psi; \chi\text{'s}; \rho\text{'s}) = \mathcal{H}_\lambda(*, \psi; \chi\text{'s}; \rho\text{'s})[1],$$

and assertions (1), (2), (3) of the theorem hold for $\mathcal{H}_\lambda(*, \psi; \chi\text{'s}; \rho\text{'s})$.

proof Duality. QED

Theorem 8.4.12 If $n = m$ and the χ's and the ρ's are identical, then

(1) the sheaf $\mathcal{H}^{-1}(\mathrm{Hyp}_\lambda(!, \psi; \chi\text{'s}; \chi\text{'s}))$ is lisse on $\mathbb{G}_m$, a successive extension of the $\mathcal{L}_\chi$'s.

(2) $\mathcal{H}^{-1}(\mathrm{Hyp}_\lambda(!, \psi; \chi\text{'s}; \chi\text{'s}))$ is isomorphic to

$$\bigoplus_{\text{distinct } \chi\text{'s}} \mathcal{L}_\chi \otimes \mathrm{Unip}(\mathrm{mult}_0(\chi)).$$

(3) the sheaf $\mathcal{H}^0(\mathrm{Hyp}_\lambda(!, \psi; \chi\text{'s}; \chi\text{'s}))$ is a punctual sheaf, concentrated at λ, of rank one.

proof By multiplicative translation, we may assume $\lambda = 1$. For $n = 1$, this is 8.4.8.1. Assertions (1) and (3) in the general case are proven inductively, using the Cancellation Theorem 8.4.7.

To prove (2), we use the fact that $\mathcal{H}^{-1}(\mathrm{Hyp}_\lambda(!, \psi; \chi\text{'s}; \chi\text{'s}))$, being lisse on $\mathbb{G}_m$ and tame, is determined by (indeed is the canonical extension of, cf [Ka-LG, 1.5]) its local monodromy at zero. We must show that, after any $\mathcal{L}_\Lambda$ twist, this local monodromy has at most one unipotent Jordan block. As explained above, this follows if we prove universally that $H^1_c(\mathbb{G}_m, \mathcal{H}^{-1}(\mathrm{Hyp}_\lambda(!, \psi; \chi\text{'s}; \chi\text{'s})))$ has dimension ≤ 1. To prove this, consider the perverse short exact sequence

$$0 \to \mathcal{H}^{-1}(\mathrm{Hyp}_\lambda(!, \psi; \chi\text{'s}; \chi\text{'s}))[1] \to \mathrm{Hyp}_\lambda(!, \psi; \chi\text{'s}; \chi\text{'s}) \to (\mathrm{pctl}) \to 0.$$

The long exact cohomology sequence for $H^i_c(\mathbb{G}_m, ?)$ gives an injection

$$H^1_c(\mathbb{G}_m, \mathcal{H}^{-1}(\mathrm{Hyp}_\lambda(!, \psi; \chi\text{'s}; \chi\text{'s}))) \hookrightarrow H^0_c(\mathbb{G}_m, \mathrm{Hyp}_\lambda(!, \psi; \chi\text{'s}; \chi\text{'s})),$$

and the target is one-dimensional (cf. 8.3.4). QED

Theorem 8.4.13 Suppose that $\mathrm{Hyp}_\lambda(!, \psi; \chi\text{'s}; \rho\text{'s})$ is defined over a finite subfield k_0 of k, that $(n, m) \neq (0, 0)$, and that the χ's and ρ's are not identical. Then the middle extension sheaf $\mathcal{H}_\lambda(!, \psi; \chi\text{'s}; \rho\text{'s})$ is pure of some weight (necessarily $n + m - 1$) if and only if the χ's and ρ's are disjoint.

proof If the χ's and ρ's are disjoint, then the middle extension sheaf $\mathcal{H}_\lambda(!, \psi; \chi\text{'s}; \rho\text{'s})$ is pure of weight $n + m - 1$ (cf 8.4.2). By Deligne's main result 3.3.1 in [De-WII], $\mathcal{H}_\lambda(!, \psi; \chi\text{'s}; \rho\text{'s})$ is always mixed of weight $\leq n + m - 1$. It suffices to show that if the χ's and ρ's are not identical, then $\mathrm{Hyp}_\lambda(!, \psi; \Lambda, \chi\text{'s}; \Lambda, \rho\text{'s})$ is **not** pure of weight $n + m + 1$, but that it has a nonzero quotient which is pure of weight $n + m + 1$. In the cancellation theorem exact sequence

$$0 \to V\otimes\mathcal{L}_\Lambda \to \mathcal{H}_\lambda(!, \psi; \Lambda, \chi\text{'s}; \Lambda, \rho\text{'s}) \to \mathcal{H}_\lambda(!, \psi; \chi\text{'s}; \rho\text{'s})(-1) \to 0,$$

V is the geometrically constant sheaf $H^1{}_c(\mathbb{G}_m, \mathcal{H}_\lambda(!, \psi; \Lambda^{-1}\chi\text{'s}; \Lambda^{-1}\rho\text{'s}))$, which is pure of some weight $\leq n + m$ (since $\mathcal{H}_\lambda(!, \psi; \Lambda^{-1}\chi\text{'s}; \Lambda^{-1}\rho\text{'s})$ is mixed of weight $\leq n + m - 1$). On the other hand, since the χ's and ρ's are not identical, we can "extract duplicates" as much as possible from the χ's and ρ's and still have some disjoint α's and β's left at the end. Iterating the above exact sequence, we see that if after extracting r pairs of duplicates from the χ's and ρ's, $\mathcal{H}_\lambda(!, \psi; \alpha\text{'s}; \beta\text{'s})(\mathbf{-1-r})$ is a nonzero quotient of $\mathcal{H}_\lambda(!, \psi; \chi\text{'s}; \rho\text{'s})(\mathbf{-1})$ which is pure of weight $n + m + 1$. QED

Corollary 8.4.13.1 Suppose that $\mathrm{Hyp}_\lambda(!, \psi; \chi\text{'s}; \rho\text{'s})$ is defined over a finite subfield k_0 of k, that $(n, m) \neq (0, 0)$, and that the χ's and ρ's are not identical. Then the canonical "forget supports" map

$$\mathcal{H}_\lambda(!, \psi; \chi\text{'s}; \rho\text{'s}) \to \mathcal{H}_\lambda(*, \psi; \chi\text{'s}; \rho\text{'s})$$

is an isomorphism if and only if the χ's and ρ's are disjoint.

proof If the χ's and ρ's are disjoint, then the map is an isomorphism. Conversely, suppose the map is an isomorphism. Then $\mathcal{H}_\lambda(!, \psi; \chi\text{'s}; \rho\text{'s})$ is pure of weight $n + m - 1$, for $\mathcal{H}_\lambda(!, \psi; \chi\text{'s}; \rho\text{'s})$ is always mixed of weight $\leq n + m - 1$, and (by duality) $\mathcal{H}_\lambda(*, \psi; \chi\text{'s}; \rho\text{'s})$ is always mixed of weight $\geq n + m - 1$. So the χ's and ρ's are disjoint. QED

8.5 Intrinsic characterization of hypergeometrics (compare 3.7)

(8.5.1) We continue to work over an algebraically closed field k of characteristic $p \neq \ell$. We fix a choice of nontrivial $\overline{\mathbb{Q}}_\ell$-valued additive character ψ of a fintite subfield k_0 of k.

Proposition 8.5.2 Let K be a perverse simple object of $D^b{}_c(\mathbb{G}_m, \overline{\mathbb{Q}}_\ell)$. Then $\chi(\mathbb{G}_m, K) \geq 0$, and $\chi(\mathbb{G}_m, K) = 0$ if and only if $K = \mathcal{L}_\Lambda[1]$ for some tame character Λ.

proof If K is punctual, then $\chi(\mathbb{G}_m, K) = 1$. If K is nonpunctual, then K of the form $\mathcal{F}[1]$ for some irreducible middle extension sheaf $\mathcal{F}$ on $\mathbb{G}_m$ with $\chi(\mathbb{G}_m, \mathcal{F}) = -\chi(\mathbb{G}_m, K)$. In the Euler-Poincare formula for an $\mathcal{F}$ without punctual sections on $\mathbb{G}_m$

$$-\chi(\mathbb{G}_m, \mathcal{F}) = \mathrm{Swan}_0(\mathcal{F}) + \mathrm{Swan}_\infty(\mathcal{F}) + \Sigma_{t \text{ in } k^\times} [\mathrm{drop}_t(\mathcal{F}) + \mathrm{Swan}_t(\mathcal{F})]$$

all of the terms on the right hand side are non-negative. QED

Theorem 8.5.3 Let K be a perverse simple object of $D^b{}_c(\mathbb{G}_m, \overline{\mathbb{Q}}_\ell)$ whose Euler characteristic $\chi(\mathbb{G}_m, K) = 1$. Then K is hypergeometric, i.e., there exist $\lambda \in k^\times$ and disjoint χ's and ρ's such that

$$K \approx \mathrm{Hyp}_\lambda(!, \psi; \chi\text{'s}; \rho\text{'s}).$$

proof If K is punctual, it is δ_λ for some $\lambda \in k^\times$, so $K = \mathrm{Hyp}_\lambda(!, \psi; \emptyset; \emptyset)$.

Suppose now that K is nonpunctual. Then K of the form $\mathcal{F}[1]$ for some irreducible middle extension sheaf $\mathcal{F}$ on $\mathbb{G}_m$ with $\chi(\mathbb{G}_m, \mathcal{F}) = -1$. We claim that

$$H^i{}_c(\mathbb{G}_m, \mathcal{F}) = 1\text{-dim'l for } i = 1,$$
$$= 0 \text{ for } i \neq 1.$$

Indeed, the $H^0{}_c$ vanishes because $\mathcal{F}$ has no punctual sections, and the $H^2{}_c$ vanishes because $\mathcal{F}$ has no $\overline{\mathbb{Q}}_\ell$ quotient (otherwise $\mathcal{F}$ would be the constant sheaf, and this is incompatible with its Euler characteristic). Dually, we have

$$H^i(\mathbb{G}_m, \mathcal{F}) = 1\text{-dim'l for } i = 1,$$
$$= 0 \text{ for } i \neq 1.$$

Now denote by $k: \mathbb{G}_m \to \mathbb{P}^1$ the inclusion. The long exact cohomology sequence attached to the short exact sequence of sheaves on $\mathbb{P}^1$

$$0 \to k_!\mathcal{F} \to k_*\mathcal{F} \to \mathcal{F}^{I(0)} \otimes \delta_0 \oplus \mathcal{F}^{I(\infty)} \otimes \delta_\infty \to 0,$$

gives an injective map

$$\mathcal{F}^{I(0)} \oplus \mathcal{F}^{I(\infty)} \to H^1_c(\mathbb{G}_m, \mathcal{F}).$$

Therefore at most one of $\mathcal{F}^{I(0)}$ or $\mathcal{F}^{I(\infty)}$ is nonzero, and if nonzero its dimension is one.

In the Euler-Poincare formula for an $\mathcal{F}$ without punctual sections on $\mathbb{G}_m$

$$-\chi(\mathbb{G}_m, \mathcal{F}) = \mathrm{Swan}_0(\mathcal{F}) + \mathrm{Swan}_\infty(\mathcal{F}) + \Sigma_{t \text{ in } k^\times} [\mathrm{drop}_t(\mathcal{F}) + \mathrm{Swan}_t(\mathcal{F})]$$

all of the terms on the right hand side are non-negative. So there are two possibilities for our $\mathcal{F}$:

(1) $\mathcal{F}$ is everywhere tame, lisse outside a single point t in $k^\times$, and has pseudoreflection local monodromy at t.

(2) $\mathcal{F}$ is lisse on $\mathbb{G}_m$, and $\mathrm{Swan}_0(\mathcal{F}) + \mathrm{Swan}_\infty(\mathcal{F}) = 1$.

We say that $\mathcal{F}$ is of type (n, m) for

n := dimension of $\mathcal{F}^{P(0)}$ = the size of the "tame at 0" part of $\mathcal{F}$,

m := dimension of $\mathcal{F}^{P(\infty)}$ = the size of the "tame at ∞" part of $\mathcal{F}$.

The generic rank of $\mathcal{F}$ is max(n, m). The two cases (1) and (2) above correspond to n = m and to n $\neq$ m respectively.

Suppose first that n = m. Tensoring with a suitable $\mathcal{L}_\Lambda$, an operation under which the theorem is invariant, we may further assume that $\mathbb{1}$ is among the characters which occur in local monodromy at zero. Denote by $j: \mathbb{G}_m \to \mathbb{A}^1$ the inclusion. Then $j_*\mathcal{F}$ is an irreducible Fourier sheaf on $\mathbb{A}^1$ (it is an irreducible middle extension, and as it is not lisse it cannot be $\mathcal{L}_{\psi(tx)}$ for any $t \in k$). By the numerology of Fourier Transform, one checks easily that $\mathrm{NFT}_\psi(j_*\mathcal{F})$ (= $\mathrm{FT}_\psi(j_*\mathcal{F})$), itself an irreducible Fourier sheaf on $\mathbb{A}^1$, is of the form $j_!\mathcal{G}$ for $\mathcal{G}$ an irreducible lisse sheaf on $\mathbb{G}_m$ of rank n-1 with $\chi(\mathbb{G}_m, \mathcal{G}) = -1$. Moreover, $\mathcal{G}$ is of type (n, n-1). By Fourier inversion, we have

$$[x \mapsto -x]_*\mathcal{F}[1](-1) = j^*\mathrm{FT}_\psi(j_!\mathcal{G}[1]) = (j^*\mathcal{L}_\psi)[1] *_! \mathrm{inv}^*\mathcal{G}[1].$$

In view of the stability of hypergeometrics, it suffices to show that $\mathcal{G}[1]$ is hypergeometric. Thus the case n=m results from the case n $\neq$ m.

We now turn to the case n $\neq$ m. We first treat the case when one of n or m vanishes. In this case, we may, by a multiplicative inversion, suppose that $m = 0$. Then $\mathcal{F}$ is tame at zero, and totally wild at ∞ withe $\mathrm{Swan}_\infty(\mathcal{F}) = 1$. Tensoring with a suitable $\mathcal{L}_\Lambda$, an operation under which the theorem is invariant, we may further assume that $\mathbb{1}$ is among the characters which occur in local monodromy at zero. If $n=1$,then by the break-depression lemma [Ka-GKM, 8.5.7] we see that $\mathcal{F}$ is $\mathcal{L}_{\psi(tx)}$ for some $t \in k^\times$. If $n \geq 2$, then $j_*\mathcal{F}$ is an irreducible Fourier sheaf on $\mathbb{A}^1$, and $\mathrm{NFT}_\psi(j_*\mathcal{F})$ (= $\mathrm{FT}_\psi(j_*\mathcal{F})$), itself an irreducible Fourier sheaf on $\mathbb{A}^1$, is of the form $j_!\mathcal{G}$ for $\mathcal{G}$ an irreducible lisse sheaf on $\mathbb{G}_m$ of rank $n-1$ with $\chi(\mathbb{G}_m, \mathcal{G}) = -1$. Moreover, $\mathcal{G}$ is of type $(0, n-1)$. By Fourier inversion, we have

$$[x \mapsto -x]_*\mathcal{F}[1](-1) = j^*\mathrm{FT}_\psi(j_!\mathcal{G}[1]) = (j^*\mathcal{L}_\psi)[1]*_!\mathrm{inv}^*\mathcal{G}[1].$$

By induction on n, $\mathrm{inv}^*\mathcal{G}[1]$ and hence $\mathcal{G}[1]$ itself is hypergeometric, whence $\mathcal{F}[1]$ is hypergeometric.

It remains to treat the case when $n \neq m$ and both n and m are nonzero. By a multiplicative inversion, we may assume that $1 \leq n < m$, or what is the same, that $\mathcal{F}$ is tame at ∞. Tensoring with a suitable $\mathcal{L}_\Lambda$, an operation under which the theorem is invariant, we may further assume that $\mathbb{1}$ is among the characters which occur in local monodromy at 0 [it is here that we use $n \geq 1$]. Then $j_*\mathcal{F}$ is an irreducible Fourier sheaf on $\mathbb{A}^1$, and $\mathrm{NFT}_\psi(j_*\mathcal{F})$ (= $\mathrm{FT}_\psi(j_*\mathcal{F})$), itself an irreducible Fourier sheaf on $\mathbb{A}^1$, is of the form $j_!\mathcal{G}$ for $\mathcal{G}$ an irreducible lisse sheaf on $\mathbb{G}_m$ of rank m with $\chi(\mathbb{G}_m, \mathcal{G}) = -1$. Moreover, $\mathcal{G}$ is of type $(m, n-1)$. By Fourier inversion, we have

$$[x \mapsto -x]_*\mathcal{F}[1](-1) = j^*\mathrm{FT}_\psi(j_!\mathcal{G}[1]) = (j^*\mathcal{L}_\psi)[1]*_!\mathrm{inv}^*\mathcal{G}[1].$$

By induction on $\min(n, m)$, $\mathrm{inv}^*\mathcal{G}[1]$ and hence $\mathcal{G}[1]$ itself is hypergeometric, whence $\mathcal{F}$ is hypergeometric. QED

Corollary 8.5.3.1 Let $\mathcal{F}$ be an irreducible middle extension sheaf on $\mathbb{G}_m$ whose Euler characteristic $\chi(\mathbb{G}_m, \mathcal{F}) = -1$. Then there exists a unique $\lambda \in k^\times$ and unique disjoint sets of χ's and ρ's such that

$$K \approx \mathcal{H}_\lambda(!, \psi; \chi\text{'s}; \rho\text{'s}).$$

proof The existence is given by the theorem. The χ's (resp. the ρ's) with their multiplicities are the precisely the tame characters which occur in the $I(0)$-semisimplification of $\mathcal{F}$ as $I(0)$-représentation (resp. in the $I(\infty)$-semisimplification of $\mathcal{F}$ as $I(\infty)$-representation). The uniqueness of λ results from the the following general lemma.

Translation Lemma 8.5.4 (compare 3.7.7 and [Ka-GKM, 4.1.6]) Let $\mathcal{F}$ be an irreducible middle extension sheaf on $\mathbb{G}_m$ whose Euler characteristic $\chi(\mathbb{G}_m, \mathcal{F})$ is nonzero. Suppose that for some $\lambda \in k^\times$ there exists an isomorphism $\mathcal{F} \approx [x \mapsto \lambda x]^*\mathcal{F}$. Then λ is a root of unity of order dividing $\chi(\mathbb{G}_m, \mathcal{F})$. In particular, if $\chi(\mathbb{G}_m, \mathcal{F}) = -1$, then $\lambda = 1$, i.e., $\mathcal{F}$ is isomorphic to no nontrivial multiplicative translate of itself.

proof We first show that λ must be a root of unity. If $\mathcal{F}$ is not lisse on $\mathbb{G}_m$, then its finite set S of points of nonlissenesss on $\mathbb{G}_m$ is stable by $s \mapsto \lambda s$, hence λ is a root of unity of order dividing Card(S). If $\mathcal{F}$ is lisse on $\mathbb{G}_m$, but λ is not a root of unity, then by Verdier's lemma [Ver, Prop. 1.1] $\mathcal{F}$ is tame at both zero and ∞, whence $\chi(\mathbb{G}_m, \mathcal{F}) = 0$, contradiction.

Once λ is a root of unity, say of order N, then because $\mathcal{F}$ is irreducible it descends through the N-fold Kummer covering, and hence $\chi(\mathbb{G}_m, \mathcal{F})$ is divisible by N. QED

Rigidity Corollary 8.5.5 Let $\mathcal{F}$ be an irreducible middle extension sheaf on $\mathbb{G}_m$ whose Euler characteristic $\chi(\mathbb{G}_m, \mathcal{F}) = -1$. Then the isomorphism class of $\mathcal{F}$ is determined up to (a unique) multiplicative translation by the isomorphism classes of the I(0) and I(∞)-semisimplifications of the tame parts of the local monodromy of $\mathcal{F}$ at zero and ∞.

Rigidity Corollary bis 8.5.6 Let $\mathcal{F}$ be an irreducible middle extension sheaf on $\mathbb{G}_m$ whose Euler characteristic $\chi(\mathbb{G}_m, \mathcal{F}) = -1$.

(1) If $\mathcal{F}$ is not lisse on $\mathbb{G}_m$, the isomorphism class of $\mathcal{F}$ is determined by the following three data:

the I(0)-semisimplification of $\mathcal{F}$,
the I(∞)-semisimplification of $\mathcal{F}$,
the unique point $\lambda \in k^\times$ where $\mathcal{F}$ is not lisse.

(2) If $\mathcal{F}$ is lisse on $\mathbb{G}_m$, the isomorphism class of $\mathcal{F}$ is determined by the following two data:

the I(0)-semisimplification of $\mathcal{F}$,
the I(∞)-semisimplification of $\mathcal{F}$.

proof The first assertion is an immediate consequence of the first

Rigidity Corollary, since fixing the unique point of nonlisseness in $\mathbb{G}_m$ rigidifies the situation entirely.

The second assertion is a bit more delicate. We know that $\mathcal{F}$ is a hypergeometric $\mathcal{H}_\lambda(!, \psi; \chi\text{'s}; \rho\text{'s})$ of type (n, m) with $n \neq m$. By inversion, we may suppose that $n > m$. Then $\mathcal{F}$ as $I(\infty)$-representation is of the form

$$\mathcal{F} \approx T \oplus W = (\text{tame of rank } m) \oplus (\text{rank } n-m, \text{ all breaks } 1/(n-m)).$$

Since W is the unique wild irreducible $I(\infty)$-constituent of $\mathcal{F}$, it is an intrinsic invariant of the the $I(\infty)$-semisimplification of $\mathcal{F}$. So it suffices to show that W "detects" multiplicative translations. This is proven in [Ka-GKM, 4.1.6, (3)]. QED

8.6 Local Rigidity

(8.6.1) We continue to work over an algebraically closed field of characteristic $p \neq \ell$, with ℓ-adic representations of $I(\infty)$.

We first note the following variant of Grothendieck's local monodromy theorem [Se-Ta, Appendix].

Theorem 8.6.2 Suppose that (W, ρ) is an irreducible $I(\infty)$-representation. Then an open subgroup of $I(\infty)$ acts as scalars. If detW is of finite order, then $\rho(I(\infty))$ is finite.

proof Clearly the first statement implies the second. To prove the first, denote by

$$t_\ell : I(\infty) \to \mathbb{Z}_\ell(1)$$

the canonical projection defined by the ℓ-power Kummer coverings. Recopying the beginning of the proof of the local monodromy theorem, one shows that there exists an endomorphism

$$N \in \mathrm{End}_{\overline{\mathbb{Q}}_\ell}(W)(-1)$$

such that on a sufficiently small open subgroup Γ of $I(\infty)$, we have

$$\rho(\gamma) = \exp(t_\ell(\gamma)N)$$

for every γ in Γ. This N is unique, and by unicity it is $I(\infty)$-equivariant. By irreducibility, N is scalar. QED

Local Rigidity Theorem 8.6.3 Let V and W be $I(\infty)$-representations, each of the same rank $d \geq 1$ with all breaks $= 1/d$. Then

(1) If $\lambda \in k^\times$ and $W \approx [x \mapsto \lambda x]^* W$, then $\lambda = 1$.

(2) If $d \geq 2$, and if there exists $\lambda \in k^\times$ with $V \approx [x \mapsto \lambda x]^* W$, then

$\det V \approx \det W$.

(3) If $\det V \approx \det W$, there exists a unique $\lambda \in k^\times$ with $V \approx [x \mapsto \lambda x]^* W$.

proof Assertion (1), as noted above, is proven in [Ka-GKM, 4.1.6, (3)]. If $d \geq 2$, then $\det W$ is tame, necessarily some $\mathcal{L}_\chi$, so its isomorphism class is invariant by multiplicative translation, whence (2). Assertion (3) is trivial for $d=1$, and in general the unicity in it results from (1).

The existence for $d \geq 2$ is more delicate. Consider the canonical extensions (cf. [Ka-LG, 1.5]) of V and W to $\mathbb{G}_m$. Both of them are necessarily hypergeometrics of type (d, 0) (by the intrinsic characterization of hypergeometrics), say

$$V_{can} = \mathcal{H}_\lambda(!, \psi; \chi\text{'s}; \emptyset), \qquad W_{can} = \mathcal{H}_\mu(!, \psi; \xi\text{'s}; \emptyset).$$

Looking at determinants, we see that

$$\prod \chi_i = \prod \xi_i.$$

Replacing W by a multiplicative translate of itself replaces W_{can} by the corresponding translate, so we may further assume that $\lambda = \mu$. By a further translation, we may suppose that $\lambda = \mu = 1$. The problem now is to show that for fixed ψ, the isomorphism class of the $I(\infty)$-representation of the Kloosterman sheaf

$$Kl(\psi; \chi\text{'s}) := \mathcal{H}_1(!, \psi, \chi\text{'s}; \emptyset)$$

depends only on $\prod \chi_i$. This is a special case of the following

Change of Characters Theorem 8.6.4 Let $\mathcal{H}_\lambda(!, \psi; \chi\text{'s}; \rho\text{'s})$ be a hypergeometric sheaf of type (n, m) with $n \neq m$. If $n > m$ (resp. if $n < m$), denote by $W_\lambda(!, \psi; \chi\text{'s}; \rho\text{'s})$ the wild part of its $I(\infty)$-representation (resp. of its $I(0)$-representation). For fixed (λ, ψ), the isomorphism class of $W_\lambda(!, \psi; \chi\text{'s}; \rho\text{'s})$ as $I(\infty)$-representation (resp. as $I(0)$-representation) depends only on the tame character $\prod_i \chi_i / \prod_j \rho_j$ and on the integer $n-m$.

proof The proof proceeds by induction on the quantity $|n - m|$. The statement is invariant under multiplicative translation, so we may assume that $\lambda = 1$. The statement is invariant under multiplicative inversion, so we may suppose that $n > m$. The statement is also invariant under $\otimes \mathcal{L}_\Lambda$. Given two $W_\lambda(!, \psi; \chi\text{'s}; \rho\text{'s})$ and $W_\lambda(!, \psi; \alpha\text{'s}; \beta\text{'s})$ with the same $n - m$, and with $\prod_i \chi_i / \prod_j \rho_j = \prod_i \alpha_i / \prod_j \beta_j$, twisting by a sufficiently general $\mathcal{L}_\Lambda$ reduces us to the case where none of the χ's and none of the α's is $\mathbb{1}$. By the Cancellation Theorem 8.4.7, the statement is invariant under "cancelling", so we may further assume that the χ's and ρ's are disjoint, and that the α's and β's are disjoint.

In this case, we have (cf. the proof of 8.4.2) equalities of irreducible Fourier sheaves on $\mathbb{A}^1$

$$j_*\mathcal{H}_1(!, \overline{\psi}; \mathbb{1}, \overline{\rho}\text{'s}; \overline{\chi}\text{'s}) \approx \mathrm{NFT}_{\overline{\psi}}(j_*\mathcal{H}_1(!, \psi; \chi\text{'s}; \rho\text{'s})),$$

$$j_*\mathcal{H}_1(!, \overline{\psi}; \mathbb{1}, \overline{\beta}\text{'s}; \overline{\alpha}\text{'s}) \approx \mathrm{NFT}_{\overline{\psi}}(j_*\mathcal{H}_1(!, \psi; \alpha\text{'s}; \beta\text{'s})).$$

The sheaves $j_*\mathcal{H}_1(!, \psi; \chi\text{'s}; \rho\text{'s})$ and $j_*\mathcal{H}_1(!, \psi; \alpha\text{'s}; \beta\text{'s})$ are lisse on $\mathbb{G}_m$, and tame at zero. If $n-m \geq 2$, all their ∞-breaks are < 1, and the sheaves $\mathcal{H}_1(!, \overline{\psi}; \mathbb{1}, \overline{\rho}\text{'s}; \overline{\chi}\text{'s})$ and $\mathcal{H}_1(!, \overline{\psi}; \mathbb{1}, \overline{\beta}\text{'s}; \overline{\alpha}\text{'s})$ are lisse on $\mathbb{G}_m$ and tame at ∞. So by Fourier inversion and stationary phase (7.4.1.1, 7.4.2) we have

$$W_1(!, \psi; \chi\text{'s}; \rho\text{'s}) \approx \mathrm{FT}_{\overline{\psi}}\mathrm{loc}(0, \infty)(W_1(!, \overline{\psi}; \mathbb{1}, \overline{\rho}\text{'s}; \overline{\chi}\text{'s})),$$

$$W_1(!, \psi; \alpha\text{'s}; \beta\text{'s}) \approx \mathrm{FT}_{\overline{\psi}}\mathrm{loc}(0, \infty)(W_1(!, \overline{\psi}; \mathbb{1}, \overline{\beta}\text{'s}; \overline{\alpha}\text{'s})).$$

By induction on $|n - m|$, $W_1(!, \overline{\psi}; \mathbb{1}, \overline{\rho}\text{'s}; \overline{\chi}\text{'s})$ and $W_1(!, \overline{\psi}; \mathbb{1}, \overline{\beta}\text{'s}; \overline{\alpha}\text{'s})$ are isomorphic, which concludes the proof in this case.

If $n-m = 1$, $j_*\mathcal{H}_1(!, \psi; \chi\text{'s}; \rho\text{'s})$ and $j_*\mathcal{H}_1(!, \psi; \alpha\text{'s}; \beta\text{'s})$ have a single nonzero ∞-break, which is 1, and the sheaves $\mathcal{H}_1(!, \overline{\psi}; \mathbb{1}, \overline{\rho}\text{'s}; \overline{\chi}\text{'s})$ and $\mathcal{H}_1(!, \overline{\psi}; \mathbb{1}, \overline{\beta}\text{'s}; \overline{\alpha}\text{'s})$ are lisse on $\mathbb{G}_m - \{1\}$, everywhere tame, with tame pseudoreflection local monodromy at 1. By stationary phase, we deduce as in 7.4.6.1 that as $I(\infty)$-representations

$$\mathcal{H}_1(!, \psi; \chi\text{'s}; \rho\text{'s})(\infty) \approx (\mathcal{L}_{\psi(x)} \otimes \mathcal{L}_{\Lambda_1}) \oplus (\text{succ. ext. of } \mathcal{L}_\rho\text{'s}),$$

$$\mathcal{H}_1(!, \psi; \alpha\text{'s}; \beta\text{'s})(\infty) \approx (\mathcal{L}_{\psi(x)} \otimes \mathcal{L}_{\Lambda_2}) \oplus (\text{succ. ext. of } \mathcal{L}_\beta\text{'s}).$$

for some tame characters Λ_1 and Λ_2. It remains only to show that

$$\Lambda_1 = \textstyle\prod_i \chi_i / \prod_j \rho_j.$$

Consider $\mathcal{L}_{\psi(-x)} \otimes \det\mathcal{H}_1(!, \psi; \chi\text{'s}; \rho\text{'s})$. It is lisse on $\mathbb{G}_m$, everywhere tame, so of the form $\mathcal{L}_\Gamma$ for some tame character Γ. Looking at zero, we see that $\Gamma = \prod_i \chi_i$, while looking at ∞ we see that $\Gamma = \Lambda_1 \prod_j \rho_j$; equationg these two expressions for Γ gives the asserted formula for Λ_1. QED

8.7 Multiplicative Translation and Change of ψ

(8.7.1) Let k_0 be a finite subfield of k over which ψ is defined. For any $\mu \in (k_0)^\times$, denote by ψ_μ the additive character of k_0 defined by

$$\psi_\mu(x) := \psi(\mu x).$$

The sheaves $\mathcal{L}_\psi$ and $\mathcal{L}_{\psi_\mu}$ are related by

$$\mathcal{L}_{\psi_\mu} = [x \mapsto \mu x]^* \mathcal{L}_\psi = [x \mapsto x/\mu]_* \mathcal{L}_\psi.$$

Similarly, their multipicative inverses on $\mathbb{G}_m$ are related by

$$\mathrm{inv}^* j^* \mathcal{L}_{\overline{\psi}_\mu} = [x \mapsto x/\mu]^* \mathrm{inv}^* j^* \mathcal{L}_{\overline{\psi}} = [x \mapsto \mu x]_* \mathrm{inv}^* j^* \mathcal{L}_{\overline{\psi}}.$$

If χ is a multiplicative character of k_0, then

$$\mathcal{L}_\chi \otimes (\chi(\mu))^{\deg} \approx [x \mapsto \mu x]^* \mathcal{L}_\chi = [x \mapsto x/\mu]_* \mathcal{L}_\chi$$

on $\mathbb{G}_m$ over k_0. For χ an arbitrary tame character, we have

$$\mathcal{L}_\chi \approx [x \mapsto \mu x]^* \mathcal{L}_\chi = [x \mapsto x/\mu]_* \mathcal{L}_\chi$$

on $\mathbb{G}_m$ over k.

Lemma 8.7.2 Over an algebraically closed field k of characteristic $p \neq \ell$, for any hypergeometric $\mathrm{Hyp}_\lambda(!, \psi; \chi\text{'s}; \rho\text{'s})$ of type (n, m), and for any μ in any finite subfield of k, we have

$$\mathrm{Hyp}_\lambda(!, \psi_\mu; \chi\text{'s}; \rho\text{'s}) \approx [x \mapsto x/\mu^{n-m}]_* \mathrm{Hyp}_\lambda(!, \psi; \chi\text{'s}; \rho\text{'s})$$
$$\approx [x \mapsto x\mu^{n-m}]^* \mathrm{Hyp}_\lambda(!, \psi; \chi\text{'s}; \rho\text{'s}).$$

Over a finite field k_0 of characteristic $p \neq \ell$, for any hypergeometric $\mathrm{Hyp}_\lambda(!, \psi; \chi\text{'s}; \rho\text{'s})$ of type (n, m) which is defined over k_0, and for any μ in k_0, if we define $\alpha \in \overline{\mathbb{Q}}_\ell^\times$ to be

$$\alpha := (\prod_i \chi_i / \prod_j \rho_j)(\mu),$$

we have

$$\mathrm{Hyp}_\lambda(!, \psi_\mu; \chi\text{'s}; \rho\text{'s}) \otimes \alpha^{\deg} \approx [x \mapsto x/\mu^{n-m}]_* \mathrm{Hyp}_\lambda(!, \psi; \chi\text{'s}; \rho\text{'s}).$$
$$\approx [x \mapsto x\mu^{n-m}]^* \mathrm{Hyp}_\lambda(!, \psi; \chi\text{'s}; \rho\text{'s}).$$

proof By the interrelation (cf. 8.1.10 (3)) of convolution and translation we reduce immediately to the case when (n, m) is either (1, 0) or (0, 1), where it is obvious. QED

Corollary 8.7.3 If p(n-m) is even, then over an algebraically closed field k of characteristic $p \neq \ell$ we have

$$\mathrm{Hyp}_\lambda(!, \overline{\psi}; \chi\text{'s}; \rho\text{'s}) \approx \mathrm{Hyp}_\lambda(!, \psi; \chi\text{'s}; \rho\text{'s}).$$

If in addition $\mathrm{Hyp}_\lambda(!, \psi; \chi\text{'s}; \rho\text{'s})$ is defined over a finite subfield k_0 of k, and $(\prod_i \chi_i/\prod_j \rho_j)(-1) = 1$ (a condition which always holds after passing to a quadratic extension), then over k_0 we have

$$\mathrm{Hyp}_\lambda(!, \overline{\psi}; \chi\text{'s}; \rho\text{'s}) \approx \mathrm{Hyp}_\lambda(!, \psi; \chi\text{'s}; \rho\text{'s}).$$

proof This is the Lemma with $\mu = -1$. QED

Corollary 8.7.4 If $n = m$, then over an algebraically closed field k of characteristic $p \neq \ell$ the isomorphism class of $\mathrm{Hyp}_\lambda(!, \psi; \chi\text{'s}; \rho\text{'s})$ is independent of ψ.

Over a finite field k_0 of characteristic $p \neq \ell$, for any hypergeometric $\mathrm{Hyp}_\lambda(!, \psi; \chi\text{'s}; \rho\text{'s})$ of type (n, n) which is defined over k_0, and for any μ in k_0, if we define $\alpha \in \overline{\mathbb{Q}}_\ell^\times$ to be

$$\alpha := (\prod_i \chi_i/\prod_j \rho_j)(\mu),$$

we have

$$\mathrm{Hyp}_\lambda(!, \psi_\mu; \chi\text{'s}; \rho\text{'s}) \otimes \alpha^{\deg} \approx \mathrm{Hyp}_\lambda(!, \psi; \chi\text{'s}; \rho\text{'s}).$$

proof This is the lemma with n=m. QED

8.8 Global and Local Duality Recognition

Duality Recognition Theorem 8.8.1 Suppose that the χ's and ρ's are disjoint. The irreducible middle extension sheaf $\mathcal{H}_\lambda(!, \psi; \chi\text{'s}; \rho\text{'s})$ is geometrically self-dual (i.e., geometrically isomorphic to its dual), if and only if the following three conditions hold:

(1) the set of χ's with multiplicity is stable under $\chi \mapsto \overline{\chi}$,

(2) the set of ρ's with multiplicity is stable under $\rho \mapsto \overline{\rho}$,

(3) the product p(n - m) is even.

proof By (8.4.2, 8.3.3), the dual of $\mathcal{H}_\lambda(!, \psi; \chi\text{'s}; \rho\text{'s})$ is $\mathcal{H}_\lambda(!, \overline{\psi}; \overline{\chi}\text{'s}; \overline{\rho}\text{'s})$, so the conditons are obviously sufficient. In order for $\mathcal{H}_\lambda(!, \psi; \chi\text{'s}; \rho\text{'s})$ and $\mathcal{H}_\lambda(!, \overline{\psi}; \overline{\chi}\text{'s}; \overline{\rho}\text{'s})$ to be geometrically isomorphic, their local monodromies at zero and ∞ must agree, whence (1) and (2). If (1) and (2) hold, then the dual of $\mathcal{H}_\lambda(!, \psi; \chi\text{'s}; \rho\text{'s})$ is its multiplicative translate by $(-1)^{n-m}$, to which it is isomorphic if and only if $(-1)^{n-m} = 1$ in the field k. QED

Parity Recognition Theorem 8.8.2 Suppose that the χ's and ρ's are disjoint, and that $\mathcal{H}_\lambda(!, \psi; \chi\text{'s}; \rho\text{'s})$ is self dual. Then on the open set of $\mathbb{G}_m$ where $\mathcal{H}_\lambda(!, \psi; \chi\text{'s}; \rho\text{'s})$ is lisse, the (unique up to a $\overline{\mathbb{Q}}_\ell{}^\times$-multiple, by irreduciblity) autodaulity pairing is alternating if and only if

$$\max(n, m) \text{ is even, } n - m \text{ is even, and } \prod_i\chi_i/\prod_j\rho_j = \mathbb{1}.$$

Otherwise (i.e., if $\max(n, m)$ is odd, or or if $n - m$ is odd, or if $\prod_i\chi_i/\prod_j\rho_j \neq \mathbb{1}$) the pairing is symmetric.

proof If the generic rank $\max(n, m)$ is odd, the pairing has no choice but to be symmetric.

Suppose now that $\max(n, m)$ is even. By multiplicative inversion, we may suppose that $n \geq m$. By multiplicative translation, we may suppose $\lambda = 1$.

If $n = m$, then the character $\Lambda := \prod_i\chi_i/\prod_j\rho_j$ is of order dividing two, because $\mathcal{L}_{\Lambda(x-1)}$ is the determinant of the (pseudoreflection) local monodromy at 1. If $\Lambda = \mathbb{1}$, local monodromy at λ is a unipotent pseudoreflection; as $O(n)$ contains no unipotent pseudoreflections (cf. the proof of 3.4), the autoduality must be alternating. If Λ is nontrivial, then the pairing must be symmetric; it cannot be alternating since $Sp(n) \subset SL(n)$.

If $n - m \geq 1$, then as $I(\infty)$-representation we have (in the notations of the proof of 8.6.4)

$$\mathcal{H}_1(!, \psi; \chi\text{'s}; \rho\text{'s}) \approx W_1(!, \psi; \chi\text{'s}; \rho\text{'s}) \oplus (\text{tame}),$$

with $W_1(!, \psi; \chi\text{'s}; \rho\text{'s})$ totally wild of Swan conductor 1, and (hence) $I(\infty)$-irreducible and Jordan-Holder disjoint from the "tame" factor. Therefore the global autoduality of $\mathcal{H}_1(!, \psi; \chi\text{'s}; \rho\text{'s})$ must induce an autoduality of $W_1(!, \psi; \chi\text{'s}; \rho\text{'s})$ as irreducible $I(\infty)$-representation. Of course this local autoduality has the same sign as the global one which induces it. By 8.6.4, $W_1(!, \psi; \chi\text{'s}; \rho\text{'s})$ depends, for fixed ψ, only upon the integer $d := n - m$ and the tame character Λ. Also, the character Λ has order dividing two, since by the Duality Recognition Thm 8.8.1 it is invariant by $\Lambda \mapsto \overline{\Lambda}$. So we are reduced to proving that the global theorem holds for the self-dual rank d Kloosterman sheaf

$$\mathrm{Kl}(\psi; \Lambda, d-1\ \mathbb{1}\text{'s}).$$

If d is odd, the duality must be symmetric.

If $d \geq 2$, then $\det\mathrm{Kl}(\psi; \Lambda, d-1\ \mathbb{1}\text{'s}) \approx \mathcal{L}_\Lambda$. So if Λ is nontrivial (or if d is odd), the pairing must be symmetric. It remains to treat the case where $d \geq 2$ is even, and Λ is trivial, i.e., the case of the Kloosterman sheaf $\mathrm{Kl}(\psi; d\ \mathbb{1}\text{'s})$. We must show that the pairing is alternating in this case. This is proven in [Ka-GKM, 4.2.1 or 5.5.1]. QED

Local Duality Recognition Theorem 8.8.3 Let W be an $I(\infty)$-representation of rank $d \geq 1$, with all breaks = $1/d$. Then
(1) W is self-dual if and only if $\det(W)$ has order dividing two and pd is even.
(2) If W is self-dual, the autoduality pairing is alternating if and only if d is even and $\det(W)$ is trivial.

proof If detW does not have order dividing two, W cannot be self dual. So we may suppose henceforth that detW has order dividing two.

If $d = 1$, then W is $\mathcal{L}_\psi \otimes \mathcal{L}_\Lambda$, which is self-dual if and only if $p = 2$ and Λ is trivial; in this case the autoduality is symmetric, as required.

If $d \geq 2$, detW is tame, say $\mathcal{L}_\Lambda$. The canonical extension of W is an $\mathcal{H}_\lambda(!, \psi; \chi\text{'s}; \emptyset)$ of type $(d, 0)$, which, being the canonical extension, is self-dual of given parity if and only if W is self-dual of the same parity. Comparing determinants, we find $\Lambda = \prod_i \chi_i$.

If pd is odd, then $\mathcal{H}_\lambda(!, \psi; \chi\text{'s}; \emptyset)$ cannot be self-dual (by 8.8.1), so we may suppose that pd is even.

By 8.5.4, W is isomorphic to the $I(\infty)$-representation attached to $\mathcal{H}_\lambda(!, \psi; \Lambda, d\text{-}1\ \mathbb{1}\text{'s})$. So it suffices to show that if

$$d \geq 2, \text{ pd is even, and } \Lambda \text{ has order 1 or 2}$$

then $\mathcal{H}_\lambda(!, \psi; \Lambda, d\text{-}1\ \mathbb{1}\text{'s})$ is self-dual, and the autoduality is alternating if and only if d is even and Λ is trivial. This is a special case of the 8.8.1 and 8.8.2 QED

8.9 Kummer Induction Formulas and Recognition Criteria

Kummer Induction Theorem 8.9.1 Let $d \geq 1$ be an integer which is prime to p, and denote by [d] the d'th power endomorphism of $\mathbb{G}_m$. Over an algebraically closed field k of characteristic $p \neq \ell$, for any hypergeometric $\mathrm{Hyp}_\lambda(!, \psi; \chi\text{'s}; \rho\text{'s})$ there exists an isomorphism

$$[d]_* \mathrm{Hyp}_\lambda(!, \psi_d; \chi\text{'s}; \rho\text{'s}) \approx$$

$$\approx \mathrm{Hyp}_{\lambda^d}(!, \psi; \text{all d'th roots of all } \chi\text{'s}; \text{all d'th roots of all } \rho\text{'s}).$$

proof By multiplicative translation, we reduce to the case $\lambda = 1$. Since [d] is a homomorphism from $\mathbb{G}_m$ to itself, we have the convolution relation

$$[d]_*(K*_!L) \approx ([d]_*K)*_!([d]_*L).$$

So we are reduced to the case where $\mathrm{Hyp}_1(!, \psi_d; \chi\text{'s}; \rho\text{'s})$ is of type (1, 0) or (0, 1). Since $[d]_*$ and inv^* (= inv_*) commute, the (0, 1) case results from the (1, 0) case.

Since every tame character χ has a d'th root, say $\chi = \xi^d$, our (1, 0) hypergeometric $\mathcal{L}_{\psi_d}\otimes\mathcal{L}_\chi$ may be rewritten $\mathcal{L}_{\psi_d}\otimes[d]^*\mathcal{L}_\xi$. Applying $[d]_*$, and using the projection formula, we get

$$[d]_*(\mathcal{L}_{\psi_d}\otimes\mathcal{L}_\chi) \approx ([d]_*(\mathcal{L}_{\psi_d}))\otimes\mathcal{L}_\xi.$$

So we are reduced to showing that, denoting by

$$\Lambda_1, \Lambda_2, \ldots, \Lambda_d$$

the d tame characters of order d, we have an isomorphism

$$[d]_*(\mathcal{L}_{\psi_d}) \approx \mathrm{Kl}(\psi; \Lambda_1, \Lambda_2, \ldots, \Lambda_d).$$

This is proven in [Ka-GKM, 5.6.2]. QED

Corollary 8.9.2 (Kummer Recognition) Let $d \geq 1$ be an integer which is prime to p, and denote by [d] the d'th power endomorphism of $\mathbb{G}_m$. Over an algebraically closed field k of characteristic $p \neq \ell$, an irreducible hypergeometric $\mathrm{Hyp}_\lambda(!, \psi; \chi\text{'s}; \rho\text{'s})$ of type (n, m) is Kummer induced of degree d, i.e., of the form $[d]_*K$ for some K in $D^b_c(\mathbb{G}_m, \overline{\mathbb{Q}}_\ell)$, if and only if the following three conditions are satisfied:
(1) d divides both n and m,
(2) there exists a set of n/d tame characters α's such that the χ's are all the d'th roots of all the α's,
(3) there exists a set of m/d tame characters β's such that the ρ's are all the d'th roots of all the β's.
Moreover, if these conditions hold, then for any $\mu \in k$ with $\mu^d = \lambda$, there exists an isomorphism

$$\mathrm{Hyp}_\lambda(!, \psi; \chi\text{'s}; \rho\text{'s}) \approx [d]_*\mathrm{Hyp}_\mu(!, \psi_d; \alpha\text{'s}; \beta\text{'s}).$$

proof If any (not necessarily irreducible) $\mathrm{Hyp}_\lambda(!, \psi; \chi\text{'s}; \rho\text{'s})$ of type (n, m) satisfies (1), (2), and (3), then by the Kummer Induction Theorem above, for any $\mu \in k$ with $\mu^d = \lambda$,there exists an isomorphism

$$\mathrm{Hyp}_\lambda(!, \psi; \chi\text{'s}; \rho\text{'s}) \approx [d]_*\mathrm{Hyp}_\mu(!, \psi_d; \alpha\text{'s}; \beta\text{'s}).$$

Conversely, suppose that an **irreducible** $\mathrm{Hyp}_\lambda(!, \psi; \chi\text{'s}; \rho\text{'s})$ is of the form $[d]_*K$ for some K in $D^b_c(\mathbb{G}_m, \overline{\mathbb{Q}}_\ell)$. Then $[d]_*K$ is perverse simple with $\chi(\mathbb{G}_m, [d]_*K) = 1$. Since $[d]$ is finite, K must be perverse simple, and $\chi(\mathbb{G}_m, K) = \chi(\mathbb{G}_m, [d]_*K) = 1$. Therefore K is itself irreducible hypergeometric, so of the form $\mathrm{Hyp}_\mu(!, \psi_d; \alpha\text{'s}; \beta\text{'s})$. Looking at the Kummer induction formula for $[d]_*\mathrm{Hyp}_\mu(!, \psi_d; \alpha\text{'s}; \beta\text{'s})$, we see that (1), (2), and (3) hold. QED

Remark 8.9.3 An alternate formulation of conditions (1), (2), and (3) is this: for some (or equivalently for every) tame character Λ of exact order d, both the χ's and the ρ's as sets with multiplicity are stable under the operation $\xi \mapsto \xi\Lambda$.

8.10 Belyi Induction Formulas and Recognition Criteria

Belyi Recognition Criterion 8.10.1 Over an algebraically closed field k of characteristic $p \neq \ell$, let $\mathrm{Hyp}_\lambda(!, \psi; \chi\text{'s}; \rho\text{'s})$ be an irreducible hypergeometric of type (n, n), and suppose that $p > n$. Then $\mathrm{Hyp}_\lambda(!, \psi; \chi\text{'s}; \rho\text{'s})$ is Belyi induced of type (a,b) for some partition of $n = a + b$ as the sum of two strictly positive integers if and only if the χ's and the ρ's are Belyi induced in the sense that there exist tame characters α and β, $\beta \neq \mathbb{1}$, such that

(1) $\{\chi\text{'s}\} = \{$all a'th roots of $\alpha\} \cup \{$ all b'th roots of $\beta\}$,

(2) $\{\rho\text{'s}\} = \{$ all a+b 'th roots of $\alpha\beta\}$.

Moreover, if (1) and (2) hold, then

$$\mathrm{Hyp}_\lambda(!, \psi; \chi\text{'s}; \rho\text{'s}) \approx [\mathrm{Bel}_{a,b,\lambda}]_*\mathrm{Hyp}_1(!, \psi; \alpha; \alpha\beta),$$

and the local monodromy at λ of $\mathcal{H}_\lambda(!, \psi; \chi\text{'s}; \rho\text{'s})$ is a reflection.

proof If an irreducible hypergeometric $\mathrm{Hyp}_\lambda(!, \psi; \chi\text{'s}; \rho\text{'s})$ of type (n, n) is of the form $[x \mapsto \mathrm{Bel}_{a,b,\lambda}(x)]_*K$ for some K in $D^b_c(\mathbb{G}_m, \overline{\mathbb{Q}}_\ell)$, then K is perverse simple, lisse on $\mathbb{G}_m - \{1\}$, and

$$\chi(\mathbb{G}_m, K) = \chi(\mathbb{G}_m, [\mathrm{Bel}_{a,b,\lambda}]_*K) = 1.$$

Therefore K is itself hypergeometric of type (1, 1) with singularity at 1, i.e., K is $\mathrm{Hyp}_1(!, \psi; \alpha; \alpha\beta)$,with $\beta \neq \mathbb{1}$ (by irreducibility). Looking at the local monodromy at zero and ∞ of $[\mathrm{Bel}_{a,b,\lambda}]_*\mathrm{Hyp}_1(!, \psi; \alpha; \alpha\beta)$, we see that (1) and (2) hold. By (7.2.6(8), its local monodromy at λ is a reflection.

Conversely, if (1) and (2) hold, we claim that

$$\mathrm{Hyp}_\lambda(!, \psi; \chi\text{'s}; \rho\text{'s}) \approx [\mathrm{Bel}_{a,b,\lambda}]_* \mathrm{Hyp}_1(!, \psi; \alpha; \alpha\beta).$$

To see this, we argue as follows. Since $\mathrm{Hyp}_1(!, \psi; \alpha; \alpha\beta)$ is perverse with $\chi(\mathbb{G}_m, \mathrm{Hyp}_1(!, \psi; \alpha, \alpha\beta)) = 1$, and $\mathrm{Bel}_{a,b,\lambda}$ is finite, the direct image $[\mathrm{Bel}_{a,b,\lambda}]_* \mathrm{Hyp}_1(!, \psi; \alpha, \alpha\beta)$ is itself perverse and has

$$\chi(\mathbb{G}_m, [\mathrm{Bel}_{a,b,\lambda}]_* \mathrm{Hyp}_1(!, \psi; \alpha; \alpha\beta)) = 1.$$

In the caterory $\mathrm{Perv}(\mathbb{G}_m)$, any object K with $\chi(\mathbb{G}_m, K) = 1$ is a successive extension of $\mathcal{L}_\Lambda$'s and of a single perverse simple L with $\chi(\mathbb{G}_m, L) = 1$. Moreover, there are no $\mathcal{L}_\Lambda$'s if and only if the tame parts of local monodromy at zero and ∞ of $\mathcal{H}^{-1}K$ are disjoint. We may apply this to the object $K := [\mathrm{Bel}_{a,b,\lambda}]_* \mathrm{Hyp}_1(!, \psi; \alpha; \alpha\beta)$, because its local monodromy at zero is $\{\chi\text{'s}\}$, and that at ∞ is $\{\rho\text{'s}\}$, and by hypothesis the $\{\chi\text{'s}\}$ and $\{\rho\text{'s}\}$ are disjoint. Therefore K is itself an irreducible hypergeometric of type (n, n). Looking at its local monodromy at zero, λ, and ∞ shows (by rigidity) that K is none other than $\mathrm{Hyp}_\lambda(!, \psi; \chi\text{'s}; \rho\text{'s})$. QED

Lemma 8.10.2 (compare 3.5.2, 3.5.7) Over an algebraically closed field k of characteristic $p \neq \ell$, let $\mathcal{H}_\lambda(!, \psi; \chi\text{'s}; \rho\text{'s})$ be an irreducible hypergeometric of type (n, n), and suppose that $p > n \geq 1$. Then either

(1) $\mathcal{H}_\lambda(!, \psi; \chi\text{'s}; \rho\text{'s}) \mid \mathbb{G}_m - \{\lambda\}$ is Lie-irreducible;

(2a) $\mathcal{H}_\lambda(!, \psi; \chi\text{'s}; \rho\text{'s}) \mid \mathbb{G}_m - \{\lambda\}$ is Kummer induced of some degree $d \geq 2$ prime to p;

(2b) Either $\mathcal{H}_\lambda(!, \psi; \chi\text{'s}; \rho\text{'s}) \mid \mathbb{G}_m - \{\lambda\}$ or $\mathrm{inv}^*\mathcal{H}_\lambda(!, \psi; \chi\text{'s}; \rho\text{'s}) \mid \mathbb{G}_m - \{\lambda\}$ is Belyi induced of type (a,b) for some partition of $n = a + b$ as the sum of two strictly positive integers;

(3) $\mathcal{H}_\lambda(!, \psi; \chi\text{'s}; \rho\text{'s}) \mid \mathbb{G}_m - \{\lambda\}$ is the tensor product $\mathcal{L}\otimes\mathcal{F}$ of a lisse rank one $\mathcal{L}$ with a lisse irreducible $\mathcal{F}$ of rank n whose G_{geom} is finite. If in addition $\det(\mathcal{H}_\lambda(!, \psi; \chi\text{'s}; \rho\text{'s}) \mid \mathbb{G}_m - \{\lambda\})$ is of finite order, then $\mathcal{H}_\lambda(!, \psi; \chi\text{'s}; \rho\text{'s}) \mid \mathbb{G}_m - \{\lambda\}$ itself has finite G_{geom} in this case (3).

proof This is a special case of 7.2.6, (1), (6), and (8). QED

Lemma 8.10.3 (Geometric Determinant Formula) Over an algebraically closed field k of characteristic $p \neq \ell$, let $\mathcal{H}_\lambda(!, \psi; \chi\text{'s}; \rho\text{'s})$ be an arbitrary hypergeometric of type (n, n). If $\prod_i\chi_i = \prod_i\rho_i = \Lambda$, then

$$\det(\mathcal{H}_\lambda(!, \psi; \chi\text{'s}; \rho\text{'s}) \mid \mathbb{G}_m - \{\lambda\}) \approx \mathcal{L}_\Lambda.$$

If $\prod_i \chi_i \neq \prod_i \rho_i$, then

$$\det(\mathcal{H}_\lambda(!, \psi; \chi\text{'s}; \rho\text{'s}) \mid \mathbb{G}_m - \{\lambda\}) \approx \mathcal{H}_\lambda(!, \psi; \prod_i \chi_i ; \prod_i \rho_i).$$

proof In the first case, local monodromy at λ is a unipotent pseudoreflection, so $\det(\mathcal{H}_\lambda(!, \psi; \chi\text{'s}; \rho\text{'s}) \mid \mathbb{G}_m - \{\lambda\})$ is unramified at λ. Since $\mathcal{H}_\lambda(!, \psi; \chi\text{'s}; \rho\text{'s})$ is everywhere tame, its determinant must be an $\mathcal{L}_\Lambda$, and we can compute Λ as the determinant of local monodromy at zero. In the second case, $\det(\mathcal{H}_\lambda(!, \psi; \chi\text{'s}; \rho\text{'s}) \mid \mathbb{G}_m - \{\lambda\})$ is everywhere tame, lisse on $\mathbb{G}_m - \{\lambda\}$ of rank one with nontrivial local monodromy at λ. So $\det(\mathcal{H}_\lambda(!, \psi; \chi\text{'s}; \rho\text{'s}) \mid \mathbb{G}_m - \{\lambda\})$ must be $\mathcal{H}_\lambda(!, \psi; \prod_i \chi_i ; \prod_i \rho_i)$. QED

Corollary 8.10.4 Over an algebraically closed field k of characteristic $p \neq \ell$, let $\mathcal{H}_\lambda(!, \psi; \chi\text{'s}; \rho\text{'s})$ be an arbitrary hypergeometric of type (n, n). Then $\det(\mathcal{H}_\lambda(!, \psi; \chi\text{'s}; \rho\text{'s}) \mid \mathbb{G}_m - \{\lambda\})$ is of finite order (resp. trivial) if and only if both of the tame characters $\prod_i \chi_i$ and $\prod_i \rho_i$ are of finite order (resp. trivial).

8.11 Calculation of G_{geom} for irreducible hypergeometrics

(8.11.1) Throughout this section we work over an algebraically closed field k of characteristic $p \neq \ell$.

Theorem 8.11.2 (compare 3.5.8) Let $\mathcal{H} := \mathcal{H}_\lambda(!, \psi; \chi\text{'s}; \rho\text{'s})$ be an irreducible hypergeometric of type (n, n). Suppose that $p > n \geq 1$ and that $\mathcal{H} \mid \mathbb{G}_m - \{\lambda\}$ is neither Kummer induced nor Belyi induced nor inverse Belyi induced. Let $\Lambda := \prod_i \chi_i / \prod_i \rho_i$. Denote by G the group $G := G_{geom}$ for $\mathcal{H} \mid \mathbb{G}_m - \{\lambda\}$. Then

(1) The group G is reductive. If both $\prod_i \chi_i$ and $\prod_i \rho_i$ are of finite order, then $G^0 = G^{0,der}$. Otherwise, $G^0 = \mathbb{G}_m G^{0,der}$.

(2) The group $G^{0,der}$ is either $\{1\}$, SL(n), SO(n), or (if n is even) Sp(n).

(3) If Λ does not have order dividing 2, $G^{0,der} = \{1\}$ or SL(n).

(4) If Λ has exact order 2, $G^{0,der} = \{1\}$ or SO(n) or SL(n).

(5) if $\Lambda = \mathbb{1}$, $G^{0,der} =$ SL(n) or (if n is even) Sp(n).

(6) If Λ is not of finite order, $G = GL(n)$.

proof Local monodromy around λ is a pseudoreflection of determinant $\mathcal{L}_{\Lambda(x - \lambda)}$. So if $\mathcal{H} \mid \mathbb{G}_m - \{\lambda\}$ is Lie irreducible, the theorem is an immediate consequence of the Pseudoreflection Theorem 1.5. In view of 8.10.2, the only other case is when $\mathcal{H} \mid \mathbb{G}_m - \{\lambda\}$ is the tensor product $\mathcal{L}\otimes\mathcal{F}$ of a lisse rank one $\mathcal{L}$ with a lisse irreducible $\mathcal{F}$ of rank n whose G_{geom} is finite. In this case G^0 is either $\{1\}$ or $\mathbb{G}_m$, depending on whether or not $\mathcal{L}$, or equivalently $\det\mathcal{H} \mid \mathbb{G}_m - \{\lambda\}$, is of finite order. So (1) through (4) hold (trivially) in this case. If Λ is either trivial or of infinite order, then we cannot be in this case, for then local monodromy around λ is a either a unipotent pseudoreflection or is $\mathrm{Diag}(\Lambda, 1, 1,\ldots, 1)$, no power of which is scalar. QED

We can be more precise about the distinguishing the various Lie-irreducible cases. (We will discuss in section 8.17 how to detect the case when $G^{0,der}$ is $\{1\}$.)

Corollary 8.11.2.1 Notations and hypotheses as above, suppose further that $G^{0,der} \neq \{1\}$. Then $G^{0,der}$ is SO(n) (respectively Sp(n)) if and only if there exists a tame character ξ such that

$$\mathcal{H}\otimes\mathcal{L}_\xi := \mathcal{H}_\lambda(!, \psi;\ \xi\chi\text{'s};\ \xi\rho\text{'s})$$

is self dual and its autoduality pairing is symmetric (resp. alternating).

proof Entirely analogous to that of 3.5.8.1. QED

In the case $n \neq m$, we have

Theorem 8.11.3 Suppose that $\mathcal{H} := \mathcal{H}_\lambda(!, \psi;\ \chi\text{'s};\ \rho\text{'s})$ is an irreducible hypergeometric of type (n,m), $n \neq m$, which is not Kummer induced. Let $N := \max(n,m)$ be the rank of $\mathcal{H}$, $d := |n-m|$, and G the group G_{geom} for $\mathcal{H}$. Suppose $p > 2N + 1$. If $d < N$, suppose also that p does not divide the integer $2N_1(d)N_2(d)$ of 7.1.1. Then

(1) G is reductive. If $\det\mathcal{H}$ is of finite order (i.e., for $n > m$, if $\prod_i\chi_i$ is of finite order; for $m > n$, if $\prod_j\rho_j$ is of finite order) , then $G^0 = G^{0,der}$; otherwise $G^0 = \mathbb{G}_m G^{0,der}$.

(2) If d is odd, $G^{0,der}$ is SL(N). If $d = 1$ then $G \supset \mu_p SL(N)$.

(3) If d is even, then $G^{0,der}$ is SL(N) or SO(N) or (if N is even) SP(N), or $|n-m|=6$, $N=7,8$ or 9, and $G^{0,der}$ is one of

N=7: the image of G_2 in its 7-dim'l irreducible representation

N=8: the image of Spin(7) in the 8-dim'l spin representation

the image of SL(3) in the adjoint representation

the image of SL(2)×SL(2)×SL(2) in std⊗std⊗std

the image of SL(2)×Sp(4) in std⊗std

the image of SL(2)×SL(4) in std⊗std

N=9: the image of SL(3)×SL(3) in std⊗std.

proof Since $p > 2N + 1$, $\mathcal{H}$ is Lie-irreducible, by 7.2.6 (4). So this theorem is just the special case $a/b = 1/d$ of the Main ℓ-adic Theorem 7.2.7. The only extra remark is that if $d = 1$, then det$\mathcal{H}$ has break $=1$ at either zero (if $n < m$) or ∞ (if $n > m$), hence det$\mathcal{H}$ has order divisible by p. QED

Proposition 8.11.4 (Ofer Gabber) For N=8, neither of the two groups

the image of SL(2)×Sp(4) in std⊗std

the image of SL(2)×SL(4) in std⊗std

occurs as $G^{0,\mathrm{der}}$ for a hypergeometric of type (8,2).

proof Entirely analogous to that of 4.0.1. QED

The discrimination among the various possible cases is aided by

Proposition 8.11.5 Hypotheses and notations as in 8.11.3 above, $G^{0,\mathrm{der}}$ is contained in SO(N) (resp. in Sp(N)) if and only if there exists a tame character ξ such that $\mathcal{H}\otimes\mathcal{L}_\xi := \mathcal{H}_\lambda(!, \psi; \xi\chi\text{'s}; \xi\rho\text{'s})$ is self dual and its autoduality pairing is symmetric (resp. alternating). Moreover, if pN is odd, then $G^{0,\mathrm{der}}$ is contained in SO(N) if and only if there exists a tame character ξ such that $\mathcal{H}\otimes\mathcal{L}_\xi := \mathcal{H}_\lambda(!, \psi; \xi\chi\text{'s}; \xi\rho\text{'s})$ has its $G_{\mathrm{geom}} \subset SO(N)$.

proof Entirely analogous to that of 3.6.1. QED

Lemma 8.11.6 (Geometric Determinant Formula) Let

$$\mathcal{H}:=\mathcal{H}_\lambda(!, \psi; \chi\text{'s}; \rho\text{'s})$$

be an irreducible hypergeometric of type (n,m), $n > m$. Let $\Lambda := \prod_i \chi_i$. Then

(1) if $n - m \geq 2$, $\det(\mathcal{H}) \approx \mathcal{L}_\Lambda$.

(2) if $n - m = 1$, $\det(\mathcal{H}) \approx \mathcal{H}_\lambda(!, \psi; \Lambda; \emptyset) := [x \mapsto \lambda x]_*(\mathcal{L}_\psi \otimes \mathcal{L}_\Lambda)$.

proof Since $n > m$, $\det(\mathcal{H})$ is lisse on $\mathbb{G}_m$, tame at zero, and $(\mathcal{L}_\Lambda)^\vee \otimes \det(\mathcal{H})$ extends to a lisse sheaf on $\mathbb{A}^1$.

If $n - m \geq 2$, then $\det(\mathcal{H})$ is tame at ∞ as well (since $\mathcal{H}$ has all its ∞-slopes $1/(n-m) < 1$), and hence $(\mathcal{L}_\Lambda)^\vee \otimes \det(\mathcal{H})$ is lisse on $\mathbb{A}^1$ and tame at ∞, hence geometrically constant.

If $n - m = 1$, then as $I(\infty)$-representation

$$\mathcal{H} \approx (\text{tame}) \oplus W_\lambda(!, \psi; \chi\text{'s}; \rho\text{'s}),$$

with $W_\lambda(!, \psi; \chi\text{'s}; \rho\text{'s})$ of rank one. By 8.6.4,

$$W_\lambda(!, \psi; \chi\text{'s}; \rho\text{'s}) \approx W_\lambda(!, \psi; \Lambda/(\textstyle\prod_j \rho_j); \emptyset) :=$$
$$:= [x \mapsto \lambda x]_*(\mathcal{L}_\psi) \otimes (\mathcal{L}_{\Lambda/\prod_j \rho_j}) \approx (\text{tame}) \otimes [x \mapsto \lambda x]_*(\mathcal{L}_\psi \otimes \mathcal{L}_\Lambda).$$

Therefore $\mathcal{H}_\lambda(!, \psi; \Lambda; \emptyset)^\vee \otimes \det(\mathcal{H})$ is lisse on $\mathbb{A}^1$ and tame at ∞, so geometrically constant. QED

Corollary 8.11.6.1 Suppose that $\mathcal{H} := \mathcal{H}_\lambda(!, \psi; \chi\text{'s}; \rho\text{'s})$ is an irreducible hypergeometric of type (n,m), $n > m$. Then $G_{geom} \subset SL(n)$ if and only if $n-m \geq 2$ and $\prod_i \chi_i = \mathbb{1}$.

8.11.7 Direct Sums and Tensor Products (compare 3.8)

We continue to work over an algebraically closed field k of characteristic $p \neq \ell$.

Lemma 8.11.7.1 (compare 3.8.1) Suppose that $\mathcal{H}$ and $\mathcal{H}'$ are irreducible hypergeometrics of types (n,m) and (n',m') respectively, whose generic ranks $\max(n,m)$ and $\max(n',m')$ are both ≥ 2. Suppose that there exists a dense open set $j: U \to \mathbb{G}_m$, a lisse rank one $\mathcal{L}$ on U, and an isomorphism $j^*\mathcal{H} \approx j^*\mathcal{H}' \otimes \mathcal{L}$ of lisse sheaves on U. Then

(1) $(n,m) = (n',m')$.

(2) If $n = m$, denoting by λ (resp. λ') the unique singularity of $\mathcal{H}$ (resp. $\mathcal{H}'$) in $\mathbb{G}_m$, we have $\lambda = \lambda'$.

(3) If (n,m) is not (2,1), (1,2) or (2,2), then $\mathcal{L}$ is $\mathcal{L}_\chi$ for some tame character χ, and $\mathcal{H} \approx \mathcal{H}' \otimes \mathcal{L}_\chi$ on $\mathbb{G}_m$.

proof Entirely analogous to the proof of 3.8.1. QED

Proposition 8.11.7.2 (compare 3.8.2) Suppose that $\mathcal{H}_1, \ldots, \mathcal{H}_n$ are $n \geq 2$ irreducible hypergeometrics, with $\mathcal{H}_i$ of rank $N_i \geq 2$. Suppose that
(1) if $N_i = 2$, $\mathcal{H}_i$ is of type (2,0) or (0,2).
(2)for each i, denote by $G_i \subset GL(N_i)$ the group G_{geom} of $\mathcal{H}_i$ (restricted to some dense open U where it is lisse), and suppose that $G_i{}^{0,der}$ is one of the groups

$SL(N_i)$, any $N_i \geq 2$,
$Sp(N_i)$, any even $N_i \geq 4$,
$SO(N_i)$, $N_i = 7$ or any $N_i \geq 9$,
$SO(3)$, if $N_i = 3$ and no $N_j = 2$,
$SO(5)$, if $N_i = 5$ and no $N_j = 4$,
$SO(6)$, if $N_i = 6$ and no $N_j = 4$,
$G_2 \subset SO(7)$, if $N_i = 7$,
$Spin(7) \subset SO(8)$ if $N_i = 8$, and no $N_j = 7$.

Suppose that for all $i \neq j$, and all $\alpha \in k^\times$, there exist no isomorphisms from $\mathcal{H}_i \otimes \mathcal{L}_\chi$ to either $\mathcal{H}_j$ or to its dual $(\mathcal{H}_j)^\vee$. Then group $G := G_{geom}$ of $\oplus \mathcal{H}_i$ has $G^{0,der} = \prod G_i{}^{0,der}$, and that of $\otimes \mathcal{H}_i$ has $G^{0,der}$ = the image of $\prod G_i{}^{0,der}$ in $\otimes \mathrm{std}_{n_i}$.

proof Entirely analogous to the proof of 3.8.2, using 8.5.4 in place of 3.7.7. QED

Corollary 8.11.7.2.1 (compare 3.8.2.1) Let $\mathcal{H} := \mathcal{H}_\lambda(!, \psi; \chi\text{'s}; \rho\text{'s})$ be an irreducible hypergeometric of rank $N \geq 2$. If $N = 2$, suppose that $\mathcal{H}$ is of type (2,0) or (0,2). Suppose that $\mathcal{H}$ is self-dual, and that G_{geom} (resp. $(G_{geom})^0$) is one of the groups G:

$Sp(N)$, if N even,
$SO(N)$, if $N \neq 4, 8$.
$G_2 \subset SO(7)$, if $N = 7$,
$Spin(7) \subset SO(8)$ if $N = 8$.

Let $d \geq 2$, and let $\mu_1, \ldots, \mu_d$ be d distinct elements of $k^\times$. Then the direct sum

$$\oplus_i \mathcal{H}_{\lambda/\mu_i}(!, \psi; \chi\text{'s}; \rho\text{'s}) \;=\; \oplus_i [x \mapsto \mu_i x]^*\mathcal{H}$$

has G_{geom} (resp. $(G_{geom})^0$) the d-fold product group G^d.

proof Entirely analogous to the proof of 3.8.2.1. QED

In the case of Kummer induction, 8.11.7.2 gives:

Theorem 8.11.7.3 (compare 3.8.3) Let $\mathcal{H} := \mathcal{H}_\lambda(!, \psi; \chi\text{'s}; \rho\text{'s})$ be an irreducible hypergeometric of rank $N \geq 2$. If $N = 2$, suppose that $\mathcal{H}$ is of type (2,0) or (0,2). Suppose that $\mathcal{H}$ has $(G_{geom})^{0,der}$ one of the groups G:

SL(N),
Sp(N), if N even,
SO(N), if $N \neq 4, 8$.
$G_2 \subset SO(7)$, if $N = 7$,
$Spin(7) \subset SO(8)$ if $N = 8$.

Fix an integer $d \geq 2$ prime to p. Let $S \subset \mu_d(k)$ be a nonempty subset of $\mu_d(k)$ which is maximal among all nonempty subsets of $\mu_d(k)$ which satisfy the following condition:

whenever ζ_1 and ζ_2 are distinct elements of S, and χ is a tame character, there exists no isomorphism from $\mathcal{H}_{\lambda\zeta_1}(!, \psi; \chi\text{'s}; \rho\text{'s})\otimes\mathcal{L}_\chi$ to either $\mathcal{H}_{\lambda\zeta_2}(!, \psi; \chi\text{'s}; \rho\text{'s})$ or to its dual $(\mathcal{H}_{\lambda\zeta_2}(!, \psi; \chi\text{'s}; \rho\text{'s}))^\vee$.

Then $[d]_*\mathcal{H}_\lambda(!, \psi; \chi\text{'s}; \rho\text{'s})$ has $(G_{gal})^{0,der} \approx G^S$.

proof Entirely analogous to the proof of 3.8.3. QED

8.12 Arithmetic Determinant Formula

(8.12.1) In this section, we work over a finite field k_0 of characteristic $p \neq \ell$ and cardinality q. We will compute the determinant of an arbitrary nonpunctual irreducible hypergeometric $\mathcal{H}_\lambda(!, \psi; \chi\text{'s}; \rho\text{'s})$ which is defined over k_0. At the expense of a multiplicative inversion and a multiplicative translation, we may and will assume that $n \geq m$ and that $\lambda = 1$.

Theorem 8.12.2 (Arithmetic Determinant Formula) Let k_0 be a finite field of characteristic $p \neq \ell$ and cardinality q, and

$$\mathcal{H} := \mathcal{H}_1(!, \psi; \chi\text{'s}; \rho\text{'s})$$

a nonpunctual irreducible hypergeometric defined over k_0 of type (n, m) with $n \geq m$ and $\lambda = 1$. Define

$$\Lambda := \prod_i \chi_i, \text{ a character of } k_0{}^\times,$$

$$\Gamma := \prod_j \rho_j, \text{ a character of } k_0{}^\times,$$

$$N := \Sigma_{\text{distinct } \chi \text{ among the } \chi_i} (1/2)\mathrm{mult}_0(\chi)(\mathrm{mult}_0(\chi) - 1)$$

$$A := q^N[\prod_{i_1,i_2} (-g(\psi, \chi_{i_1}/\chi_{i_2}))]\times[\prod_{i,j} (-g(\bar{\psi}, \bar{\rho}_j/\bar{\chi}_i))]$$

$$= \Lambda((-1)^{n-1})q^{n(n-1)/2}\prod_{i,j} (-g(\bar{\psi}, \bar{\rho}_j/\bar{\chi}_i)) \in \bar{\mathbb{Q}}_\ell.$$

Then

(1a) if $n=m$ and $\Lambda = \Gamma$, $\det(\mathcal{H})\otimes(\mathcal{L}_\Lambda)^\vee \approx A^{\deg}$.

(1b) if $n=m$ and $\Lambda \neq \Gamma$, $\det(\mathcal{H})\otimes(\mathcal{L}_{\Lambda(x)}\otimes\mathcal{L}_{(\Gamma/\Lambda)(1-x)})^\vee \approx A^{\deg}$.

(2) if $n - m = 1$, $\det(\mathcal{H})\otimes(\mathcal{L}_\psi\otimes\mathcal{L}_\Lambda)^\vee \approx A^{\deg}$.

(3) if $n - m \geq 2$, $\det(\mathcal{H})\otimes(\mathcal{L}_\Lambda)^\vee \approx A^{\deg}$.

proof In all cases, the left hand side is a lisse sheaf of rank one which is geometrically constant (by the geometric determinant formulas 8.10.3 and 8.11.6), so it is necessarily of the form $\alpha^{\deg}$ for some unit α in $\bar{\mathbb{Q}}_\ell$. Since the sheaves $\mathcal{L}_{(\Gamma/\Lambda)(1-x)}$ and $\mathcal{L}_\psi$ are both canonically trivial at zero, it follows that α is the determinant of $\det(\mathcal{H})\otimes(\mathcal{L}_\Lambda)^\vee$ **as D(0)-representation**. The verification that $\alpha = A$ is a straightforward if tedious modification of that given in [Ka-GKM, 7.0.8, 7.2, 7.3, 7.4] in the case of Kloosterman sheaves, with 7.3.1 there replaced by 8.2.10 in the general case. That the two expressions for A coincide is [Ka-GKM, 7.4.1.1, 7.4.1.2 and 7.4.1.4]. QED

8.13 Sato-Tate Law for Hypergeometric Sums; Nonexceptional Cases

(8.13.1) Let k_0 be a finite field of characteristic $p \neq \ell$, and

$$\mathcal{H} := \mathcal{H}_1(!, \psi; \chi\text{'s}; \rho\text{'s})$$

a nonpunctual irreducible hypergeometric defined over k_0 of type (n, m) with $n \geq m$ and $\lambda = 1$. Denote by G the group G_{geom} for $\mathcal{H}$. Recall that since $\mathcal{H}$ is pure (of weight $n + m - 1$), G_{geom} is semisimple, so $G^{0,der} = G^0$. As above, put

$$\Lambda := \prod_i \chi_i, \text{ a character of } k_0{}^\times,$$

$\Gamma := \prod_j \rho_j$, a character of $k_0{}^{\times}$,

$A := \Lambda((-1)^{n-1})q^{n(n-1)/2}\prod_{i,j} (-g(\overline{\psi}, \overline{\rho}_j/\overline{\chi}_i)) \in \overline{\mathbb{Q}}_\ell$.

We also pick a square root of $q := \mathrm{Card}(k_0)$ so as to be able to speak of the Tate twist $\mathcal{H}(\mathbf{(n+m-1)/2})$ of $\mathcal{H}$, which is pure of weight zero.

Lemma 8.13.2 Notations and hypotheses as above, suppose that $\mathcal{H}$ is geometrically self-dual. If $(\Lambda/\Gamma)(-1) = 1$ (a condition which is always satisfied if either the autoduality is symplectic, or if $p = 2$, or if p is odd and we replace k_0 by its quadratic extension) then the Tate twist

$$\mathcal{H}(\mathbf{(n+m-1)/2})$$

is arithmetically self-dual with values in $\overline{\mathbb{Q}}_\ell$.

proof $\mathcal{H} := \mathcal{H}_1(!, \psi; \chi\text{'s}; \rho\text{'s})$ and $\mathcal{H}_1(!, \overline{\psi}; \overline{\chi}\text{'s}; \overline{\rho}\text{'s})(\mathbf{n+m-1})$ are dual as lisse sheaves on $\mathbb{G}_m - \{1\}$ (indeed on $\mathbb{G}_m$ if $n > m$), in virtue of the duality formulas 8.2.12 and 8.4.2, (5). By the Duality Recognition Theorem 8.8.1, $\mathcal{H} := \mathcal{H}_1(!, \psi; \chi\text{'s}; \rho\text{'s})$ is geometrically self dual if and only if $p(n-m)$ is even and the sets of the χ's and of the ρ's are each stable by inversion. So the arithmetic dual of $\mathcal{H}$ is

$$\mathcal{H}_1(!, \overline{\psi}; \chi\text{'s}; \rho\text{'s})(\mathbf{n+m-1}).$$

By 8.7.3, if $(\Lambda/\Gamma)(-1) = 1$, then

$$\mathcal{H}_1(!, \overline{\psi}; \chi\text{'s}; \rho\text{'s})(\mathbf{n+m-1}) \approx \mathcal{H}_1(!, \psi; \chi\text{'s}; \rho\text{'s})(\mathbf{n+m-1}). \quad \text{QED}$$

Theorem 8.13.3 Hypotheses and notations as above, suppose that $G := G_{\text{geom}}$ for $\mathcal{H}$ is either $O(n)$ or, if n is even, $Sp(n)$. Suppose also that

$$(\Lambda/\Gamma)(-1) = 1.$$

Then the Tate twist

$$\mathcal{H}(\mathbf{(n+m-1)/2})$$

is pure of weight zero, and all of its Frobenii land in G_{geom}. Fix an embedding of $\overline{\mathbb{Q}}_\ell$ into $\mathbb{C}$, and a maximal compact subgroup K of the Lie group $G_{\text{geom}}(\mathbb{C})$. The conjugacy class of the semisimple part of each $\rho(\mathrm{Frob}_{E,t})$ for $\mathcal{H}(\mathbf{(n+m-1)/2})$ meets K in a single conjugacy class, denoted $\vartheta(E, t)$. The conjugacy classes $\vartheta(E, t)$ are equidistributed in the space $K^\natural$ of conjugacy classes of K with respect to normalized Haar measure, in any of the three senses of equidistrbiution of ([Ka-GKM, 3.5]).

proof Since $\mathcal{H}$ is pure of weight $n+m-1$, $\mathcal{H}(\mathbf{(n+m-1)/2})$ is certainly pure of weight zero. That all the Frobenii of $\mathcal{H}((\mathbf{n+m-1})/2)$ land in G is precisely the content of the previous Lemma. The equidistribution results from Deligne's Weil II, as recalled in 7.11.1. QED

Theorem 8.13.4 Hypotheses and notations as above, suppose that $G := G_{geom}$ for $\mathcal{H}$ is SO(n), and that

$$(\Lambda/\Gamma)(-1) = 1.$$

Then after replacing k_0 by a quadratic extension, we have:
The Tate twist

$$\mathcal{H}((\mathbf{n+m-1})/2)$$

is pure of weight zero and all of its Frobenii land in G_{geom}. Fix an embedding of $\overline{\mathbb{Q}}_\ell$ into $\mathbb{C}$, and a maximal compact subgroup K of the Lie group $G_{geom}(\mathbb{C})$. The conjugacy class of the semisimple part of each $\rho(\mathrm{Frob}_{E,t})$ for $\mathcal{H}((\mathbf{n+m-1})/2)$ meets K in a single conjugacy class, denoted $\vartheta(E, t)$. The conjugacy classes $\vartheta(E, t)$ are equidistributed in the space $K^\natural$ of conjugacy classes of K with respect to normalized Haar measure, in any of the three senses of equidistrbiution of ([Ka-GKM, 3.5]).

proof By the above lemma, all the Frobenii of $\mathcal{H}((\mathbf{n+m-1})/2)$ lie in O(n). The obstruction to their lying in SO(n) is a character of order dividing two which is geometrically constant, hence is trivialized by any constant field extension of even degree. QED

Theorem 8.13.5 Hypotheses and notations as above, suppose that G_{geom} is SL(n). Let $\alpha \in \overline{\mathbb{Q}}_\ell$ be any solution of

$$\alpha^{-n} = A.$$

Then $\mathcal{H}\otimes\alpha^{deg}$ is pure of weight zero, and has all its Frobenii in G_{geom}. Fix an embedding of $\overline{\mathbb{Q}}_\ell$ into $\mathbb{C}$, and a maximal compact subgroup K of the Lie group $G_{geom}(\mathbb{C})$. The conjugacy class of the semisimple part of each $\rho(\mathrm{Frob}_{E,t})$ for $\mathcal{H}\otimes\alpha^{deg}$ meets K in a single conjugacy class, denoted $\vartheta(E, t)$. The conjugacy classes $\vartheta(E, t)$ are equidistributed in the space $K^\natural$ of conjugacy classes of K with respect to normalized Haar measure, in any of the three senses of equidistrbiution of ([Ka-GKM, 3.5]).

proof This results from the arithmetic determinant formula (and Weil II). QED

In a similar vein, we have the following slightly less precise result.

Theorem 8.13.6 Hypotheses and notations as above, suppose that **either**

(1) $(G_{geom})^0$ is $SL(n)$ or $Sp(n)$

or

(2) n is odd and $(G_{geom})^0$ is $SO(n)$.

Then $G := G_{geom}$ is of the form $\mu_d G^0$ for some d, and there exists a constant α in $\overline{\mathbb{Q}}_\ell^\times$ with

$$\alpha^{-n} = (\text{root of unity of order dividing order of } \det G_{geom}) \times A,$$

such that $\mathcal{H} \otimes \alpha^{\deg}$ is pure of weight zero, and has all its Frobenii in G_{geom}. Fix an embedding of $\overline{\mathbb{Q}}_\ell$ into $\mathbb{C}$, and a maximal compact subgroup K of the Lie group $G_{geom}(\mathbb{C})$. The conjugacy class of the semisimple part of each $\rho(\mathrm{Frob}_{E,t})$ for $\mathcal{H} \otimes \alpha^{\deg}$ meets K in a single conjugacy class, denoted $\vartheta(E, t)$. The conjugacy classes $\vartheta(E, t)$ are equidistributed in the space $K^\natural$ of conjugacy classes of K with respect to normalized Haar measure, in any of the three senses of equidistrbiution of ([Ka-GKM, 3.5]).

proof In all the cases listed, the normalizer of G^0 in $GL(n)$ is $\mathbb{G}_m G^0$. The rest of the proof proceeds as in [Ka-MG, Cor. 16]. QED

8.14 Criteria for finite monodromy

(8.14.1) Let C be a smooth geometrically connected curve over a finite field k of characteristic $p \neq \ell$, and $\mathcal{F}$ a lisse $\overline{\mathbb{Q}}_\ell$ sheaf on C. Fix a geometric point ξ of $C \otimes_k \bar{k}$, and denote by

$$\pi_1^{geom} := \pi_1(C \otimes_k \bar{k}, \xi) \subset \pi_1^{arith} := \pi_1(C, \xi)$$

the geometric and arithmetic fundamental groups respectively of C, by

$$\rho: \pi_1^{arith} \to GL(\mathcal{F}_\xi)$$

the ℓ-adic representation that $\mathcal{F}$ "is", and by

$$G_{geom} := \text{the Zariski closure of } \rho(\pi_1{}^{geom}),$$

$$G_{arith} := \text{the Zariski closure of } \rho(\pi_1{}^{arith}).$$

(8.14.2) One knows that the radical of $(G_{geom})^0$ is unipotent (this is Grothendieck's global version of the local monodromy theorem, cf. [De-WII, 1.3.8]). Thus if $\mathcal{F}$ is geometrically semisimple, its G_{geom} is a semisimple group. Applying this to $\det(\mathcal{F})$, we recover the fact that $\det(\mathcal{F})$ is geometrically of finite order. Therefore a suitable twist $\mathcal{F}\otimes\alpha^{deg}$ has $\det(\mathcal{F}\otimes\alpha^{deg})$ arithmetically of finite order.

Proposition 8.14.3 Suppose that $\mathcal{F}$ is geometrically semisimple. Consider the following conditions:

(1) G_{geom} is finite (i.e., $\rho(\pi_1{}^{geom})$ is finite).

(2) the image of $\pi_1{}^{arith}$ (or equivalently of G_{arith}) in $PGL(\mathcal{F}_{\xi})$ is finite.

(3) for every finite extension E of k, and every point $t \in C(E)$, some strictly positive power of $\rho(Frob_{E,t})$ is a scalar.

(4) for every finite extension E of k, and every point $t \in C(E)$, some strictly positive power of $\rho(Frob_{E,t})$ has all its eigenvalues equal.

We have the implications

$$(2) \Rightarrow (3) \Rightarrow (4) \Rightarrow (1),$$

and if $\mathcal{F}$ is geometrically irreducible, these conditions are all equivalent:

$$(1) \Leftrightarrow (2) \Leftrightarrow (3) \Leftrightarrow (4)$$

proof We first show that (1) $\Rightarrow$ (2) if $\mathcal{F}$ is geometrically irreducible. Any element γ of $\pi_1{}^{arith}$ normalizes $\pi_1{}^{geom}$, so $\rho(\gamma)$ normalizes G_{geom}. Since G_{geom} is finite, its automorphism group is finite, say of order N. Thus $\rho(\gamma)^N$ centralizes G_{geom}; as G_{geom} is an irreducible subgroup of $GL(\mathcal{F}_{\xi})$, $\rho(\gamma)^N$ is a scalar. Therefore the image of $\pi_1{}^{arith}$ in $PGL(\mathcal{F}_{\xi})$ has every element of order dividing N. By Zariski density, the image of G_{arith} has the same property. Therefore the Lie algebra of this image is killed by N, so this image is finite.

That (2) $\Rightarrow$ (3) $\Rightarrow$ (4) is obvious.

It remains to show that (4) $\Rightarrow$ (1). Since $\det(\mathcal{F})$ is geometrically of finite order, the intersection of G_{geom} with the scalars is finite.

Therefore the restriction to G_{geom} of the adjoint representation of $GL(\mathcal{F}_\xi)$ has a finite kernel. So it suffices to show that if (4) holds for $\mathcal{F}$, then $End(\mathcal{F})$ has **its** G_{geom} finite. But if (4) holds for $\mathcal{F}$, then for $End(\mathcal{F})$ the following condition is satisfied:

for every finite extension E of k, and every point $t \in C(E)$, some strictly positive power of $(Ad\circ\rho)(Frob_{E,t})$ is unipotent.

We claim that this implies that there exists an $N \geq 1$ such that

for every finite extension E of k, and every point $t \in C(E)$, $(Ad\circ\rho)(Frob_{E,t})^N$ is unipotent.

Indeed, $\mathcal{F}$ is definable over some finite extension E_λ of $\mathbb{Q}_\ell$, so the eigenvalues of $(Ad\circ\rho)(Frob_{E,t})$ are roots of unity which are algebraic of degree at most the rank of $End(\mathcal{F})$ over E_λ; as E_λ has only finitely many extensions of any given degree, all the eigenvalues of $(Ad\circ\rho)(Frob_{E,t})$ lie in a fixed finite extension F_λ of $\mathbb{Q}_\ell$, and in any such field F_λ there are only finitely many roots of unity.

By Chebataroff, it follows now that $(Ad\circ\rho)(\gamma)^N$ is unipotent for **every** element γ in π_1^{arith}, so a fortiori for every element γ in π_1^{geom}. By Zariski density, it follows that g^N is unipotent for every element in the group $G := G_{geom}$ for $End(\mathcal{F})$. But $\mathcal{F}$ and hence $End(\mathcal{F})$ is geometrically semisimple, so G is a semisimple group. Thus G^0 is a connected semisimple group in which the N'th power of every element is unipotent; looking at elements of a maximal torus, we infer that its rank is zero, hence that $G^0 = \{e\}$, and hence that G itself is finite. QED

Corollary 8.14.3.1 Suppose that $\mathcal{F}$ is geometrically irreducible and that $\det(\mathcal{F})$ is arithmetically of finite order. Then the following conditions are equivalent:

(1) G_{geom} is finite.

(2) G_{arith} is finite.

(3) for every finite extension E of k, and every point $t \in C(E)$, $\rho(Frob_{E,t})$ is quasi-unipotent, i.e., all its eigenvalues are roots of unity.

Theorem 8.14.4 Suppose that $\mathcal{F}$ is geometrically irreducible, that its determinant is arithmetically of finite order, and that $\mathcal{F}$ is pure of weight zero for all embeddings of $\overline{\mathbb{Q}}_\ell$ into $\mathbb{C}$. Then the following conditions are equivalent:

(1) G_{geom} is finite.

(2) G_{arith} is finite.

(3) for every finite extension E of k, and every point $t \in C(E)$, $\rho(Frob_{E,t})$ is quasi-unipotent, i.e., all its eigenvalues are roots of unity.

(4)for every finite extension E of k, and every point $t \in C(E)$, all the eigenvalues of $\rho(Frob_{E,t})$ are algebraic integers.

(5) for every finite extension E of k, every point $t \in C(E)$, and every integer $N \geq 1$, $Trace(\rho(Frob_{E,t})^N)$ is an algebraic integer.

(6) for every finite extension E of k, and every point $t \in C(E)$, $Trace(\rho(Frob_{E,t}))$ is an algebraic integer.

proof We already know (1) $\Leftrightarrow$ (2) $\Leftrightarrow$ (3), and trivially (3) $\Rightarrow$ (4) $\Rightarrow$ (5) $\Rightarrow$ (6). We have (4) $\Rightarrow$ (3), simply because roots of unity are characterized among all algebraic integers as those all of whose archimedean absolute values are one. We have (5) $\Rightarrow$ (4) by [Ax], and we have (6) $\Rightarrow$ (5) for (E, $t \in C(E)$) by applying (6) to each (E_N, t), where E_N denotes the extension of degree N of E. QED

Here is a mild variant, which will be needed for dealing with hypergeometrics of type (n, n).

Theorem 8.14.5. Let C be a smooth geometrically connected curve over a finite field k of characteristic $p \neq \ell$, and $\mathcal{F}$ a middle extension $\overline{\mathbb{Q}}_\ell$ sheaf on C. Let $U \subset C$ be any nonempty open set on which $\mathcal{F}$ is lisse. Fix a geometric point ξ of $U \otimes_k \overline{k}$, and denote by

$$\pi_1^{geom} := \pi_1(U \otimes_k \overline{k}, \xi) \subset \pi_1^{arith} := \pi_1(U, \xi)$$

the geometric and arithmetic fundamental groups respectively of U, by

$$\rho: \pi_1^{arith} \rightarrow GL(\mathcal{F}_\xi)$$

the ℓ-adic representation that $\mathcal{F} \mid U$ "is", and by

$$G_{geom} := \text{the Zariski closure of } \rho(\pi_1^{geom}),$$

$$G_{arith} := \text{the Zariski closure of } \rho(\pi_1^{arith}).$$

Suppose that $\mathcal{F} \mid U$ is geometrically irreducible, that its determinant is arithmetically of finite order, and that $\mathcal{F} \mid U$ is pure of weight zero for all embeddings of $\overline{\mathbb{Q}}_\ell$ into $\mathbb{C}$. Then the following conditions are equivalent:

(1) G_{geom} is finite.
(2) G_{arith} is finite.
(3) for every finite extension E of k, and every point $t \in C(E)$, $Frob_{E,t} \mid \mathcal{F}_{\bar{t}}$ is quasi-unipotent, i.e., all its eigenvalues are roots of unity.
(4) for every finite extension E of k, and every point $t \in C(E)$, all the eigenvalues of $Frob_{E,t} \mid \mathcal{F}_{\bar{t}}$ are algebraic integers.
(5) for every finite extension E of k, every point $t \in C(E)$, and every integer $N \geq 1$, $Trace((Frob_{E,t})^N \mid \mathcal{F}_{\bar{t}})$ is an algebraic integer.
(6) for every finite extension E of k, and every point $t \in C(E)$, $Trace(Frob_{E,t} \mid \mathcal{F}_{\bar{t}})$ is an algebraic integer.

proof Restricting to t's in U, we see by the previous theorem that any of (3), (4), (5), (6) implies the equivalent conditions (1) and (2). Trivially, (3) $\Rightarrow$ (4) $\Rightarrow$ (5) $\Rightarrow$ (6). So it suffices to show that (2) $\Rightarrow$ (3). This is only a problem at points t in C - U. By Deligne's result on integrality [SGA7 Part II, Expose XXI, Appendice, Cor. 5.3] we see that if (3) holds at all point of U, then the eigenvalues of $Frob_{E,t} \mid \mathcal{F}_{\bar{t}}$ are algebraic integers, and by [De-WII, 1.8.1] we see that all their archimedean absolute values are ≤ 1. QED

Theorem 8.14.6 (p-adic criterion for finite monodromy). Let k be a finite field of characteristic $p \neq \ell$, and $\mathcal{H}$ a middle extension $\overline{\mathbb{Q}}_\ell$-sheaf on $\mathbb{G}_m$ over k. Let $U \subset \mathbb{G}_m$ be any nonempty open set on which $\mathcal{F}$ is lisse. Fix a geometric point ξ of $U \otimes_k \bar{k}$, and denote by

$$\pi_1{}^{geom} := \pi_1(U \otimes_k \bar{k}, \xi) \subset \pi_1{}^{arith} := \pi_1(U, \xi)$$

the geometric and arithmetic fundamental groups respectively of U, by

$$\rho: \pi_1{}^{arith} \to GL(\mathcal{H}_\xi)$$

the ℓ-adic representation that $\mathcal{H} \mid U$ "is", and by

$$G_{geom} := \text{the Zariski closure of } \rho(\pi_1{}^{geom}),$$

$$G_{arith} := \text{the Zariski closure of } \rho(\pi_1{}^{arith}).$$

Suppose that
(1) $\mathcal{H} \mid U$ is geometrically irreducible of generic rank $n \geq 2$,
(2) $\mathcal{H} \mid U$ is pure of some weight w_ι for every embedding ι of $\overline{\mathbb{Q}}_\ell$ into $\mathbb{C}$.
(3) for every finite extension E of k, and every point $t \in E^\times$, $Trace(Frob_{E,t} \mid \mathcal{H}_{\bar{t}})$ is an algebraic number which is integral at all places of $\overline{\mathbb{Q}}$ of residue characteristic $\ell' \neq p$.

(4) "the" number α in $\overline{\mathbb{Q}}_\ell^\times$ such that $\det(\mathcal{H}\otimes\alpha^{-\deg} \mid U)$ is arithmetically of finite order (strictly speaking, α is only well defined up to multiplication by a root of unity) is an algebraic integer which is a unit at all places of $\overline{\mathbb{Q}}$ of residue characteristic $\ell' \neq p$.

Then the following conditions are equivalent:
(1) G_{geom} is finite.
(2) for every finite extension E of k, every multiplicative character χ of $E^\times$, and every p-adic valuation "ord" of $\overline{\mathbb{Q}}$, we have the inequality

$$\mathrm{ord}(\mathrm{Trace}(\mathrm{Frob}_E \mid H^1{}_c(\mathbb{G}_m\otimes_k\overline{k}, \mathcal{H}\otimes\mathcal{L}_\chi))) \geq \deg(E/k)\mathrm{ord}(\alpha).$$

proof Extending the finite field k if necessary, we may suppose that the open set U where $\mathcal{H}$ is lisse contains a rational point u. Then we may take for α any n'th root of $\det(\mathrm{Frob}_{k,u} \mid \mathcal{H}_{\overline{u}})$. Since $\mathcal{H}$ is pure of some weight for every complex embedding, we see that $\mathcal{H}\otimes\alpha^{-\deg}$ is pure (necessarily of weight zero, since its determinant is pure of weight zero, being arithmetically of finite order) for every complex embedding. In view of the previous theorem (applied to $\mathcal{H}\otimes\alpha^{-\deg}$), G_{geom} is finite if and only if for every finite extension E of k, and every point $t \in E^\times$,

$$\mathrm{Trace}(\mathrm{Frob}_{E,t} \mid (\mathcal{H}\otimes\alpha^{-\deg})_{\overline{t}}) := \mathrm{Trace}(\mathrm{Frob}_{E,t} \mid (\mathcal{H})_{\overline{t}})/\alpha^{\deg(E/k)}$$

is an algebraic integer. Since this trace is by hypothesis integral outside of p, the condition is that for every p-adic valuation "ord" of $\overline{\mathbb{Q}}$, we have the inequality

$$\mathrm{ord}(\mathrm{Trace}(\mathrm{Frob}_{E,t} \mid (\mathcal{H})_{\overline{t}})) \geq \deg(E/k)\mathrm{ord}(\alpha).$$

Fix the choice of the p-adic valuation "ord", and the field E, and allow t to vary over $E^\times$. Since $\mathrm{Card}(E^\times)$ is **prime to p**, it is equivalent (by multiplicative Fourier inversion on the finite group $E^\times$) to show that for every multiplicative character χ of $E^\times$, we have

$$\mathrm{ord}(\Sigma_{t \text{ in } E^\times}\, \chi(t)\mathrm{Trace}(\mathrm{Frob}_{E,t} \mid (\mathcal{H})_{\overline{t}})) \geq \deg(E/k)\mathrm{ord}(\alpha).$$

By the Lefschetz Trace Formula (applied to $\mathcal{H}\otimes\mathcal{L}_\chi$), and the vanishing of $H^i{}_c$ for a geometrically irreducible middle extension of generic rank ≥ 2 on an open curve, we see that

$$-\mathrm{Trace}(\mathrm{Frob}_E \mid H^1{}_c(\mathbb{G}_m\otimes_k\overline{k}, \mathcal{H}\otimes\mathcal{L}_\chi)) =$$
$$= \Sigma_{t \text{ in } E^\times}\, \chi(t)\mathrm{Trace}(\mathrm{Frob}_{E,t} \mid (\mathcal{H})_{\overline{t}}). \qquad \text{QED}$$

8.15 Irreducible Hypergeometrics with finite G_{geom}

Theorem 8.15.1 (finite monodromy criterion for hypergeometrics) Let k be a finite field of characteristic $p \neq \ell$, $q := \mathrm{Card}(k)$, and

$$\mathcal{H} := \mathcal{H}_1(!, \psi; \chi\text{'s}; \rho\text{'s})$$

a nonpunctual, geometrically irreducible hypergeometric defined over k of type (n, m) with $n \geq m$, $n \geq 2$ and $\lambda = 1$. Put

$$\Lambda := \prod_i \chi_i, \text{ a character of } k_0^{\times},$$

$$A := \Lambda((-1)^{n-1})q^{n(n-1)/2}\prod_{i,j} (-g(\overline{\psi}, \overline{\rho}_j/\overline{\chi}_i)) \in \overline{\mathbb{Q}}_\ell.$$

$$\alpha := \text{any n'th root of } A.$$

Then G_{geom} for $\mathcal{H}$ is finite if and only if for every finite extension E of k, every multiplicative character η of $E^{\times}$, and every p-adic valuation "ord" of $\overline{\mathbb{Q}}$, we have the inequality

$$\Sigma_i \, \mathrm{ord}(g(\psi_E, \eta\chi_{i,E}) + \Sigma_j \, \mathrm{ord}(g(\overline{\psi}_E, \overline{\eta}\overline{\rho}_{j,E}) \geq \deg(E/k)\mathrm{ord}(\alpha).$$

proof This is immediate from the p-adic criterion 8.14.6, the fact (8.2.11) that

$$\mathrm{Trace}(\mathrm{Frob}_E \mid H^1_c(\mathbb{G}_m \otimes_k \overline{k}, \mathcal{L}_\eta \otimes \mathcal{H}_1(!, \psi; \chi\text{'s}; \rho\text{'s}))) =$$
$$= (\prod_i(-g(\psi_E, \eta\chi_{i,E})))(\prod_j(-g(\overline{\psi}_E, \overline{\eta}\overline{\rho}_{j,E}))),$$

and the fact that Gauss sums are algebraic integers which are units outside of p. QED

The next proposition shows that in searching for hypergeometrics with finite G_{geom}, we "lose" nothing by looking only at those defined over finite fields.

Proposition 8.15.2 Let k be an algebraically closed field of characteristic $p \neq \ell$, $q := \mathrm{Card}(k)$, and

$$\mathcal{H} := \mathcal{H}_1(!, \psi; \chi\text{'s}; \rho\text{'s})$$

a nonpunctual, geometrically irreducible hypergeometric defined over k of type (n, m) with $n \geq m$ and $\lambda = 1$. Suppose that G_{geom} is finite. Then

(1) the n χ_i's are all distinct and all of finite order (i.e., local monodromy at zero is of finite order).

(2) the m ρ_j's are all distinct and all of finite order (i.e., local monodromy at ∞ is of finite order).

(3) $\mathcal{H}$ is defined over a finite subfield of k.

proof If G_{geom} is finite, then the local monodromy at both zero and ∞ must be of finite order, whence (1) and (2). Once (1) and (2) hold, (3) is tautologous. QED

8.16 Explicitation via Stickelberger

(8.16.1) We now explicate the finite monodromy criterion 8.15.1 with the aid of the Stickelberger formula for the p-adic valuations of Gauss sums. For $n \geq 1$, we denote by ζ_n a primitive n'th root of unity in $\overline{\mathbb{Q}}_\ell$. We denote by $K_p \subset \overline{\mathbb{Q}}_\ell$ the subfield

$$K_p := \mathbb{Q}(\zeta_p, \text{ all } \zeta_N \text{ with N prime to p}),$$

by $\mathcal{O}_p \subset K_p$ the subring

$$\mathcal{O}_p := \mathbb{Z}[\zeta_p, \text{ all } \zeta_N \text{ with N prime to p}]$$

of all algebraic integers in K_p,

by $K_{p,nr} \subset K_p$ the subfield

$$K_{p,nr} := \mathbb{Q}(\text{all } \zeta_N \text{ with N prime to p}),$$

and by $\mathcal{O}_{p,nr} \subset K_{p,nr}$ the subring

$$\mathcal{O}_{p,nr} := \mathbb{Z}[\text{all } \zeta_N \text{ with N prime to p}]$$

of all algebraic integers in $K_{p,nr}$.

All multiplicative characters χ of finite fields E of characteristic p take values in $\mathcal{O}_{p,nr}$, and their Gauss sums $g(\psi, \chi)$ lie in $\mathcal{O}_p$.

(8.16.2) Fix an embedding of fields

$$\iota : K_{p,nr} \subset \overline{\mathbb{Q}}_p.$$

For any integer N prime to p, reduction mod $\mathcal{P}$ defines an isomorphism of groups $\mu_N(\overline{\mathbb{Q}}_p) \approx \mu_N(\overline{\mathbb{F}}_p)$, so we have

$$\mu_N(\overline{\mathbb{Q}}_\ell) = \mu_N(\mathcal{O}_{p,nr}) = \mu_N(\overline{\mathbb{Q}}_p) \approx \mu_N(\overline{\mathbb{F}}_p).$$

Taking $N := q - 1$ where $q := \text{Card}(E)$ for a finite subfield E of $\overline{\mathbb{F}}_p$, we find

$$\mu_{q-1}(\mathcal{O}_{p,nr}) \approx \mu_{q-1}(\overline{\mathbb{Q}}_p) \approx E^\times.$$

The inverse of this isomorphism is the construction $x \mapsto \text{Teich}(x)$, where Teich(x) denotes the "Teichmuller representative" of x.

Passing to the inverse limit, and recalling that the numbers $q - 1$ are cofinal among the N's prime to p, we obtain an isomorphism

$$T_{\text{not } p}(\overline{\mathbb{Q}}_\ell) = T_{\text{not } p}(\mathcal{O}_{p,nr}) \approx T_{\text{not } p}(\overline{\mathbb{Q}}_p) \approx \varprojlim_{\text{Norm}} E^\times.$$

By means of this identification, $\overline{\mathbb{Q}}_\ell$-valued characters of finite order of the group $\varprojlim_{\text{Norm}} E^\times$ are elements of the discrete group $(\mathbb{Q}/\mathbb{Z})_{\text{not } p}$. Concretely, an element x of $(\mathbb{Q}/\mathbb{Z})_{\text{not } p}$ corresponds to the

$\overline{\mathbb{Q}}_p$-valued character $\chi_{x,E}$ of any finite E of cardinality q such that

$$(q-1)x \in \mathbb{Z}$$

defined by

$$\chi_{x,E}(t) := (\mathrm{Teich}(t))^{(q-1)x} \quad \text{for } t \in E^\times.$$

(8.16.3) We denote by ord_q the p-adic valuation of $\overline{\mathbb{Q}}_p$ normalized by $\mathrm{ord}_q(q) = 1$. For any real number x, we denote by $\langle x \rangle$ its "fractional part", defined to be the unique real number in [0, 1) such that $x \equiv \langle x \rangle$ mod $\mathbb{Z}$. Since $\langle x + n \rangle = \langle x \rangle$ for any $n \in \mathbb{Z}$, we may speak of $\langle x \rangle$ for x in $\mathbb{R}/\mathbb{Z}$, so in particular for x in $(\mathbb{Q}/\mathbb{Z})_{\text{not } p}$.

(8.16.4) Given x in $(\mathbb{Q}/\mathbb{Z})_{\text{not } p}$, we define $\int_p \langle x \rangle$ to be the rational number in [0, 1) defined as follows: pick an integer $f \geq 1$ such that

$$(p^f - 1)x \in \mathbb{Z},$$

and define

$$\int_p \langle x \rangle := (1/f)\Sigma_{i \bmod f} \langle p^i x \rangle.$$

It is immediate that this definition is independent of the auxiliary choice of the integer f.

With these notations, we can state the classical Stickelberger theorem.

Theorem 8.16.5 (Stickelberger) Fix an embedding $\iota : K_{p,nr} \subset \overline{\mathbb{Q}}_p$. Let

$$x \in (\mathbb{Q}/\mathbb{Z})_{\text{not } p},$$

and let E be a finite subfield of $\overline{\mathbb{F}}_p$ of cardinality q such that

$$(q-1)x \in \mathbb{Z},$$

so that we can speak of the $\overline{\mathbb{Q}}_p$-valued multiplicative character $\chi_{x,E}$ of E,

$$\chi_{x,E}(t) := (\mathrm{Teich}(t))^{(q-1)x} \quad \text{for } t \in E^\times.$$

For any nontrivial $\overline{\mathbb{Q}}_p$-valued additive character ψ of E, we have the formula

$$\mathrm{ord}_q(g(\psi, \chi_{x,E})) = \int_p \langle -x \rangle.$$

(8.16.6) The extension $K_{p,nr}$ of $\mathbb{Q}$ is Galois with group

$$\mathrm{Gal}(K_{p,nr}/\mathbb{Q}) \approx \prod_{\ell \neq p} \mathbb{Z}_\ell^\times \approx \mathrm{Aut}((\mathbb{Q}/\mathbb{Z})_{\text{not } p}).$$

Given an element

$$\alpha \in \prod_{\ell \neq p} \mathbb{Z}_\ell^\times \approx \mathrm{Aut}((\mathbb{Q}/\mathbb{Z})_{\text{not } p}),$$

we denote by σ_α the unique element of $\mathrm{Gal}(K_{p,nr})$ such that

$$\sigma_\alpha \circ \chi_{x,E} = \chi_{\alpha x,E}$$

for every x in $(\mathbb{Q}/\mathbb{Z})_{\text{not } p}$ and every finite subfield E of $\overline{\mathbb{F}}_p$ of cardinality q such that $(q - 1)x \in \mathbb{Z}$. We denote by $\text{ord}_{\alpha,q}$ the p-adic valuation of $K_{p,nr}$ defined by

$$\text{ord}_{\alpha,q}(z) := \text{ord}_q(\sigma_\alpha(z)).$$

Every p-adic valuation of $K_{p,nr}$ with $\text{ord}(q) = 1$ is $\text{ord}_{\alpha,q}$ for some α.

Theorem bis 8.16.7 (Stickelberger) Fix an embedding $\iota : K_{p,nr} \subset \overline{\mathbb{Q}}_p$. Let

$$x \in (\mathbb{Q}/\mathbb{Z})_{\text{not } p},$$

and let E be a finite subfield of $\overline{\mathbb{F}}_p$ of cardinality q such that

$$(q - 1)x \in \mathbb{Z},$$

so that we can speak of the $\overline{\mathbb{Q}}_p$-valued multiplicative character $\chi_{x,E}$ of E,

$$\chi_{x,E}(t) := (\text{Teich}(t))^{(q-1)x} \quad \text{for } t \in E^\times.$$

For any nontrivial $\overline{\mathbb{Q}}_p$-valued additive character ψ of E, and any element

$$\alpha \in \prod_{\ell \neq p} \mathbb{Z}_\ell^\times \approx \text{Aut}((\mathbb{Q}/\mathbb{Z})_{\text{not } p}),$$

we have the formula

$$\text{ord}_{\alpha,q}(g(\psi, \chi_{x,E})) = \textstyle\int_p \langle -\alpha x \rangle.$$

Theorem 8.16.8 (numerical criterion for finite monodromy of hypergeometrics) Let E be a finite field of characteristic $p \neq \ell$, and

$$\mathcal{H} := \mathcal{H}_1(!, \psi; \chi\text{'s}; \rho\text{'s})$$

a nonpunctual, geometrically irreducible hypergeometric defined over E of type (n, m) with $n \geq m$ and $\lambda = 1$. Fix an embedding $\iota : K_{p,nr} \subset \overline{\mathbb{Q}}_p$. Let

$$x_1, \ldots, x_n, y_1, \ldots, y_m \in (\mathbb{Q}/\mathbb{Z})_{\text{not } p}$$

be the unique elements of $(\mathbb{Q}/\mathbb{Z})_{\text{not } p}$ such that

$$\iota \circ \chi_i = \chi_{x_i, E} \quad \text{for } i = 1, \ldots, n,$$

$$\iota \circ \rho_j = \chi_{y_j, E} \quad \text{for } j = 1, \ldots, m.$$

Then $\mathcal{H}$ has finite G_{geom} if and only if the following condition holds for every $\alpha \in \prod_{\ell \neq p} \mathbb{Z}_\ell^\times \approx \text{Aut}((\mathbb{Q}/\mathbb{Z})_{\text{not } p})$:

For every $z \in (\mathbb{Q}/\mathbb{Z})_{\text{not } p}$, we have the inequality

$$\textstyle\sum_i \int_p \langle \alpha z - \alpha x_i \rangle \;+\; \sum_j \int_p \langle \alpha y_j - \alpha z \rangle \;\geq$$

$\geq (1/n)[n(n-1)/2 \;+\; \Sigma_{i,\,j} \int_p \langle \alpha y_j - \alpha x_i \rangle]$.

proof This is just the Stickelberger spelling out of the finite monodromy criterion 8.15.1, when the variable multiplicative character η of a variable finite extension F, card(F) := q, of E is written as $\overline{\chi}_{z,\,F}$, and the p-adic valuation tested is $\mathrm{ord}_{\alpha,q}$. QED

8.17 Finite monodromy for type (n, n), intertwining, and specialization

(8.17.1) In this section we will show that the very same intertwining conditions which for irreducible hypergeometric $\mathcal{D}$-modules of type (n, n) are equivalent to having finite G_{gal} (cf 5.5.3) are equivalent to having finite G_{geom} for irreducible ℓ-adic hypergeometrics of type (n, n).

Theorem 8.17.2 Suppose that $x_1, \ldots, x_n, y_1, \ldots, y_n$ are 2n distinct elements of $(\mathbb{Q}/\mathbb{Z})_{\text{not }p}$ such that for every

$$\alpha \in \textstyle\prod_{\ell \neq p} \mathbb{Z}_\ell^\times \approx \mathrm{Aut}((\mathbb{Q}/\mathbb{Z})_{\text{not }p}),$$

the two subsets

$$X_\alpha := \{\alpha x_1, \ldots, \alpha x_n\} \quad \text{and} \quad Y_\alpha := \{\alpha y_1, \ldots, \alpha y_n\}$$

of $(\mathbb{Q}/\mathbb{Z})_{\text{not }p}$ are **intertwined** in $(\mathbb{Q}/\mathbb{Z})_{\text{not }p}$, in the sense that if we display their images under $x \mapsto \exp(2\pi i x)$ on the unit circle, then as we walk counterclockwise around the unit circle we alternately encounter one from each subset. Then for every $z \in (\mathbb{Q}/\mathbb{Z})_{\text{not }p}$, we have the inequality

$$\Sigma_i \int_p \langle \alpha z - \alpha x_i \rangle \;+\; \Sigma_j \int_p \langle \alpha y_j - \alpha z \rangle \;\geq$$
$$\geq (1/n)[n(n-1)/2 \;+\; \Sigma_{i,\,j} \int_p \langle \alpha y_j - \alpha x_i \rangle].$$

proof If the two subsets

$$X_\alpha := \{\alpha x_1, \ldots, \alpha x_n\} \quad \text{and} \quad Y_\alpha := \{\alpha y_1, \ldots, \alpha y_n\}$$

of $(\mathbb{Q}/\mathbb{Z})_{\text{not }p}$ are **intertwined** in $(\mathbb{Q}/\mathbb{Z})_{\text{not }p}$, then so are their additive translates by αz,

$$\{\alpha x_1 + \alpha z, \ldots, \alpha x_n + \alpha z\} \quad \text{and} \quad \{\alpha y_1 + \alpha z, \ldots, \alpha y_n + \alpha z\}.$$

Since the right-hand side of the asserted inequalities are invariant by such additive translation, it suffices to treat universally the case in which z = 0. We must show universally that

$$\Sigma_i \int_p \langle -\alpha x_i \rangle \;+\; \Sigma_j \int_p \langle \alpha y_j \rangle \;\geq$$

$$\geq (1/n)[n(n-1)/2 \quad + \quad \Sigma_{i,\,j} \textstyle\int_p \langle \alpha y_j - \alpha x_i \rangle].$$

Now pick a common denominator N for all the x_i's and y_j's, and an integer f such that $p^f \equiv 1 \bmod N$. In view of the definition of $\int_p$, it suffices to show that for every integer d = 0, 1, ... , f-1, we have

$$\Sigma_i \langle -p^d \alpha x_i \rangle \quad + \quad \Sigma_j \langle p^d \alpha y_j \rangle \quad \geq$$
$$\geq (1/n)[n(n-1)/2 \quad + \quad \Sigma_{i,\,j} \langle p^d \alpha y_j - p^d \alpha x_i \rangle].$$

Now $p^d\alpha$ is simply another element α' of $\prod_{\ell \neq p} \mathbb{Z}_\ell^\times \approx \mathrm{Aut}((\mathbb{Q}/\mathbb{Z})_{\text{not } p})$.

Since the hypotheses are $\mathrm{Aut}((\mathbb{Q}/\mathbb{Z})_{\text{not } p})$-stable, it suffices to prove universally that

$$\Sigma_i \langle -x_i \rangle \; + \; \Sigma_j \langle y_j \rangle \; \geq \; (1/n)[n(n-1)/2 \quad + \quad \Sigma_{i,\,j} \langle y_j - x_i \rangle]$$

whenever the two subsets

$$X := \{x_1, \ldots, x_n\} \quad \text{and} \quad Y := \{y_1, \ldots, y_n\}$$

of $(\mathbb{Q}/\mathbb{Z})_{\text{not } p}$ are **intertwined** in $(\mathbb{Q}/\mathbb{Z})_{\text{not } p}$.

The verification of this is straightforward. The only properties of the function $\langle x \rangle$ which will be used in the proof are

$\langle x \rangle = x$ for x in [0, 1),

$\langle -x \rangle = 1 - \langle x \rangle$ if x is not in $\mathbb{Z}$.

By renumbering, we may suppose that we are in one of the three following cases:

(Case 1) $0 = x_1 < y_1 < x_2 < y_2 < \ldots < x_n < y_n < 1$,

(Case 2) $0 < x_1 < y_1 < x_2 < y_2 < \ldots < x_n < y_n < 1$,

(Case 3) $0 \leq y_1 < x_1 < y_2 < x_2 < \ldots < y_n < x_n < 1$.

In cases 1 and 2, we have

$$\begin{aligned}\Sigma_{i,\,j}\langle y_j - x_i \rangle &= \Sigma_{i \leq j}\langle y_j - x_i \rangle \; + \; \Sigma_{i > j}\langle y_j - x_i \rangle \\ &= \Sigma_{i \leq j}(y_j - x_i) \; + \; \Sigma_{i > j}[\,1 - (x_i - y_j)] \\ &= \Sigma_{i,\,j}(y_j - x_i) \; + \; \Sigma_{i > j}\, 1 \\ &= n\Sigma_j y_j \; - \; n\Sigma_i x_i \; + \; n(n-1)/2.\end{aligned}$$

In case 3, we have

$$\begin{aligned}\Sigma_{i,\,j}\langle y_j - x_i \rangle &= \Sigma_{i \geq j}\langle y_j - x_i \rangle \; + \; \Sigma_{i < j}\langle y_j - x_i \rangle \\ &= \Sigma_{i \geq j}[\,1 - (x_i - y_j)] \; + \; \Sigma_{i < j}(y_j - x_i) \\ &= n\Sigma_j y_j \; - \; n\Sigma_i x_i \; + \; n(n+1)/2.\end{aligned}$$

In case 1, we have

$$\Sigma_i\langle -x_i\rangle + \Sigma_j\langle y_j\rangle = \Sigma_{i\geq 2}(1 - x_i) + \Sigma_j y_j = n - 1 + \Sigma_j y_j - \Sigma_i x_i.$$

In cases 2 and 3, we have

$$\Sigma_i\langle -x_i\rangle + \Sigma_j\langle y_j\rangle = \Sigma_i (1 - x_i) + \Sigma_j y_j = n + \Sigma_j y_j - \Sigma_i x_i.$$

Comparing, we see see that the asserted inequality is in fact an equality in cases 1 and 3, and that it holds with a margin of 1 in case 2. QED

Combining this last result with the numerical criterion 8.16.8, we obtain

Corollary 8.17.2.1 Let E be a finite field of characteristic $p \neq \ell$, and

$$\mathcal{H} := \mathcal{H}_1(!, \psi; \chi\text{'s}; \rho\text{'s})$$

a nonpunctual, geometrically irreducible hypergeometric defined over E of type (n, n) with $n \geq 1$ and $\lambda = 1$, whose local monodromy at both zero and ∞ is of finite order. Fix an embedding $\iota : K_{p,nr} \subset \overline{\mathbb{Q}}_p$. Let

$$x_1, \ldots, x_n, y_1, \ldots, y_n \in (\mathbb{Q}/\mathbb{Z})_{\text{not } p}$$

be the 2n distinct elements of $(\mathbb{Q}/\mathbb{Z})_{\text{not } p}$ such that

$$\iota\circ\chi_i = \chi_{x_i, E} \text{ for } i = 1, \ldots, n,$$

$$\iota\circ\rho_j = \chi_{y_j, E} \text{ for } j = 1, \ldots, n.$$

Then $\mathcal{H}$ has finite G_{geom} if for every $\alpha \in \prod_{\ell\neq p} \mathbb{Z}_\ell^\times \approx \mathrm{Aut}((\mathbb{Q}/\mathbb{Z})_{\text{not } p})$, the two subsets

$$X_\alpha := \{\alpha x_1, \ldots, \alpha x_n\} \quad \text{and} \quad Y_\alpha := \{\alpha y_1, \ldots, \alpha y_n\}$$

of $(\mathbb{Q}/\mathbb{Z})_{\text{not } p}$ are **intertwined** in $(\mathbb{Q}/\mathbb{Z})_{\text{not } p}$.

(8.17.3) We will now establish the converse to this corollary: if an irreducible hypergeometric of type (n, n) has G_{geom} finite, then the intertwining condition holds. In view of the numerical criterion 8.16.8, this amounts to a purely combinatorial statement. However, we do not know a combinatorial proof; our proof is based upon Grothendieck's theory of specialization of the fundamental group.

(8.17.4) Fix an integer $N \geq 1$. Let

$\Phi_N(x)$:= the N'th cyclotomic polynomial,

R_N := the ring $\mathbb{Z}[1/N\ell, X]/(\Phi_N(X))$,

μ_N := the cyclic group $\mu_N(R_N)$ of order N,

S_N := $\mathrm{Spec}(R_N)$.

The group $\mathrm{Hom}(\mu_N, \mu_N)$ is canonically $(1/N)\mathbb{Z}/\mathbb{Z}$, with x in $(1/N)\mathbb{Z}/\mathbb{Z}$ corresponding to the character $\zeta \mapsto \zeta^{Nx}$. If we fix an embedding

$$\iota_\ell : R_N \subset \overline{\mathbb{Q}}_\ell,$$

we have an induced isomorphism $\mu_N \approx \mu_N(\overline{\mathbb{Q}}_\ell)$; this allows us to identify $\overline{\mathbb{Q}}_\ell$-valued characters of μ_N with elements of $(1/N)\mathbb{Z}/\mathbb{Z}$. Over the base S_N, the N'th power endomorphism of $\mathbb{G}_m/S_N$ is a finite etale μ_N-torsor. For any character

$$\chi : \mu_N \to \overline{\mathbb{Q}}_\ell^\times$$

we can speak of the lisse $\overline{\mathbb{Q}}_\ell$-sheaf $\mathcal{L}_\chi$ on $\mathbb{G}_m/S_N$ obtained from this μ_N-torsor by pushing out via $\overline{\chi}$ (sic). Similarly, for any nontrivial character $\Lambda: \mu_N \to \overline{\mathbb{Q}}_\ell^\times$, we can speak of the $\overline{\mathbb{Q}}_\ell$-sheaf $\mathcal{L}_{\Lambda(1-x)}$ on $\mathbb{G}_m/S_N$; this sheaf $\mathcal{L}_{\Lambda(1-x)}$ is lisse of rank one on the complement of the unit section "1" of $\mathbb{G}_m/S_N$, extended by zero to all of $\mathbb{G}_m/S_N$. Both $\mathcal{L}_\chi$ and $\mathcal{L}_{\Lambda(1-x)}$ are tame along both zero and ∞ in the ambient $\mathbb{P}^1/S_N$.

(8.17.5) Given two distinct $\overline{\mathbb{Q}}_\ell^\times$-valued characters χ and ρ of μ_N, we can speak of the $\overline{\mathbb{Q}}_\ell$-sheaf $\mathcal{L}_{\chi(x)}\otimes\mathcal{L}_{(\rho/\chi)(1-x)}$ on $\mathbb{G}_m/S_N$; it is lisse of rank one on the complement of the unit section "1" of $\mathbb{G}_m/S_N$, extended by zero to all of $\mathbb{G}_m/S_N$. This sheaf is also tame along both zero and ∞.

(8.17.6) We denote by

$$\mathcal{H}(\chi;\rho) := \text{the } \overline{\mathbb{Q}}_\ell\text{-sheaf } \mathcal{L}_{\chi(x)}\otimes\mathcal{L}_{(\rho/\chi)(1-x)} \text{ on } \mathbb{G}_m/S_N.$$

For any prime number p which is prime to $N\ell$, and for any ring homomorphism

$$R_N \to \overline{\mathbb{F}}_p,$$

the induced map on N'th roots of unity is an isomorphism $\mu_N \approx \mu_N(\overline{\mathbb{F}}_p)$. This allows us to view both χ and ρ as $\overline{\mathbb{Q}}_\ell^\times$-valued characters of $\mu_N(\overline{\mathbb{F}}_p)$. So viewing them, it makes sense to form the sheaf $\mathcal{H}_1(!,\psi;\chi;\rho)$ on $\mathbb{G}_m/\overline{\mathbb{F}}_p$. Clearly the restriction of $\mathcal{H}(\chi;\rho)$ to $\mathbb{G}_m/\overline{\mathbb{F}}_p$ is geometrically isomorphic to $\mathcal{H}_1(!,\psi;\chi;\rho)$.

(8.17.7) We can use the multiplication morphism

$$\pi : (\mathbb{G}_m\times\mathbb{G}_m)_{S_N} \to (\mathbb{G}_m)_{S_N}$$

to define ! convolution (relative to S_N); for K and L in $D^b_c(\mathbb{G}_m/S_N, \overline{\mathbb{Q}}_\ell)$,

$$K*_!L := R\pi_!(pr_1{}^*K\otimes pr_2{}^*L).$$

(8.17.8) Let us say that a $\overline{\mathbb{Q}}_\ell$-sheaf $\mathcal{F}$ on $\mathbb{G}_m/S$ is "tame and adapted to the unit section" if it satisfies the following three conditions:
(1) $\mathcal{F}$ is lisse on the complement of the unit section "1" of $\mathbb{G}_m/S$,
(2) the restriction of $\mathcal{F}$ to the unit section is lisse on S,
(3) $\mathcal{F}$ is tame along each of the three sections "0", "1", and ∞ of the ambient $\mathbb{P}^1/S$.

Let us say that an object K of $D^b{}_c(\mathbb{G}_m/S, \overline{\mathbb{Q}}_\ell)$ is "tame and adapted to the unit section" if each of its cohomology sheaves is "tame and adapted to the unit section" in the above sense.

Theorem 8.17.9 Let S be any irreducible noetherian $\mathbb{Z}[1/\ell]$-scheme whose generic point has characteristic zero. The subcategory of $D^b{}_c(\mathbb{G}_m/S, \overline{\mathbb{Q}}_\ell)$ consisting of those objects K which are "tame and adapted to the unit section" is stable by ! convolution.

proof Let K and L be "tame and adapted to the unit section". That the cohomology sheaves of $K*_!L$ are lisse outside the unit section and that their restriction to the unit section is lisse on S both result directly from the "trivial" case of Deligne's semicontinuity theorem [Lau-SCS]). Once these cohomology sheaves are lisse outside the unit section, they are automatically tame along the three sections "0", "1", and ∞ simply because S is irreducible with generic point of characteristic zero (cf [SGAI, Exposé XIII, 5.5]). QED

Two Variants 8.17.10 In these variants, S is any irreducible noetherian $\mathbb{Z}[1/\ell]$-scheme whose generic point has characteristic zero.
(1) Let Γ be a finite etale subgroupscheme of $\mathbb{G}_m/S$. Thus Γ is μ_M for some integer $M \geq 1$ which is invertible on S. In the ambient $\mathbb{P}^1/S$, Γ, "0" and ∞ are three disjoint smooth/S divisors. We say that an object K of $D^b{}_c(\mathbb{G}_m/S, \overline{\mathbb{Q}}_\ell)$ is "tame and adapted to Γ" if each of its cohomology sheaves is lisse on $\mathbb{G}_m - \Gamma$, lisse on Γ, and tame along each of the divisors Γ, "0" and ∞. For $\Gamma = \mu_M$, K is "tame and adapted to Γ" if and only if $[M]_*K$ is "tame and adapted to the unit section". The subcategory of $D^b{}_c(\mathbb{G}_m/S, \overline{\mathbb{Q}}_\ell)$ consisting of those objects K which are "tame and adapted to Γ" is stable by ! convolution. (Indeed, since [M] is a homomorphism, $[M]_*(K*_!L) \approx ([M]_*K)*_!([M]_*L)$, so this is immediate from the theorem.)

(2) Let E/S be an elliptic curve over S, and $\Gamma \subset E$ a finite etale subgroupscheme of E/S. Then Γ is a smooth/S divisor in E. We say that an object K of $D^b_c(E/S, \overline{\mathbb{Q}}_\ell)$ is "tame and adapted to Γ" if each of its cohomology sheaves is lisse on $E - \Gamma$, lisse on Γ, and tame along Γ. If we denote by E_1/S the quotient of E by Γ, and by $\pi : E \to E_1$ the isogeny with kernel Γ, then K is tame and adapted to Γ if and only if π_*K is tame and adapted to the unit section on E_1/S. The subcategory of $D^b_c(E/S, \overline{\mathbb{Q}}_\ell)$ consisting of those objects K which are tame and adapted to Γ is stable by ! convolution. (Since π is a homomorphism, $\pi_*(K*_!L) \approx (\pi_*K)*_!(\pi_*L)$, so may we reduce to the case when Γ is the zero-section. Now apply Deligne's semicontinuity theorem [Lau-SCS].)

(8.17.11) Given an integer $n \geq 1$, a set $\{\chi_1, \dots, \chi_n\}$ of n not necessarily distinct characters χ_i of μ_N, and a disjoint set $\{\chi_1, \dots, \chi_n\}$ of n not necessarily distinct characters ρ_j of μ_N, we can define the multiple ! convolution (relative to S_N)

$$\mathcal{H}(\chi_1; \rho_1)[1] *_! \mathcal{H}(\chi_2; \rho_2)[1] *_! \dots *_! \mathcal{H}(\chi_n; \rho_n)[1].$$

as an object of $D^b_c(\mathbb{G}_m/S_N, \overline{\mathbb{Q}}_\ell)$. In view of the above theorem, this object is tame and adapted to the unit section. Looking fibre by fibre over S_N (permissible by proper base change), we see that this object is of the form

$$\mathcal{H}(\chi\text{'s}; \rho\text{'s})[1]$$

for some $\overline{\mathbb{Q}}_\ell$-sheaf $\mathcal{H}(\chi\text{'s}; \rho\text{'s})$ on $\mathbb{G}_m/S_N$ which is tame and adapted to the unit section. Furthermore, we see that the restriction of $\mathcal{H}(\chi\text{'s}; \rho\text{'s})$ to each geometric fibre $\mathbb{G}_m/\overline{\mathbb{F}}_p$ is geometrically isomorphic to the hypergeometric sheaf $\mathcal{H}_1(!, \psi; \chi\text{'s}; \rho\text{'s})$.

Pick a complex embedding

$$R_N \subset \mathbb{C}.$$

The restriction of $\mathcal{H}(\chi\text{'s}; \rho\text{'s})$ to $((\mathbb{G}_m - \{1\})_\mathbb{C})^{an}$ is an ℓ-adic local system $\mathcal{M}_\ell$ on $\mathbb{C}^\times - \{1\}$ whose local monodromy at zero (resp.at ∞) is (via the $t \mapsto \{\exp(2\pi it/N)\}_N$ embedding $\mathbb{Z} \to \hat{\mathbb{Z}}(1)$) a successive extension of the χ's (resp. of the ρ's), and whose local monodromy at 1 is a pseudoreflection. Since the χ's and the ρ's are disjoint, this local system is irreducible.

Pick any embedding of $\mathbb{C}$ into $\overline{\mathbb{Q}}_\ell$ such that the composite embedding

$$R_N \subset \mathbb{C} \subset \overline{\mathbb{Q}}_\ell$$

is the fixed embedding

$$\iota_\ell : R_N \subset \overline{\mathbb{Q}}_\ell.$$

By 3.5.4 (Rigidity), the local system $\mathcal{M}_\ell$ must be the (extension of scalars by the embedding $\mathbb{C} \subset \overline{\mathbb{Q}}_\ell$ of the) complex local system attached to the hypergeometric $\mathcal{D}$-module $\mathcal{H}_1(x_1, \ldots, x_n; y_1, \ldots, y_n)$, where the x_i and the y_j are the unique elements of $(1/N)\mathbb{Z}/\mathbb{Z}$ to which the characters $\chi_1, \ldots, \chi_n; \rho_1, \ldots, \rho_n$ correspond.

Comparison Theorem 8.17.12 Hypotheses and notations as above, suppose given an integer $n \geq 1$, a set $\{\chi_1, \ldots, \chi_n\}$ of n not necessarily distinct characters χ_i of μ_N, and a disjoint set $\{\rho_1, \ldots, \rho_n\}$ of n not necessarily distinct characters ρ_j of μ_N. Let the x_i and the y_j be the unique elements of $(1/N)\mathbb{Z}/\mathbb{Z}$ to which correspond the characters $\chi_1, \ldots, \chi_n; \rho_1, \ldots, \rho_n$. Then the following $\overline{\mathbb{Q}}_\ell$-algebraic subgroups of $GL(n)$ are all conjugate:

(1) for any complex embedding $R_N \subset \mathbb{C}$, the group $G_{gal} \otimes_{\mathbb{C}} \overline{\mathbb{Q}}_\ell$, where G_{gal} is the differential galois group of the hypergeometric $\mathcal{D}$-module $\mathcal{H}_1(x_1, \ldots, x_n; y_1, \ldots, y_n)$.

(2) for any prime number p which is prime to $N\ell$, and for any ring homomorphism $R_N \to \overline{\mathbb{F}}_p$, the group G_{geom} for $\mathcal{H}_1(!, \psi; \chi\text{'s}; \rho\text{'s})$.

proof Since the $\mathcal{D}$-module $\mathcal{H}_1(x_1, \ldots, x_n; y_1, \ldots, y_n)$ has regular singular points, its G_{gal} is the Zariski closure of the image of its monodromy representation. Thus $G_{gal} \otimes_{\mathbb{C}} \overline{\mathbb{Q}}_\ell$ is the group G_{geom} for the restriction of $\mathcal{H}(\chi\text{'s}; \rho\text{'s})$ to the complex fibre of $(\mathbb{G}_m - \{1\})/S_N$ given by $R_N \subset \mathbb{C}$. The sheaf $\mathcal{H}(\chi\text{'s}; \rho\text{'s})$ on $(\mathbb{G}_m - \{1\})/S_N$ is lisse, and tame along "0", "1", and ∞.

For such a sheaf on $(\mathbb{G}_m - \{1\})/S_N$ (lisse, and tame along "0", "1", and ∞), the groups G_{geom} for its restrictions to the various geometric fibres of $(\mathbb{G}_m - \{1\})/S_N$ are all conjugate in GL(n). This is a special case of:

Tame Specialization Theorem 8.17.13 Let S be a normal irreducible noetherian scheme with generic point η, X/S a proper smooth morphism with geometrically connected fibres, $D \subset X$ a divisor with normal crossings relative to S, U := X - D, and $u \in U(S)$ a section of U/S. Let ℓ be a prime number, and $\mathcal{F}$ a lisse $\overline{\mathbb{Q}}_\ell$-sheaf of rank n on U such that $\mathcal{F}_{\overline{\eta}} \mid U_{\overline{\eta}}$ is tamely ramified at all the maximal points of $D_{\overline{\eta}}$. Then for any geometric point s of S, the image of $\pi_1(U_s, u_s)$ in $GL(n, \overline{\mathbb{Q}}_\ell)$ is (conjugate to) the image of $\pi_1(U_{\overline{\eta}}, u_{\overline{\eta}})$ in $GL(n, \overline{\mathbb{Q}}_\ell)$.

proof For this, it suffices to prove that the images of the π_1's of these geometric fibres in $GL(n, \overline{\mathbb{Q}}_\ell)$ are all conjugate. For this, we may reduce successively to the case of E_λ-sheaves, then to $\mathcal{O}_\lambda$-sheaves, and finally to the case of $\mathcal{O}_\lambda/\ell^\nu\mathcal{O}_\lambda$-sheaves.

Given a finite ring A (e.g., $A = \mathcal{O}_\lambda/\ell^\nu\mathcal{O}_\lambda$) and a lisse sheaf of free A-modules $\mathcal{F}$ of rank n on U, denote by

$$E \rightarrow U$$

the associated finite etale GL(n, A) torsor.

The theorem is now reduced to the following finite variant (compare [De-WII, 1.11.1 and 1.11.2]):

Tame Specialization Theorem bis 8.17.14 Let G be a finite group. Let S be a normal irreducible noetherian scheme with generic point η, X/S a proper smooth morphism with geometrically connected fibres, $D \subset X$ a divisor with normal crossings relative to S, U := X - D, and $u \in U(S)$ a section of U/S. Let $E \rightarrow U$ be a finite etale G-torsor such that $E_{\overline{\eta}} \rightarrow U_{\overline{\eta}}$ is tamely ramified at all the maximal points of $D_{\overline{\eta}}$. Then for any geometric point s of S, the image of $\pi_1(U_s, u_s)$ in G is (conjugate to) the image of $\pi_1(U_{\overline{\eta}}, u_{\overline{\eta}})$ in G.

proof For any geometric point s of S, we have a diagram of homomorphisms of π_1's induced by the evident inclusions

$$\begin{array}{ccccccc} & & & & & & \pi_1(U_s, u_s) \\ & & & & & & \downarrow \\ \pi_1(U_{\bar\eta}, u_{\bar\eta}) & \to & \pi_1(U_\eta, u_{\bar\eta}) & \twoheadrightarrow & \pi_1(U, u_{\bar\eta}) & \approx & \pi_1(U, u_s). \end{array}$$

Because $E \to U$ is finite etale, there exists a finite galois extension L/η such that $\pi_1(U_{\bar\eta}, u_{\bar\eta})$ and $\pi_1(U_\eta \otimes L, u_{\bar\eta})$ have the same image in G. Replacing S by its normalization in L, we reduce to the case when $\pi_1(U_{\bar\eta}, u_{\bar\eta})$ and $\pi_1(U, u_{\bar\eta})$ have the same image in G.

In this case, the above diagram shows that for every geometric point s of S, the image of $\pi_1(U_s, u_s)$ in G is (conjugate to) a subgroup of the image of $\pi_1(U_{\bar\eta}, u_{\bar\eta})$.

So to show that these two images are conjugate, it suffices to show that both images have the same index in G. But these indices are precisely the number of connected components of the geometric fibres E_s and $E_{\bar\eta}$ of E/S. By the reduction already performed, the connected components of E are in bijection with those of $E_{\bar\eta}$. So replacing E by one of its connected components,say E_1, and G by the subgroup G_1 which stabilizes E_1, we are reduced to showing universally that (for a tame covering G-torsor $E \to U$) if $E_{\bar\eta}$ is connected, then E_s is connected. For this, we may reduce to the case where S is local and strictly henselian (replace S by its strict henselization at s). Since $E_{\bar\eta}$ is connected, the total space E is connected. By [SGA1, XIII, Thm. 2.4, 1] (applied to F := G, S' := s and h := the inclusion of s into S), the functor "pullback from U to U_s" on the categories of tame G-torsors is an equivalence. In particular, if E_s is disconnected, then E itself is disconnected. QED

Corollary 8.17.15 Let E be a finite field of characteristic $p \neq \ell$, and

$$\mathcal{H} := \mathcal{H}_1(!, \psi; \chi\text{'s}; \rho\text{'s})$$

a nonpunctual, geometrically irreducible hypergeometric defined over E of type (n, n) with $n \geq 1$ and $\lambda = 1$, whose local monodromy at both zero and ∞ is of finite order. Fix an embedding $\iota : K_{p,nr} \subset \bar{\mathbb{Q}}_p$. Let

$$x_1, \ldots, x_n, y_1, \ldots, y_n \in (\mathbb{Q}/\mathbb{Z})_{\text{not } p}$$

be the 2n distinct elements of $(\mathbb{Q}/\mathbb{Z})_{\text{not } p}$ such that

$$\iota \circ \chi_i = \chi_{x_i, E} \text{ for } i = 1, \ldots, n,$$

$$\iota \circ \rho_j = \chi_{y_j, E} \text{ for } j = 1, \dots, n.$$

Then the following conditions are equivalent.

(1) for every $\alpha \in \prod_{\ell \neq p} \mathbb{Z}_\ell^\times \approx \mathrm{Aut}((\mathbb{Q}/\mathbb{Z})_{\text{not } p})$, the two subsets

$$X_\alpha := \{\alpha x_1, \dots, \alpha x_n\} \quad \text{and} \quad Y_\alpha := \{\alpha y_1, \dots, \alpha y_n\}$$

of $(\mathbb{Q}/\mathbb{Z})_{\text{not } p}$ are **intertwined** in $(\mathbb{Q}/\mathbb{Z})_{\text{not } p}$.

(2) $\mathcal{H}$ has finite G_{geom}.

(3) the hypergeometric $\mathcal{D}$-module $\mathcal{H}_1(x_1, \dots, x_n; y_1, \dots, y_n)$ has G_{gal} finite.

(4) the hypergeometric $\mathcal{D}$-module $\mathcal{H}_1(x_1, \dots, x_n; y_1, \dots, y_n)$ has p-curvature zero for almost all primes p.

proof We have (1) ⇒ (2) by 8.17.2.1, we have (2) ⇒ (3) by the previous theorem, (3) ⇒ (4) is elementary (formation of the p-curvature commutes with etale localization, cf [Ka-AS, Intro]), and (4) ⇔ (1) is the Beukers-Heckman lemma 5.5.2. QED

Remark 8.17.16 The proof of the equivalence (3) ⇔ (4) in the above corollary provides a third proof of Grothendieck's p-curvature conjecture for hypergeometric $\mathcal{D}$-modules of type (n, n), (apparently) independent of both the Beukers-Heckman "signature of a hermitian form" proof (cf. [B-H, 4.8]) and of the "reduction to the Gauss-Manin case" proof in 5.5.1.

8.18 Appendix : Semicontinuity and Specialization for G_{geom} d'apres R. Pink

(8.18.1) In this appendix, we will consider the following situation. S is a normal connected noetherian scheme with generic point η, X/S is a smooth S-scheme with geometrically connected fibres, ℓ is a prime number (which is **not** assumed invertible on S) and $\mathcal{F}$ is a lisse $\overline{\mathbb{Q}}_\ell$-sheaf of rank $n \geq 1$ on X. For each geometric point s in S, the group

$$\Gamma(s) := \text{the image of } \pi_1(X_s, \text{any base point } x_s) \text{ in } GL(n, \overline{\mathbb{Q}}_\ell)$$

is a closed subgroup, whose conjugacy class in $GL(n, \overline{\mathbb{Q}}_\ell)$ is independent of the auxiliary choice of base point x_s.

Specialization Theorem 8.18.2 (compare [De-WII, 1.11.5]) Let S be a normal connected noetherian scheme with generic point η, X/S a smooth S-scheme with geometrically connected fibres, ℓ a prime number (which is **not** assumed invertible on S) and $\mathcal{F}$ a lisse $\overline{\mathbb{Q}}_\ell$-sheaf of rank $n \geq 1$ on X. For each geometric point s in S, define

$\Gamma(s)$:= the image of $\pi_1(X_s$, any base point $x_s)$ in $GL(n, \overline{\mathbb{Q}}_\ell)$.

Then

(1) the group $\Gamma(s)$ decreases under specialization, in the sense that if t is a specialization of s, then $\Gamma(t)$ is conjugate in $GL(n, \overline{\mathbb{Q}}_\ell)$ to a subgroup of $\Gamma(s)$.

(2) there exists an open neighborhood V of η in S such that for any geometric point s in V, $\Gamma(s)$ is conjugate in $GL(n, \overline{\mathbb{Q}}_\ell)$ to $\Gamma(\overline{\eta})$.

proof We first reduce to the case of lisse $\mathcal{O}_\lambda$-sheaves which are free of rank n. In order to prove (1), it suffices to show universally that $\Gamma(s)$ is conjugate to a subgroup of $\Gamma(\overline{\eta})$. For this, one can reduce further to the case of $\mathcal{O}_\lambda/\ell^\nu\mathcal{O}_\lambda$ -sheaves, and then to the case of G-torsors for a finite group G. This case is then treated by repeating verbatim the first two paragraphs of the proof of 8.17.14. In the notations used there, the constructibility on S of the function "$s \mapsto$ number of irreducible components of E_s" [EGA IV, 9.7.8] shows that (2) holds for any lisse $\mathcal{O}_\lambda/\ell^\nu\mathcal{O}_\lambda$ -sheaf.

To prove (2) for a lisse $\mathcal{O}_\lambda$-sheaf $\mathcal{F}$, for each integer $\nu \geq 1$ let V_ν be an open neighborhood η in S on which (2) holds for the sheaf $\mathcal{F}/\ell^\nu\mathcal{F}$. We will show that for $\nu >> 0$, (2) for $\mathcal{F}$ itself holds on V_ν. In view of part (1), this results from the following lemma, applied to the group $K := \Gamma(\overline{\eta})$ inside $GL(n, \mathcal{O}_\lambda)$.

Key Lemma 8.18.3 (R. Pink) Let $\mathcal{O}_\lambda$ be the ring of integers in a finite extension of $\mathbb{Q}_\ell$, $n \geq 1$ an integer, and K a closed subgroup of $GL(n, \mathcal{O}_\lambda)$. There exists an integer ν, depending only on K, with the following property:

for any closed subgroup H of K, H = K if (and only if) H and K have the same image in $GL(n, \mathcal{O}_\lambda/\ell^\nu\mathcal{O}_\lambda)$.

proof Denote by M a free $\mathcal{O}_\lambda$-module of rank n, and by G the group $\mathrm{Aut}_{\mathcal{O}_\lambda}(M) \approx GL(n, \mathcal{O}_\lambda)$. For each integer $d \geq 2$, define

$$G_d := \text{Kernel of } \mathrm{Aut}_{\mathcal{O}_\lambda}(M) \to \mathrm{Aut}_{\mathcal{O}_\lambda}(M/\ell^d M),$$

$$K_d := K \cap G_d.$$

Then we have injective group homomorphisms

$$K_d/K_{d+1} \subset G_d/G_{d+1} \subset \mathrm{End}_{\mathcal{O}_\lambda/\ell\mathcal{O}_\lambda}(M/\ell M)$$

$$X \mapsto (X-1)/\ell^d.$$

For each $d \geq 2$, we define

$$L_d(K) := \text{the image of } K_d/K_{d+1} \text{ in } \mathrm{End}_{\mathcal{O}_\lambda/\ell\mathcal{O}_\lambda}(M/\ell M).$$

Notice that all the $L_d(K)$ are subgroups of the **same finite group** $\mathrm{End}_{\mathcal{O}_\lambda/\ell\mathcal{O}_\lambda}(M/\ell M)$. The key observation is that the ℓ-th power map $X \mapsto X^\ell$ defines an injective group homomorphism of K_d/K_{d+1} into K_{d+1}/K_{d+2} for any $d \geq 2$, which gives inclusions

$$L_d(K) \subset L_{d+1}(K) \subset L_{d+2}(K) \subset \ldots \subset \mathrm{End}_{\mathcal{O}_\lambda/\ell\mathcal{O}_\lambda}(M/\ell M).$$

Therefore for some D, we have

$$L_d(K) = L_{d+1}(K) = L_\infty(K) \quad \text{for all } d \geq D.$$

We claim that we can take $\nu := D + 1$ in the lemma. Indeed, let $H \subset K$ be a closed subgroup. Then $L_d(H) \subset L_d(K)$ for every $d \geq 2$. Suppose that H and K have the same image mod ℓ^{D+1}, i.e., suppose that

$$H/H_{D+1} \approx K/K_{D+1}.$$

Then H and K must have the same image mod any lower power of ℓ, in particular mod ℓ^D. So we have short exact sequences

$$\begin{array}{ccccccccc} 0 & \to & L_D(H) & \to & H/H_{D+1} & \to & H/H_D & \to & 0 \\ & & \cap & & \| & & \| & & \\ 0 & \to & L_D(K) & \to & K/K_{D+1} & \to & K/K_D & \to & 0, \end{array}$$

which show that $L_D(H) = L_D(K)$. For any $d \geq D$, we have

$$\begin{array}{ccc} L_D(H) & = & L_D(K) \\ \cap & & \| \\ L_d(H) & \subset & L_d(K), \end{array}$$

which shows that $L_d(H) = L_d(K)$ for every $d \geq D$. For any $d \geq D$, the short exact sequences

$$\begin{array}{ccccccccc} 0 & \to & L_d(H) & \to & H/H_{d+1} & \to & H/H_d & \to & 0 \\ & & \| & & \cap & & \cap & & \\ 0 & \to & L_d(K) & \to & K/K_{d+1} & \to & K/K_d & \to & 0, \end{array}$$

allow us to show inductively that $\mathrm{Card}(H/H_{d+1}) = \mathrm{Card}(K/K_{d+1})$, and hence that the inclusion $H/H_{d+1} \subset K/K_{d+1}$ is an isomorphism. Taking the inverse limit over d, we deduce that $H = K$, as required. QED

CHAPTER 9

G_2 Examples, Fourier Transforms, and Hypergeometrics

9.1 Another G_2 Example

In Theorem 2.10.5, we proved that the rank seven D.E. on $\mathbb{A}^1$

$$\partial^7 - x\partial - 1/2,$$

whose FT defines $x^{-1/2}\exp(-x^7/7)$, has differential galois group G_{gal} the subgroup G_2 of SO(7). In this section, we will give a diophantine proof of the following ℓ-adic analogue of that result.

G_2 Theorem 9.1.1 Let k be an algebraically closed field of characteristic $p > 15$. Denote by $\Lambda_{1/2}$ the unique tame $\overline{\mathbb{Q}}_\ell$-valued character of $\pi_1(\mathbb{G}_m \otimes k)$ of exact order two. Denote by $j: \mathbb{G}_m \to \mathbb{A}^1$ the inclusion. Let ψ be any nontrivial $\overline{\mathbb{Q}}_\ell$-valued additive character of a finite subfield k_0 of k. Then $NFT_\psi(\mathcal{L}_{\psi(x^7)} \otimes j_*\mathcal{L}_{\Lambda_{1/2}(x)})$ has $G_{geom} = G_2$.

proof Let us define $\mathcal{F} := \mathcal{L}_{\psi(x^7)} \otimes j_*\mathcal{L}_{\Lambda_{1/2}(x)}$, $\mathcal{G} := NFT_\psi(\mathcal{F})$. Since $\mathcal{F}$ is irreducible Fourier, $\mathcal{G}$ is irreducible Fourier. By stationary phase, $\mathcal{G}$ is lisse on $\mathbb{A}^1$ of rank seven, with ∞-slopes {0 once, 7/6 six times}. Since $D(\mathcal{F}) \approx [-1]^*\mathcal{F}$, $\mathcal{G}$ is self dual. Since $\mathcal{G}$ has rank seven, the autoduality must be orthogonal. Since $\mathcal{G}$ is lisse irreducible on $\mathbb{A}^1$ and

$$p > 2(\mathrm{rank}\mathcal{G} + 1) = 15,$$

$\mathcal{G}$ is Lie-irreducible. Since $\det\mathcal{G}$ has order dividing two (being self-dual), the group G_{geom} is semisimple and connected. By Theorem 1.6 on prime-dimensional representations, the only possibilities for G_{geom} are the image PSL(2) of SL(2) in $\mathrm{Sym}^6(\mathrm{std}_2)$, or G_2 or SO(7) or SL(7).

Of these, SL(7) is ruled out by the existence of the autoduality. We can rule out PSL(2) by a slope argument. Indeed, if G_{geom} were PSL(2), then by the lifting lemma 7.2.5 there exists a lisse sheaf $\mathcal{K}$ on $\mathbb{A}^1$ of rank two whose G_{geom} is SL(2), such that $\mathcal{G} \approx \mathrm{Sym}^6(\mathcal{K})$. By the highest slope lemma 7.2.4, $\mathcal{G}$ and $\mathcal{K}$ have the highest ∞-slope. Therefore $\mathcal{K}$ has rank two but has highest ∞-slope 7/6, contradicting the fundamental integrality property of slopes (cf [Ka-GKM, 1.9]). Thus $\mathcal{G}$ has G_{geom} either SO(7) or G_2.

To distinguish these two possibilities, recall that for the standard representation std_7 of SO(7), the tensor cube $(std_7)^{\otimes 3}$ has no nonzero SO(7)-invariants. [This amounts to the statement that std_7 does not occur in $(std_7)^{\otimes 2}$. But for any $n \geq 4$, the decomposition of $(std_n)^{\otimes 2}$ as SO(n)-representation is

$$(std_n)^{\otimes 2} = \mathbb{1} \oplus \text{Spherical Harmonics of deg. 2} \oplus \mathrm{Lie}(SO(n)),$$

and none of three irreducible constituents has dimension n.]

On the other hand, one knows that under G_2, already the subspace $\Lambda^3(std_7)$ has a one-dimensional space of G_2-invariants. [In fact these are all the G_2-invariants in $(std_7)^{\otimes 3}$, but will not use this.]

Therefore the following conditions are equivalent:

$G_{geom} = SO(7)$

$\Leftrightarrow$ $\mathcal{G}^{\otimes 3}$ has no nonzero $\pi_1(\mathbb{A}^1\otimes k, \bar{\eta})$-invariants,

$\Leftrightarrow$ $\mathcal{G}^{\otimes 3}$ has no nonzero $\pi_1(\mathbb{A}^1\otimes k, \bar{\eta})$-coinvariants,

$\Leftrightarrow$ $H_c^2(\mathbb{A}^1\otimes k, \mathcal{G}^{\otimes 3}) = 0$.

Since the sheaf $\mathcal{F}$ is "defined over k_0", so is $\mathcal{G}$. By proper base change, we may replace k by the algebraic closure of k_0 in k without changing the cohomology group $H_c^2(\mathbb{A}^1\otimes k, \mathcal{G}^{\otimes 3})$ in question.

Since $\mathcal{G}$ is pure of weight one, $\mathcal{G}^{\otimes 3}$ is pure of weight three. So if $H_c^2(\mathbb{A}^1\otimes k, \mathcal{G}^{\otimes 3})$ is nonzero, it is pure of weight five. On the other hand, by Weil II we know that $H_c^1(\mathbb{A}^1\otimes k, \mathcal{G}^{\otimes 3})$ is mixed of weight ≤ 4. As $\mathcal{G}$ is lisse on the open curve $\mathbb{A}^1\otimes k$, these are the only two possibly nonvanishing cohomology groups. By the Lefschetz Trace Formula, for any finite overfield E of k_0, we have

$$\Sigma_{t \text{ in } E} (\mathrm{Trace}(\mathrm{Frob}_{E,t} \mid \mathcal{G}))^3 =$$
$$= \mathrm{Trace}(\mathrm{Frob}_E \mid H_c^2(\mathbb{A}^1\otimes k, \mathcal{G}^{\otimes 3})) - \mathrm{Trace}(\mathrm{Frob}_E \mid H_c^1(\mathbb{A}^1\otimes k, \mathcal{G}^{\otimes 3})).$$

Now let us denote by

$$h^i = \dim H_c^i(\mathbb{A}^1\otimes k, \mathcal{G}^{\otimes 3}) \text{ for } i = 1, 2.$$

By a standard argument (cf [Ka-SE, 2.2.2.1]) we have

$$\limsup_{\text{all } E/k_0} \mathrm{Card}(E)^{-5/2} | \Sigma_{t \text{ in } E} (\mathrm{Trace}(\mathrm{Frob}_{E,t} \mid \mathcal{G}))^3 | = h^2.$$

Therefore we have only to compute the sums

$$S := \Sigma_{t \text{ in } E} (\mathrm{Trace}(\mathrm{Frob}_{E,t} \mid \mathcal{G}))^3$$

in order to decide whether G_{geom} is SO(7) (the case $h^2 = 0$) or G_2 (the case $h_2 > 0$).

Let us fix attention on a single E. We denote Card(E) by q. We wish to compute, for ψ any nontrivial additive $\mathbb{C}$-valued character of E, and for Λ the quadratic character of $E^\times$, extended by zero to E, the sum

$$S := \Sigma_{t \text{ in } E} \ (\mathrm{Trace}(\mathrm{Frob}_{E,t} \mid \mathcal{G}))^3$$
$$= \Sigma_{t \text{ in } E} \ (-\Sigma_{x \text{ in } E} \psi(tx + x^7)\Lambda(x))^3$$
$$= - \Sigma_{t,x,y,z \text{ in } E} \psi(t(x + y + z))\psi(x^7 + y^7 + z^7)\Lambda(xyz).$$

Summing first over t we see that only the terms with $x + y + z = 0$ survive:

$$S = -q\Sigma_{x,y \text{ in } E} \psi(x^7 + y^7 - (x + y)^7)\Lambda(-xy(x + y)).$$

At this point, we extract the rabbit from the hat: the universal identity

$$(x + y)^7 - x^7 - y^7 =$$
$$= 7x^6y + 21x^5y^2 + 35x^4y^3 + 35x^3y^4 + 21x^2y^5 + 7xy^6$$
$$= 7xy(x + y)(x^4 + 2x^3y + 3x^2y^2 + 2xy^3 + y^4)$$
$$= 7xy(x + y)(x^2 + xy + y^2)^2.$$

Substituting into our sum, we find

$$S = -q\Sigma_{x,y \text{ in } E} \psi(-7xy(x + y)(x^2 + xy + y^2)^2)\Lambda(-xy(x + y)).$$

Since Λ is the quadratic character of $E^\times$ extended by zero, we have

$$-S/q = A + B,$$
$$A := \ \Sigma_{x,y \text{ in } E} \psi(-7xy(x + y)(x^2 + xy + y^2)^2)\Lambda(-xy(x + y)(x^2 + xy + y^2)^2)$$
$$B := \ \Sigma_{x,y \text{ in } E \text{ with } x^2 + xy + y^2 = 0} \ \Lambda(-xy(x + y)).$$

We first remark that the sum B is trivially bounded by 2q, since it is the sum of at most 2q terms, each of which is ± 1. So B will not affect our limsup.

What about the sum A? Consider the one parameter family

$$\pi : \mathcal{C} \to \mathbb{G}_m$$

of affine curves C_u, $u \neq 0$, over the $\mathbb{G}_m/\mathbb{F}_p$ of u's defined by the equation

$$C_u : -xy(x + y)(x^2 + xy + y^2)^2 = u.$$

Because $xy(x + y)(x^2 + xy + y^2)^2$ is homogeneous of degree 7 but is not a 7'th power, these curves C_u, $u \neq 0$, are smooth and geometrically irreducible over any field of characteristic $\neq 7$ (cf. [Ka-PES, proof of Cor. 6.5]). On the other hand, this family becomes constant after extracting the seventh root of the parameter "u". Therefore the sheaves $R^i\pi_!\overline{\mathbb{Q}}_\ell$ on $\mathbb{G}_m\otimes k$ are all lisse on $\mathbb{G}_m$, and everywhere tame (since they become constant after [7]*). Moreover, we have

$$R^i\pi_!\overline{\mathbb{Q}}_\ell = 0 \text{ for } i \neq 1, 2,$$
$$R^2\pi_!\overline{\mathbb{Q}}_\ell = \overline{\mathbb{Q}}_\ell(-1) \quad \text{for } i = 2,$$
$$R^1\pi_!\overline{\mathbb{Q}}_\ell \text{ is mixed of weight} \leq 1.$$

In terms of these curves C_u,

$$A = \Sigma_{a \text{ in } E^\times}\ \psi(7a)\Lambda(a)\text{Card}(C_a(E))$$
$$= \Sigma_{a \text{ in } E^\times}\ \psi(7a)\Lambda(a)[q - \text{Trace}(\text{Frob}_{a,E} \mid R^1\pi_!\overline{\mathbb{Q}}_\ell)]$$
$$= q\Sigma_{a \text{ in } E^\times}\ \psi(7a)\Lambda(a) \ - \Sigma_{a \text{ in } E^\times}\ \text{Trace}(\text{Frob}_{a,E} \mid \mathcal{L}_{\tilde{\psi}}\otimes\mathcal{L}_\Lambda\otimes R^1\pi_!\overline{\mathbb{Q}}_\ell).$$

where we write $\tilde{\psi}(x) := \psi(7x)$.
The sum

$$\Sigma_{a \text{ in } E^\times}\ \psi(7a)\Lambda(a)$$

is a Gauss sum, so it has absolute value $q^{1/2}$. What about the sum

$$D := \Sigma_{a \text{ in } E^\times}\ \text{Trace}(\text{Frob}_{a,E} \mid \mathcal{L}_{\tilde{\psi}}\otimes\mathcal{L}_\Lambda\otimes R^1\pi_!\overline{\mathbb{Q}}_\ell)?$$

Using the Lefschetz Trace Formula, we get

$$D = \text{Trace}(\text{Frob}_E \mid H_c^2(\mathbb{G}_m\otimes k,\ \mathcal{L}_{\tilde{\psi}}\otimes\mathcal{L}_\Lambda\otimes R^1\pi_!\overline{\mathbb{Q}}_\ell))$$
$$- \text{Trace}(\text{Frob}_E \mid H_c^1(\mathbb{G}_m\otimes k,\ \mathcal{L}_{\tilde{\psi}}\otimes\mathcal{L}_\Lambda\otimes R^1\pi_!\overline{\mathbb{Q}}_\ell)).$$

Because $\mathcal{L}_\Lambda\otimes R^1\pi_!\overline{\mathbb{Q}}_\ell$ is tame on $\mathbb{G}_m$, $\mathcal{L}_{\tilde{\psi}}\otimes\mathcal{L}_\Lambda\otimes R^1\pi_!\overline{\mathbb{Q}}_\ell$ is totally wild at ∞, and consequently its $H_c^2(\mathbb{G}_m\otimes k,\ \mathcal{L}_{\tilde{\psi}}\otimes\mathcal{L}_\Lambda\otimes R^1\pi_!\overline{\mathbb{Q}}_\ell) = 0$. On the other hand, $\mathcal{L}_{\tilde{\psi}}\otimes\mathcal{L}_\Lambda\otimes R^1\pi_!\overline{\mathbb{Q}}_\ell$ is mixed of weight ≤ 1, so $H_c^1(\mathbb{G}_m\otimes k,\ \mathcal{L}_{\tilde{\psi}}\otimes\mathcal{L}_\Lambda\otimes R^1\pi_!\overline{\mathbb{Q}}_\ell)$ is mixed of weight ≤ 2. So if we denote by K the dimension of $H_c^1(\mathbb{G}_m\otimes k,\ \mathcal{L}_{\tilde{\psi}}\otimes\mathcal{L}_\Lambda\otimes R^1\pi_!\overline{\mathbb{Q}}_\ell)$, we get

$$|D| \leq Kq.$$

So all in all we have

$$-S/q = A + B$$
$$|B| \leq 2q$$
$$A = q\Sigma_{a \text{ in } E^\times}\ \psi(7a)\Lambda(a) + D$$
$$|q\Sigma_{a \text{ in } E^\times}\ \psi(7a)\Lambda(a)| = q^{3/2}$$
$$|D| \leq Kq,$$

whence

$$| \, |S| - q^{5/2} \, | \le (2 + K)q^2.$$

Taking the limsup of $|S|/q^{5/2}$ over larger and larger E's, we see that $h^2 := \dim H_c^2(\mathbb{A}^1 \otimes k, \mathcal{G}^{\otimes 3}) = 1$. In particular, $G_{geom} = G_2$. QED

9.2 Relation of Simple Fourier Transforms to Hypergeometrics

(9.2.1) What is the relation of the sheaf $NFT_\psi(\mathcal{L}_{\psi(x^7)} \otimes j_*\mathcal{L}_{\Lambda_{1/2}(x)})$, whose G_{geom} we have proven to be G_2, to hypergeometrics of type (7, 1), some of which also have $G_{geom} = G_2$? We will see below that it is the Kummer pullback by $[7]^*$ of just such a hypergeometric. Indeed this is a special case of a general phenomenon.

Proposition 9.2.2 Suppose that $n \ge 2$, and that p does not divide n. Let k be an algebraically closed field of characteristic p. Let Λ_1 be a nontrivial tame $\overline{\mathbb{Q}}_\ell$-valued character of $\pi_1(\mathbb{G}_m \otimes k)$ of finite order prime to n. Let Λ_2 be any tame $\overline{\mathbb{Q}}_\ell$-valued character of $\pi_1(\mathbb{G}_m \otimes k)$ which satisfies

$$(\Lambda_2)^n = \overline{\Lambda}_1.$$

Denote by $\{\rho_1, \ldots, \rho_n\}$ all the characters of $\pi_1(\mathbb{G}_m \otimes k)$ of order dividing n. Denote by $j: \mathbb{G}_m \to \mathbb{A}^1$ the inclusion. Let ψ be any nontrivial $\overline{\mathbb{Q}}_\ell$-valued additive character of a finite subfield k_0 of k. Then for some λ in $k^\times$ there exists an isomorphism of lisse sheaves on $\mathbb{G}_m \otimes k$

$$j^*NFT_\psi(\mathcal{L}_{\psi(x^n)} \otimes j_*\mathcal{L}_{\Lambda_1(x)}) \approx [n]^*\mathcal{H}_\lambda(!, \psi; \rho_1, \ldots, \rho_n; \Lambda_2).$$

proof Define $\mathcal{G} := NFT_\psi(\mathcal{L}_{\psi(x^n)} \otimes j_*\mathcal{L}_{\Lambda_1(x)})$. By stationary phase, $\mathcal{G}$ is lisse of rank n on $\mathbb{A}^1$, irreducible, and its I_∞-representation is

$$\mathcal{L}_{\overline{\Lambda}_1(x)} \oplus (\text{rank } n-1, \text{ all slopes } n/(n-1)).$$

Enlarging k_0 if necessary, we may assume that k_0 contains a primitive n'th root of unity, say ζ, and that Λ_2 is defined over k_0. For any finite extension E of k_0, it is obvious that the trace function of $j^*\mathcal{G}$ on $E^\times$

$$t \in E^\times \mapsto -\Sigma_{x \text{ in } E}\, \psi_E(tx + x^n)\Lambda_{2,E}(x^{-n})$$

is invariant under the multiplicative translation $t \mapsto \zeta t$: simply replace x by $\zeta^{-1}x$ in the sum. Therefore $j^*\mathcal{G}$ and $T_\zeta{}^*j^*\mathcal{G}$ have the same trace function. Therefore they have isomorphic semisimplifications as representations of $\pi_1(\mathbb{G}_m \otimes k_0)$. Since they are $\pi_1(\mathbb{G}_m \otimes k_0)$-irreducible, being irreducible for the subgroup $\pi_1(\mathbb{G}_m \otimes k)$, it follows that $j^*\mathcal{G} \approx T_\zeta{}^* j^*\mathcal{G}$ as lisse sheaves on $\mathbb{G}_m \otimes k_0$. Since $j^*\mathcal{G}$ is irreducible, we can descend $j^*\mathcal{G}$ through $[n]$. Thus there exists a lisse $\mathcal{H}$ on $\mathbb{G}_m \otimes k_0$ of rank n with $[n]^*\mathcal{H} \approx j^*\mathcal{G}$. Any such $\mathcal{H}$ is irreducible on $\mathbb{G}_m \otimes k$ (since it has an irreducible pullback).

The I_∞-representation of $\mathcal{H}$ must be of the form

$$\mathcal{L}_{\chi(x)} \oplus (\text{rank } n-1, \text{ all slopes } 1/(n-1)),$$

for some χ with $\chi^n = \overline{\Lambda}_1$. So if we twist $\mathcal{H}$ by a suitable tame $\mathcal{L}_\Lambda$ of order dividing n, we may assume that

$$[n]^*\mathcal{H} \approx j^*\mathcal{G}$$
$$\mathcal{H} \mid I_\infty \approx \mathcal{L}_{\Lambda_2(x)} \oplus (\text{rank } n-1, \text{ all slopes } 1/(n-1)).$$

In particular, $\mathcal{H}$ has $\mathrm{Swan}_\infty(\mathcal{H}) = 1$.

Since $[n]^*\mathcal{H}$ is lisse on $\mathbb{A}^1$, $\mathcal{H}$ must be tame at zero. Being irreducible lisse on $\mathbb{G}_m$ with $\mathrm{Swan}_\infty = 1$ and tame at zero, it must be a hypergeometric of type (n, 1), so on $\mathbb{G}_m$ there exists an isomorphism

$$\mathcal{H} \approx \mathcal{H}_\lambda(!, \psi; \chi_1, \dots, \chi_n; \Lambda_2)$$

for some λ in $k^\times$.

Since $[n]^*\mathcal{H}$ is lisse on $\mathbb{A}^1$, each χ_i has order dividing n, and there can be no repetition of the χ_i which occur (since local monodromy at zero is of finite order dividing n). Therefore the χ_i are precisely all the characters of order dividing n. QED

In an entirely similar fashion, one proves:

Proposition 9.2.3 Suppose that $n \geq 2$, and that p does not divide n. Let k be an algebraically closed field of characteristic p. Let $\{\rho_1, \dots, \rho_{n-1}\}$ be all but one of the characters of $\pi_1(\mathbb{G}_m \otimes k)$ of order dividing n. Denote by $j: \mathbb{G}_m \to \mathbb{A}^1$ the inclusion. Let ψ be any nontrivial $\overline{\mathbb{Q}}_\ell$-valued additive character of a finite subfield k_0 of k. Then for some λ in $k^\times$ there exists an isomorphism of lisse sheaves on $\mathbb{G}_m \otimes k$

$$j^*\mathrm{NFT}_\psi(\mathcal{L}_{\psi(x^n)}) \approx [n]^*\mathcal{H}_\lambda(!, \psi; \rho_1, \dots, \rho_{n-1}; \emptyset).$$

9.3 Fourier Transforms of Kummer Pullbacks of

Hypergeometrics: a remarkable stability (compare 6.2, 6.3, 6.4)

(9.3.1) The results of the previous section are themselves special cases of a quite general and remarkable stability property, to which this section is devoted.

We work over an algebraically closed field of characteristic p, with hypergeometrics of arbitrary type (n, m), including (0, 0). In order to formulate the main result of this section, it will be convenient to introduce the operator **Cancel** on hypergeometrics which "cancels" the characters common to numerator and denominator. Given a hypergeometric

$$\mathrm{Hyp} := \mathrm{Hyp}_\lambda(!, \psi; \chi_1, \ldots, \chi_n; \rho_1, \ldots, \rho_m)$$

of type (n, m), look to see how many of the χ_i's are also ρ_j's. If there are r such common characters, renumber so that

$$\chi_{n-k} = \rho_{m-k} \quad \text{for } k < r,$$
$$\chi_i \neq \rho_j \text{ if} \quad i \leq n-r \text{ and } j \leq m-r,$$

and define

$$\mathbf{Cancel}(\mathrm{Hyp}_\lambda(!, \psi; \chi_1, \ldots, \chi_n; \rho_1, \ldots, \rho_m)) :=$$
$$:= \mathrm{Hyp}_\lambda(!, \psi; \chi_1, \ldots, \chi_{n-r}; \rho_1, \ldots, \rho_{m-r}).$$

Thus **Cancel**($\mathrm{Hyp}_\lambda(!, \psi; \chi_1, \ldots, \chi_n; \rho_1, \ldots, \rho_m)$) is an irreducible hypergeometric of type (n-r, m-r).

Theorem 9.3.2 Over an algebraically closed field of characteristic p, let

$$\mathrm{Hyp}_\lambda(!, \psi; \chi_i\text{'s}; \rho_j\text{'s}) := \mathrm{Hyp}_\lambda(!, \psi; \chi_1, \ldots, \chi_n; \rho_1, \ldots, \rho_m)$$

be an irreducible (i.e., no χ_i is a ρ_j) hypergeometric of type (n, m). Let $d \geq 1$ be an integer which is prime to p. Denote by $\{\Lambda_1, \ldots, \Lambda_d\}$ all the characters of $\pi_1(\mathbb{G}_m)$ of order dividing d. Then we have isomorphisms of perverse objects on $\mathbb{A}^1$

(1) $$\mathrm{FT}_\psi(j_*[d]^*\mathrm{Hyp}_\lambda(!, \psi; \chi_1, \ldots, \chi_n; \rho_1, \ldots, \rho_m)) \approx$$
$$\approx j_*[d]^*\mathbf{Cancel}(\mathrm{Hyp}_{(-1)^{m-n}(d)^d/\lambda}(!, \psi; \Lambda_1, \ldots, \Lambda_d, \overline{\rho}_j\text{'s}; \overline{\chi}_i\text{'s})).$$

(2) $$j_*[d]^*\mathrm{Hyp}_\lambda(!, \psi; \chi_i\text{'s}; \rho_j\text{'s}) \approx$$
$$\approx \mathrm{FT}_{\overline{\psi}}(j_*[d]^*\mathbf{Cancel}(\mathrm{Hyp}_{(-1)^{n-m}(d)^d/\lambda}(!, \psi; \Lambda_1, \ldots, \Lambda_d, \overline{\rho}_j\text{'s}; \overline{\chi}_i\text{'s}))$$
$$\approx \mathrm{FT}_\psi(j_*[d]^*\mathbf{Cancel}(\mathrm{Hyp}_{(-1)^{d+n-m}(d)^d/\lambda}(!, \psi; \Lambda_1, \ldots, \Lambda_d, \overline{\rho}_j\text{'s}; \overline{\chi}_i\text{'s})).$$

proof The isomorphism (2) is obtained from (1) by Fourier inversion. In order to prove (1), we will first establish the following

Lemma 9.3.3 Over an algebraically closed field of characteristic p, for any hypergeometric $\mathrm{Hyp}_\lambda(!, \psi; \chi_i\text{'s}; \rho_j\text{'s})$, and any integer $d \geq 1$ prime to p, we have an isomorphism of perverse objects on $\mathbb{G}_m$

$$j^*\mathrm{FT}_\psi(j_![d]^*\mathrm{Hyp}_\lambda(!, \psi; \chi_i\text{'s}; \rho_j\text{'s})) \approx$$
$$\approx [d]^*\mathrm{Hyp}_{(-1)^{m-n}(d)^d/\lambda}(!, \psi; \Lambda_1, \dots, \Lambda_d, \overline{\rho}_j\text{'s}; \overline{\chi}_i\text{'s}).$$

proof As recalled in 8.1.12, for any object K we have

$$(j^*\mathcal{L}_\psi)[1]*_!K \approx j^*\mathrm{FT}_\psi(j_!\mathrm{inv}^*K).$$

We apply this to the object

$$K := [d]^*\mathrm{inv}^*\mathrm{Hyp}_\lambda(!, \psi; \chi_i\text{'s}; \rho_j\text{'s}) = \mathrm{inv}^*[d]^*\mathrm{Hyp}_\lambda(!, \psi; \chi_i\text{'s}; \rho_j\text{'s})$$

and find

$$j^*\mathrm{FT}_\psi(j_![d]^*\mathrm{Hyp}_\lambda(!, \psi; \chi_i\text{'s}; \rho_j\text{'s})) \approx$$
$$\approx (j^*\mathcal{L}_\psi)[1]*_![d]^*\mathrm{inv}^*\mathrm{Hyp}_\lambda(!, \psi; \chi_i\text{'s}; \rho_j\text{'s})$$

By the base change formula for convolution (8.1.10, 2(b)), for any two objects K and L on $\mathbb{G}_m$, and any nonzero integer d, we have

$$K*_!([d]^*L) \approx [d]^*(([d]_*K)*_!L).$$

Thus we have

$$j^*\mathrm{FT}_\psi(j_![d]^*\mathrm{Hyp}_\lambda(!, \psi; \chi_i\text{'s}; \rho_j\text{'s})) \approx$$
$$\approx [d]^*(([d]_*(j^*\mathcal{L}_\psi)[1])*_!\mathrm{inv}^*\mathrm{Hyp}_\lambda(!, \psi; \chi_i\text{'s}; \rho_j\text{'s})).$$

It remains only to simplify the convolvees.

By the inversion property (8.3.3) and the change of ψ formula (8.7.2), we have

$$\mathrm{inv}^*\mathrm{Hyp}_\lambda(!, \psi; \chi_i\text{'s}; \rho_j\text{'s}) \approx \mathrm{Hyp}_{1/\lambda}(!, \overline{\psi}; \overline{\rho}_j\text{'s}; \overline{\chi}_i\text{'s})$$
$$\approx \mathrm{Hyp}_{(-1)^{n-m}/\lambda}(!, \psi; \overline{\rho}_j\text{'s}; \overline{\chi}_i\text{'s}).$$

By the Kummer Induction Theorem 8.9.1, we have

$$[d]_*(j^*\mathcal{L}_\psi)[1]) := [d]_*(\mathrm{Hyp}_1(!, \psi; \mathbb{1}, \varnothing) \approx$$
$$\approx \mathrm{Hyp}_1(!, \psi_{1/d}; \Lambda_1, \dots, \Lambda_d; \varnothing),$$

and by 8.7.2, we have

$$\mathrm{Hyp}_{(d)^d}(!, \psi; \Lambda_1, \dots, \Lambda_d; \varnothing) \approx \mathrm{Hyp}_1(!, \psi_{1/d}; \Lambda_1, \dots, \Lambda_d; \varnothing).$$

Combining all this, we find

$$j^*\mathrm{FT}_\psi(j_![d]^*\mathrm{Hyp}_\lambda(!, \psi; \chi_i\text{'s}; \rho_j\text{'s})) \approx$$
$$\approx [d]^*(\mathrm{Hyp}_1(!, \psi_{1/d}; \Lambda_1, \dots, \Lambda_d; \varnothing)*_!\mathrm{Hyp}_{(-1)^{n-m}/\lambda}(!, \psi; \overline{\rho}_j\text{'s}; \overline{\chi}_i\text{'s}))$$

$\approx [d]^*\mathrm{Hyp}_{(-1)^{m-n(d)}d/\lambda}(!, \psi; \Lambda_1, \ldots, \Lambda_d, \bar{\rho}_j\text{'s}; \bar{\chi}_i\text{'s})$.

QED for lemma 9.3.3

We now return to the proof of the theorem 9.3.2.

We claim that $j_*[d]^*\mathrm{Hyp}_\lambda(!, \psi; \chi_i\text{'s}; \rho_j\text{'s})$ is the direct sum of perverse irreducibles on $\mathbb{A}^1$, none of which is $j_*\mathcal{L}_\Lambda[1]$ for any tame character Λ of $\pi_1(\mathbb{G}_m)$. Indeed, since $\mathrm{Hyp}_\lambda(!, \psi; \chi_i\text{'s}; \rho_j\text{'s})$ is perverse irreducible on $\mathbb{G}_m$, and $[d]$ is finite etale, $[d]^*\mathrm{Hyp}_\lambda(!, \psi; \chi_i\text{'s}; \rho_j\text{'s})$ is semisimple as a perverse object on $\mathbb{G}_m$, and hence its middle extension $j_*[d]^*\mathrm{Hyp}_\lambda(!, \psi; \chi_i\text{'s}; \rho_j\text{'s})$ is a direct sum of perverse irreducibles on $\mathbb{A}^1$. To show that no $j_*\mathcal{L}_\Lambda[1]$ is a direct factor, it suffices to show that on $\mathbb{G}_m$, no $\mathcal{L}_\Lambda[1]$ is a direct factor of $[d]^*\mathrm{Hyp}_\lambda(!, \psi; \chi_i\text{'s}; \rho_j\text{'s})$. But **all** the irreducible constituents of $[d]^*\mathrm{Hyp}_\lambda(!, \psi; \chi_i\text{'s}; \rho_j\text{'s})$ are μ_d-translates of each other (since $\mathrm{Hyp}_\lambda(!, \psi; \chi_i\text{'s}; \rho_j\text{'s})$ is irreducible). So if $\mathcal{L}_\Lambda[1]$ were a direct factor of $[d]^*\mathrm{Hyp}_\lambda(!, \psi; \chi_i\text{'s}; \rho_j\text{'s})$, then

$$[d]^*\mathrm{Hyp}_\lambda(!, \psi; \chi_i\text{'s}; \rho_j\text{'s})$$

would be a direct sum of copies of $\mathcal{L}_\Lambda[1]$. This in turn would imply that

$$\chi(\mathbb{G}_m, [d]^*\mathrm{Hyp}_\lambda(!, \psi; \chi_i\text{'s}; \rho_j\text{'s})) = 0.$$

But this is nonsense, because

$$\chi(\mathbb{G}_m, [d]^*\mathrm{Hyp}_\lambda(!, \psi; \chi_i\text{'s}; \rho_j\text{'s})) = d\chi(\mathbb{G}_m, \mathrm{Hyp}_\lambda(!, \psi; \chi_i\text{'s}; \rho_j\text{'s})) = d.$$

Therefore $\mathrm{FT}_\psi(j_*[d]^*\mathrm{Hyp}_\lambda(!, \psi; \chi_i\text{'s}; \rho_j\text{'s}))$ is a sum of perverse irreducibles on $\mathbb{A}^1$, none of which is either the delta sheaf δ_0 at the origin, or $j_*\mathcal{L}_\Lambda[1]$ for any nontrivial tame character Λ. Since any perverse irreducible M on $\mathbb{A}^1$ other than δ_0 satisfies $M \approx j_*j^*M$, we have

$$\mathrm{FT}_\psi(j_*[d]^*\mathrm{Hyp}_\lambda(!, \psi; \chi_i\text{'s}; \rho_j\text{'s})) \approx j_*j^*\mathrm{FT}_\psi(j_*[d]^*\mathrm{Hyp}_\lambda(!, \psi; \chi_i\text{'s}; \rho_j\text{'s})).$$

So to prove the theorem it suffices to prove that on $\mathbb{G}_m$ we have

$$j^*\mathrm{FT}_\psi(j_*[d]^*\mathrm{Hyp}_\lambda(!, \psi; \chi_i\text{'s}; \rho_j\text{'s})) \approx$$
$$\approx [d]^*\mathbf{Cancel}(\mathrm{Hyp}_{(-1)^{m-n(d)}d/\lambda}(!, \psi; \Lambda_1, \ldots, \Lambda_d, \bar{\rho}_j\text{'s}; \bar{\chi}_i\text{'s})).$$

Since both of these perverse objects are semisimple, it suffices to show that they have isomorphic semisimplifications. For this, we argue as follows.

We have a short exact sequence of perverse objects on $\mathbb{A}^1$ of the form

$$0 \to V \otimes \delta_0 \to j_![d]^*\mathrm{Hyp}_\lambda(!, \psi; \chi_i\text{'s}; \rho_j\text{'s}) \to j_*[d]^*\mathrm{Hyp}_\lambda(!, \psi; \chi_i\text{'s}; \rho_j\text{'s}) \to 0,$$

for some punctual sheaf $V \otimes \delta_0$ at zero. In view of the known structure of the local monodromy at zero of $\mathrm{Hyp}_\lambda(!, \psi; \chi_i\text{'s}; \rho_j\text{'s})$, we see that V has dimension

$$r := \mathrm{Card}(R),\ R := \{k \text{ in } \{1, \ldots, d\} \text{ such that } \Lambda_k \text{ is among the } \chi_i\}.$$

Taking the Fourier Transform of the above exact sequence, applying j^*, and using the lemma, we find a short exact sequence

$$0 \to (V \otimes \overline{\mathbb{Q}}_\ell[1]) \to [d]^*\mathrm{Hyp}_{(-1)^{m-n}(d)^d/\lambda}(!, \psi; \Lambda_1, \ldots, \Lambda_d, \overline{\rho}_j\text{'s}; \overline{\chi}_i\text{'s}) \to$$
$$\to j^*\mathrm{FT}_\psi(j_*[d]^*\mathrm{Hyp}_\lambda(!, \psi; \chi_1, \ldots, \chi_n; \rho_1, \ldots, \rho_m)) \to 0.$$

On the other hand, by the Semisimplification Theorem 8.4.10, the semisimplifiation of

$$\mathrm{Hyp}_{(-1)^{m-n}(d)^d/\lambda}(!, \psi; \Lambda_1, \ldots, \Lambda_d, \overline{\rho}_j\text{'s}; \overline{\chi}_i\text{'s})$$

is

$$\oplus_{i \text{ in } R}\, \mathcal{L}_{\Lambda_i}[1] \oplus \mathbf{Cancel}(\mathrm{Hyp}_{(-1)^{m-n}(d)^d/\lambda}(!, \psi; \Lambda_1, \ldots, \Lambda_d, \overline{\rho}_j\text{'s}; \overline{\chi}_i\text{'s})).$$

Pulling back by $[d]^*$, we find that the semisimplification of

$$[d]^*\mathrm{Hyp}_{(-1)^{m-n}(d)^d/\lambda}(!, \psi; \Lambda_1, \ldots, \Lambda_d, \overline{\rho}_j\text{'s}; \overline{\chi}_i\text{'s})$$

is

$$(\overline{\mathbb{Q}}_\ell[1])^r \oplus [d]^*\mathbf{Cancel}(\mathrm{Hyp}_{(-1)^{m-n}(d)^d/\lambda}(!, \psi; \Lambda_1, \ldots, \Lambda_d, \overline{\rho}_j\text{'s}; \overline{\chi}_i\text{'s})).$$

Comparing this with the above short exact sequence, we conclude that the two perverse objects on $\mathbb{G}_m$

$$[d]^*\mathbf{Cancel}(\mathrm{Hyp}_{(-1)^{m-n}(d)^d/\lambda}(!, \psi; \Lambda_1, \ldots, \Lambda_d, \overline{\rho}_j\text{'s}; \overline{\chi}_i\text{'s})),$$

$$j^*\mathrm{FT}_\psi(j_*[d]^*\mathrm{Hyp}_\lambda(!, \psi; \chi_1, \ldots, \chi_n; \rho_1, \ldots, \rho_m)),$$

have isomorphic semisimplifications, hence, both being semisimple, are themselves isomorphic. QED

Corollary 9.3.4 Hypotheses as in 9.3.2, suppose in addition that for all i, $(\chi_i)^d$ is nontrivial.

Then we have isomorphisms of perverse objects on $\mathbb{A}^1$

(1) $\mathrm{FT}_\psi(Rj_*[d]^*\mathrm{Hyp}_\lambda(!, \psi; \chi_1, \ldots, \chi_n; \rho_1, \ldots, \rho_m)) \approx$

$\approx j_*[d]^*\mathrm{Hyp}_{(-1)^{m-n}(d)^d/\lambda}(!, \psi; \Lambda_1, \ldots, \Lambda_d, \overline{\rho}_j\text{'s}; \overline{\chi}_i\text{'s}).$

(2) $Rj_*[d]^*\mathrm{Hyp}_\lambda(!, \psi; \chi_i\text{'s}; \rho_j\text{'s}) \approx$

$\approx \mathrm{FT}_{\overline{\psi}}(j_*[d]^*\mathrm{Hyp}_{(-1)^{n-m}(d)^d/\lambda}(!, \psi; \Lambda_1, \ldots, \Lambda_d, \overline{\rho}_j\text{'s}; \overline{\chi}_i\text{'s}))$

$\approx \mathrm{FT}_\psi(j_*[d]^*\mathrm{Hyp}_{(-1)^{d+n-m}(d)^d/\lambda}(!, \psi; \Lambda_1, \ldots, \Lambda_d, \overline{\rho}_j\text{'s}; \overline{\chi}_i\text{'s})).$

proof If no χ_i has order dividing d, then

$$\mathrm{Hyp} := \mathrm{Hyp}_\lambda(!, \psi; \chi_1, \ldots, \chi_n; \rho_1, \ldots, \rho_m)$$

has

$$j_![d]^*\mathrm{Hyp} \approx j_*[d]^*\mathrm{Hyp} \approx Rj_*[d]^*\mathrm{Hyp},$$

and $\mathrm{Hyp}_{(-1)^{n-m}(d)^d/\lambda}(!, \psi; \Lambda_1, \ldots, \Lambda_d, \overline{\rho}_j\text{'s}; \overline{\chi}_i\text{'s})$ is its own **Cancel**. QED

Corollary 9.3.5 Hypotheses as in 9.3.2, suppose in addition that

$$\mathrm{Hyp}_\lambda(!, \psi; \chi_1, \ldots, \chi_n; \rho_1, \ldots, \rho_m)$$

is not Kummer induced of any degree d_1 which divides d. Then

$$j_*[d]^*\mathrm{Hyp}_\lambda(!, \psi; \chi_1, \ldots, \chi_n; \rho_1, \ldots, \rho_m)$$

is perverse irreducible on $\mathbb{A}^1$, and consequently the isomorphism

$\mathrm{FT}_\psi(j_*[d]^*\mathrm{Hyp}_\lambda(!, \psi; \chi_1, \ldots, \chi_n; \rho_1, \ldots, \rho_m)) \approx$

$\approx j_*[d]^*\mathbf{Cancel}\mathrm{Hyp}_{(-1)^{m-n}(d)^d/\lambda}(!, \psi; \Lambda_1, \ldots, \Lambda_d, \overline{\rho}_j\text{'s}; \overline{\chi}_i\text{'s})$

is an isomorphism of perverse irreducibles on $\mathbb{A}^1$.

proof Indeed, since $H := \mathrm{Hyp}_\lambda(!, \psi; \chi_1, \ldots, \chi_n; \rho_1, \ldots, \rho_m)$ is perverse irreducible, and [d] is finite etale galois, either $[d]^*H$ is isotypical or H is induced from an intermediate covering. So the hypothesis insures that $[d]^*H$ is isotypical. It remains only to show that if $[d]^*H$ is isotypical, then it is irreducible.

If $[d]^*H$ is isotypical, say $k \geq 1$ copies of an irreducible K, then since the isomorphism class of K is μ_d-invariant, K itself descends through the cyclic covering [d], to a perverse irreducible K_0. Therefore H is of the form $K_0 \otimes M$, with M the k-dimensional representation of $\pi_1(\mathbb{G}_m)$ given by $\mathcal{H}om(K_0, H)$. This M must be irreducible if H is to be irreducible. But this M becomes trivial after $[d]^*$, so it is a sum of $\mathcal{L}_\Lambda$'s.

Therefore we have $k = 1$, and hence $[d]^*H$ is perverse irreducible on $\mathbb{G}_m$. Taking its middle extension $j_*[d]^*H$ to $\mathbb{A}^1$, we find that $j_*[d]^*H$ and with it $FT_\psi(j_*[d]^*H)$ are perverse irreducible on $\mathbb{A}^1$. QED

9.4 Reduction to the Tame Case

(9.4.1) In the case when $n > m$ and $d = n-m$ is prime to p, we obtain a striking relation between hypergeometrics of "wild" type (n, m) and those of "tame" type (n-r, n-r). The above results give, in this case:

Corollary 9.4.2 Hypotheses as in the theorem, suppose that $n > m$ and that $d = n - m$ is prime to p. Then

(a) we have isomorphisms of perverse sheaves on $\mathbb{A}^1$

(1) $FT_\psi(j_*[d]^*Hyp_\lambda(!, \psi; \chi_1, \ldots, \chi_n; \rho_1, \ldots, \rho_m)) \approx$

$$\approx j_*[d]^*\mathbf{Cancel}(Hyp_{(-1)^{m-n}(d)^d/\lambda}(!, \psi; \Lambda_1, \ldots, \Lambda_d, \overline{\rho}_j\text{'s}; \overline{\chi}_i\text{'s})).$$

(2) $j_*[d]^*Hyp_\lambda(!, \psi; \chi_i\text{'s}; \rho_j\text{'s}) \approx$

$$\approx FT_{\overline{\psi}}(j_*[d]^*\mathbf{Cancel}(Hyp_{(-1)^{n-m}(d)^d/\lambda}(!, \psi; \Lambda_1, \ldots, \Lambda_d, \overline{\rho}_j\text{'s}; \overline{\chi}_i\text{'s}))$$

$$\approx FT_\psi(j_*[d]^*\mathbf{Cancel}(Hyp_{(-1)^{d+n-m}(d)^d/\lambda}(!, \psi; \Lambda_1, \ldots, \Lambda_d, \overline{\rho}_j\text{'s}; \overline{\chi}_i\text{'s})).$$

(b) If $Hyp_\lambda(!, \psi; \chi_i\text{'s}; \rho_j\text{'s})$ is not Kummer induced, these are isomorphisms of perverse irreducibles.

(c) If none of the χ_i satisfies $(\chi_i)^d = \mathbb{1}$, we may rewrite these isomorphisms:

(1) $FT_\psi(Rj_*[d]^*Hyp_\lambda(!, \psi; \chi_1, \ldots, \chi_n; \rho_1, \ldots, \rho_m)) \approx$

$$\approx j_*[d]^*Hyp_{(-1)^{m-n}(d)^d/\lambda}(!, \psi; \Lambda_1, \ldots, \Lambda_d, \overline{\rho}_j\text{'s}; \overline{\chi}_i\text{'s}).$$

(2) $Rj_*[d]^*Hyp_\lambda(!, \psi; \chi_i\text{'s}; \rho_j\text{'s}) \approx$

$$\approx FT_{\overline{\psi}}(j_*[d]^*Hyp_{(-1)^{n-m}(d)^d/\lambda}(!, \psi; \Lambda_1, \ldots, \Lambda_d, \overline{\rho}_j\text{'s}; \overline{\chi}_i\text{'s}))$$

$$\approx FT_\psi(j_*[d]^*(Hyp_{(-1)^{d+n-m}(d)^d/\lambda}(!, \psi; \Lambda_1, \ldots, \Lambda_d, \overline{\rho}_j\text{'s}; \overline{\chi}_i\text{'s})).$$

proof This is just rewriting the previous results for $d = n - m$. QED

CHAPTER 10
ℓ-adic Exceptional Cases

10.0 Introduction

This chapter is devoted to the exceptional possibilities for the group G_{geom} of an irreducible ℓ-adic hypergeometric on $\mathbb{G}_m$ in characteristic p of type (n,m), $n \neq m$, which is not Kummer induced. Let $N:=\max(n,m)$, $d := |n - m|$. Suppose that $p > 2N + 1$ and that p does not divide the integer $2N_1(d)N_2(d)$ of 7.1.1. Recall (8.11.2-4) that the exceptional possibilities for $G^{0,der}$ can occur only for $|n-m|=6$, N=7,8 or 9:

N=7: the image of G_2 in its 7-dim'l irreducible representation

N=8: the image of Spin(7) in the 8-dim'l spin representation
 the image of SL(3) in the adjoint representation
 the image of SL(2)×SL(2)×SL(2) in std⊗std⊗std

N=9: the image of SL(3)×SL(3) in std⊗std.

We will show that the cases in which these exceptional groups occur are "the same" as they were for hypergeometric $\mathcal{D}$-modules. Indeed, the proofs in the two cases are quite analogous. We will largely content ourselves with indicating these analogies, rather that giving the ℓ-adic proofs in complete detail.

10.1 The G_2 and Spin(7) Cases

G_2 Recognition Theorem 10.1.1 Let k be an algebraically closed field of characteristic $p > 15$. Suppose that p does not divide the integer $2N_1(6)N_2(6)$ of 7.1.1. Let χ, ρ be two tame $\overline{\mathbb{Q}}_\ell$-valued characters of $\pi_1(\mathbb{G}_m \otimes k)$ such that none of χ, ρ, or $\chi\rho$ is the unique character $\Lambda_{1/2}$ of exact order two. Then for any $\lambda \in k^\times$, and ψ any nontrivial $\overline{\mathbb{Q}}_\ell$-valued additive character of a finite subfield k_0 of k

$$\mathcal{H} := \mathcal{H}_\lambda(!, \psi; \mathbb{1}, \chi, \overline{\chi}, \rho, \overline{\rho}, \chi\rho, \overline{\chi}\overline{\rho}; \Lambda_{1/2})$$

has $G_{geom} = G_2$. These are all the hypergeometric of type (7,1) with $G_{geom} = G_2$. The hypergeometrics of type (7,1) with $G^{0,der} = G_2$ are precisely the tame $\mathcal{L}_\Lambda$ twists of these.

proof Exactly as in the differential galois case (4.1), but using 8.11.5 instead of 3.6.1, the only nonobvious point is that such an $\mathcal{H}$ has $G_{geom} = G_2$. To show this, we argue as follows. $\mathcal{H}$ is irreducible, and, being of type (7, 1), it is not Kummer induced. Since $p > 15$, $\mathcal{H}$ is Lie-irreducible. Visibly $\mathcal{H}$ is self-dual with trivial determinant. Exactly as in 9.1.1, we see that the only possibilities for $(G_{geom})^0$ are PSL(2), G_2, or SO(7). In all of these casess, every automorphism of $(G_{geom})^0$ is inner, so $G_{geom} \subset \mathbb{G}_m(G_{geom})^0$. Since $G_{geom} \subset SO(7)$, it contains no nontrivial scalars, so $G_{geom} = (G_{geom})^0$. By the same slope argument as in 9.1.1, we can rule out the PSL(2) possibility. So G_{geom} is either G_2 or SO(7).

To rule out the SO(7) possibility, it suffices to show that, denoting by $j: \mathbb{G}_m \to \mathbb{P}^1$ the inclusion, we have

$$\chi(\mathbb{P}^1, j_*\Lambda^3(\mathcal{H})) \geq 2 > 0.$$

By the Euler-Poincare formula,

$$\chi(\mathbb{P}^1, j_*\Lambda^3(\mathcal{H})) =$$
$$= -\mathrm{Swan}_0(\Lambda^3(\mathcal{H})) - \mathrm{Swan}_\infty(\Lambda^3(\mathcal{H})) + \dim(\Lambda^3(\mathcal{H}))^{I_0} + \dim(\Lambda^3(\mathcal{H}))^{I_\infty}.$$

So it suffices to show that

(1) $\dim(\Lambda^3(\mathcal{H}))^{I_0} \geq 5$.

(2) $\mathrm{Swan}_0(\Lambda^3(\mathcal{H})) = 0$.

(3) $\dim(\Lambda^3(\mathcal{H}))^{I_\infty} = 2$.

(4) $\mathrm{Swan}_\infty(\Lambda^3(\mathcal{H})) = 5$.

The proofs of these four assertions are entirely analogous to those of their differential galois theoretic avatars, using 8.6.4 and 8.9.1 instead of 3.4.1.1. Assertions (1) and (2) hold in any characteristic p. In proving (3) and (4), one needs that p is prime to 6, and that the relations of the form $\zeta_1 + \zeta_2 + \zeta_3 = 0$, with the ζ_i three distinct sixth roots of unity in characteristic p are the two cases

$\{\zeta_1, \zeta_2, \zeta_3\}$ = {all the cube roots of 1},

$\{-\zeta_1, -\zeta_2, -\zeta_3\}$ = {all the cube roots of 1}.

We will now show that this is the case so long as $p \neq 2, 3, 7$. Indeed, suppose we have any relation

$$\zeta_1 + \zeta_2 + \zeta_3 = 0$$

where each ζ_i is a sixth root of unity. Since either $\pm\zeta_i$ is a cube root of unity, we can rewrite this relation in the form

$$\pm\omega_1 \pm \omega_2 \pm \omega_3 = 0$$

where the ω_i are cube roots of unity. Dividing by $\pm\omega_3$, we get a relation of the form

$$1 = \pm \omega_1 \pm \omega_2,$$

where the ω_i are cube roots of unity.

We now analyze the possible cases. If $\omega_1 = 1$, then its sign must be minus (lest $\omega_2 = 0$), and the relation is $2 = \pm \omega_2$, whence 2 is a sixth root of unity. But $2^6 - 1 = 63 \neq 0$ (since $p \neq 2, 3, 7$). If $\omega_1 = \omega_2$, then the relation is $1 = \pm 2\omega_1$, and again 2 would be a sixth root of 1. So the only possible relation has ω_1 and ω_2 the two nontrivial cube roots of unity, say ω and ω^2. The relation is one of

$$1 = \omega + \omega^2,$$
$$1 = -\omega + \omega^2,$$
$$1 = \omega - \omega^2,$$
$$1 = -\omega - \omega^2.$$

Of these, we claim that only the last one (which always holds so long as $p \neq 3$) holds in characteristic $p \neq 2, 3, 7$. Using the last one, the first three become

$$-\omega - \omega^2 = \omega + \omega^2,$$
$$-\omega - \omega^2 = -\omega + \omega^2,$$
$$-\omega - \omega^2 = \omega - \omega^2,$$

each of which trivially implies that $p = 2$. QED

Remark 10.1.2 We can summarize the result of the preceeding calculation as the statement that $N_2(6)$ (cf 7.1.1) is divisible only by the primes 2, 3, 7.

Spin(7) Recognition Theorem 10.1.3 Let k be an algebraically closed field of characteristic $p > 17$. Let $\lambda \in k^\times$, and ψ any nontrivial $\overline{\mathbb{Q}}_\ell$-valued additive character of a finite subfield k_0 of k. Suppose that p does not divide the integer $2N_1(6)N_2(6)$ of 7.1.1. Let χ, ρ, ξ be three tame $\overline{\mathbb{Q}}_\ell$-valued characters of $\pi_1(\mathbb{G}_m \otimes k)$ such that

$$\mathcal{H} := \mathcal{H}_\lambda(!, \psi; \chi, \overline{\chi}, \rho, \overline{\rho}, \xi, \overline{\xi}, \chi\rho\xi, \overline{\chi}\overline{\rho}\overline{\xi}; \mathbb{1}, \Lambda_{1/2})$$

is irreducible and not Kummer induced. Then $\mathcal{H}$ has G_{geom} equal to (the image in SO(8) of) Spin(7). These are all the hypergeometric of type (8,2) with G_{geom} = Spin(7). The hypergeometrics of type (8,2) with $G^{0,der}$ = Spin(7) are precisely the tame $\mathcal{L}_\Lambda$ twists of these.

proof Again the only hard point is that such an $\mathcal{H}$ has $G_{geom} = \mathrm{Spin}(7)$. For this it suffices to show that, denoting by $j: \mathbb{G}_m \to \mathbb{P}^1$ the inclusion, we have

$$\chi(\mathbb{P}^1, j_*\Lambda^4(\mathcal{H})) \geq 2 > 0.$$

By the Euler-Poincare formula,

$$\chi(\mathbb{P}^1, j_*\Lambda^4(\mathcal{H})) =$$
$$= -\mathrm{Swan}_0(\Lambda^4(\mathcal{H})) - \mathrm{Swan}_\infty(\Lambda^4(\mathcal{H})) + \dim(\Lambda^4(\mathcal{H}))^{I_0} + \dim(\Lambda^4(\mathcal{H}))^{I_\infty}.$$

So it suffices to show that

(1) $\dim(\Lambda^4(\mathcal{H}))^{I_0} \geq 8$.

(2) $\mathrm{Swan}_0(\Lambda^4(\mathcal{H})) = 0$.

(3) $\dim(\Lambda^4(\mathcal{H}))^{I_\infty} = 4$.

(4) $\mathrm{Swan}_\infty(\Lambda^4(\mathcal{H})) = 10$.

The proofs of these four assertions are entirely analogous to those of their differential galois theoretic avatars. Assertions (1) and (2) hold in any characteristic p.

In proving (3) and (4), one needs that p is prime to 6, and that all relations of the two forms

$$\zeta_1 + \zeta_2 + \zeta_3 = 0, \qquad \zeta_1 + \zeta_2 + \zeta_3 + \zeta_4 = 0$$

with the ζ_i three (resp. four) distinct sixth roots of unity in characteristic p are exactly the same as in characteristic zero. Since -1 is a sixth root of unity, the relations in question can be rewritten to be of the forms

$$\zeta_1 + \zeta_2 = \zeta_3, \qquad \zeta_1 - \zeta_2 = \zeta_3 - \zeta_4.$$

So this is guaranteed by the hypothesis that p does not divide the integer $2N_1(6)N_2(6)$. QED

10.2 The PSL(3), SL(2)×SL(2)×SL(2), and SL(3)×SL(3) Cases, via Tensor Induction

(10.2.1) We will give a unified treatment of these three cases by thinking systematically about tensor induction (cf. [C-R-MRT, 13], [Ev]) of ℓ-adic hypergeometrics. I am indebted to Ofer Gabber for making me aware of this point of view. [This same method could also be used in the differential galois case, where it would obviate the use of the specialization theorem.]

10.3 Short Review of Tensor Induction

(10.3.1) Let us recall the basic setup (cf. [C-R-MRT, 13], [Ev]). Given a group G and a subgroup H of finite index n, consider the "wreath product" $(H)^n \ltimes S_n$, where $\pi \in S_n$ acts on $(H)^n$ by

$$\pi^{-1}(h_1, h_2, \ldots, h_n)\pi = (h_{\pi(1)}, h_{\pi(2)}, \ldots, h_{\pi(n)}).$$

If we pick an ordered set $\gamma_1, \ldots, \gamma_n$ of left coset representatives for G/H, we can, following Frobenius, define an injective group homomorphism

$$G \to (H)^n \ltimes S_n$$
$$g \mapsto \pi\cdot(h_1, h_2, \ldots, h_n)$$

by writing

$$g\gamma_i = \gamma_{\pi(i)}h_i, \text{ with each } h_i \text{ in H, and } \pi \text{ in } S_n.$$

If we change the ordered set of coset representatives, this homomorphism changes by an inner automorphism of the target.

Now suppose we are given a ring A, and a representation of H on a free A module V of finite rank r. The wreath product $(H)^n \ltimes S_n$ acts on $V^{\otimes n}$ as follows: $(H)^n$ acts as

$$(h_1, h_2, \ldots, h_n): (v_1 \otimes v_2 \otimes \ldots \otimes v_n) \mapsto (h_1v_1 \otimes h_2v_2 \otimes \ldots \otimes h_nv_n),$$

and S_n acts as

$$\pi^{-1} : (v_1 \otimes v_2 \otimes \ldots \otimes v_n) \mapsto (v_{\pi(1)} \otimes v_{\pi(2)} \otimes \ldots \otimes v_{\pi(n)}).$$

Restricting this representation of $(H)^n \ltimes S_n$ to the subgroup G, we obtain a representation of G on $V^{\otimes n}$, a free A-module of rank r^n. The isomorphism class of this representation is independent of the auxiliary choice of ordered set of left coset representatives used to define it. We call it the tensor induction of V from H to G, and denote it

$$\otimes\text{-Ind}_{H\subset G}(V).$$

(10.3.2) Here are some of the basic properties of tensor induction, all of which result directly from the definitions:

(1) (additivity) If V and W are two representations of H in free A-modules of finite rank, we have an isomorphism of G-representations

$$\otimes\text{-Ind}_{H\subset G}(V\otimes W) \approx (\otimes\text{-Ind}_{H\subset G}(V))\otimes(\otimes\text{-Ind}_{H\subset G}(W)).$$

(2) (Jordan-Holder compatibility) Suppose that V is is a representation of H in a free A-module of finite rank, and that we are given a finite filtration

$$0 = \mathrm{Fil}^{d+1}V \subset \mathrm{Fil}^{d}V \subset \ldots \subset \mathrm{Fil}^{0}V = V$$

of V by subrepresentations such that each gr^iV is A-free. Consider the induced filtration of $V^{\otimes n}$ defined by

$$\mathrm{Fil}^k(V^{\otimes n}) := \Sigma_{a_1 + a_2 + \ldots + a_n \geq k} \mathrm{Fil}^{a_1}(V)\otimes \mathrm{Fil}^{a_2}(V)\otimes \ldots \otimes \mathrm{Fil}^{a_n}(V).$$

This is a filtration of $\otimes\text{-Ind}_{H\subset G}(V)$ by G-subrepresentations whose gr^i are each A-free.

Moreover, if we filter the H-representation $\oplus_i \mathrm{gr}^i V$ by the submodules

$$\mathrm{Fil}^k(\oplus_i \mathrm{gr}^i V) := \oplus_{i \geq k} \mathrm{gr}^i V,$$

the induced filtration on $\otimes\text{-Ind}_{H\subset G}(\oplus_i \mathrm{gr}^i V)$ is split, and has an isomorphic associated graded:

$$\mathrm{gr}^j(\otimes\text{-Ind}_{H\subset G}(\oplus_i \mathrm{gr}^i V)) \approx \mathrm{gr}^j(\otimes\text{-Ind}_{H\subset G}(V)).$$

In particular, we have an isomorphism of G-representations

$$\otimes\text{-Ind}_{H\subset G}(\oplus_i \mathrm{gr}^i V) \approx \oplus_j \mathrm{gr}^j(\otimes\text{-Ind}_{H\subset G}(V)).$$

(2bis) (inclusion) In the case of a two step filtration $\mathrm{Fil}^1 V \subset V$, the subobject $\mathrm{Fil}^n(\otimes\text{-Ind}_{H\subset G}(V))$ of $\otimes\text{-Ind}_{H\subset G}(V)$ is $\otimes\text{-Ind}_{H\subset G}(\mathrm{Fil}^1 V)$, and the quotient $\mathrm{gr}^0(\otimes\text{-Ind}_{H\subset G}(V))$ is $\otimes\text{-Ind}_{H\subset G}(V/\mathrm{Fil}^1 V)$.

(2ter) (inclusion) In the case of a direct sum decomposition $V = \oplus_i V_i$, there is a canonical inclusion

$$\oplus_i (\otimes\text{-Ind}_{H\subset G}(V_i)) \subset \otimes\text{-Ind}_{H\subset G}(V)$$

with A-free quotient.

(3) (transfer) If V is a representation of H in a free A-module of rank one, i.e., a character $\chi: H \to A^\times$, then $\otimes\text{-Ind}_{H\subset G}(V)$ is the character of G defined by $g \mapsto \chi(V_{G\subset H}(g))$, where $V_{G\subset H} : G^{ab} \to H^{ab}$ is the transfer.

(4) (inflation) Suppose that $K \subset H$ is a subgroup, and that K is normal in G. For any representation V of H/K, we have an isomorphism of G-representations

$$\mathrm{Infl}_{G/K \text{ to } G}(\otimes\text{-Ind}_{H/K \subset G/K}(V)) \approx \otimes\text{-Ind}_{H\subset G}(\mathrm{Infl}_{H/K \text{ to } H}(V)).$$

The other basic properties of tensor induction which we will need are almost all immediate consequences of the tensor version of the Mackey Subgroup Theorem. Here is the statement (cf [Ev]).

Mackey Subgroup Theorem for Tensor Induction 10.3.3 Suppose that K is an arbitrary subgroup of G, and H a subroup of G of finite index n. Then G is a disjoint union of $d \le n$ double cosets Kg_iH. Let $g_1, \ldots, g_d$ be double coset representatives. For each g_i, let $H_i := g_iHg_i^{-1}$. Given an H-representation V, say $\rho: H \to \mathrm{Aut}(V)$, denote by V_i the H_i-representation $\rho_i: H \to \mathrm{Aut}(V)$ defined by $\rho_i(g_ihg_i^{-1}) := \rho(h)$. Then

$$\otimes\text{-Ind}_{H\subset G}(V) \mid K \approx \bigotimes_{\text{reps } g_i \text{ of } K\backslash G/H} (\otimes\text{-Ind}_{H_i\cap K\subset K}(V_i)).$$

Corollary 10.3.4 Hypotheses and notations as in the theorem above, if A is a field of characteristic zero, and if V is semisimple (e.g., irreducible) as an H-representation, then $\otimes\text{-Ind}_{H\subset G}(V)$ is semisimple as a G-representation.

proof Since we are in characteristic zero, it suffices to show that the restriction of $\otimes\text{-Ind}_{H\subset G}(V)$ to some subgroup K of G of finite index is semisimple as a K-representation. Take for K the intersection of all the G-conjugates of H. Since K is normal in each H_i, each $V_i \mid K$ is K-semisimple, and hence so is their tensor product. Since $H_i \cap K = K$, the theorem gives

$$\otimes\text{-Ind}_{H\subset G}(V) \mid K \approx \bigotimes_{\text{reps } g_i \text{ of } K\backslash G/H} (V_i \mid K). \qquad \text{QED}$$

Corollary 10.3.5 Hypotheses and notations as in the theorem above, suppose in addition that H is normal in G. Then

(1) $$\otimes\text{-Ind}_{H\subset G}(V) \mid K \approx \bigotimes_{\text{reps } g_i \text{ of } K\backslash G/H} (\otimes\text{-Ind}_{H\cap K\subset K}(V_i)).$$

(2) If K maps onto G/H (e.g., if K has finite index m in G and if $\gcd(n, m) = 1$), then

$$\otimes\text{-Ind}_{H\subset G}(V) \mid K \approx \otimes\text{-Ind}_{H\cap K\subset K}(V).$$

(3) If $K = H$, then

$$\otimes\text{-Ind}_{H\subset G}(V) \mid H \approx \bigotimes_{\text{reps } g_i \text{ of } G/H} (V_i).$$

10.4 A Basic Example; tensor induction of polynomials

(10.4.1) In this section we will consider the following situation: G is the group $\mathbb{Z}$, H is the subgroup $n\mathbb{Z}$, V is a free A-module of rank $r \geq 1$ on which the canonical generator γ_n := "n" of H acts by an automorphism φ. We take the integers $\{1, 2, \ldots, n\}$ as ordered set of coset representatives of G/H. By definition, the canonical generator γ_1 := "1" of G acts on $V^{\otimes n}$ by the automorphism

$$\gamma_1(v_1 \otimes v_2 \otimes \ldots \otimes v_n) := \varphi(v_n) \otimes v_1 \otimes v_2 \otimes \ldots \otimes v_{n-1}.$$

Let us define

$$P(T) := \det(T - \gamma_n \mid V),$$
$$P_{\otimes n}(T) := \det(T - \gamma_1 \mid \otimes\text{-}\mathrm{Ind}_{n\mathbb{Z} \subset \mathbb{Z}}(V)).$$

By using the Jordan-Holder compatibility and "reduction to the universal case", one sees that $P_{\otimes n}(T)$ depends only on P(T), and that this dependence is itself by means of universal formulas.

Definition 10.4.2 We say that $P_{\otimes n}(T)$ is the n-fold tensor induction of P(T), and that the roots of $P_{\otimes n}(T)$ are the "$\otimes$n'th roots of the roots of P(T)".

(10.4.3) Here is a concrete way to make this explicit. Suppose that V admits an A-eigenbasis $e_1, \ldots, e_r$, with

$$\varphi(e_i) = \lambda_i e_i.$$

(Thus P(T) is $\prod_i(T - \lambda_i)$). Then $V^{\otimes n}$ admits as A-basis the r^n vectors

$$e_{a_1} \otimes e_{a_2} \otimes \ldots \otimes e_{a_n} := e[a_1, \ldots, a_n],$$

indexed by the set $E := \{1, 2, \ldots, r\}^n$. Consider the action of "cyclic permutation of the n factors" on this set, and the corresponding decomposition of E into orbits. Given an orbit Z, we can attach to it the integer Card(Z), and the quantity $\lambda(Z)$ in $A^\times$ defined as

$$\lambda(Z) := \lambda_{a_1} \times \lambda_{a_2} \times \ldots \times \lambda_{a_{\mathrm{Card}(Z)}},$$

where $[a_1, \ldots, a_n]$ is any element of E which lies in the orbit Z. For each orbit Z, we denote by $\mathrm{Span}(Z) \subset V^{\otimes n}$ the A-span of the corresponding basis vectors. Then we have a G-stable direct sum decomposition

$$V^{\otimes n} = \bigoplus_{\text{orbits } Z} \mathrm{Span}(Z).$$

One sees directly that the characteristic polynomial of γ_1 on Span(Z) is

$$\det(T - \gamma_1 \mid \mathrm{Span}(Z)) = T^{\mathrm{Card}(Z)} - \lambda(Z).$$

Thus we obtain the following formula:

$$P_{\otimes n}(T) = \prod_{\text{orbits } Z} (T^{\mathrm{Card}(Z)} - \lambda(Z)).$$

(10.4.4) In the special case when V has rank one, this specializes to $P_{\otimes n}(T) = P(T)$.

(10.4.5) In the special case when the index n is a prime q (but rank(V) is arbitrary), all orbits have Card either 1 or q, and the formula becomes

$$P_{\otimes n}(T) = P(T) \times H(T^q),$$

where H(T) is the monic polynomial defined (universally) by

$$\det(T - \varphi^{\otimes q} \mid V^{\otimes q})/\det(T - \varphi^q \mid V) = H(T)^q.$$

An alternate description of the polynomial H(T) is this: since q is a prime, when we expand $(X_1 + X_2 + \ldots + X_r)^q$ by the binomial theorem, we get

$$(X_1 + X_2 + \ldots + X_r)^q = \Sigma_i (X_i)^q + q\Sigma_W a(W)X^W \quad \text{in } \mathbb{Z}[\text{the } X_i],$$

with nonnegative integers a(W). Then H(T) is given by

$$H(T) = \prod_W (T - \lambda^W)^{a(W)}.$$

10.5 The Geometric Incarnation

(10.5.1) Suppose given connected schemes X and Y, and a finite etale map $f: Y \to X$ of degree $n \geq 1$. If we pick geometric points y of Y and $x := f(y)$ of X, then $H := \pi_1(Y, y)$ is an open subgroup of index n in $G := \pi_1(X, x)$. Let A be a coefficient ring, and $\mathcal{F}$ a lisse sheaf of free A-modules of finite rank $r \geq 1$ on Y. View $\mathcal{F}$ as a representation of $\pi_1(Y, y)$ on the free A-module $V := \mathcal{F}_y$. Then we can form the tensor induction $\otimes\mathrm{Ind}_{H \subset G}(V)$, and then interpret it as the fibre at x of a lisse sheaf on X of free A-modules of rank r^n. We will denote this sheaf

$$f_{\otimes *}\mathcal{F},$$

and call it interchangeably the "tensor direct image", or the "tensor induction", of the lisse sheaf $\mathcal{F}$ by the finite etale map f.

Descent Proposition 10.5.2 Let Γ be a finite group, X a connected scheme, and $f : Y \to X$ a finite etale connected Γ-torsor over X. For any coefficient ring A, and any lisse sheaf $\mathcal{F}$ of free A-modules of finite rank $r \geq 1$ on Y, we have an isomorphism of lisse sheaves of free A-modules

$$f^*(f_{\otimes *}\mathcal{F}) \approx \bigotimes_{\gamma \in \Gamma} \gamma^*(\mathcal{F}).$$

proof This is just the geometric transcription of 10.3.5 (3). QED

Base Change Proposition 10.5.3 Let Γ be a finite group, X a connected scheme, and $f : Y \to X$ a finite etale connected Γ-torsor over X. Let Z be a connected scheme, $\pi : Z \to X$ a morphism. Suppose that the fibre product $Z\times_X Y$ is connected. Consider the cartesian diagram

$$\begin{array}{ccccc} \pi_Y{}^*\mathcal{F} & & & & \mathcal{F} \\ & \searrow & \pi_Y & \swarrow & \\ & Z\times_X Y & \longrightarrow & Y & \\ & f_Z \downarrow & & \downarrow f & \\ & Z & \xrightarrow{\pi} & X & \end{array}$$

Then on Z we have a "base-change" isomorphism of lisse sheaves

$$\pi^*(f_{\otimes *}\mathcal{F}) \approx (f_Z)_{\otimes *}(\pi_Y{}^*\mathcal{F}).$$

proof This is just the geometric transcription of 10.3.5 (2). QED

10.6 Tensor Induction on $\mathbb{G}_m$

(10.6.1) In this section, we work over an algebraically closed field k of characteristic p on $\mathbb{G}_m := \mathrm{Spec}(k[t, 1/t])$. We denote by I_0 (resp. I_∞) a choice of inertia groups at 0 and ∞ respectively. For each integer $n \geq 1$ prime to p, we denote by $I_0(n)$ (resp. $I_\infty(n)$) their unique subgroups of index n, We fix a prime $\ell \neq p$, and consider the tensor induction of lisse $\overline{\mathbb{Q}}_\ell$-sheaves $\mathcal{F}$ with respect to the Kummer coverings

$$[n] : \mathbb{G}_m \to \mathbb{G}_m$$

of degrees $n \geq 1$ prime to p. We will also consider the corresponding "local" tensor inductions of I_0 (resp. I_∞)-representations with respect to the same Kummer coverings [n] of $\mathrm{Spec}(k((t)))$ (resp. of $\mathrm{Spec}(k((t^{-1}))))$ by itself.

For each $\alpha \in k^\times$, we denote by

$$T_\alpha : \mathbb{G}_m \to \mathbb{G}_m$$

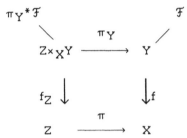

$$T_\alpha(x) := \alpha x$$

the multiplicative translation by α.

Lemma 10.6.2 Let $\mathcal{F}$ be a lisse sheaf on $\mathbb{G}_m$, and $n \geq 1$ an integer prime to p. Then its local and global tensor inductions with respect to any Kummer covering [n] with $n \geq 1$ prime to p are related by

$$[n]_{\otimes *}(\mathcal{F}) \mid I_0 \approx [n]_{\otimes *}(\mathcal{F} \mid I_0),$$
$$[n]_{\otimes *}(\mathcal{F}) \mid I_\infty \approx [n]_{\otimes *}(\mathcal{F} \mid I_\infty).$$

proof This is the base-change proposition 10.5.3, with $f := [n]$ and π the inclusion of Spec(k((t))) (resp. Spec(k((t^{-1})))) into $\mathbb{G}_m$. QED

Lemma 10.6.3 Suppose that $\mathcal{F}$ is lisse of rank r on $\mathbb{G}_m$ and tame at zero. If $(\mathcal{F} \mid I_0)^{ss} \approx \oplus_{i=1, \ldots, r} \mathcal{L}_{\chi_i}$, then for any integer $n \geq 1$ prime to p,

$$([n]_{\otimes *}(\mathcal{F}) \mid I_0)^{ss} \approx \oplus_{j=1, \ldots, r^n} \mathcal{L}_{\rho_j},$$

where the ρ_j are the r^n tame characters (with multiplicity) such that for any topological generator γ of $I_0{}^{tame}$, if we define

$$P(T) := \prod_i (T - \chi_i(\gamma))$$

then we have

$$P_{\otimes n}(T) = \prod_j (T - \rho_j(\gamma)).$$

proof This is obvious from the previous lemma, the inflation compatibility 10.3.2 (4) applied to $K := P_0 \subset G := I_0$, and 10.3.5 (2) applied to the inclusion of $\mathbb{Z}\gamma$ into $I_0{}^{tame} \approx \prod_{\ell \neq p} \mathbb{Z}_\ell(1)$. QED

Definition 10.6.4 Given $r \geq 1$ not necessarily distinct tame characters $\chi_1, \ldots, \chi_r$, and an integer $n \geq 1$ prime to p, we will refer to the r^n characters ρ_j in the above lemma as the "$\otimes$n'th roots of the {χ_i's}".

Examples 10.6.5 (1) The $\otimes$2'nd roots of $\{\chi_1, \ldots, \chi_r\}$ are the r^2 characters

$$\{\chi_1, \ldots, \chi_r\} \cup \bigcup_{1 \leq i < j \leq r} \{\text{both square roots of } \chi_i\chi_j\}.$$

(2) The $\otimes$3'rd roots of $\{\chi_1, \chi_2\}$ are the 8 characters

$$\{\chi_1, \chi_2\} \cup \{\text{all cube roots of } (\chi_1)(\chi_2)^2\} \cup \{\text{all cube roots of } (\chi_1)^2(\chi_2)\}.$$

Lemma 10.6.6 Let χ be a tame $\overline{\mathbb{Q}}_\ell$-valued characters of $\pi_1(\mathbb{G}_m \otimes k)$, ψ a nontrivial additive character of a finite subfield k_0 of k, and $n \geq 1$ an integer prime to p. Then we have isomorphisms of lisse sheaves on $\mathbb{G}_m$

(1) $\quad [n]_{\otimes *}(\mathcal{L}_\chi) \approx \mathcal{L}_\chi$,

(2) $\quad [n]_{\otimes *}(\mathcal{L}_\psi) \approx \overline{\mathbb{Q}}_\ell$ if $n \geq 2$,

(3) $\quad [n]_{\otimes *}(\mathcal{L}_\psi \otimes \mathcal{L}_\chi) \approx \mathcal{L}_\chi$ if $n \geq 2$.

proof (1) follows from 10.6.2, 10.6.3 and 10.4.4, since $[n]_{\otimes *}(\mathcal{L}_\chi)$ is lisse of rank one on $\mathbb{G}_m$ and everywhere tame. For (2), the transfer interpretation of tensor induction of characters shows that $[n]_{\otimes *}(\mathcal{L}_\psi)$ has order dividing p. But if $n \geq 2$ is prime to p, then $[n]_{\otimes *}(\mathcal{L}_\psi)$ is tame, because by 10.5.2 we have

$$[n]^*([n]_{\otimes *}(\mathcal{L}_\psi)) \approx \otimes_{\alpha \in \mu_n(k)} T_\alpha{}^*(\mathcal{L}_\psi) = \otimes_{\alpha \in \mu_n(k)} (\mathcal{L}_{\psi(\alpha x)}) \approx$$
$$\approx \mathcal{L}_{\psi((\Sigma\alpha)x)} \approx \overline{\mathbb{Q}}_\ell.$$

Therefore $[n]_{\otimes *}(\mathcal{L}_\psi)$ is trivial. Finally (3) follows from (1), (2), and the "additivity" of tensor induction. QED

Lemma 10.6.7 Let $n \geq 1$ and $m \geq 1$ be integers which are both prime to p and which are relatively prime: $(n, m) = 1$. For any lisse $\overline{\mathbb{Q}}_\ell$-sheaf $\mathcal{F}$ on $\mathbb{G}_m$, we have an isomorphism

(1) $\quad [n]^*([m]_{\otimes *}(\mathcal{F})) \approx [m]_{\otimes *}([n]^*(\mathcal{F}))$,

and an injective homomorphism

(2) $\quad [n]_*([m]_{\otimes *}(\mathcal{F})) \hookrightarrow [m]_{\otimes *}([n]_*(\mathcal{F}))$.

proof The first assertion is a special case of the base-change isomorphism, applied to the cartesian (because $(n, m) = 1$) diagram

$$\begin{array}{ccccc} [n]^*\mathcal{F} & & & & \mathcal{F} \\ & \searrow & [n] & \swarrow & \\ & \mathbb{G}_m & \longrightarrow & \mathbb{G}_m & \\ [m] & \downarrow & [n] & \downarrow & [m] \\ & \mathbb{G}_m & \longrightarrow & \mathbb{G}_m. & \end{array}$$

We next define the define the map (2) when $(n, m) = 1$. For this, it suffices by adjunction to define a map

$$[m]_{\otimes *}(\mathcal{F}) \to [n]^*([m]_{\otimes *}([n]_*(\mathcal{F}))) \approx [m]_{\otimes *}([n]^*[n]_*(\mathcal{F}))$$
$$\approx [m]_{\otimes *}(\oplus_{\alpha \in \mu_n(k)} T_\alpha{}^*(\mathcal{F})).$$

By 10.3.2 (2ter), we have an inclusion

$$\oplus_{\alpha \in \mu_n(k)} [m]_{\otimes *}(T_\alpha{}^*(\mathcal{F})) \subset [m]_{\otimes *}(\oplus_{\alpha \in \mu_n(k)} T_\alpha{}^*(\mathcal{F})),$$

and the desired map is the restriction of this one to the direct summand ($\alpha = 1$) which is $[m]_{\otimes *}(\mathcal{F})$.

To show that this map

(2) $$[n]_*([m]_{\otimes *}(\mathcal{F})) \to [m]_{\otimes *}([n]_*(\mathcal{F}))$$

is injective, it suffices to show that it is injective after pullback by $[n]$. But after this pullback, this map is none other than the above inclusion

$$\oplus_{\alpha \in \mu_n(k)} [m]_{\otimes *}(T_\alpha{}^*(\mathcal{F})) \subset [m]_{\otimes *}(\oplus_{\alpha \in \mu_n(k)} T_\alpha{}^*(\mathcal{F})).$$ QED

Proposition 10.6.8 Suppose that p is odd. Let χ be a tame $\overline{\mathbb{Q}}_\ell$-valued character of $\pi_1(\mathbb{G}_m \otimes k)$, ψ a nontrivial additive character of a finite subfield k_0 of k, and $n \geq 1$ an odd integer prime to p. For any hypergeometric of type (n, 0)

$$\mathcal{H} := \mathcal{H}_\lambda(!, \psi; \chi_1, \ldots, \chi_n; \varnothing) \text{ with } \chi = \prod_{i=1,\ldots,n} \chi_i,$$

the I_∞-representation $[2]_{\otimes *}(\mathcal{H}) \mid I_\infty$ is a direct sum

$$[2]_{\otimes *}(\mathcal{H}) \mid I_\infty \approx [n]_*(\mathcal{L}_\chi) \oplus (\text{rank } n^2 - n, \text{ all } \infty\text{-slopes } 1/2n).$$

proof By 10.6.2, $[2]_{\otimes *}(\mathcal{H}) \mid I_\infty$ depends only on $\mathcal{H} \mid I_\infty$. By the change of characters theorem 8.6.4, for fixed (λ, ψ), $\mathcal{H} \mid I_\infty$ is the same for any n tame characters whose product is χ. Since n is odd, we may take the n n'th roots of χ as these characters, and prove the assertion for that particular $\mathcal{H}$.

In this case, we have (for some μ in $k^\times$) a global isomorphism (cf. 8.9.1 or [Ka-GKM, 5.6.2])

$$\mathcal{H} \approx [n]_*(\mathcal{L}_{\psi(\mu x)} \otimes \mathcal{L}_{\chi(x)}).$$

Let us first compute the ∞-slopes of $[2]_{\otimes *}(\mathcal{H})$. We must show that $[2]^*[n]^*([2]_{\otimes *}(\mathcal{H}))$ has $n^2 - n$ slopes of 1, and n slopes of zero. But this is clear, since

$$[2]^*[n]^*([2]_{\otimes *}(\mathcal{H})) = [2]^*[n]^*([2]_{\otimes *}([n]_*(\mathcal{L}_{\psi(\mu x)} \otimes \mathcal{L}_{\chi(x)})))$$
$$\approx [2]^*([2]_{\otimes *}([n]^*[n]_*(\mathcal{L}_{\psi(\mu x)} \otimes \mathcal{L}_{\chi(x)})))$$

$$\approx ([n]^*[n]_*(\mathcal{L}_{\psi(\mu x)}\otimes\mathcal{L}_{\chi(x)}))\otimes T_{-1}{}^*([n]^*[n]_*(\mathcal{L}_{\psi(\mu x)}\otimes\mathcal{L}_{\chi(x)}))$$
$$\approx (\oplus_{\alpha\in\mu_n}\mathcal{L}_{\psi(\mu\alpha x)}\otimes\mathcal{L}_{\chi(x)})\otimes(\oplus_{\alpha\in\mu_n}\mathcal{L}_{\psi(-\mu\alpha x)}\otimes\mathcal{L}_{\chi(x)})$$
$$\approx \oplus_{\alpha,\beta\in\mu_n\times\mu_n}\mathcal{L}_{\psi(\mu(\alpha-\beta)x)}\otimes\mathcal{L}_{\chi^2(x)}.$$

It remains to see that the n-dimensional tame part, corresponding to the pairs (α, β) with $\alpha = \beta$, is in fact $[n]_*\mathcal{L}_\chi$. For this, we use part (2) of the previous lemma, according to which

$$[n]_*([2]_{\otimes *}(\mathcal{L}_{\psi(\mu x)}\otimes\mathcal{L}_{\chi(x)})) \subset [2]_{\otimes *}([n]_*(\mathcal{L}_{\psi(\mu x)}\otimes\mathcal{L}_{\chi(x)})),$$

and 10.6.6, (3), according to which

$$[2]_{\otimes *}(\mathcal{L}_{\psi(\mu x)}\otimes\mathcal{L}_{\chi(x)}) \approx \mathcal{L}_{\chi(x)}.$$ QED

We now exploit the special case n = 3 of the above proposition, the only case in which $[2]_{\otimes *}(\mathcal{H})$ has $\mathrm{Swan}_\infty = 1$.

Corollary 10.6.9 Suppose that p is prime to 6. Let χ be a tame $\overline{\mathbb{Q}}_\ell$-valued character of $\pi_1(\mathbb{G}_m\otimes k)$, ψ a nontrivial additive character of a finite subfield k_0 of k. For any hypergeometric of type (3, 0)

$$\mathcal{H} := \mathcal{H}_\lambda(!, \psi; \chi_1, \chi_2, \chi_3; \varnothing) \text{ with } \chi = \textstyle\prod_{i=1,\ldots,3}\chi_i,$$

denote by

$$\{\rho_1, \ldots, \rho_9\} := \text{the } \otimes 2\text{'nd roots of } \{\chi_1, \chi_2, \chi_3\},$$
$$=\{\chi_1, \chi_2, \chi_3\} \cup \{\text{both square roots of } \chi_1\chi_2, \text{ of } \chi_1\chi_3, \text{ of } \chi_2\chi_3\},$$

and by

$$\{\Lambda_1, \Lambda_2, \Lambda_3\} := \text{the cube roots of } \chi.$$

Then for some $\mu \in k^\times$, there exists an isomorphism of lisse sheaves on $\mathbb{G}_m$

$$[2]_{\otimes *}(\mathcal{H}) \approx \mathcal{H}_\mu(!, \psi; \rho_1, \ldots, \rho_9; \Lambda_1, \Lambda_2, \Lambda_3)^{ss},$$

where "ss" means "semisimplification as lisse sheaf on $\mathbb{G}_m$".

proof Since $\mathcal{H}$ is irreducible, $[2]_{\otimes *}(\mathcal{H})$ is certainly semisimple. It has $\mathrm{Swan}_\infty = 1$, with six slopes 1/6 and three slopes zero, and it is tame at zero. By the intrinsic characterization of hypergeometrics (8.5.3) and (8.5.2), either $[2]_{\otimes *}(\mathcal{H})$ is itself an irreducible hypergeometric, necessarily of type (9, 3), or for some integer $1 \le k \le 3$ it is the direct sum of an irreducible hypergeometric of type (9 - k, 3 - k) with k sheaves among the $\mathcal{L}_{\Lambda_i}$. Looking at the tame parts of $[2]_{\otimes *}(\mathcal{H})$ and of

$\mathcal{H}_\mu(!, \psi; \rho_1, \ldots, \rho_g; \Lambda_1, \Lambda_2, \Lambda_3)$ at both zero and ∞, the result is immediate from the cancellation theorem 8.4.7 and the rigidity corollary bis 8.5.6 (2). QED

Lemma 10.6.10 Let q be a prime. There exists a nonzero integer D(q) such that for any algebraically closed field k of characteristic not dividing 2qD(q), the following statement (*) holds:

(*)for ζ any primitive q'th root of unity in $k^\times$, of the 2^q sums in k,

$$(\pm 1) + (\pm \zeta) + (\pm \zeta^2) + \ldots + (\pm \zeta^{q-1}),$$

where the signs $\pm$ are chosen independently, the only ones that vanish are the two corrseponding to the choices (all +) and (all -).

Moreover, for q = 3 or q = 5, we have D(q) = 1.

proof Let k be an algebraically closed field of charcteristic neither two nor q. Fix a primitive q'th root of unity ζ in $k^\times$, i.e., ζ is a root in k of the polynomial

$$\Phi_q(X) := 1 + X + X^2 + \ldots + X^{q-1} \text{ in } \mathbb{Z}[X].$$

That (*) hold in k amounts to the statement that if we partition {0, 1, 2, ..., q-1} into two nonempty disjoint subsets S and T, then

$$\Sigma_{n\in S}\, \zeta^n \neq \Sigma_{m\in T}\, \zeta^m \text{ in k.}$$

Since we always have

$$\Sigma_{n\in S}\, \zeta^n = -\Sigma_{m\in T}\, \zeta^m \text{ in k,}$$

and k has odd characteristic, we can only have $\Sigma_{n\in S}\, \zeta^n = \Sigma_{m\in T}\, \zeta^m$ in k if in fact $\Sigma_{n\in S}\, \zeta^n = 0 = \Sigma_{m\in T}\, \zeta^m$ in k.

Thus we must show that if S is any proper nonempty subset of {0, 1, 2, ..., q-1}, we have $\Sigma_{n\in S}\, \zeta^n \neq 0$. Replacing S by its complement if necessary, we may assume in addition that Card(S) $\leq$ (q - 1)/2. Thus the cases q = 3 and q = 5 (lest -1 be a fifth root of unity) are trivially okay in any characteristic not dividing 2q.

For general q, we argue as follows. Multiplying the putative relation by a power of ζ, we may further suppose that S does not contain q-1. Define

$$F_S(X) := \Sigma_{n\in S}\, X^n \in \mathbb{Z}[X].$$

Then if (k, ζ) "fails" for S, then $X \mapsto \zeta$ is a k-valued point of the finite $\mathbb{Z}$-scheme Spec(A_S),

$$A_S := \mathbb{Z}[X]/(\Phi_q(X), F_S(X)).$$

But over $\mathbb{Q}$ the polynomial $\Phi_q(X)$ is irreducible, and as $F_S(X)$ is a nonzero polynomial of lower degree, we must have

$$\text{g.c.d.}(\Phi_q(X), F_S(X)) = 1 \text{ in } \mathbb{Q}[X].$$

Clearing denominators, we find that the ideal $(\Phi_q(X), F_S(X))$ contains some nonzero integer $D_S(q)$. Therefore $A_S[1/D_S(q)]$ is the zero ring, and hence $\mathrm{Spec}(A_S)$ has no k-valued points with values in fields where $D_S(q)$ is invertible. So one can take for $D(q)$ the product of the $D_S(q)$ over all nonempty S not containing q-1. QED

Remark 10.6.10.1 Here is a method, due to J. Conway, to show that in general $D(q) \neq 1$. Suppose that q and p are distinct odd primes, that $r \geq 2$ is the least integer such that

$$p^r \equiv 1 \bmod q,$$

and that

$$p^r - 1 = nq \text{ for some integer n with}$$
$$p \equiv 1 \bmod n.$$

Examples: $(p = 3, q = 13, r = 3, n = 2)$, $(p = 5, q = 31, r = 3, n = 4)$.

Then certainly $r \leq q - 1$ by Fermat. We claim that there exists a primitive q'th root of unity ζ in $\mathbb{F}_{p^r}$ whose trace from $\mathbb{F}_{p^r}$ to $\mathbb{F}_p$ vanishes. But this trace is the sum of r distinct (since $\mathbb{F}_p(\zeta) = \mathbb{F}_{p^r}$ by the choice of r) primitive q'th roots of 1, so p is a "bad" characteristic for q.

To verify the claim, we argue by contradiction. The key point is that

$$(\mathbb{F}_{p^r})^\times = \mu_q \times \mu_n.$$

[Indeed all the n'th roots of unity lie in the ground field $\mathbb{F}_p$, while any element $\zeta \neq 1$ of μ_q generates $\mathbb{F}_{p^r}$ over $\mathbb{F}_p$, so $\mu_q \cap \mu_n = \{1\}$.]

Suppose that each of the q-1 nontrivial q'th roots of unity in $\mathbb{F}_{p^r}$ has nonzero trace. Since all the n'th roots of unity lie in the ground field $\mathbb{F}_p$, the linearity of the trace shows that only the elements of $\{0\} \cup \mu_n$ of $\mathbb{F}_{p^r}$ can possibly have trace zero. Since there exist $p^{r-1} > 1$ elements of trace zero, some element of μ_n must have trace zero, and by linearity every element of μ_n has trace zero. Therefore 1 has trace zero, so $p|r$. and there are exactly $1 + n$ elements of trace zero. Thus $p^{r-1} = 1 + n$. But if $p|r$, then every element of $\mathbb{F}_p$ has trace zero, so $p \leq 1 + n$. Therefore $p = 1 + n$, whence $p^{r-1} = p$, so $r = 2$. But $p|r$, so

$p = 2$, contradiction.

Proposition 10.6.11 Let q be an odd prime. Suppose that p is odd and does not divide $2qD(q)$. Let χ be a tame $\overline{\mathbb{Q}}_\ell$-valued character of $\pi_1(\mathbb{G}_m \otimes k)$, and ψ a nontrivial additive character of a finite subfield k_0 of k. Denote by $\Lambda_{1/2}$ the quadratic character. For any hypergeometric of type (2, 0)

$$\mathcal{H} := \mathcal{H}_\lambda(!, \psi; \chi_1, \chi_2; \emptyset) \text{ with } \chi = \chi_1\chi_2\Lambda_{1/2},$$

the I_∞-representation $[q]_{\otimes *}(\mathcal{H}) \mid I_\infty$ is a direct sum

$$[q]_{\otimes *}(\mathcal{H}) \mid I_\infty \approx [2]_*(\mathcal{L}_\chi) \oplus (\text{rank } 2^q - 2, \text{ all } \infty\text{-slopes } 1/2q).$$

proof The proof is entirely analogous to that of the proposition above. One first reduces to the case when $\mathcal{H}$ is $[2]_*(\mathcal{L}_{\psi(\mu x)} \otimes \mathcal{L}_{\chi(x)})$. One shows that $[2]_*(\mathcal{L}_\chi)$ is a subrepresentation of $[q]_{\otimes *}(\mathcal{H})$ exactly as above.

The only nonobvious point is that

$$[2q]^*([q]_{\otimes *}(\mathcal{H})) \approx [q]^*([q]_{\otimes *}([2]^*\mathcal{H})) \approx \otimes_{\alpha \in \mu_q} T_\alpha{}^*([2]^*\mathcal{H})$$

$$\approx \otimes_{\alpha \in \mu_q} T_\alpha{}^*([2]^*[2]_*(\mathcal{L}_{\psi(\mu x)} \otimes \mathcal{L}_{\chi(x)}))$$

$$\approx \otimes_{\alpha \in \mu_q} T_\alpha{}^*(\mathcal{L}_{\psi(\mu x)} \otimes \mathcal{L}_{\chi(x)} \oplus \mathcal{L}_{\psi(-\mu x)} \otimes \mathcal{L}_{\chi(x)})$$

has all but two of its ∞-slopes $=1$. This amounts to the statement that if ζ is a primitive q'th root of unity in $k^\times$, then of the 2^q sums in k,

$$(\pm 1) + (\pm \zeta) + (\pm \zeta^2) + \ldots + (\pm \zeta^{q-1}),$$

where the signs $\pm$ are chosen independently, the only ones that vanish are the two corrseponding to the choices (all +) and (all -). This holds precisely because we are in characteristic p not dividing $2qD(q)$. QED

We now exploit the special case $q = 3$, the only case in which $[q]_{\otimes *}(\mathcal{H})$ has $\text{Swan}_\infty = 1$ for p not dividing $2qD(q)$.

Corollary 10.6.12 Suppose that p is prime to 6. Let χ be a tame $\overline{\mathbb{Q}}_\ell$-valued character of $\pi_1(\mathbb{G}_m \otimes k)$, ψ a nontrivial additive character of a finite subfield k_0 of k. Denote by $\Lambda_{1/2}$ the quadratic character. For any hypergeometric of type (2, 0)

$$\mathcal{H} := \mathcal{H}_\lambda(!, \psi; \chi_1, \chi_2; \emptyset) \text{ with } \chi = \chi_1\chi_2\Lambda_{1/2},$$

denote by

$$\{\rho_1, \ldots, \rho_8\} := \text{the } \otimes 3\text{'rd roots of } \{\chi_1, \chi_2\}$$

$= \{\chi_1, \chi_2\} \cup \{$all cube roots of $(\chi_1)(\chi_2)^2$, of $(\chi_1)^2(\chi_2)\}$,

and by

$\{\Lambda_1, \Lambda_2\} :=$ the square roots of χ.

Then for some $\mu \in k^\times$, there exists an isomorphism of lisse sheaves on $\mathbb{G}_m$

$[3]_{\otimes *}(\mathcal{H}) \approx \mathcal{H}_\mu(!, \psi; \rho_1, \ldots, \rho_8; \Lambda_1, \Lambda_2)^{ss}$,

where "ss" means "semisimplification as lisse sheaf on $\mathbb{G}_m$".

proof The proof is entirely analogous to that of 10.6.9. QED

10.7 Return to the PSL(3), SL(2)×SL(2)×SL(2), and SL(3)×SL(3) cases

PSL(3) Recognition Theorem 10.7.1 Let k be an algebraically closed field of characteristic $p > 7$. Let $\lambda \in k^\times$, and ψ any nontrivial $\overline{\mathbb{Q}}_\ell$-valued additive character of a finite subfield k_0 of k. Let χ be a tame $\overline{\mathbb{Q}}_\ell$-valued character of $\pi_1(\mathbb{G}_m \otimes k)$ which is not of exact order three. Denote by

$\Lambda_{1/2} :=$ the unique tame character of exact order 2,

$\Lambda_{1/3}, \Lambda_{2/3} :=$ the two characters of exact order 3,

$\{\rho_1, \ldots, \rho_8, \mathbb{1}\} :=$ the $\otimes$2'nd roots of $\{\chi, \overline{\chi}, \mathbb{1}\}$

$= \{\chi, \overline{\chi}, \mathbb{1}\} \cup \{$both square roots of $\mathbb{1}$, of χ, of $\overline{\chi}\}$.

(1) Any hypergeometric of type (3, 0) of the form

$\mathcal{H} := \mathcal{H}_\lambda(!, \psi; \chi, \overline{\chi}, \mathbb{1}; \varnothing)$

has $G_{geom} = SL(3)$, and its dual is isomorphic to to $T_{-1}{}^*\mathcal{H}$.

(2) The tensor direct image $[2]_{\otimes *}(\mathcal{H})$ admits, for some μ in $k^\times$, a direct sum decomposition

$[2]_{\otimes *}(\mathcal{H}) \approx \overline{\mathbb{Q}}_\ell \oplus \mathcal{H}_\mu(!, \psi; \rho_1, \ldots, \rho_8; \Lambda_{1/3}, \Lambda_{2/3})$.

(3) There exists an isomorphism of lisse sheaves

$End^0(\mathcal{H}) \approx [2]^*\mathcal{H}_\mu(!, \psi; \rho_1, \ldots, \rho_8; \Lambda_{1/3}, \Lambda_{2/3})$.

(4) For any μ in $k^\times$, $\mathcal{H}_\mu(!, \psi; \rho_1, \ldots, \rho_8; \Lambda_{1/3}, \Lambda_{2/3})$ has $(G_{geom})^0 =$ PSL(3) in its adjoint representation, and PSL(3) has index two in G_{geom}.

(5) The hypergeometrics of type (8, 2) with $(G_{geom})^{0,der}$ = PSL(3) in its adjoint representation are precisely the tame $\mathcal{L}_\Lambda$-twists of these.

proof In (1), the assertion about G_{geom} is a special case of 8.11.3, and the duality assertions 8.4.2 and 8.3.3. Assertions (2), (3), and (4) follow from (1), via 10.6.9 and the semisimplification theorem 8.4.10. Assertion (5) is proven exactly as its differential-galois analogue 4.3.6. QED

10.8 The SL(2)×SL(2)×SL(2) Case

SL(2)×SL(2)×SL(2) Recognition Theorem 10.8.1 Let k be an algebraically closed field of characteristic p > 5. Let $\lambda \in k^\times$, and ψ any nontrivial $\overline{\mathbb{Q}}_\ell$-valued additive character of a finite subfield k_0 of k. Let χ be a tame $\overline{\mathbb{Q}}_\ell$-valued character of $\pi_1(\mathbb{G}_m \otimes k)$ which is not of exact order four. Denote by

ζ_1, ζ_2 := the primitive cube roots of unity in k,

$\Lambda_{1/2}$:= the unique tame character of exact order 2,

$\Lambda_{1/4}, \Lambda_{3/4}$:= the two characters of exact order 4,

$\{\rho_1, \ldots, \rho_8\}$:= the ⊗3'rd roots of $\{\chi, \overline{\chi}\}$

$= \{\chi, \overline{\chi}\} \cup$ {all cube roots of χ, of $\overline{\chi}$}.

(1) Any hypergeometric of type (2, 0) of the form

$$\mathcal{H} := \mathcal{H}_\lambda(!, \psi; \chi, \overline{\chi}; \varnothing)$$

has G_{geom} = SL(2), and $\mathcal{H} \otimes (T_{\zeta_1}{}^*\mathcal{H}) \otimes (T_{\zeta_2}{}^*\mathcal{H})$ has G_{geom} the image of SL(2)×SL(2)×SL(2) in SL(8).

(2) There exists, for some μ in $k^\times$, an isomorphism of lisse sheaves

$$[3]_{\otimes *}(\mathcal{H}) \approx \mathcal{H}_\mu(!, \psi; \rho_1, \ldots, \rho_8; \Lambda_{1/4}, \Lambda_{3/4}).$$

(3) There exists an isomorphism of lisse sheaves

$$\mathcal{H} \otimes (T_{\zeta_1}{}^*\mathcal{H}) \otimes (T_{\zeta_2}{}^*\mathcal{H}) \approx [3]^*\mathcal{H}_\mu(!, \psi; \rho_1, \ldots, \rho_8; \Lambda_{1/3}, \Lambda_{2/3}).$$

(4) For any μ in $k^\times$, $\mathcal{H}_\mu(!, \psi; \rho_1, \ldots, \rho_8; \Lambda_{1/4}, \Lambda_{3/4})$ has G_{geom} = the image in Sp(8) of the semidirect product $(SL(2) \times SL(2) \times SL(2)) \ltimes A_3$.

(5) The hypergeometrics of type (8, 2) with $(G_{geom})^{0,der}$ = the image in Sp(8) of SL(2)×SL(2)×SL(2) are precisely the tame $\mathcal{L}_\Lambda$-twists of these.

proof Similar to the case above, using 8.11.7 and 10.6.12, and imitating

4.5.3. QED

10.9 The SL(3)×SL(3) Case

SL(3)×SL(3) Recognition Theorem 10.9.1

Let k be an algebraically closed field of characteristic $p > 7$. Let $\lambda \in k^\times$, and ψ any nontrivial $\overline{\mathbb{Q}}_\ell$-valued additive character of a finite subfield k_0 of k. Let χ_1, χ_2, χ_3 be tame $\overline{\mathbb{Q}}_\ell$-valued characters of $\pi_1(\mathbb{G}_m \otimes k)$ none of which has order dividing three, and whose product is trivial:

$$\chi_1\chi_2\chi_3 = \mathbb{1}.$$

Denote by

$\Lambda_{1/2}$:= the unique tame character of exact order 2,

$\Lambda_{1/3}, \Lambda_{2/3}$:= the two characters of exact order 3,

$\{\rho_1, \dots, \rho_9\}$:= the $\otimes$2'nd roots of $\{\chi_1, \chi_2, \chi_3\}$

$= \{\chi_1, \chi_2, \chi_3\} \cup$ {both square roots of $\overline{\chi}_1$, of $\overline{\chi}_2$, of $\overline{\chi}_3$}.

(1) Any hypergeometric of type (3, 0) of the form

$$\mathcal{H} := \mathcal{H}_\lambda(!, \psi; \chi_1, \chi_2, \chi_3; \emptyset)$$

has G_{geom} = SL(3), and $\mathcal{H}\otimes T_{-1}{}^*\mathcal{H}$ has G_{geom} = the image in SL(9) of SL(3)×SL(3).

(2) There exists, for some μ in $k^\times$, an isomorphism of lisse sheaves

$$[2]_{\otimes *}(\mathcal{H}) \approx \mathcal{H}_\mu(!, \psi; \rho_1, \dots, \rho_9; \mathbb{1}, \Lambda_{1/3}, \Lambda_{2/3}).$$

(3) There exists an isomorphism of lisse sheaves

$$\mathcal{H}\otimes T_{-1}{}^*\mathcal{H} \approx [2]^*\mathcal{H}_\mu(!, \psi; \rho_1, \dots, \rho_9; \mathbb{1}, \Lambda_{1/3}, \Lambda_{2/3}).$$

(4) For any μ in $k^\times$, $\mathcal{H}_\mu(!, \psi; \rho_1, \dots, \rho_9; \mathbb{1}, \Lambda_{1/3}, \Lambda_{2/3})$ has $(G_{geom})^0$ = the image in SL(9) of SL(3)×SL(3), and $(G_{geom})^0$ has index two in G_{geom}.

(5) The hypergeometrics of type (9, 3) with $(G_{geom})^{0,der}$ = the image in SL(9) of SL(3)×SL(3) are precisely the tame $\mathcal{L}_\Lambda$-twists of these.

proof Exactly like those of the previous two theorems, now using 10.6.9 and imitating 4.6.10. QED

CHAPTER 11
Reductive Tannakian Categories

11.1 Homogeneous Space Recovery of a Reductive Group

(11.1.1) In this section we work over an algebraically closed field K of characteristic zero. Suppose that we are given an integer $n \geq 1$, an n-dimensional K-vector space V, and an algebraic subgroup G of GL(V). We know that we can recover G from the the Tannakian category $\mathcal{C} := \mathrm{Rep}(G)$ of its finite-dimensional K-representations and the fibre functor

$$\omega := \text{"forget the G-action"} : \mathrm{Rep}(G) \to (\text{fin.-dim'l K-spaces}).$$

(11.1.2) Suppose now that G is **reductive**. Following Ofer Gabber, we will explain how to recover G (more precisely, the conjugacy class in GL(n, K) of G(K)) from the Tannakian category $\mathcal{C} := \mathrm{Rep}(G)$ using only the functor

$$\text{"G-invariants"} : \mathcal{C} \to (\text{fin.-dim'l K-spaces})$$
$$W \mapsto W^G := \mathrm{Hom}_{\mathcal{C}}(\mathbb{1}, W),$$

but **without** using a fibre functor.

(11.1.3) Let us first describe the **idea**. Consider the space

$$X := \mathit{Isom}_K(K^n, \omega(V)).$$

This is a left $GL(\omega(V))$-torsor by postcomposition, and a right GL(n, K)-torsor by precomposition. The quotient space $Y := G\backslash X$ is then a right homogeneous space under GL(n, K). In terms of this homogeneous space we recover G (up to GL(n, K)-conjugacy) as the **stabilizer** in GL(n, K) of any chosen point $y \in Y$.

(11.1.4) How are we to turn this idea into a proof? And where will the hypothesis that G is reductive enter? First of all, we can construct the space X using the fibre functor ω as follows. If we view X as the space of ordered bases $(v_1, v_2, \dots, v_n)$ of ω(V), then X sits as an **open** set in V^n. We will instead view X as the **closed** subscheme of $V^n \times (V^\vee)^n$ consisting of those tuples

$$(v_1, v_2, \dots, v_n; v_1^\vee, \dots, v_n^\vee)$$

satisfying the n^2 equations

$$v_j^\vee(v_i) = \delta_{i,j}.$$

(This is just a longwinded way of saying that the v_i are a basis of V and the $v_j^\vee$ are the dual basis of $V^\vee$. For n=1, it amounts to defining $\mathbb{G}_m$ by the equation xy=1 rather than by the condition "x invertible".)

This description has the merit of exhibiting X as an affine K-scheme of finite type, say $X = \mathrm{Spec}(A)$. The group G, being a subgroup of $GL(V)$, certainly acts freely on X; indeed, for any K-algebra R, the group $G(R)$ acts freely on the set $X(R)$. Because G is **reductive**, the quotient $Y := G\backslash X$ exists and is affine, with coordinate ring $B = A^G$. On K-valued points, we have

$$Y(K) = G(K)\backslash X(K).$$

Since $X(K) = \mathrm{Isom}(K^n, \omega(V))$ is a left $GL(\omega(V))$-torsor and a right $GL(n, K)$-torsor we see that $Y(K)$ is a right homogeneous space under $GL(n, K)$, and the stabilizer in $GL(n, K)$ of any point $y \in Y(K)$ is (a conjugate of) $G(K)$.

(11.1.5) To conclude this discussion, it remains to construct the coordinate ring B of the affine scheme Y and the left action of $GL(n, K)$ on B, using only the Tannakian structure of $\mathcal{C}$ and the functor "G-invariants". To clarify the discussion which follows, let us denote by $V_1, \ldots, V_n$ n copies of V, and by $V_1^\vee, \ldots, V_n^\vee$ n copies of $V^\vee$. We can construct X as the closed subscheme of the spectrum of the symmetric algebra

$$S := \bigotimes_{i=1,\ldots,n} \mathrm{Symm}^*(\omega(V_i^\vee)) \otimes \mathrm{Symm}^*(\omega(V_i))$$

defined by the ideal I generated by the n^2 elements

$$f_{i,j} - \delta_{i,j},$$

where $f_{i,j}$ is the G-invariant element of

$$\omega(V_i^\vee) \otimes \omega(V_j) = \omega(V^\vee) \otimes \omega(V) = \omega(V^\vee \otimes V)$$

which is the canonical map $\mathbb{1} \to V^\vee \otimes V$. This $f_{i,j}$ is just the function

$$(v_1, v_2, \ldots, v_n; v_1^\vee, \ldots, v_n^\vee) \mapsto v_j^\vee(v_i),$$

described in an invariant way.

(11.1.6) Since G is reductive, and the ideal I is generated by the invariants $f_{i,j} - \delta_{i,j}$, we have

$$B = (S/I)^G = S^G/(\text{the ideal in } S^G \text{ generated by all the } f_{i,j} - \delta_{i,j}),$$

and S^G is the $\mathbb{N}^{2n}$-graded algebra

$$\bigoplus_{(a_1,\ldots,a_n,b_1,\ldots,b_n)} \Big(\bigotimes_i \mathrm{Symm}^{a_i}(\omega(V_i^\vee)) \otimes \mathrm{Symm}^{b_i}(\omega(V_i))\Big)^G$$

$$= \bigoplus_{(a_1,\ldots,a_n,b_1,\ldots,b_n)} \omega\Big(\bigotimes_i \mathrm{Symm}^{a_i}(V_i^\vee) \otimes \mathrm{Symm}^{b_i}(V_i)\Big)^G$$

$$= \bigoplus_{(a_1,\ldots,a_n,b_1,\ldots,b_n)} \mathrm{Hom}_{\mathcal{C}}\Big(\mathbb{1}, \bigotimes_i \mathrm{Symm}^{a_i}(V_i^\vee) \otimes \mathrm{Symm}^{b_i}(V_i)\Big)$$

$$= \mathrm{Hom}_{\mathrm{Ind}\text{-}\mathcal{C}}(\mathbb{1}, \bigotimes_i \mathrm{Symm}^*(V_i^\vee)\otimes \mathrm{Symm}^*(V_i))$$

$$= \mathrm{Hom}_{\mathrm{Ind}\text{-}\mathcal{C}}(\mathbb{1}, \mathrm{Symm}^*((K^n\otimes V)^\vee)\otimes \mathrm{Symm}^*(K^n\otimes V)).$$

In this last description, we see the left action of GL(n, K), through its standard left action on K^n, the induced ($A \mapsto A\otimes \mathrm{id}_V$) left action on $K^n\otimes V$, and the contragredient left action on $(K^n\otimes V)^\vee$. This left action of GL(n, K) respects the $\mathbb{N}^2$-grading.

(11.1.7) In order to simplify the bookkeeping which is about to follow, we perform one final rewriting of S^G as

$$S^G = \mathrm{Hom}_{\mathrm{Ind}\text{-}\mathcal{C}}(\mathbb{1}, \mathrm{Symm}^*((K^n\otimes V)^\vee \oplus (K^n\otimes V))).$$

This allows us to view S^G as an $\mathbb{N}$-graded algebra:

$$S^G = \bigoplus_{d\geq 0} S^G(d), \text{ with}$$

$$S^G(d) := \mathrm{Hom}_{\mathcal{C}}(\mathbb{1}, \mathrm{Symm}^d((K^n\otimes V)^\vee \oplus (K^n\otimes V))).$$

In this picture, the left action of GL(n, K) respects the $\mathbb{N}$-grading.

11.2 First analysis of finiteness properties

(11.2.1) Since G is reductive, and we are in characteristic zero, the graded ring of invariants S^G is **finitely generated** as a graded K-algebra. To keep track systematically of generators and relations, it will be convenient to introduce the following ad hoc terminology.

(11.2.2) For any finitely generated graded commutative K-algebra

$$A = \oplus_{d\geq 0} A_d,$$

and any integer $D \geq 1$, we define an $\mathbb{N}$-graded commutative K-algebra A{D} by setting

$$A\{D\} := \mathrm{Symm}^*(\oplus_{0\leq d\leq D} A_d) \approx \otimes_{0\leq d\leq D} \mathrm{Symm}^*(A_d),$$

with the grading that makes A_d isobaric of weight d. We denote by $A\{D\}_m$ the part of A{D} which is isobaric of weight m.

There is a unique homomorphism of graded rings

$$\alpha_D : A\{D\} \rightarrow A$$

which is the identity on $\oplus_{0\leq d\leq D} A_d$. Since A is finitely generated, this homomorphism α_D is surjective if D is sufficiently large, say $D \geq D_0$.

For each integer $m \geq 0$, we denote by

$$\mathrm{Rel}\{D, m\} := \mathrm{Ker}(\alpha_D)\cap A\{D\}_m$$

the relations among the A_i with $i \leq D$ which are isobaric of degree m.

For each integer $N \geq 1$, we denote by

$$\mathrm{Rel}\{D, \leq N\} \subset A\{D\}$$

the graded ideal generated by all the $\mathrm{Rel}\{D, m\}$ with $m \leq N$.

(11.2.3) For any (D, N) with $D \geq 1$ and $N \geq D$, we denote by $A\{D, N\}$ the quotient ring

$$A\{D, N\} := A\{D\}/\mathrm{Rel}\{D, \leq N\},$$

and by

$$\alpha_{D,N} : A\{D, N\} \to A$$

the canonical graded homomorphism induced by α_D.

(11.2.4) The key point is this: for a fixed D which is $\geq D_0$, this map $\alpha_{D,N}$ is surjective, and for N sufficiently large, say $N \geq N_0(D)$, $\alpha_{D,N}$ is an isomorphism, simply because $A\{D\}$ is noetherian.

(11.2.5) Let us say that the graded ring A is **(D, N)-determined** if $\alpha_{D,N}$ is an isomorphism. Clearly if A is (D, N)-determined, then A is determined by the following **finite** amount of data:

the element 1 in A_0

the graded vector space $\oplus_{0 \leq d \leq N} A_d$

the multiplication maps $A_i \otimes A_j \to A_{i+j}$, for each (i, j) with $i+j \leq N$.

Indeed, the ring $A\{D, N\}$ is always determined by this data, whether or not A is (D, N)-determined.

(11.2.6) We will now apply this to the situation $A = S^G$. Notice that the graded action of GL(n, K) on S^G induces by functoriality a graded action on each of the graded rings $S^G\{D\}$, and this action stabilizes the ideals $\mathrm{Rel}\{D, \leq N\}$. So each of the approximations $S^G\{D, N\}$ carries a graded left action of GL(n, K), and all the maps $\alpha_{D,N}$ are GL(n, K)-equivariant.

(11.2.7) For any integer $D \geq 2$, $S^G\{D\}$ contains the n^2 elements $f_{i,j}$ as isobaric elements of degree 2. So for any (D, N) with $N \geq D \geq 2$, it makes sense to form the quotient ring

$$B\{D,N\} := S^G\{D, N\}/(\text{ideal gen. by the } f_{i,j} - \delta_{i,j}).$$

This is a (no longer graded) ring on which GL(n, K) acts, and the canonical ring homomorphism

$$\beta_{D,N}: B\{D,N\} \to B := S^G/(\text{ideal gen. by the } f_{i,j} - \delta_{i,j})$$

deduced from $\alpha_{D,N}$ is GL(n, K)-equivariant. If S^G is (D, N)-determined, then this map $\beta_{D,N}$ is an isomorphism.

(11.2.8) How much data determines $B\{D,N\}$ as a K-algebra with GL(n,

K) action? Clearly, it is determined by the following **finite** amount of data:

the element 1 in $S^G(0)$

the elements $f_{i,j}$ in $S^G(2)$

the graded vector space $\oplus_{0 \le d \le N} S^G(d)$

the action of GL(n, K) on $S^G(d)$, for each $0 \le d \le N$

the multiplication maps $S^G(i) \otimes S^G(j) \to S^G(i+j)$, for each (i, j) with $i+j \le N$.

11.3 Transition away from Tannakian categories

(11.3.1) We now make one further simplification in the presentation of this data. Since the field K is of characteristic zero, for any object W of $\mathcal{C}$, the symmetric algebra Symm*(W) in Ind-$\mathcal{C}$ may be recovered as a graded subring of the corresponding tensor algebra which in each degree d is the image of the appropriate symmetrizing idempotent in the rational group ring $\mathbb{Q}[\mathfrak{S}_d]$. Therefore the invariants $S^G(d)$ may be viewed as the image of the symmetrizing idempotent on

$$T^G(d) := \mathrm{Hom}_{\mathcal{C}}(\mathbb{1}, \bigotimes{}^d((K^n \otimes V)^\vee \oplus (K^n \otimes V))).$$

The elements $f_{i,j}$ in $T^G(2)$ are the images of the canonical element δ in $\mathrm{Hom}_{\mathcal{C}}(\mathbb{1}, V^\vee \otimes V)$ via the n^2 "mixed crossterms"coordinate inclusions of $V^\vee \otimes V$ into $\bigotimes{}^2((K^n \otimes V)^\vee \oplus (K^n \otimes V))$ in the Tannakian category $\mathcal{C}$. These inclusions depend only on the fact that $\mathcal{C}$ is a K-linear ACU $\otimes$-category whose $\otimes$ is K-bilinear. [But the notion of the canonical element δ in $\mathrm{Hom}_{\mathcal{C}}(\mathbb{1}, V^\vee \otimes V)$ depends on the fact that $\mathcal{C}$ is Tannakian.]

(11.3.2) The action of GL(n, K) on $T^G(d)$ is deduced by functoriality from the dual actions of GL(n, K) on the objects $(K^n \otimes V)^\vee := K^n \otimes V^\vee$ and $K^n \otimes V$ of $\mathcal{C}$. These actions exist for any n and any objects V and $V^\vee$ of any K-linear additive category $\mathcal{C}$.

(11.3.3) The multiplications $S^G(i) \otimes S^G(j) \to S^G(i+j)$ may be recovered by symmetrization from the multiplications

$$T^G(i) \otimes T^G(j) \to T^G(i+j),$$

and these in turn depend only on the fact that V and $V^\vee$ are two objects of a K-linear ACU $\otimes$-category whose $\otimes$ is K-bilinear.

11.4 Mock Tannakian Categories

(11.4.1) Exactly how much depends on having a Tannakian category? We continue to work over our algebraically closed field K of characteristic zero. Suppose we are given a K-linear additive category $\mathcal{M}$ with an ACU $\otimes$-operation in the sense of [Saa], with unit object $\mathbb{1}$, whose $\otimes$ is K-bilinear. Suppose we are given

two objects of $\mathcal{M}$, denoted V and $V^\vee$,

a morphism $\delta_V: \mathbb{1} \to V^\vee \otimes V$

an integer $n \geq 1$.

(11.4.2) For each integer $N \geq 1$, denote by $\mathcal{M}^{\leq N}(V, V^\vee)$ the full subcategory of $\mathcal{M}$ consisting of all finite direct sums of the objects

$$W_1 \otimes W_2 \otimes \ldots \otimes W_r,\ r \leq N,$$

where each object W_i is either $\mathbb{1}$ or V or $V^\vee$. Clearly the two objects

$$(K^n \otimes V)^\vee := K^n \otimes V^\vee, \text{ and } K^n \otimes V$$

both lie in $\mathcal{M}^{\leq 1}(V, V^\vee)$. The $\otimes$ operation in $\mathcal{M}$ defines for each (i, j) a bifunctor

$$\mathcal{M}^{\leq i}(V, V^\vee) \times \mathcal{M}^{\leq j}(V, V^\vee) \to \mathcal{M}^{\leq i+j}(V, V^\vee).$$

(11.4.3) So for each integer $d \geq 0$ the object

$$\bigotimes^d((K^n \otimes V)^\vee \oplus (K^n \otimes V))$$

makes sense as an object of $\mathcal{M}^{\leq d}(V, V^\vee)$ on which $GL(n, K) \times \mathfrak{S}_d$ acts. For $d = 0$, we define this to be the unit object $\mathbb{1}$, with trivial action.

(11.4.4) For each d, we define

$$T^G(d) := \mathrm{Hom}_{\mathcal{M}}(\mathbb{1}, \bigotimes^d((K^n \otimes V)^\vee \oplus (K^n \otimes V))).$$

This is a K-vector space on which $GL(n, K) \times \mathfrak{S}_d$ acts, and their direct sum is a (noncommutative) graded K-algebra, with multiplication given by tensor product of morphisms in $\mathcal{M}$. In each degree d we define $S^G(d)$ to be the image of the symmetrization idempotent for $\mathfrak{S}_d$ on $T^G(d)$. The direct sum S^G of the $S^G(d)$ becomes a graded commutative K-algebra, taking for product the symmetrization of the product in the ambient tensor algebra, on which we have a graded action of GL(n, K).

(11.4.5) We can now define the graded algebras $S^G(D)$, their quotients $S^G(D, N)$, and the maps $\alpha_{D,N}$; everything will be GL(n, K)-equivariant.

Scholie 11.4.6 The approximation $S^G(D, N)$ with $N \geq D \geq 1$ is determined entirely by the full subcategory $\mathcal{M}^{\leq N}(V, V^\vee)$ and all the tensor product bifunctors

$$\mathcal{M}^{\le i}(V, V^\vee) \times \mathcal{M}^{\le j}(V, V^\vee) \to \mathcal{M}^{\le i+j}(V, V^\vee),$$

for all (i, j) with $i+j \le N$.

(11.4.7) There are n^2 visible "mixed crossterms" maps

$$V^\vee \otimes V \to \bigotimes{}^2((K^n \otimes V)^\vee \oplus (K^n \otimes V)).$$

By composition with $\delta_V: \mathbb{1} \to V^\vee \otimes V$, we obtain n^2 elements $f_{i,j}$ in $S^G(2)$. If we **assume** that these elements $f_{i,j}$ are GL(n, K)-invariant, we can define the quotient ring

$$B := S^G/(\text{ideal gen. by the } f_{i,j} - \delta_{i,j})$$

on which GL(n, K) acts, its approximations

$$B\{D,N\} := S^G\{D, N\}/(\text{ideal gen. by the } f_{i,j} - \delta_{i,j}),$$

on which GL(n, K) also acts, and the GL(n, K)-equivariant maps

$$\beta_{D,N} : B\{D,N\} \to B.$$

We can then define the K-schemes

$$Y := \mathrm{Spec}(B), \quad Y\{D,N\} := \mathrm{Spec}(B\{D,N\})$$

on which GL(n,K) acts on the right, and the GL(n, K)-equivariant maps

$$\mathrm{Spec}(\beta_{D,N}) : Y \to Y\{D,N\}.$$

Proposition 11.4.8 If $\mathcal{M}$ above is a neutralizable Tannakian category $\mathcal{C}$, if V and $V^\vee$ are dual n-dimensional objects of $\mathcal{C}$, and if δ_V is the canonical map δ, and if the Tannakian galois group G of V is reductive, then

(1) Y(K) is a right homogeneous space for GL(n, K), and the stabilizer in GL(n, K) of any point $y \in Y(K)$ is (a conjugate of) G(K).

(2) there exists an integer D_0 such that for any $N \ge D \ge D_0$, the map

$$\mathrm{Spec}(\beta_{D,N}) : Y \to Y\{D,N\}$$

is a closed immersion.

(3) for each $D \ge D_0$ there exists an integer $N_0(D)$ such that $Y \cong Y\{D,N\}$ for all $N \ge N_0(D)$.

proof The hypotheses, that V be n-dimensional and that its Tannakian group G be reductive, are invariant under change of K-valued fibre functor, as is the conjugacy class of G(K) in GL(n, K). Once we we pick a fibre functor on the Tannakian subcategory <V> of $\mathcal{C}$ generated by V, this proposition reduces to the previous discussion (11.1, 11.2). QED

Definition 11.4.9 Hypotheses as in the above proposition, we say that the object V of $\mathcal{C}$ is (D, N)-determined if $Y \cong Y\{D,N\}$.

11.5 Statement of the reductive specialization theorem

Reductive SpecializationTheorem 11.5.1 Let K be an algebraically closed field of characteristic zero. Let $\mathcal{M}$ be a K-linear additive category $\mathcal{M}$ with an ACU $\otimes$-operation in the sense of [Saa], with unit object $\mathbb{1}$, whose $\otimes$ is K-bilinear. Suppose we are given

two objects of $\mathcal{M}$, denoted V and $V^\vee$,

a morphism $\delta_V: \mathbb{1} \to V^\vee \otimes V$

an integer $n \geq 1$.

For each integer $d \geq 1$, denote by $\mathcal{M}^{\leq d}(V, V^\vee)$ the full subcategory of $\mathcal{M}$ consisting of all finite direct sums of the objects

$$W_1 \otimes W_2 \otimes \ldots \otimes W_r,\ r \leq d,$$

where each object W_i is either $\mathbb{1}$ or V or $V^\vee$.

Suppose in addition we are given two neutralizable Tannakian categories $\mathcal{C}_{\mathbb{C}}$ and $\mathcal{C}_{\mathbb{F}}$ over K, and K-linear additive ACU $\otimes$-functors

$$\mathcal{M} \to \mathcal{C}_{\mathbb{C}} \qquad \mathcal{M} \to \mathcal{C}_{\mathbb{F}}$$
$$M \mapsto M_{\mathbb{C}} \qquad M \mapsto M_{\mathbb{F}}.$$

Fix an integer $N \geq 2$ and suppose that the following conditions hold:

(1$\mathbb{C}$ dual) The object $V_{\mathbb{C}}$ of $\mathcal{C}_{\mathbb{C}}$ is n-dimensional, its dual is $(V^\vee)_{\mathbb{C}}$, and under this identification of $(V^\vee)_{\mathbb{C}}$ with $(V_{\mathbb{C}})^\vee$, $(\delta_V)_{\mathbb{C}}$ is the canonical map $\delta: \mathbb{1} \to (V_{\mathbb{C}})^\vee \otimes V_{\mathbb{C}}$ in $\mathcal{C}_{\mathbb{C}}$.

(2$\mathbb{C}$ red) The Tannakian group $G_{\mathbb{C}}$ of $V_{\mathbb{C}}$ is reductive.

(3$\mathbb{C}$ $\leq$N) The induced functor $\mathcal{M}^{\leq N}(V, V^\vee) \to \mathcal{C}_{\mathbb{C}}^{\leq N}(V_{\mathbb{C}}, (V^\vee)_{\mathbb{C}})$ is an equivalence of categories.

(1$\mathbb{F}$ dual) The object $V_{\mathbb{F}}$ of $\mathcal{C}_{\mathbb{F}}$ is n-dimensional, its dual is $(V^\vee)_{\mathbb{F}}$, and under this identification of $(V^\vee)_{\mathbb{F}}$ with $(V_{\mathbb{F}})^\vee$, $(\delta_V)_{\mathbb{F}}$ is the canonical map $\delta: \mathbb{1} \to (V_{\mathbb{F}})^\vee \otimes V_{\mathbb{F}}$ in $\mathcal{C}_{\mathbb{F}}$.

(2$\mathbb{F}$ red) The Tannakian group $G_{\mathbb{F}}$ of $V_{\mathbb{F}}$ is reductive.

(3$\mathbb{F}$ $\leq$N) The induced functor $\mathcal{M}^{\leq N}(V, V^\vee) \to \mathcal{C}_{\mathbb{F}}^{\leq N}(V_{\mathbb{F}}, (V^\vee)_{\mathbb{F}})$ is an equivalence of categories.

Suppose that $V_{\mathbb{C}}$ is (D, N)-determined for some integer D with

$1 \le D \le N$. Then $G_{\mathbb{F}}(K)$ is (conjugate in $GL(n, K)$ to) a subgroup of $G_{\mathbb{C}}(K)$.

11.6 Proof of the reductive specialization theorem

For any D with $1 \le D \le N$, the $\{D, N\}$ approximations of the spaces Y, $Y_{\mathbb{C}}$, and $Y_{\mathbb{F}}$ are related by $GL(n, K)$-equivariant isomorphisms

$$\begin{array}{ccc} & Y\{D, N\} & \\ \overset{\approx}{\swarrow} & & \overset{\approx}{\searrow} \\ Y_{\mathbb{F}}\{D, N\} & & Y_{\mathbb{C}}\{D, N\}. \end{array}$$

If in addition $V_{\mathbb{C}}$ is (D, N)-determined, the canonical map

$$Y_{\mathbb{C}} \to Y_{\mathbb{C}}\{D, N\}$$

is an isomorphism, so we obtain a $GL(n, K)$-equivariant diagram

$$\begin{array}{ccc} & Y\{D, N\} & \\ \overset{\approx}{\swarrow} & & \overset{\approx}{\searrow} \\ Y_{\mathbb{F}}\{D, N\} & & Y_{\mathbb{C}}\{D, N\}. \\ \uparrow & & \uparrow \wr\!\wr \\ Y_{\mathbb{F}} & & Y_{\mathbb{C}}. \end{array}$$

So there exists a (unique) $GL(n, K)$-equivariant morphism

$$Y_{\mathbb{F}} \to Y_{\mathbb{C}}$$

which makes the diagram commute. On K-valued points, this is a $GL(n, K)$-equivariant morphism of right $GL(n, K)$-**homogeneous spaces**. Fix a point $y_{\mathbb{F}}$ in $Y_{\mathbb{F}}(K)$, with image $y_{\mathbb{C}}$ in $Y_{\mathbb{C}}(K)$. Their stabilizers $\mathrm{Stab}(y_{\mathbb{F}})$ and $\mathrm{Stab}(y_{\mathbb{C}})$ in $GL(n, K)$ are (conjugates of) the groups $G_{\mathbb{F}}(K)$ and $G_{\mathbb{C}}(K)$ respectively. Since the map is $GL(n, K)$-equivariant, we have the inclusion of stabilizers $\mathrm{Stab}(y_{\mathbb{F}}) \subset \mathrm{Stab}(y_{\mathbb{C}})$, which means precisely that $G_{\mathbb{F}}(K)$ is (conjugate to) a subgroup of $G_{\mathbb{C}}(K)$. QED

11.7 A minor variation on the reductive specialization theorem

Here is a minor variation, in which we onit mention of δ_V but insist that $N \ge 3$.

Reductive SpecializationTheorem bis 11.7.1 Let K be an

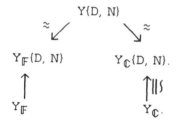

algebraically closed field of characteristic zero. Let $\mathcal{M}$ be a K-linear additive category with an ACU $\otimes$-operation in the sense of [Saa], with unit object $\mathbb{1}$, whose $\otimes$ is K-bilinear. Suppose we are given

two objects of $\mathcal{M}$, denoted V and $V^\vee$,
an integer $n \geq 1$.

For each integer $d \geq 1$, denote by $\mathcal{M}^{\leq d}(V, V^\vee)$ the full subcategory of $\mathcal{M}$ consisting of all finite direct sums of the objects

$$W_1 \otimes W_2 \otimes \ldots \otimes W_r,\ r \leq d,$$

where each object W_i is either $\mathbb{1}$ or V or $V^\vee$.

Suppose in addition we are given two neutralizable Tannakian categories $\mathcal{C}_{\mathbb{C}}$ and $\mathcal{C}_{\mathbb{F}}$ over K, and K-linear additive ACU $\otimes$-functors

$$\mathcal{M} \to \mathcal{C}_{\mathbb{C}} \qquad \mathcal{M} \to \mathcal{C}_{\mathbb{F}}$$
$$M \mapsto M_{\mathbb{C}} \qquad M \mapsto M_{\mathbb{F}}.$$

Fix an integer $N \geq 3$ and suppose that the following conditions hold:

(1$\mathbb{C}$ dual) The object $V_{\mathbb{C}}$ of $\mathcal{C}_{\mathbb{C}}$ is n-dimensional, and its dual is $(V^\vee)_{\mathbb{C}}$.

(2$\mathbb{C}$ red) The Tannakian group $G_{\mathbb{C}}$ of $V_{\mathbb{C}}$ is reductive.

(3$\mathbb{C}$ $\leq$N) The induced functor $\mathcal{M}^{\leq N}(V, V^\vee) \to \mathcal{C}_{\mathbb{C}}{}^{\leq N}(V_{\mathbb{C}}, (V^\vee)_{\mathbb{C}})$ is an equivalence of categories.

(1$\mathbb{F}$ dual) The object $V_{\mathbb{F}}$ of $\mathcal{C}_{\mathbb{F}}$ is n-dimensional, and its dual is $(V^\vee)_{\mathbb{F}}$.

(2$\mathbb{F}$ red) The Tannakian group $G_{\mathbb{F}}$ of $V_{\mathbb{F}}$ is reductive.

(3$\mathbb{F}$ $\leq$N) The induced functor $\mathcal{M}^{\leq N}(V, V^\vee) \to \mathcal{C}_{\mathbb{F}}{}^{\leq N}(V_{\mathbb{F}}, (V^\vee)_{\mathbb{F}})$ is an equivalence of categories.

Suppose that $V_{\mathbb{C}}$ is (D, N)-determined for some integer D with $1 \leq D \leq N$. Then $G_{\mathbb{F}}(K)$ is (conjugate in GL(n, K) to) a subgroup of $G_{\mathbb{C}}(K)$.

proof The point is to show that the map δ_V in the hypotheses of the reductive specialization theorem is already uniqely determined by the other data. As explained in [De-CT, 2.1.2], given an object X in a Tannakian category $\mathcal{C}$, the data consisting of its dual $X^\vee$ together with the maps

$$\delta : \mathbb{1} \to X^\vee \otimes X, \qquad ev : X \otimes X^\vee \to \mathbb{1}$$

(which morally correspond to the identity mapping id_X in End(X) and to the trace form on End($X^\vee$)) is entirely characterized by the sole requirement that the two composites

$$X \xrightarrow{X\otimes\delta} X\otimes X^\vee \otimes X \xrightarrow{ev\otimes X} X$$

$$X^\vee \xrightarrow{\delta\otimes X^\vee} X^\vee \otimes X\otimes X^\vee \xrightarrow{X^\vee\otimes ev} X^\vee$$

be the respective identities.

The key point is that these conditions can be stated entirely in terms of the category $\mathcal{C}^{\leq 3}(X, X^\vee)$ and the tensor product bifunctors

$$\mathcal{C}^{\leq 1}(X, X^\vee) \times \mathcal{C}^{\leq 2}(X, X^\vee) \rightarrow \mathcal{C}^{\leq 3}(X, X^\vee),$$
$$\mathcal{C}^{\leq 2}(X, X^\vee) \times \mathcal{C}^{\leq 1}(X, X^\vee) \rightarrow \mathcal{C}^{\leq 3}(X, X^\vee).$$

Take $\mathcal{C} := \mathcal{C}_{\mathbb{C}}$ and $X := V_{\mathbb{C}}$. Then by axioms (1$\mathbb{C}$ dual) and (3$\mathbb{C}$ $\leq$N) we may "back up" the maps δ and ev in $\mathcal{C}_{\mathbb{C}}$ **to** maps δ_V and ev_V in $\mathfrak{M}$,

$$\delta_V : \mathbb{1} \rightarrow V^\vee \otimes V, \qquad ev_V : V\otimes V^\vee \rightarrow \mathbb{1}$$

and these are the **unique** maps in $\mathfrak{M}$ for which the two composites

$$V \xrightarrow{V\otimes\delta_V} V\otimes V^\vee \otimes V \xrightarrow{ev_V\otimes V} V$$

$$V^\vee \xrightarrow{\delta_V\otimes V^\vee} V^\vee \otimes V\otimes V^\vee \xrightarrow{V^\vee\otimes ev_V} V^\vee$$

are the respective identities.

Replace $\mathcal{C}_{\mathbb{C}}$ and $V_{\mathbb{C}}$ in the above paragraph by $\mathcal{C}_{\mathbb{F}}$ and $V_{\mathbb{F}}$, and repeat the above "backing up" argument. By uniqueness, the backups to $\mathfrak{M}$ of the maps δ and ev in $\mathcal{C}_{\mathbb{F}}$ must coincide with the above δ_V and ev_V in $\mathfrak{M}$.

Thus the map δ_V can be reconstructed from the the other data.

QED

CHAPTER 12
Fourier Universality

12.1 The situation over $\mathbb{C}$

In sections 12.2 through 12.8, we work over $\mathbb{C}$.

12.2 Additive Convolution, Exotic Tensor Product, and Fourier Transform on $\mathbb{A}^1$ over $\mathbb{C}$

(12.2.1) To avoid confusion between convolution of $\mathcal{D}$-modules on $\mathbb{G}_m$ and convolution of $\mathcal{D}$-modules on $\mathbb{A}^1$, we will continue to denote the former by $K*L$, and we will denote the latter by $K*_+L$.

(12.2.2) We first establish the basic notations. We denote by

$\pi : \mathbb{A}^1 \to \mathrm{Spec}(\mathbb{C})$ the structural map,

$i_\alpha : \mathrm{Spec}(\mathbb{C}) \to \mathbb{A}^1$ the inclusion of the point $\alpha \in \mathbb{C} = \mathbb{A}^1(\mathbb{C})$,

$\Delta: \mathbb{A}^1 \to \mathbb{A}^1 \times_{\mathbb{C}} \mathbb{A}^1$ the diagonal embedding,

sum: $\mathbb{A}^1 \times_{\mathbb{C}} \mathbb{A}^1 \to \mathbb{A}^1$ the addition map $(x, y) \mapsto x+y$,

$j : \mathbb{G}_m \to \mathbb{A}^1$ the inclusion.

Key Lemma 12.2.3 For K, L objects of $D^{b,\mathrm{holo}}(\mathbb{A}^1)$, and $\alpha \in \mathbb{C}$, we have canonical isomorphisms in $D^{b,\mathrm{holo}}(\mathrm{Spec}(\mathbb{C})) = D^b(\mathbb{C}\text{-vector spaces})$

(1) $\pi_*(K \otimes e^{\alpha x}) \approx i_\alpha{}^!(FT(K))[1]$,

(2) $\pi_*(FT(K) \otimes e^{-\alpha x}) \approx i_\alpha{}^!(K)[1]$,

(3) $\pi_*(K*_+L) \approx \pi_*(K) \otimes \pi_*(L)$,

(4) $i_\alpha{}^!(K \otimes^! L) \approx i_\alpha{}^!(K) \otimes i_\alpha{}^!(L)$,

and a canonical isomorphism in $D^{b,\mathrm{holo}}(\mathbb{A}^1)$

(5) $FT(K*_+L) \approx FT(K) \otimes^! FT(L)[1]$.

proof Assertion (1) is base change for the cartesian diagram

$$\begin{array}{ccc}
\mathbb{A}^1 & \xrightarrow{\mathrm{id} \times i_\alpha} & \mathbb{A}^1 \times_{\mathbb{C}} \mathbb{A}^1 \quad pr_1{}^!(K) \otimes e^{xy} \\
\pi \downarrow & i_\alpha & \downarrow pr_2 \\
\mathrm{Spec}(\mathbb{C}) & \longrightarrow & \mathbb{A}^1.
\end{array}$$

Assertion (2) is just (1) at $-\alpha$ applied to FT(K).

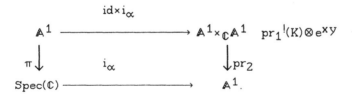

Assertion (3), "mise pour memoire", has already been pointed out in (5.1.9 1(a)). Assertion (4) is the transitivity of "upper shriek", together with the observation that the composite

$$\mathrm{Spec}(\mathbb{C}) \xrightarrow{\;i_\alpha\;} \mathbb{A}^1 \xrightarrow{\;\Delta\;} \mathbb{A}^1\times_{\mathbb{C}}\mathbb{A}^1$$

is the product map

$$\mathrm{Spec}(\mathbb{C})\times\mathrm{Spec}(\mathbb{C}) \xrightarrow{\;i_\alpha\times\, i_\alpha\;} \mathbb{A}^1\times_{\mathbb{C}}\mathbb{A}^1.$$

Since formation of $f^!$ is compatible with products, we have

$$i_\alpha{}^!(K\otimes^! L) := i_\alpha{}^!\Delta^!(K\times L) = (i_\alpha\times i_\alpha)^!(K\times L) =$$
$$= i_\alpha{}^!(K)\times i_\alpha{}^!(L) = i_\alpha{}^!(K)\otimes_{\mathbb{C}} i_\alpha{}^!(L).$$

Assertion (5), in the equivalent form

$$FT(K*_+L)[1] \approx FT(K)[1]\otimes^! FT(L)[1],$$

is base change for the following diagram, whose outer square is cartesian:

$$\begin{array}{ccc}
\mathbb{A}^1\times_{\mathbb{C}}\mathbb{A}^1\times_{\mathbb{C}}\mathbb{A}^1 & \xrightarrow{(x,y,z)\,\mapsto\,(x,z,y,z)} & \mathbb{A}^2\times_{\mathbb{C}}\mathbb{A}^2 \quad (pr_1{}^!(K)\otimes e^{xy})\times(pr_1{}^!(L)\otimes e^{xy}) \\
\downarrow \mathrm{sum}\times \mathrm{id} & & \\
\mathbb{A}^1\times_{\mathbb{C}}\mathbb{A}^1 & & \downarrow pr_2\times pr_2 \\
\downarrow pr_2 & & \\
\mathbb{A}^1 & \xrightarrow{\;\Delta\;} & \mathbb{A}^1\times_{\mathbb{C}}\mathbb{A}^1.
\end{array}$$

QED

Remark 12.2.3.1 The use of the full $\mathcal{D}$-module formalism in the above proof should not obscure the entirely elementary nature of the result. Here is an entirely elementary proof of (1), (2), and (5) in the case when K and L are themselves holonomic $\mathcal{D}$-modules.

If we think of an $\mathcal{O}$-quasicoherent $\mathcal{D}$-module K on $\mathbb{A}^1$ as (the sheaf associated to) a module over $\mathbb{C}[x, \partial]$, then K "is" a triple (V, A, B), with V a $\mathbb{C}$-vector space (namely the global sections of K as $\mathcal{O}$-module) endowed with an ordered pair (A, B) of $\mathbb{C}$-linear endomorphisms A and B (i.e., the effects of x and ∂ respectively) whose commutator satisfies

$$[A, B] = -1.$$

The Fourier Transform FT(K) of K = (V, A, B) is just (V, -B, A).

The complexes $\pi_*(K \otimes e^{\alpha x})$ and $i_\alpha{}^!(FT(K))[1]$ are respectively

$$\underset{\text{deg } -1}{K} \xrightarrow{\partial+\alpha} \underset{\text{deg } 0}{K} \qquad \underset{\text{deg } -1}{FT(K)} \xrightarrow{x-\alpha} \underset{\text{deg } 0}{FT(K).}$$

From the (V, A, B) point of view, these two complexes are

$$V \xrightarrow{B+\alpha} V \qquad V \xrightarrow{-B-\alpha} V,$$

so (1) is obvious. Similarly for (2).

For (5), think of K and L as given by data (V, A, B) and (W, C, D) respectively as above. Then $K *_+ L := \mathrm{sum}_*(K \times L)$ is the two-term complex

$$\underset{\text{deg } -1}{V \otimes_{\mathbb{C}} W} \xrightarrow{B\otimes 1 - 1 \otimes D} \underset{\text{deg } 0}{V \otimes_{\mathbb{C}} W,}$$

with operators $(A\otimes 1 + 1\otimes C, (1/2)(B\otimes 1 + 1\otimes D))$.
Applying FT, we see that $FT(K *_+ L)$ is the **same** two term complex

$$\underset{\text{deg } -1}{V \otimes_{\mathbb{C}} W} \xrightarrow{B\otimes 1 - 1 \otimes D} \underset{\text{deg } 0}{V \otimes_{\mathbb{C}} W,}$$

but with with operators $(-(1/2)(B\otimes 1 + 1\otimes D), A\otimes 1 + 1\otimes C)$.

What about $FT(K) \otimes^! FT(L)[1]$? This is the complex

$$\underset{\text{deg } -1}{FT(K) \otimes_{\mathbb{C}} FT(L)} \xrightarrow{x-y} \underset{\text{deg } 0,}{FT(K) \otimes_{\mathbb{C}} FT(L)}$$

with operators $(x+y, (1/2)(\partial_x + \partial_y))$.
If we trace this through, it means the complex

$$\underset{\text{deg } -1}{V \otimes_{\mathbb{C}} W} \xrightarrow{-B\otimes 1 + 1 \otimes D} \underset{\text{deg } 0}{V \otimes_{\mathbb{C}} W,}$$

with operators $(-(1/2)(B\otimes 1 + 1\otimes D), A\otimes 1 + 1\otimes C)$. So (5) is proven.QED

Corollary 12.2.4 For $\mathcal{M}$, $\mathcal{N}$ holonomic $\mathcal{D}$-modules on $\mathbb{A}^1$, we have
(1) $FT(\mathcal{M}) \approx j_*j^*FT(\mathcal{M})$ if and only if $\pi_*\mathcal{M} = 0$.
(2) if either $\pi_*\mathcal{M} = 0$ or $\pi_*\mathcal{N} = 0$, then $\pi_*(\mathcal{M}*_+\mathcal{N}) = 0$.
(3) If either of $\mathcal{A} := j^*FT(\mathcal{M})$ or $\mathcal{B} := j^*FT(\mathcal{N})$ is a D.E. on $\mathbb{G}_m$, and if $\pi_*(\mathcal{M}*_+\mathcal{N}) = 0$, then denoting by $\mathcal{A}\otimes\mathcal{B}$ the usual $\mathcal{O}_{\mathbb{G}_m}$-tensor product,

$$FT(\mathcal{M}*_+\mathcal{N}) \approx j_*(\mathcal{A}\otimes\mathcal{B}),$$

and $\mathcal{M}*_+\mathcal{N}$ is a single $\mathcal{D}$-module, concentrated in degree zero.

proof (1):The canonical map $FT(\mathcal{M}) \to j_*j^*FT(\mathcal{M})$ is an isomorphism if and only if the map Left(x): $FT(\mathcal{M}) \to FT(\mathcal{M})$ is an isomorphism, i.e., if and only if $i_0{}^!(FT(\mathcal{M}))[1]$ ($= \pi_*\mathcal{M}$) vanishes.

(2): This is obvious from $\pi_*(K*_+L) \approx \pi_*(K)\otimes\pi_*(L)$.

(3): Since $\pi_*(\mathcal{M}*_+\mathcal{N}) = 0$, part (1) above and 12.2.3 (5) give

$$FT(\mathcal{M}*_+\mathcal{N}) \approx j_*j^*(FT(\mathcal{M}*_+\mathcal{N})) \approx j_*j^*(FT(\mathcal{M})\otimes^!FT(\mathcal{N})[1]) := j_*(\mathcal{A}\otimes^!\mathcal{B}[1]).$$

If say $\mathcal{A}$ is a D.E. on $\mathbb{G}_m$, then as $\mathcal{O}$-module $\mathcal{A}$ is locally free of finite rank, in which case $\mathcal{A}\otimes^!\mathcal{B}[1]$ on $\mathbb{G}_m$ is visibly the usual $\mathcal{O}$-tensor product. Thus we have $FT(\mathcal{M}*_+\mathcal{N}) \approx j_*(\mathcal{A}\otimes\mathcal{B})$ is a single $\mathcal{D}$-module in degree zero. So by Fourier inversion, $\mathcal{M}*_+\mathcal{N}$ is a single $\mathcal{D}$-module in degree zero. QED

12.3 The Tannakian Category $D_{A,B}$

Definition 12.3.1 Given a holonomic $\mathcal{D}$-module $\mathcal{M}$ on $\mathbb{A}^1$, consider the following conditions (A) and (B):
Condition(A) : $\pi_*\mathcal{M} = 0$ (or equivalently, $FT\mathcal{M} \approx j_*j^*FT\mathcal{M}$).
Condition(B) : $j^*FT\mathcal{M}$ is a D.E. on $\mathbb{G}_m$ (i.e., $j^*FT\mathcal{M}$ is $\mathcal{O}_{\mathbb{G}_m}$-locally free of finite rank).

Remark 12.3.2 In virtue of 2.10.16 (1), a holonomic $\mathcal{M}$ satisfies Condition (B) if and only if all of its ∞-slopes are $\neq 1$. In particular, if $\mathcal{M}$ is RS then $\mathcal{M}$ satisfies Condition (B).

Proposition 12.3.3 Suppose that $\mathcal{M}$ is a holonomic $\mathcal{D}$-module on $\mathbb{A}^1$ which satisfies both Conditions (A) and (B). Then for any holonomic $\mathcal{D}$-module $\mathcal{N}$, $\mathcal{M}*_+\mathcal{N}$ is a single $\mathcal{D}$-module which satsifies (A). If in addition

$\mathcal{M}$ is nonconstant (i.e., if $j^*FT(\mathcal{M}) \neq 0$), then $\mathcal{M}*_+\mathcal{N}$ satisfies (B) if and only if $\mathcal{N}$ itself satisfies (B).

proof The first assertion is just a restatement of the previous corollary. If in addition $\mathcal{M}$ is nonconstant, then $j^*FT\mathcal{M}$ is $\mathcal{O}$-locally free of **nonzero** finite rank. Therefore $j^*FT\mathcal{N}$ is $\mathcal{O}$-coherent if and only if $j^*FT\mathcal{N}\otimes j^*FT\mathcal{M}$ is $\mathcal{O}$-coherent. QED

(12.3.4) Recall (2.10.1 (1)) that $FT(j_!\mathcal{O}_{\mathbb{G}_m}) = j_*\mathcal{O}_{\mathbb{G}_m}$. So $j_!\mathcal{O}_{\mathbb{G}_m}$ satisfies both conditions (A) and (B). Using $j_!\mathcal{O}_{\mathbb{G}_m}$ as the $\mathcal{M}$ above, we find

Proposition 12.3.5 For $\mathcal{N}$ a holonomic $\mathcal{D}$-modules on $\mathbb{A}^1$, $\mathcal{N}*_+(j_!\mathcal{O}_{\mathbb{G}_m})$ is a single holonomic $\mathcal{D}$-module on $\mathbb{A}^1$ which satisfies Condition (A), and we have a canonical isomorphism

$$FT(\mathcal{N}*_+(j_!\mathcal{O}_{\mathbb{G}_m})) \approx j_*j^*FT(\mathcal{N}).$$

The operator $\mathcal{N} \mapsto \mathcal{N}*_+(j_!\mathcal{O}_{\mathbb{G}_m})$ is idempotent; it is the projector onto those holonomic $\mathcal{D}$-modules $\mathcal{N}$ which satisfy Condition (A).

proof All save the last assertion is a formal consequence of the preceeding two results, and the fact that $j^*FT(j_!\mathcal{O}_{\mathbb{G}_m}) = \mathcal{O}_{\mathbb{G}_m}$.

If $\mathcal{N}$ already satisfies Condition (A), then $FT(\mathcal{N}) \approx j_*j^*FT(\mathcal{N})$, so

$$FT(\mathcal{N}*_+(j_!\mathcal{O}_{\mathbb{G}_m})) \approx j_*j^*FT(\mathcal{N}) = FT(\mathcal{N}),$$

and by Fourier inversion we have $\mathcal{N}*_+(j_!\mathcal{O}_{\mathbb{G}_m}) \approx \mathcal{N}$. QED

Theorem 12.3.6 Denote by $\mathcal{D}\mathrm{Mod}^{\mathrm{holo}}(\mathbb{A}^1)$ the abelian catergory of all holonomic $\mathcal{D}$-modules on $\mathbb{A}^1$, and by $D_{A,B}$ the full subcategory of $\mathcal{D}\mathrm{Mod}^{\mathrm{holo}}(\mathbb{A}^1)$ consisting of those objects which satisfy both Conditions (A) and (B). Then

(1) $D_{A,B}$ is stable by convolution.

(2) The exact functor

$$\mathcal{D}\mathrm{Mod}^{\mathrm{holo}}(\mathbb{A}^1) \to \mathcal{D}\mathrm{Mod}^{\mathrm{holo}}(\mathbb{G}_m)$$
$$\mathcal{M} \mapsto j^*FT(\mathcal{M})$$

induces an equivalence of categories

$$D_{A,B} \approx D.E.(\mathbb{G}_m/\mathbb{C}),$$

whose quasi-inverse $D.E.(\mathbb{G}_m/\mathbb{C}) \approx D_{A,B}$ is given by

$$V \mapsto [x \mapsto -x]^*FT(j_*V).$$

(3) This equivalence carries convolution product of objects of $D_{A,B}$ to usual tensor product of D.E.'s on $\mathbb{G}_m$.

proof We have already seen (12.3.3) that $D_{A,B}$ is stable by convolution. That $\mathcal{M} \mapsto j^*FT(\mathcal{M})$ carries $D_{A,B}$ to $D.E.(\mathbb{G}_m/\mathbb{C})$ is built into condition (B). That $V \mapsto [x \mapsto -x]^*FT(j_*V)$ carries $D.E.(\mathbb{G}_m/\mathbb{C})$ to $D_{A,B}$ and is a two-sided quasi-inverse is Fourier inversion. That these inverse equivalences interchange convolution in $D_{A,B}$ and usual tensor product in $D.E.(\mathbb{G}_m/\mathbb{C})$ is 12.2.4 (3). QED

Corollary 12.3.7 $D_{A,B}$ is abelian, and kernels and cokernels of morphisms in it are the same as in the ambient $\mathcal{D}\mathrm{Mod}^{\mathrm{holo}}(\mathbb{A}^1)$.

proof Denote by $FTD_{A,B}$ the full subcategory of $\mathcal{D}\mathrm{Mod}^{\mathrm{holo}}(\mathbb{A}^1)$ consisting of objects of the form j_*V, for V a D.E. on $\mathbb{G}_m$. Fourier Transform makes $D_{A,B}$ equivalent to $FTD_{A,B}$, so it suffices to prove the same assertion for $FTD_{A,B}$.

Suppose we are given any $\mathcal{O}$-quasicoherent $\mathcal{D}$-modules V and W on $\mathbb{G}_m$, and a $\mathcal{D}$-module map $\varphi: j_*V \to j_*W$. Then $\varphi = j_*j^*(\varphi)$. Denote by S and T the kernel and cokernel of $j^*(\varphi): V \to W$. Since j_* is exact, it follows that j_*S and j_*T are the kernel and cokernel of φ. If now V and W are D.E.'s on $\mathbb{G}_m$, i.e., if V and W are each $\mathcal{O}_{\mathbb{G}_m}$-coherent, then S and T are $\mathcal{O}_{\mathbb{G}_m}$-coherent, so S and T are D.E.'s on $\mathbb{G}_m$. QED

Corollary 12.3.8 $D_{A,B}$ is a Tannakian category, with

(1) "tensor product" given by convolution $*_+$,

(2) "unit object" $\mathbb{1}$ given by $j_!\mathcal{O}_{\mathbb{G}_m}$,

(3) "dual" given by $\mathcal{M}^\vee := ([x \mapsto -x]^*(\mathcal{M}^*)) *_+ (j_!\mathcal{O}_{\mathbb{G}_m})$.

proof This is just the inverse Fourier Transform of the standard structure of Tannakian category on $D.E.(\mathbb{G}_m/\mathbb{C})$. QED

Remark 12.3.9 For any $\alpha \in \mathbb{C}^\times$, the functor

$$D_{A,B} \to (\text{fin. dim'l. } \mathbb{C}\text{-vector spaces})$$

$$\mathcal{M} \mapsto H^0(\pi_*(\mathcal{M}\otimes e^{\alpha x})) := H^1{}_{DR}(\mathbb{G}_m/\mathbb{C}, \mathcal{M}\otimes e^{\alpha x})$$

is a fibre functor. [It is the Fourier Transform of the usual fibre functor ω_α := fibre at α on $D.E.(\mathbb{G}_m/\mathbb{C})$.] We do **not** know how to construct explicit fibre functors on $D_{A,B}$ without invoking Fourier Transform.

12.4 The Tannakian Category $D_{A,RS}$

(12.4.1) Recall (cf. [Ber], [Bor]) that for X a smooth separated $\mathbb{C}$-scheme of finite type, one has the notion of RS ("regular singular") $\mathcal{D}$-modules, and of the full subcategory $D^{RS}(X)$ consisting of the RS objects of $D^{b,holo}(X)$. One knows that $D^{RS}(X)$ is stable under the "six operations". In particular, if X is a smooth $\mathbb{C}$-groupscheme G, then $D^{RS}(G)$ is stable under convolution.

(12.4.2) We now return to the $\mathbb{A}^1$ setting. We define $D_{A,RS}$ to be the full subcategory of $D_{A,B}$ consisting of those objects which are RS. Given a short exact sequence

$$0 \to \mathcal{M}_1 \to \mathcal{M}_2 \to \mathcal{M}_3 \to 0$$

in $D_{A,B}$, $\mathcal{M}_2$ is RS if and only if both $\mathcal{M}_1$ and $\mathcal{M}_3$ are RS. The subcategory $D_{A,RS}$ of $D_{A,B}$ is stable under convolution and under "dual", and contains the unit object $\mathbb{1} := j_!\mathcal{O}_{\mathbb{G}_m}$. Thus we find

Theorem 12.4.3 $D_{A,RS}$ is itself a Tannakian category with

(1) "tensor product" given by convolution $*_+$,

(2) "unit object" $\mathbb{1}$ given by $j_!\mathcal{O}_{\mathbb{G}_m}$,

(3) "dual" given by $\mathcal{M}^\vee := ([x \mapsto -x]^*(\mathcal{M}^*))*_+(j_!\mathcal{O}_{\mathbb{G}_m})$.

Proposition 12.4.4 Let $\mathcal{M}$ be an object of $D_{A,RS}$. Then $\mathcal{M}$ and $FT\mathcal{M}$ have the same generic rank. In other words, the "dimension" of $\mathcal{M}$ as an object of the Tannakian category $D_{A,RS}$ is its generic rank.

proof Since $FT\mathcal{M}$ is a D.E. on $\mathbb{G}_m$, say of rank r, for any $\alpha \in \mathbb{C}^\times$ the complex

$$i_\alpha^!(\mathrm{FT}\mathcal{M}) := \mathrm{FT}\mathcal{M} \xrightarrow{x-\alpha} \mathrm{FT}\mathcal{M},$$

placed in degree 0 and 1, has $H^0 = 0$, $\dim_{\mathbb{C}} H^1 = r$. Thus the generic rank of $\mathrm{FT}\mathcal{M}$ is $-\chi(i_\alpha^!(\mathrm{FT}\mathcal{M}))$ for any $\alpha \neq 0$. By 12.2.3, we have

$$\pi_*(\mathcal{M}\otimes e^{\alpha x}) \approx i_\alpha^!(\mathrm{FT}(\mathcal{M}))[1],$$

so the generic rank r of $\mathrm{FT}\mathcal{M}$ is

$$r = \chi(\pi_*(\mathcal{M}\otimes e^{\alpha x})) := -\chi_{DR}(\mathbb{A}^1, \mathcal{M}\otimes e^{\alpha x}).$$

Because $\mathcal{M}$ satisfies Condition (A), $\pi_*(\mathcal{M}) = 0$, so we may rewrite this as

$$r = \chi_{DR}(\mathbb{A}^1, \mathcal{M}) - \chi_{DR}(\mathbb{A}^1, \mathcal{M}\otimes e^{\alpha x}).$$

At this point, we need the following numerological lemma.

Lemma 12.4.5 For any $\alpha \in \mathbb{C}$, and any holonomic $\mathcal{M}$ on $\mathbb{A}^1$ we have

$$\chi_{DR}(\mathbb{A}^1, \mathcal{M}) - \chi_{DR}(\mathbb{A}^1, \mathcal{M}\otimes e^{\alpha x}) = \mathrm{Irr}_\infty(\mathcal{M}\otimes e^{\alpha x}) - \mathrm{Irr}_\infty(\mathcal{M}).$$

proof Both sides are additive in $\mathcal{M}$, and the assertion is obvious for $\mathcal{M}$ punctual. So it suffices to treat the case when $\mathcal{M} \approx k_*k^*\mathcal{M}$, $k : U \to \mathbb{A}^1$ the inclusion of a nonempty open set on which $\mathcal{M}$ is a D.E. Looking at Deligne's Euler-Poincare formula for $\mathcal{M}$ (cf 2.9.8.2)

$$\chi_{DR}(\mathbb{A}^1, \mathcal{M}) = \chi_{DR}(U, k^*\mathcal{M}) =$$

$$\mathrm{rank}(k^*\mathcal{M})\chi_{DR}(U, \mathcal{O}_U) - \Sigma_{x \in \mathbb{A}^1-U} \mathrm{Irr}_x(\mathcal{M}) - \mathrm{Irr}_\infty(\mathcal{M})$$

and comparing it term by term with that for $\mathcal{M}\otimes e^{\alpha x}$, only the last terms can differ (since $e^{\alpha x}\mathbb{C}[x]$ is a rank one D.E. on all of $\mathbb{A}^1$). QED

Returning now to the proof of the Proposition, the lemma gives

$$r = \mathrm{Irr}_\infty(\mathcal{M}\otimes e^{\alpha x}) - \mathrm{Irr}_\infty(\mathcal{M}),$$

for any $\alpha \in \mathbb{C}^\times$. Since $\mathcal{M}$ is RS at ∞, all its ∞-slopes are zero, (so $\mathrm{Irr}_\infty(\mathcal{M}) = 0$), and $\mathcal{M}\otimes e^{\alpha x}$ has all its ∞-slopes $=1$, so $\mathrm{Irr}_\infty(\mathcal{M}\otimes e^{\alpha x})$ is the generic rank of $\mathcal{M}$. QED

12.5 A minor variant: ! convolution of $\mathcal{D}$-modules

(12.5.1) In the world of $\mathcal{D}$-modules, the "natural" operations are f_* and $f^!$; $f_!$ and f^* are **defined** by duality. Since it is these latter operations which are more natural in the "topological" world, we will make a transition to this point of view.

(12.5.2) For a holonomic $\mathcal{D}$-module $\mathcal{M}$ on a smooth $\mathbb{C}$-scheme X, let us write $D_X\mathcal{M}$, or simply $D\mathcal{M}$ if no confusion is likely, for its adjoint $\mathcal{M}^*$. Making use of Beilinson's theorem [Bei] that the natural functor

$$D^b(\mathcal{D}Mod^{holo}(X)) \to D^{b,holo}(X)$$

is an **equivalence** of the bounded derived category of the abelian category $\mathcal{D}Mod^{holo}(X)$ of holonomic left $\mathcal{D}$-modules on X with the subcategory $D^{b,holo}(X)$ of the bounded derived category of all left $\mathcal{D}$-modules on X whose cohomology sheaves are holonomic, we can extend the duality functor D_X to an involutive autoduality of $D^{b,holo}(X)$.

(12.5.3) For X and Y smooth C-schemes, and any map $f : X \to Y$, the functors

$$f_! : D^{b,holo}(X) \to D^{b,holo}(Y), \quad f^* : D^{b,holo}(Y) \to D^{b,holo}(X)$$

are defined by

$$f_!(K) := D_Y \circ f_* \circ D_X, \qquad f^* := D_X \circ f^! \circ D_Y.$$

(12.5.4) Given two objects K, L in $D^{b,holo}(X)$, their "naive" tensor product $K\otimes L$ is defined in terms of the diagonal embedding

$$\Delta : X \to X\times X$$

by

$$K\otimes L := \Delta^*(K\times L).$$

Since duality is compatible with cartesian product, we have the alternative description

$$K\otimes L := D(DK\otimes^! DL).$$

(12.5.5) For any smooth group-scheme G over $\mathbb{C}$, with product map

$$\text{product} : G\times G \to G,$$

we define the ! convolution of two objects K, L in $D^{b,holo}(G)$ by

$$K*_!L := \text{sum}_!(K\times L).$$

Since duality is compatible with cartesian product, we have the alternative description

$$K*_!L := D(DK*DL).$$

(12.5.6) We now apply this general setup to $\mathbb{A}^1$. In order to prevent any confusion about which convolution we have in mind, we will denote by

$$K*_{!+}L$$

the additive ! convolution on $\mathbb{A}^1$.

(12.5.7) For $j : \mathbb{G}_m \to \mathbb{A}^1$, duality interchanges $j_!\mathcal{O}_{\mathbb{G}_m}$ and $j_*\mathcal{O}_{\mathbb{G}_m}$, and it tautologically interchanges the functors π_* and $\pi_!$. It respects the conditions B (no ∞-slope $=1$) and RS (no slopes >0 anywhere), since $\mathcal{M}$ and $D\mathcal{M}$ have the same slopes everywhere. So it is natural to consider the following condition (A!), which is satisfied by $\mathcal{M}$ if and only if $D\mathcal{M}$ satisfes our original condition (A):

Condition(A!): $\pi_!\mathcal{M} = 0$ (or equivalently, $FT\mathcal{M} \approx j_!j^*\mathcal{M}$).

Proposition 12.5.8 (dual to 12.3.5) For $\mathcal{N}$ a holonomic $\mathcal{D}$-module on $\mathbb{A}^1$, $\mathcal{N}*_{!+}(j_*\mathcal{O}_{\mathbb{G}_m})$ is a single holonomic $\mathcal{D}$-module on $\mathbb{A}^1$ which satisfies Condition (A!), and we have a canonical isomorphism

$$FT(\mathcal{N}*_{!+}(j_*\mathcal{O}_{\mathbb{G}_m})) \approx j_!j^*FT(\mathcal{N}).$$

The operator $\mathcal{N} \mapsto \mathcal{N}*_{!+}(j_*\mathcal{O}_{\mathbb{G}_m})$ is idempotent; it is the projector onto those holonomic $\mathcal{D}$-modules $\mathcal{N}$ which satisfy Condition (A!).

Theorem 12.5.9 (dual to 12.3.6) Denote by $\mathcal{D}Mod^{holo}(\mathbb{A}^1)$ the abelian catergory of all holonomic $\mathcal{D}$-modules on $\mathbb{A}^1$, and by $D_{A!,B}$ the full subcategory of $\mathcal{D}Mod^{holo}(\mathbb{A}^1)$ consisting of those objects which satisfy both Conditions (A!) and (B). Then

(1) $D_{A!,B}$ is stable by additive ! convolution.

(2) The exact functor

$$\mathcal{D}Mod^{holo}(\mathbb{A}^1) \to \mathcal{D}Mod^{holo}(\mathbb{G}_m)$$
$$\mathcal{M} \mapsto j^*FT(\mathcal{M})$$

induces an equivalence of categories

$$D_{A!,B} \approx D.E.(\mathbb{G}_m/\mathbb{C}),$$

whose quasi-inverse $D.E.(\mathbb{G}_m/\mathbb{C}) \approx D_{A,B}$ is given by

$$V \mapsto [x \mapsto -x]^*FT(j_!V).$$

(3) This equivalence carries additive ! convolution product of objects of $D_{A,B}$ to usual tensor product of D.E.'s on $\mathbb{G}_m$.

Corollary 12.5.10 (dual to 12.3.8) $D_{A!,B}$ is a Tannakian category, with

(1) "tensor product" given by additive ! convolution $*_{!+}$,

(2) "unit object" $\mathbb{1}$ given by $j_*\mathcal{O}_{\mathbb{G}_m}$,

(3) "dual" given by $\mathcal{M}^\vee := ([x \mapsto -x]^*(D\mathcal{M}))*_{!+}(j_*\mathcal{O}_{\mathbb{G}_m})$.

(12.5.11) We define $D_{A!,RS}$ to be the full subcategory of $D_{A!,B}$ consisting of those objects which are RS.

Theorem 12.5.12 (dual to 12.4.3) $D_{A!,RS}$ is itself a Tannakian category with
(1) "tensor product" given by additive ! convolution $*_{!+}$,
(2) "unit object" $\mathbb{1}$ given by $j_*\mathcal{O}_{\mathbb{G}_m}$,
(3) "dual" given by $\mathcal{M}^\vee := ([x \mapsto -x]^*(D\mathcal{M}))*_{!+}(j_*\mathcal{O}_{\mathbb{G}_m})$.

Proposition 12.5.13 (dual to 12.4.4) Let $\mathcal{M}$ be an object of $D_{A!,RS}$. Then $\mathcal{M}$ and $FT\mathcal{M}$ have the same generic rank. In other words, the "dimension" of $\mathcal{M}$ as an object of the Tannakian category $D_{A!,RS}$ is its generic rank.

12.6 Brief Review of Riemann-Hilbert; Transition from $\mathcal{D}$-modules to $D^b{}_c$.

(12.6.1) Let X be a smooth $\mathbb{C}$-scheme of finite type. Denote by $X(\mathbb{C})^{an}$ the underlying complex manifold. For any field F, denote by $D^b{}_c(X(\mathbb{C})^{an}, F)$ the full subcategory of the bounded derived category of sheaves of F-spaces on $X(\mathbb{C})^{an}$ whose cohomology sheaves are lisse of finite rank on each piece of an algebraic stratification. By "Riemann-Hilbert"(cf. [Ber], [Bor, VIII, 9.6], [Me-HR]), the DR functor defines an equivalence of categories

$$D^b{}_{holo,\,RS}(X, \mathcal{D}) \cong D^b{}_c(X(\mathbb{C})^{an}, \mathbb{C})$$

which for variable X is compatible with the "six operations", and induces an equivalence

$$\mathcal{D}Mod^{holo,RS}(X) \approx Perv(X(\mathbb{C})^{an}, \mathbb{C}).$$

For $\mathcal{M}$ a D.E. on X, the corresponding perverse object on $X(\mathbb{C})^{an}$ is $\mathcal{F}[\dim X]$, for $\mathcal{F}$ the local system on $X(\mathbb{C})^{an}$ of germs of horizontal sections of $\mathcal{M}^{an}$.

(12.6.2) We now return to $\mathbb{A}^1$. Here the DR functor sets up the following particular correspondences:

$\mathcal{O}_{\mathbb{A}^1}$ .... $\mathbb{C}[1]$

$j_*\mathcal{O}_{\mathbb{G}_m}$ .... $Rj_*j^*\mathbb{C}[1]$

$j_!\mathcal{O}_{\mathbb{G}_m}$ .... $j_!j^*\mathbb{C}[1]$.

So the translation through Riemann-Hilbert of 12.5.8 is

Proposition 12.6.3 For K in $\mathrm{Perv}(\mathbb{A}^1(\mathbb{C})^{an}, \mathbb{C})$ corresponding to a holonomic RS $\mathcal{D}$-module $\mathcal{M}$, $K*_{!+}(Rj_*j^*\mathbb{C}[1])$ is perverse, satisfies Condition (A!), and corresponds to a holonomic RS $\mathcal{D}$-module $\mathcal{N}$ for which we have a canonical isomorphism of holonomic $\mathcal{D}$-modules

$$\mathrm{FT}(\mathcal{N}) \approx j_!j^*\mathrm{FT}(\mathcal{M}).$$

The operator on $\mathrm{Perv}(\mathbb{A}^1(\mathbb{C})^{an}, \mathbb{C})$

$$K \mapsto K*_{!+}(Rj_*j^*\mathbb{C}[1])$$

is idempotent ; it is the projector onto those objects which satisfy Condition (A!).

(12.6.4) Translating the succeeding $\mathcal{D}$-module results through Riemann-Hilbert, we find the two following results, which are interchanged by duality.

Theorem 12.6.5 Denote by $\mathrm{Perv}_{A!}(\mathbb{A}^1(\mathbb{C})^{an}, \mathbb{C})$ the full subcategory of $\mathrm{Perv}(\mathbb{A}^1(\mathbb{C})^{an}, \mathbb{C})$ consisting of those K with $R\pi_!K = 0$. Then $\mathrm{Perv}_{A!}(\mathbb{A}^1(\mathbb{C})^{an}, \mathbb{C})$ is a Tannakian category with

(1) "tensor product" given by additive ! convolution $*_{!+}$,

(2) "unit object" $\mathbb{1}$ given by $Rj_*\mathbb{C}[1]$,

(3) "dual" given by $K^\vee := ([x \mapsto -x]^*(DK))*_{!+}(Rj_*\mathbb{C}[1])$.

Moreover, for any K in $\mathrm{Perv}_{A!}(\mathbb{A}^1(\mathbb{C})^{an}, \mathbb{C})$, its "dimension" is the generic rank of $\mathcal{H}^{-1}(K)$.

Theorem 12.6.6 Denote by $\mathrm{Perv}_{A}(\mathbb{A}^1(\mathbb{C})^{an}, \mathbb{C})$ the full subcategory of $\mathrm{Perv}(\mathbb{A}^1(\mathbb{C})^{an}, \mathbb{C})$ consisting of those K with $R\pi_*K = 0$. Then $\mathrm{Perv}_{A}(\mathbb{A}^1(\mathbb{C})^{an}, \mathbb{C})$ is a Tannakian category with

(1) "tensor product" given by additive * convolution $*_+$,

(2) "unit object" $\mathbb{1}$ given by $j_!\mathbb{C}[1]$,

(3) "dual" given by $K^{\vee} := ([x \mapsto -x]^*(DK)) *_+ (j_!\mathbb{C}[1])$.

Moreover, for any K in $\mathrm{Perv}_A(\mathbb{A}^1(\mathbb{C})^{an}, \mathbb{C})$, its "dimension" is the generic rank of $\mathcal{H}^{-1}(K)$.

12.7 Transition to $\overline{\mathbb{Q}}_\ell$ coefficients

(12.7.1) Let X be a $\mathbb{C}$-scheme of finite type. Fix an isomorphism

$$\iota: \overline{\mathbb{Q}}_\ell \approx \mathbb{C}$$

of fields, and use it to identify $D^b{}_c(X(\mathbb{C})^{an}, \overline{\mathbb{Q}}_\ell)$ with $D^b{}_c(X(\mathbb{C})^{an}, \mathbb{C})$. The continuous map

$$\varepsilon: X(\mathbb{C})^{an} \to X_{et}$$

induces a fully faithful functor ε^*

$$D^b{}_c(X, \overline{\mathbb{Q}}_\ell) \to D^b{}_c(X(\mathbb{C})^{an}, \overline{\mathbb{Q}}_\ell),$$

which for variable X is compatible with the "six operations", and induces an exact fully faithful functor between the abelian categories of perverse objects

$$\mathrm{Perv}(X, \overline{\mathbb{Q}}_\ell) \to \mathrm{Perv}(X(\mathbb{C})^{an}, \overline{\mathbb{Q}}_\ell).$$

Notice that an object K of $D^b{}_c(X, \overline{\mathbb{Q}}_\ell)$ is perverse if and only if its image in $D^b{}_c(X(\mathbb{C})^{an}, \overline{\mathbb{Q}}_\ell)$ is perverse; this is obvious from looking at the cohomology sheaves of K and DK in the two contexts.

Lemma 12.7.1.1 An object K of $\mathrm{Perv}(X, \overline{\mathbb{Q}}_\ell)$ is irreducible (resp. is a direct sum of irreducibles) in $\mathrm{Perv}(X, \overline{\mathbb{Q}}_\ell)$ if and only if its image $\varepsilon^*(K)$ is irreducible (resp. a direct sum of irreducibles) in $\mathrm{Perv}(X(\mathbb{C})^{an}, \overline{\mathbb{Q}}_\ell)$.

proof By [B-B-D, 4.3.1], given an irreducible K in $\mathrm{Perv}(X, \overline{\mathbb{Q}}_\ell)$, there exists a locally closed smooth irreducible subscheme $j: U \to X$, and an irreducible local system $\mathcal{F}$ on U such that $K \approx j_{!*}\mathcal{F}[\dim U]$ is the middle extension of $\mathcal{F}[\dim U]$, and every object of this form is irreducible in $\mathrm{Perv}(X, \overline{\mathbb{Q}}_\ell)$. Similarly, Riemann-Hilbert and the classification of irreducible holonomic $\mathcal{D}$-modules (cf [Bor, 10.5 and 10.6]) shows that given an irreducible K in $\mathrm{Perv}(X(\mathbb{C})^{an}, \overline{\mathbb{Q}}_\ell)$, there exists a locally closed smooth irreducible subscheme $j: U \to X$, and an irreducible local system $\mathcal{F}$ on $U(\mathbb{C})^{an}$ such that $K \approx j_{!*}\mathcal{F}[\dim U]$ is the middle extension of $\mathcal{F}[\dim U]$, and every object of this form is irreducible in $\mathrm{Perv}(X(\mathbb{C})^{an}, \overline{\mathbb{Q}}_\ell)$.

The functor ε^* commutes with formation of middle extensions, and for $\mathcal{F}$ an irreducible local system on U, $\varepsilon^*\mathcal{F}$ is an irreducible local system on $U(\mathbb{C})^{an}$. Therefore ε^* maps irreducibles in $\mathrm{Perv}(X, \overline{\mathbb{Q}}_\ell)$ to irreducibles in $\mathrm{Perv}(X(\mathbb{C})^{an}, \overline{\mathbb{Q}}_\ell)$, and hence it maps direct sums of irreducibles to direct sums of irreducibles.

Conversely, suppose that K in $\mathrm{Perv}(X, \overline{\mathbb{Q}}_\ell)$ has ε^*K a direct sum of irreducibles L_i in $\mathrm{Perv}(X(\mathbb{C})^{an}, \overline{\mathbb{Q}}_\ell)$. By the full faithfullness of ε^*, each projector π_i of ε^*K onto L_i comes from a unique projector $\tilde{\pi}_i$ on K, and so K is the direct sum of the objects $\tilde{\pi}_i(K)$, and $\varepsilon^*\tilde{\pi}_i(K) \approx L_i$ is irreducible. Since ε^* is fully faithful, the irreducibility of $\varepsilon^*\tilde{\pi}_i(K) \approx L_i$ implies the irreducibility of $\tilde{\pi}_i(K)$. QED

(12.7.2) We now return to the case of $\mathbb{A}^1$. The translation of 12.6.3 through this change of coefficients is

Proposition 12.7.3 For K in $\mathrm{Perv}(\mathbb{A}^1_{\mathbb{C}}, \overline{\mathbb{Q}}_\ell)$, corresponding to a holonomic RS $\mathcal{D}$-module $\mathcal{M}$, $K*_{!+}(Rj_*j^*\overline{\mathbb{Q}}_\ell(1)[1])$ is perverse, satisfies Condition (A!), and corresponds to a holonomic RS $\mathcal{D}$-module $\mathcal{N}$ for which we have a canonical isomorphism of holonomic $\mathcal{D}$-modules

$$\mathrm{FT}(\mathcal{N}) \approx j_!j^*\mathrm{FT}(\mathcal{M}).$$

The operator on $\mathrm{Perv}(\mathbb{A}^1_{\mathbb{C}}, \overline{\mathbb{Q}}_\ell)$

$$K \mapsto K*_{!+}(Rj_*j^*\overline{\mathbb{Q}}_\ell(1)[1])$$

is idempotent ; it is the projector onto those objects which satisfy Condition (A!).

(12.7.4) We have the following two results, which are interchanged by duality.

Theorem 12.7.5 Denote by $\mathrm{Perv}_{A!}(\mathbb{A}^1_{\mathbb{C}}, \overline{\mathbb{Q}}_\ell)$ the full subcategory of $\mathrm{Perv}(\mathbb{A}^1_{\mathbb{C}}, \overline{\mathbb{Q}}_\ell)$ consisting of those K with $R\pi_!K = 0$. Then

$\mathrm{Perv}_{A!}(\mathbb{A}^1_{\mathbb{C}}, \overline{\mathbb{Q}}_\ell)$ is a Tannakian category with
(1) "tensor product" given by additive ! convolution $*_{!+}$,
(2) "unit object" $\mathbb{1}$ given by $Rj_*\overline{\mathbb{Q}}_\ell(1)[1]$,
(3) "dual" given by $K^\vee := ([x \mapsto -x]^*(DK)) *_{!+} (Rj_*\overline{\mathbb{Q}}_\ell(1)[1])$.
Moreover, for any K in $\mathrm{Perv}_{A!}(\mathbb{A}^1_{\mathbb{C}}, \overline{\mathbb{Q}}_\ell)$, its "dimension" is the generic rank of $\mathcal{H}^{-1}(K)$.

proof The only tricky point is that the subcategory $\mathrm{Perv}_{A!}(\mathbb{A}^1_{\mathbb{C}}, \overline{\mathbb{Q}}_\ell)$ of $\mathrm{Perv}(\mathbb{A}^1(\mathbb{C})^{an}, \overline{\mathbb{Q}}_\ell)$ is itself a Tannakian category. But using Deligne's version [De-CT, 2.1] of the axioms defining a Tannakian category, this results from the full faithfulness of

$$\mathrm{Perv}(\mathbb{A}^1_{\mathbb{C}}, \overline{\mathbb{Q}}_\ell) \to \mathrm{Perv}(\mathbb{A}^1(\mathbb{C})^{an}, \overline{\mathbb{Q}}_\ell)$$

and the stability of $\mathrm{Perv}_{A!}(\mathbb{A}^1_{\mathbb{C}}, \overline{\mathbb{Q}}_\ell)$ under the operation "dual". QED

Theorem 12.7.6 Denote by $\mathrm{Perv}_{A}(\mathbb{A}^1_{\mathbb{C}}, \overline{\mathbb{Q}}_\ell)$ the full subcategory of $\mathrm{Perv}(\mathbb{A}^1_{\mathbb{C}}, \overline{\mathbb{Q}}_\ell)$ consisting of those K with $R\pi_*K = 0$. Then $\mathrm{Perv}_{A}(\mathbb{A}^1_{\mathbb{C}}, \overline{\mathbb{Q}}_\ell)$ is a Tannakian category with
(1) "tensor product" given by additive * convolution $*_+$,
(2) "unit object" $\mathbb{1}$ given by $j_!\overline{\mathbb{Q}}_\ell[1]$,
(3) "dual" given by $K^\vee := ([x \mapsto -x]^*(DK)) *_+ (j_!\overline{\mathbb{Q}}_\ell[1])$.
Moreover, for any K in $\mathrm{Perv}_{A}(\mathbb{A}^1_{\mathbb{C}}, \overline{\mathbb{Q}}_\ell)$, its "dimension" is the generic rank of $\mathcal{H}^{-1}(K)$.

12.8 Recapitulation of the situation over $\mathbb{C}$

(12.8.1) To help the reader keep track of the where we are so far, it may be useful to keep in mind the following diagrams of Tannakian categories, in which ε^* and j^*FT are exact, fully faithful $\otimes$-functors, and in which ι and DR are equivalences of Tannakian categories:

$$\begin{array}{ccccc}
\mathrm{Perv}_{A!}(\mathbb{A}^1_{\mathbb{C}}, \overline{\mathbb{Q}}_\ell) & \xrightarrow{\varepsilon^*} & \mathrm{Perv}_{A!}(X(\mathbb{C})^{an}, \overline{\mathbb{Q}}_\ell) & \overset{\iota}{\approx} & \mathrm{Perv}_{A!}(X(\mathbb{C})^{an}, \mathbb{C}) \\
 & & & & \uparrow \wr\, DR \\
 & & & & D_{A!,RS} \\
 & & & & \downarrow j^*FT \\
 & & & & D.E.(\mathbb{G}_m/\mathbb{C})
\end{array}$$

$$\begin{array}{ccccc}
\mathrm{Perv}_{A}(\mathbb{A}^1_{\mathbb{C}}, \overline{\mathbb{Q}}_\ell) & \xrightarrow{\varepsilon^*} & \mathrm{Perv}_{A}(X(\mathbb{C})^{an}, \overline{\mathbb{Q}}_\ell) & \overset{\iota}{\approx} & \mathrm{Perv}_{A}(X(\mathbb{C})^{an}, \mathbb{C}) \\
 & & & & \uparrow \wr\, DR \\
 & & & & D_{A,RS} \\
 & & & & \downarrow j^*FT \\
 & & & & D.E.(\mathbb{G}_m/\mathbb{C})
\end{array}$$

Lemma 12.8.2 Fix a fibre functor ω on $\mathrm{Perv}_{A!}(X(\mathbb{C})^{an}, \overline{\mathbb{Q}}_\ell)$, and let K in $\mathrm{Perv}_{A!}(\mathbb{A}^1_{\mathbb{C}}, \overline{\mathbb{Q}}_\ell)$ be a semisimple object, i.e., K is a direct sum of irreducibles in $\mathrm{Perv}_{A!}(\mathbb{A}^1_{\mathbb{C}}, \overline{\mathbb{Q}}_\ell)$. Then K and ε^*K have the "same" Tannakian galois group, i.e., the natural inclusion of Tannakian galois groups $G_{\varepsilon^*K,\,\omega} \subset G_{K,\,\omega\circ\varepsilon^*}$ induced by ε^* is an isomorphism.

proof K and ε^*K are semisimple objects of $\mathrm{Perv}_{A!}(\mathbb{A}^1_{\mathbb{C}}, \overline{\mathbb{Q}}_\ell)$ and of $\mathrm{Perv}_{A!}(X(\mathbb{C})^{an}, \overline{\mathbb{Q}}_\ell)$ respectively, by 12.7.1.1. Therefore they are faithful, completely reducible representations of the groups $G_{K,\,\omega\circ\varepsilon^*}$ and $G_{\varepsilon^*K,\,\omega}$ respectively, and hence, as we are over a field of characteristic zero, these groups are both reductive. So the lemma results from the explicit recovery of these groups, via invariants in symmetric powers, given in 11.1, and the full faithfulness of the exact $\otimes$-functor ε^*. QED

Lemma 12.8.3 Suppose that K in $\mathrm{Perv}_{A!}(\mathbb{A}^1_{\mathbb{C}}, \overline{\mathbb{Q}}_\ell)$ is a semisimple object, i.e., K is a direct sum of irreducibles in $\mathrm{Perv}_{A!}(\mathbb{A}^1_{\mathbb{C}}, \overline{\mathbb{Q}}_\ell)$. Denote by $\mathcal{M}$ the RS $\mathcal{D}$-module on $\mathbb{A}^1_{\mathbb{C}}$ which corresponds, via Riemann-Hilbert and ι, to the object $\varepsilon^*(K)$ of $\mathrm{Perv}(\mathbb{A}^1(\mathbb{C})^{an}, \overline{\mathbb{Q}}_\ell)$. Fix a fibre functor ω on $\mathrm{D.E.}(\mathbb{G}_m/\mathbb{C})$, and denote by

$\omega_1 := \omega\circ j^*FT$, a fibre functor on $D_{A!,RS}$,

$\omega_2 :=$ the unique fibre functor on $\mathrm{Perv}(\mathbb{A}^1(\mathbb{C})^{an}, \mathbb{C})$ for which

$$\omega_1 = \omega_2\circ DR,$$

$\omega_3 := \omega_2\circ\iota$, a fibre functor on $\mathrm{Perv}_A(\mathbb{A}^1(\mathbb{C})^{an}, \overline{\mathbb{Q}}_\ell)$,

$\omega_4 := \omega_3\circ\varepsilon^*$, a fibre functor on $\mathrm{Perv}_{A!}(\mathbb{A}^1_{\mathbb{C}}, \overline{\mathbb{Q}}_\ell)$.

With respect to these fibre functors, the objects

$j^*FT(\mathcal{M})$ in $\mathrm{D.E.}(\mathbb{G}_m/\mathbb{C})$

$\mathcal{M}$ in $D_{A!,RS}$

$\iota\varepsilon^*(K)$ in $\mathrm{Perv}(\mathbb{A}^1(\mathbb{C})^{an}, \mathbb{C})$

$\varepsilon^*(K)$ in $\mathrm{Perv}(\mathbb{A}^1(\mathbb{C})^{an}, \overline{\mathbb{Q}}_\ell)$

K in $\mathrm{Perv}_{A!}(\mathbb{A}^1_{\mathbb{C}}, \overline{\mathbb{Q}}_\ell)$

all have the same Tannakian galois group.

proof This is immediate from 12.5.9, 12.7.1.1, and 12.8.2. QED

12.9 The situation in characteristic p

(12.9.1) In this section, we fix an algebraically closed field k of positive characteristic $p \neq \ell$ and a nontrivial additive $\overline{\mathbb{Q}}_\ell^\times$-valued character ψ of a finite subfield of k. We will work on $\mathbb{A}^1_k$. We denote by

$\pi : \mathbb{A}^1_k \to \mathrm{Spec}(k)$ the structural map,

$i_\alpha : \mathrm{Spec}(k) \to \mathbb{A}^1$ the inclusion of the point $\alpha \in k = \mathbb{A}^1(k)$,

$\Delta : \mathbb{A}^1 \to \mathbb{A}^1\times_k\mathbb{A}^1$ the diagonal embedding,

sum: $\mathbb{A}^1\times_k\mathbb{A}^1 \to \mathbb{A}^1$ the addition map $(x, y) \mapsto x+y$,

$j : \mathbb{G}_m \to \mathbb{A}^1$ the inclusion.

We will denote additive $*$ and ! convolution by $*_{*+}$ and $*_{!+}$ respectively.

(12.9.2) Recall (8.1.7) that the Fourier Transform

$$\mathrm{FT}_\psi : D^b_c(\mathbb{A}^1_k, \overline{\mathbb{Q}}_\ell) \to D^b_c(\mathbb{A}^1_k, \overline{\mathbb{Q}}_\ell)$$

is defined by

$$\mathrm{FT}_\psi(K) := R(pr_2)_!(pr_1{}^*(K)\otimes\mathcal{L}_{\psi(xy)})[1].$$

It commutes with duality up to an additive inversion,

$$D\circ\mathrm{FT}_\psi = \mathrm{FT}_\psi\circ[x \mapsto -x]^*\circ D,$$

and is involutive up to a Tate twist of (-1) and an additive inversion:

$$\mathrm{FT}_\psi\circ\mathrm{FT}_\psi(K) = [x \mapsto -x]^*(K)(-1).$$

One knows (cf [Br, 9.6]) that $\mathrm{FT}_\psi[-1]$ interchanges tensor product and additive ! convolution:

$$\mathrm{FT}_\psi(K*_{!+}L)[-1] \approx \mathrm{FT}_\psi(K)[-1]\otimes\mathrm{FT}_\psi(L)[-1].$$

Key Lemma 12.9.3 (compare Key lemma 12.2.3) For K, L objects in $D^b_c(\mathbb{A}^1_k, \overline{\mathbb{Q}}_\ell)$, and any $\alpha \in k$, we have canonical isomorphisms in $D^b_c(\mathrm{Spec}(k), \overline{\mathbb{Q}}_\ell)$

(1) $\qquad R\pi_!(K\otimes\mathcal{L}_{\psi(\alpha x)}) \approx i_\alpha{}^*(\mathrm{FT}_\psi(K))[-1],$

(2) $\qquad R\pi_!(\mathrm{FT}_\psi(K)\otimes\mathcal{L}_{\psi(-\alpha x)}) \approx i_\alpha{}^*(K)(-1)[-1],$

(3) $\qquad R\pi_!(K*_{!+}L) \approx R\pi_!(K)\otimes R\pi_!(L),$

(4) $\qquad i_\alpha{}^*(K\otimes L) \approx i_\alpha{}^*(K)\otimes i_\alpha{}^*(L),$

and a canonical isomorphism in $D^b_c(\mathbb{A}^1_k, \overline{\mathbb{Q}}_\ell)$

(5) $\qquad \mathrm{FT}_\psi(K*_{!+}L)[-1] \approx \mathrm{FT}_\psi(K)[-1]\otimes\mathrm{FT}_\psi(L)[-1].$

proof Assertion (1) is proper base change, (2) results from (1) by Fourier inversion, (3) is the special case $\varphi = \pi$ of 8.1.10(2a), (4) is tautological, and (5), recalled just above, is "mise pour memoire". QED

Exactly as in the $\mathcal{D}$-module setting, this immediately gives

Corollary 12.9.4 (compare 12.2.4) For K, L in $\mathrm{Perv}(\mathbb{A}^1_k, \overline{\mathbb{Q}}_\ell)$, we have

(1) $\mathrm{FT}_\psi(K) \approx j_!j^*\mathrm{FT}_\psi(K)$ if and only if $R\pi_!K = 0$.

(2) If either $R\pi_!K = 0$ or $R\pi_!L = 0$, then $R\pi_!(K*_{!+}L) = 0$.

(3) If $R\pi_!(K*_{!+}L) = 0$, and if $j^*\mathrm{FT}_\psi(K)[-1]$ and $j^*\mathrm{FT}_\psi(L)[-1]$ are both lisse sheaves, on $\mathbb{G}_m$ placed in degree zero, say $\mathcal{F}$ and $\mathcal{G}$, then

$$\mathrm{FT}_\psi(K*_{!+}L)[-1] \approx j_!(\mathcal{F}\otimes\mathcal{G}),$$

and $K*_{!+}L$ is perverse.

12.10 The Tannakian Category $\mathrm{Perv}_{A!,B}(\mathbb{A}^1, \overline{\mathbb{Q}}_\ell)$

Definition 12.10.1 Given K in $\mathrm{Perv}(\mathbb{A}^1{}_k, \overline{\mathbb{Q}}_\ell)$, consider the following conditions (A!) and (B):

Condition(A!) : $R\pi_!K = 0$ (or equivalently, $FT_\psi(K) \approx j_!j^*FT_\psi(K)$).

Condition(B) : $j^*FT_\psi(K)[-1]$ is a lisse sheaf on $\mathbb{G}_m$ placed in degree zero.

Remark 12.10.2 In virtue of [Ka-GKM, 8.5.8 (2)], K in $\mathrm{Perv}(\mathbb{A}^1{}_k, \overline{\mathbb{Q}}_\ell)$ satisfies Condition (B) if and only if all of its ∞-slopes (i.e., all the ∞-slopes of its only nonpunctual cohomology sheaf $\mathcal{H}^{-1}(K)$) are $\neq 1$. In particular, if K is everywhere tame, then K satisfies Condition (B).

(12.10.3) According to [Ka-PES, A2], we have (cf 2.10.1 (1) for the $\mathcal{D}$-module analogue)

$$FT_\psi(j_!j^*\overline{\mathbb{Q}}_\ell[1]) \approx Rj_*j^*\overline{\mathbb{Q}}_\ell[1],$$

which by Fourier inversion gives

$$FT_\psi(Rj_*j^*\overline{\mathbb{Q}}_\ell(1)[1]) \approx j_!j^*\overline{\mathbb{Q}}_\ell[1].$$

Therefore the object $Rj_*j^*\overline{\mathbb{Q}}_\ell[1]$ satisfies Condition (B).

Lemma 12.10.4 The object $Rj_*j^*\overline{\mathbb{Q}}_\ell(1)[1]$ satisfies Condition (A!).

proof (compare [Ka-LG, proof of 1.6.8], [Ka-GKM,2.2.1-3]) We must show that $R\pi_!(Rj_*j^*\overline{\mathbb{Q}}_\ell) = 0$. By interchanging the points 0 and ∞ in $\mathbb{P}^1$, this becomes the statement $R\pi_*(j_!j^*\overline{\mathbb{Q}}_\ell) = 0$. [Alternately, these two statements are interchanged by duality.] This amounts to the vanishing of the ordinary cohomology groups $H^i(\mathbb{A}^1{}_k, j_!j^*\overline{\mathbb{Q}}_\ell)$. The H^i for $i \neq 0,1$ vanish for reasons of cohomological dimension, the H^0 vanishes by inspection, and the remaining group H^1 vanishes by the Euler-Poincare formula. QED

Exactly as in the $\mathcal{D}$-module setting, this gives

Proposition 12.10.5 For K in $\mathrm{Perv}(\mathbb{A}^1{}_k, \overline{\mathbb{Q}}_\ell)$, $K*_{!+}(Rj_*j^*\overline{\mathbb{Q}}_\ell(1)[1])$ is perverse and satisfies Condition (A!), and we have a canonical isomorphism

$$FT_\psi(K*_{!+}(Rj_*j^*\overline{\mathbb{Q}}_\ell(1)[1])) \approx j_!j^*FT_\psi(K).$$

The operator on $\mathrm{Perv}(\mathbb{A}^1{}_k, \overline{\mathbb{Q}}_\ell)$

$$K \mapsto K *_{!+}(Rj_*j^*\overline{\mathbb{Q}}_\ell(1)[1])$$

is idempotent ; it is the projector onto those objects which satisfy Condition (A!).

Proceeding as in the $\mathcal{D}$-module case, we now prove

Theorem 12.10.6 Denote by $\mathrm{Perv}_{A!,B}(\mathbb{A}^1{}_k, \overline{\mathbb{Q}}_\ell)$ the full subcategory of $\mathrm{Perv}(\mathbb{A}^1{}_k, \overline{\mathbb{Q}}_\ell)$ consisting of those K satisfying conditions (A!) and (B). Then $\mathrm{Perv}_{A!,B}(\mathbb{A}^1{}_k, \overline{\mathbb{Q}}_\ell)$ is a Tannakian category with

(1) "tensor product" given by additive ! convolution $*_{!+}$,

(2) "unit object" $\mathbb{1}$ given by $Rj_*j^*\overline{\mathbb{Q}}_\ell(1)[1]$,

(3) "dual" given by $K^\vee := ([x \mapsto -x]^*(DK)) *_{!+}(Rj_*j^*\overline{\mathbb{Q}}_\ell(1)[1])$.

The exact functor

$$j^*FT_\psi[-1] : \mathrm{Perv}_{A!,B}(\mathbb{A}^1{}_k, \overline{\mathbb{Q}}_\ell) \to (\text{lisse } \overline{\mathbb{Q}}_\ell\text{-Sheaves on } (\mathbb{G}_m)_k)$$

carries additive ! convolution to usual tensor product of lisse sheaves on $\mathbb{G}_m$, and thus defines an equivalence of Tannakian categories. The quasi-inverse is

$$\mathcal{F} \mapsto [x \mapsto -x]^*FT_\psi(j_!\mathcal{F}(1)[1]).$$

Moreover, for any K in $\mathrm{Perv}_{A!,B}(\mathbb{A}^1{}_k, \overline{\mathbb{Q}}_\ell)$ with all ∞-slopes < 1, its "dimension", i.e., the rank of the lisse sheaf $j^*FT_\psi(K)[-1]$ on $\mathbb{G}_m$, is the generic rank of $\mathcal{H}^{-1}(K)$.

12.11 The Tannakian Category $\mathrm{Perv}_{A!,\mathrm{tame}}(\mathbb{A}^1, \overline{\mathbb{Q}}_\ell)$

(12.11.1) Denote by $\mathrm{Perv}_{A!,\mathrm{tame}}(\mathbb{A}^1{}_k, \overline{\mathbb{Q}}_\ell)$ the full subcategory of $\mathrm{Perv}_{A!,B}(\mathbb{A}^1{}_k, \overline{\mathbb{Q}}_\ell)$ consisting of those K which are everywhere tame.

(12.11.2) Denote by $\mathcal{T}$ the full subcategory of the category of all lisse $\overline{\mathbb{Q}}_\ell$-sheaves on $\mathbb{G}_m$ consisting of those objects $\mathcal{F}$ which satisfy the following two supplementary conditions:

($\mathcal{T}$1) $\mathcal{F}$ is tame at 0.

($\mathcal{T}$2) $\mathcal{F}$ as I_∞-representation is a direct sum of representations of the form $\mathcal{L}_{\psi(\alpha x)}\otimes(\text{tame})$, $\alpha \in k$.

(12.11.3) Clearly $\mathcal{T}$ is a full thick abelian subcategory of the category of all lisse $\overline{\mathbb{Q}}_\ell$-sheaves on $\mathbb{G}_m$, which is stable by tensor product. By Laumon's analysis of the local monodromy of Fourier Transforms (cf. 7.4, 7.5), the exact functor

$$j^*FT_\psi[-1] : \mathrm{Perv}_{A!,B}(\mathbb{A}^1{}_k, \overline{\mathbb{Q}}_\ell) \to (\text{lisse } \overline{\mathbb{Q}}_\ell\text{-Sheaves on } (\mathbb{G}_m)_k)$$

induces an equivalence $\mathrm{Perv}_{A!,\mathrm{tame}}(\mathbb{A}^1{}_k, \overline{\mathbb{Q}}_\ell) \approx \mathcal{T}$. Thus we obtain

Theorem 12.11.4 $\mathrm{Perv}_{A!,\mathrm{tame}}(\mathbb{A}^1{}_k, \overline{\mathbb{Q}}_\ell)$ is a Tannakian category with
(1) "tensor product" given by additive ! convolution $*_{!+}$,
(2) "unit object" $\mathbb{1}$ given by $Rj_*j^*\overline{\mathbb{Q}}_\ell(1)[1]$,
(3) "dual" given by $K^\vee := ([x \mapsto -x]^*(DK))*_{!+}(Rj_*j^*\overline{\mathbb{Q}}_\ell(1)[1])$.
The exact functor

$$j^*FT_\psi[-1] : \mathrm{Perv}_{A!,\mathrm{tame}}(\mathbb{A}^1{}_k, \overline{\mathbb{Q}}_\ell) \to \mathcal{T}$$

carries additive ! convolution to usual tensor product, and defines an equivalence of Tannakian categories. The quasi-inverse is

$$\mathcal{F} \mapsto [x \mapsto -x]^*FT_\psi(j_!\mathcal{F}(1)[1]).$$

For any K in $\mathrm{Perv}_{A!,\mathrm{tame}}(\mathbb{A}^1{}_k, \overline{\mathbb{Q}}_\ell)$, its "dimension", i.e., the rank of the lisse sheaf $j^*FT_\psi(K)[-1]$ on $\mathbb{G}_m$, is the generic rank of $\mathcal{H}^{-1}(K)$.

Remark 12.11.5 One could dually develop the $*$ version of this theory, everywhere interchanging $*_{!+}$ with $*_{*+}$, $R\pi_!$ with $R\pi_*$, $Rj_*j^*\overline{\mathbb{Q}}_\ell(1)[1]$ with $j_!j^*\overline{\mathbb{Q}}_\ell[1]$, and $j_!$ with Rj_*, to be exactly consonant with the theory developed in the $\mathcal{D}$-module context.

CHAPTER 13
Stratifications and Convolution

13.1 Generalities on Stratifications and Convolution

(13.1.1) In this section, we recall some of the basic definitions and results about stratifications (cf. [Ka-Lau, section 3], [Ka-PES, section 1], [B-B-D, 2.1, 2.2, 6.1.9]). Schemes are always understood to be separated and noetherian. A good scheme S is one which admits a map f of finite type to a scheme T which is regular of dimension at most one. [Thus a good $\mathbb{Z}[1/\ell]$-scheme is a good scheme on which ℓ is invertible, but it need not be of finite type over $\mathbb{Z}[1/\ell]$, e.g., Spec($\mathbb{C}$).] A good ring R is one whose Spec is a good scheme. For variable good schemes X, and ℓ any fixed prime number, we can speak of the triangulated categories $D^b{}_c(X[1/\ell], \overline{\mathbb{Q}}_\ell)$, which admit the full Grothendieck formalism of the "six operations" (cf [De-TF], [De-WII], [Ek], [Me-SO]).

(13.1.2) Given a good $\mathbb{Z}[1/\ell]$-scheme S, an S-scheme

$$f : X \to S$$

of finite type, and an object K in $D^b{}_c(X, \overline{\mathbb{Q}}_\ell)$, we define its S-dual $D_{X/S}(K)$ in $D^b{}_c(X, \overline{\mathbb{Q}}_\ell)$ by

$$D_{X/S}(K) := RHom(K, f^!\overline{\mathbb{Q}}_\ell).$$

We say (cf. [K-L, 1.1]) that K is S-semireflexive if the formation of $D_{X/S}(K)$ commutes with arbitrary change of base on S to a good scheme S'. We say that K is S-reflexive if both K and $D_{X/S}(K)$ are S-semireflexive. If K is reflexive, the canonical map

$$K \to D_{X/S}(D_{X/S}(K))$$

is an isomorphism (check fibre by fibre) whose formation commutes with arbitrary change of base on S to a good scheme.

(13.1.3) A stratification $\mathcal{X} := \{X_\alpha\}$ of a scheme X is a finite partition of X^{red} into a disjoint union of locally closed subschemes X_α. An object K of $D^b{}_c(X[1/\ell], \overline{\mathbb{Q}}_\ell)$ is said to be adapted to $\mathcal{X}$ if on each connected component of each stratum $X_\alpha[1/\ell]$, each of the cohomology sheaves $\mathcal{H}^i(K)$ is lisse (in the sense of corresponding to a finite-dimensional $\overline{\mathbb{Q}}_\ell$-representation of the profinite fundamental group which is definable over a finite extension of $\mathbb{Q}_\ell$).

(13.1.4) Suppose that S is a good $\mathbb{Z}[1/\ell]$-scheme, and $\pi : G \to S$ is a commutative group scheme over S which is separated and of finite type. Denote by 0_G the zero-section of G, and by $\delta_{0,\ell}$ the constant sheaf

$\overline{\mathbb{Q}}_\ell$ on the zero-section 0_G, extended by zero to all of G. Using $*_!$ convolution as the "tensor" operation, the $\overline{\mathbb{Q}}_\ell$-linear triagulated category $D^b{}_c(G[1/\ell], \overline{\mathbb{Q}}_\ell)$ is an ACU $\otimes$-category in the sense of [Saa], with $\delta_{0,\ell}$ as unit object, in which $\otimes$ is $\overline{\mathbb{Q}}_\ell$-bilinear and bi-exact ("exact" in the sense of triangulated categories).

(13.1.5) For any good scheme X, and any morphism $f : X \to S$, we denote $G_X := G\times_S X$ the pullback of G/S to X. The pullback functor

$$f^* : D^b{}_c(G[1/\ell], \overline{\mathbb{Q}}_\ell) \to D^b{}_c(G_X[1/\ell], \overline{\mathbb{Q}}_\ell)$$

is an exact $\overline{\mathbb{Q}}_\ell$-linear ACU $\otimes$-functor (the $\otimes$-compatibility because, by proper base change, the formation of $K*_!L$ commutes with change of base on S).

(13.1.6) Let $\mathcal{X}$ be a stratification of G. By [Ka-Lau, section 3], there exists an integer $N \geq 1$, a dense open set U in S[1/N], and a stratification $\mathcal{Y}$ of G_U which refines $\mathcal{X}_U$ such that

(1) If K on G_U is adapted to $\mathcal{X}_U$, then K is U-reflexive and DK is adapted to $\mathcal{Y}$ [Ka-Lau, 3.2.2 applied to G/S and $\mathcal{X}$]

(2) If K and L on G_U are adapted to $\mathcal{X}_U$, then $K*_!L$ is adapted to $\mathcal{Y}$ and its formation commutes with arbitrary change of base on S to a good scheme [Ka-Lau, 3.1.2 applied to sum : $G\times_S G \to G$ and $\mathcal{X}\times_S\mathcal{X}$]

(3) If K on G_U is adapted to $\mathcal{X}_U$, then $R\pi_!K$ and $R\pi_*K$ are lisse on U, and their formation commutes with arbitrary change of base on S to a good scheme [Ka-Lau, 3.3.3 applied to G/S and $\mathcal{X}$].

(13.1.7) So if we do it twice (i.e., apply this result to $\mathcal{Y}$ on G_U) then we get $\mathcal{Y}_2$ and U_2 such that if K and L on G_{U_2} are adapted to $\mathcal{X}$, then

(1) *RHom*(K, L) is adapted to $\mathcal{Y}_2$ and its formation commutes with arbitrary change of base on S to a good scheme [since *RHom*(K, DM) = $D(K\otimes M)$, so taking M := DL gives

$$RHom(K, L) = RHom(K, DDL) = D(K\otimes DL)],$$

(2) $R\pi_*$*RHom*(K, L) is lisse on U_2, and its formation commutes with arbitrary change of base on S to a good scheme [since*RHom*(K, L) is adapted to $\mathcal{Y}_2$ and its formation commutes with arbitrary change of base on S to a good scheme].

(13.1.8) Thus, if we start with G/S and $\mathcal{X}$, and then we have shrinking opens

$$S = U_0 \supset U_1 \supset U_2 \supset U_3 \supset U_4 \supset U_5 \ldots \supset S\otimes\mathbb{Q}$$

with each U_{i+1} open dense in $U_i[1/N_i]$ for some $N_i \geq 1$, and successively refining stratifications $\mathcal{Y}_i$ of G_{U_i}, such that if K and L on G_{U_i} are both adapted to $\mathcal{Y}_{i-2}$, then

(1) $K*_!L$ is adapted to $\mathcal{Y}_{i-1}$ and its formation commutes with arbitrary change of base on S to a good scheme.

(2) $R\pi_*RHom(K, L)$ is lisse on U_i and its formation commutes with arbitrary change of base on S to a good scheme.

(3) the derived category homs along the geometric fibres, $s \mapsto \mathrm{Hom}(K_s, L_s)$, form a local system on U_i. [Take the zeroeth cohomolology sheaf of $R\pi_*RHom(K, L)$, and apply (2).]

13.2 Interlude: Review of Elementary Stratification Facts about Normal Crossings

(13.2.1) Let $S := \mathrm{Spec}(R)$ be a good $\mathbb{Z}[1/\ell]$-scheme which is normal and connected and whose generic point has characteristic zero. Let X/S be proper and smooth, and let $D = \cup_{i\in I} D_i \subset X$ a union of R-smooth divisors D_i with relative normal crossings in X/R (e.g., $\mathbb{P}^1{}_R$, $\{0, 1, \infty\}$). Recall that there is a natural "normal crossings stratification" of X attached to this situation. For each subset S of the index set I, let

$$D_S := \cap_{i\in S} D_i, \text{ with the convention } D_\emptyset = X,$$
$$U_S := D_S - \cup_{S<T} D_T.$$

Thus D_S is proper and smooth over R, and U_S is the complement in D_S of a union of smooth divisors in D_S with normal crossings relative to S (namely the $D_S \cap D_i$ for each i **not** in S).

(13.2.2) Consider the stratification of X by the subschemes U_n, $0 \leq n \leq \mathrm{Card}(I)$,

$$U_n := \coprod_{\mathrm{Card}(S)=n} U_S.$$

The closure of U_n is

$$\overline{U}_n := \cup_{\mathrm{Card}(S)=n} D_S = \coprod_{m \geq n} U_m.$$

We denote by

$$j_n : U_n \to \overline{U}_n$$

the inclusion of U_n into $\overline{U}_n$.

(13.2.3) We endow each $\overline{U}_n$ with the stratification $\coprod_{m \geq n} U_m$. In view of the definition of each U_m as the disjoint union

$$U_m := \coprod_{\mathrm{Card}(S)=m} U_S,$$

an object K_n in $D^b{}_c(\overline{U}_n, \overline{\mathbb{Q}}_\ell)$ is adapted to the stratification $\coprod_{m \geq n} U_m$ if and only if it is adapted to the finer stratification $\coprod_{\mathrm{Card}(S) \geq n} U_S$ of $\overline{U}_n$.

Proposition 13.2.4 Hypotheses and notations as above, let $0 \leq n \leq \mathrm{Card}(I)$, and suppose K_n in $D^b{}_c(\overline{U}_n, \overline{\mathbb{Q}}_\ell)$ is adapted to the stratification $\coprod_{m \geq n} U_m$. Then

(1) $Rj_{n*}j_n{}^*K_n$ is adapted to $\coprod_{m \geq n} U_m$, and its formation commutes with arbitrary change of base on R to a good scheme.

(2) K_n is R-reflexive, and $D_{\overline{U}_n/R}(K_n)$ is adapted to the **same** stratification $\coprod_{m \geq n} U_m$.

(3) In particular, if K in $D^b{}_c(X, \overline{\mathbb{Q}}_\ell)$ is adapted to the normal crossing stratification of X, then K is reflexive and $D_{X/R}(K)$ is adapted to the **same** stratification.

proof That (1) holds for $n = 0$ is standard (cf. [Ill-ATF, A.1.1.3], [Ka-Lau, 3.4.3]). We first reduce (1) for general n to this case. The inclusion j_n of U_n into $\overline{U}_n$ factors through $\coprod_{\mathrm{Card}(S)=n} D_S$, as

$$\begin{array}{ccc} U_n = \coprod_{\mathrm{Card}(S)=n} U_S & \xrightarrow{\alpha} & \coprod_{\mathrm{Card}(S)=n} D_S \\ & & \downarrow \beta \\ & j_n \searrow & \overline{U}_n = \bigcup_{\mathrm{Card}(S)=n} D_S = \coprod_{\mathrm{Card}(T) \geq n} U_T. \end{array}$$

In this factorization, α is the disjoint union of the inclusions of U_S into D_S, and β is finite. Over each stratum U_T of the target, β is just several copies (one for each $S \subset T$ with $\mathrm{Card}(S) = n$) of the identity map. Using this factorization to compute $Rj_{n*}j_n{}^*K_n$ as $\beta_! R\alpha_*(j_n{}^*K_n)$, and applying (1) with $n=0$ to α, (1) is obvious.

To prove (2), we proceed by descending induction on n (cf. [Ka-Lau, 3.4.3]). Denote by

$$i_n : \overline{U}_{n+1} = \overline{U}_n - U_n \to \overline{U}_n$$

the inclusion. Thus i_n is a closed immersion. The "exact sequence"

$$0 \to j_{n!}j_n{}^*K_n \to K_n \to i_{n*}i_n{}^*K_n \to 0$$

gives under D a triangle

$$i_{n*} D_{\bar{U}_{n+1}/R}(i_n{}^*K_n) \to D_{\bar{U}_n/R}(K_n) \to Rj_{n*}D_{U_n/R}(j_n{}^*K_n) \to ,$$

whose first term is handled by the induction, and whose third term is handled by (1). This proves that K is R-semirelexive, and that $D_{\bar{U}_n/R}(K_n)$ is adapted to the same stratification. To get (2), simply apply this same argument to $D_{\bar{U}_n/R}(K_n)$. Assertion (3) is just the special (but most important) case n=0 of (2). QED

Corollary13.2.5 Let S := Spec(R) be a good $\mathbb{Z}[1/\ell]$-scheme which is normal and connected and whose generic point has characteristic zero. Let (X_1, D_1) and (X_2, D_2) be two normal crossing situations over R as above. Suppose K_1 in $D^b{}_c(X_1, \bar{\mathbb{Q}}_\ell)$ and K_2 in $D^b{}_c(X_2, \bar{\mathbb{Q}}_\ell)$ are adapted to the corresponding stratifications. On the fibre product $X_1 \times_R X_2$ with its "product" divisor with normal crossings $(D_1 \times_R X_2) \cup (X_1 \times_R D_2)$, consider the tensor product complex $pr_1{}^*(K_1) \otimes pr_2{}^*(K_2)$, which is adapted to the normal crossing stratification. This object is reflexive, its dual is adapted to the same stratification, and the canonical map

$$D(pr_1{}^*(K_1)) \otimes D(pr_2{}^*(K_2)) \to D(pr_1{}^*(K_1) \otimes pr_2{}^*(K_2))$$

is an isomorphism whose formation commutes with arbitrary change of base on R to a good scheme.

proof By the above 13.2.4 applied to the product, the above morphism, source and target are all of formation compatible with arbitrary change of base on R to a good scheme, so it suffices to check over geometric points of R, where it becomes the compatibility of duality with products. QED

Corollary 13.2.6 Let S :=Spec(R) be a good $\mathbb{Z}[1/\ell]$-scheme which is normal and connected and whose generic point has characteristic zero. Let (X, D) be a normal crossing situations over R as above. Let U be an open subscheme of X whose closed complement Z is a partial union of the subschemes D_S. Denote by $\mathcal{U}$ the stratification $\mathcal{U} := \coprod$(those U_T in U) of U. Then

(1) If K in $D^b{}_c(U, \bar{\mathbb{Q}}_\ell)$ is adapted to $\mathcal{U}$, then K is R-reflexive and $D_{U/R}(K)$ is adapted to $\mathcal{U}$.

(2) If K and L in $D^b{}_c(U, \bar{\mathbb{Q}}_\ell)$ are each adapted to $\mathcal{U}$, then on $U \times_R U$, $pr_1{}^*(K) \otimes pr_2{}^*(L)$ is reflexive, it and its dual are adapted to $\mathcal{U} \times \mathcal{U}$, and the canonical map

$$D(pr_1{}^*(K)) \otimes D(pr_2{}^*(L)) \to D(pr_1{}^*(K) \otimes pr_2{}^*(L))$$

is an isomorphism whose formation commutes with arbitrary change of base on R to a good scheme.

proof Let $j:U \to X$ denote the inclusion. Then $j_!K$ on X is adapted to the normal crossing stratification of X. Since j is open, we have

$$j^*D_{X/R}(j_!K) = D_{U/R}(j^!j_!K) = D_{U/R}(j^*j_!K) = D_{U/R}(K),$$

so (1) follows from 13.2.4 (3). Similarly, (2) follows from the above Corollary. QED

13.3 The special case of $\mathbb{A}^1$

(13.3.1) The general G/S discussion above has the merit of applying quite generally, but the attendant disadvantage of not being very explicit. However, in the special case when G is $\mathbb{A}^1$, the theory becomes very explicit. [It does so also if G is either $\mathbb{G}_m$ or an elliptic curve E over S. The common element is that G is, as a scheme, the complement in a proper smooth S-curve $\overline{G}$ of a disjoint union Z of sections. We leave to the reader the task of adapting the following discussion of $\mathbb{A}^1$ to these cases.] This allows us to apply the normal crossing results of the previous section.

(13.3.2) For any ring R, we denote by $\mathbb{A}^1{}_R := \mathrm{Spec}(R[x])$ the affine line over R.

(13.3.3) Fix a prime number ℓ, and an isomorphism of fields

$$\iota:\overline{\mathbb{Q}}_\ell \approx \mathbb{C}.$$

Let R be a subring of $\mathbb{C}$ which is a finitely generated $\mathbb{Z}[1/\ell]$-algebra. Let K be an object in $D^b{}_c(\mathbb{A}^1{}_R, \overline{\mathbb{Q}}_\ell)$. At the expense of replacing R by R[1/r] for some nonzero element $r \in R$, we may further assume that R is normal, and that K is adapted to the stratification $(\mathbb{A}^1{}_R - D, D)$ where $D \subset \mathbb{A}^1{}_R$ is a divisor which is finite etale over R of some degree $d \geq 1$, defined by a monic polynomial $f(x) \in R[x]$ of degree d whose discriminant Δ is a unit in R.

Recall that in this situation, we have:

Proposition 13.3.4 ([Ka-Lau,3.4]) Let R be a normal integral domain in which ℓ is invertible, whose fraction field has characteristic zero, and such that $S := \mathrm{Spec}(R)$ is a good scheme. Let $D \subset \mathbb{A}^1{}_R$ be a divisor which is finite etale over R of some degree $d \geq 1$, defined by a monic polynomial $f(x) \in R[x]$ of degree d whose discriminant Δ is a unit in R. Suppose that K in $D^b_c(\mathbb{A}^1_R, \overline{\mathbb{Q}}_\ell)$ is adapted to the stratification

$$(\mathbb{A}^1_R - D, D).$$

Denote by $\pi : X := \mathbb{A}^1_R \to S := \mathrm{Spec}(R)$ the structural morphism. Then

(1) For $j: \mathbb{A}^1_R - D \to \mathbb{A}^1_R$ the inclusion, Rj_*j^*K is adapted to the stratification $(\mathbb{A}^1_R - D, D)$, of formation compatible with arbitrary change of base on S to a good scheme.

(2) $R\pi_!K$ is lisse on S, of formation compatible with arbitrary change of base on S to a good scheme.

(3) K is S-reflexive, and $D_{X/S}(K)$ is adapted to the **same** stratification $(\mathbb{A}^1_R - D, D)$.

(4) $R\pi_*K$ is lisse on S, of formation compatible with arbitrary change of base on S to a good scheme, and we have canonical isomorphisms

$$D(R\pi_*K) \approx R\pi_!DK, \quad D(R\pi_!K) \approx R\pi_*DK,$$

of formation compatible with arbitrary change of base on S to a good scheme.

(5) For any algebraically closed field k, and any ring homomorphism $\varphi: R \to k$, the inverse image K_φ of K in $D^b_c(\mathbb{A}^1_k, \overline{\mathbb{Q}}_\ell)$ is adapted to the stratification $(\mathbb{A}^1_k - D_k, D_k)$, and K_φ is tamely ramified at all points of $D_k \cup \{\infty\}$.

Remark 13.3.5 The key point of the above proposition is that "duality costs us nothing" for objects K adapted to such a stratification. The earlier discussion (13.1) thus gives:

Corollary 13.3.6 Hypotheses and notations as in 13.3.4 above, suppose that K and L in $D^b_c(\mathbb{A}^1_R, \overline{\mathbb{Q}}_\ell)$ are both adapted to the stratification $(\mathbb{A}^1_R - D, D)$. Then

(1) $RHom(K, L) = D(K \otimes DL)$ is adapted to the stratification $(\mathbb{A}^1_R - D, D)$, and of formation compatible with arbitrary change of base on S to a good scheme.

(2) $R\pi_*RHom(K, L)$ is lisse on S, of formation compatible with arbitrary change of base on S to a good scheme.

(3) The zeroeth cohomology sheaf $\mathcal{H}^0(R\pi_*RHom(K, L))$ is lisse on S, and

its fibre at each geometric point s in S is the $\overline{\mathbb{Q}}_\ell$-space of hom's from K_s to L_s in $D^b_c(\mathbb{A}^1_s, \overline{\mathbb{Q}}_\ell)$.

(13.3.7) In order to state the next result, it will be convenient to introduce the following temporary notation. For any good R-scheme T, denote by

$$\mathcal{C}(T) \subset D^b_c(\mathbb{A}^1_T, \overline{\mathbb{Q}}_\ell)$$

the strictly full subcategory of $D^b_c(\mathbb{A}^1_T, \overline{\mathbb{Q}}_\ell)$ consisting of those objects which are adapted to the stratification $(\mathbb{A}^1_T - D_T, D_T)$, and whose restriction to each geometric fibre $\mathbb{A}^1_t$ is tamely ramified at all points of $D_t \cup \{\infty\}$.

(13.3.8) Notice that $\mathcal{C}(T)$ is a triangulated subcategory of $D^b_c(\mathbb{A}^1_T, \overline{\mathbb{Q}}_\ell)$; if two vertices of a triangle lie in $\mathcal{C}(T)$, then so does the third, as is immediate from the long exact sequence of cohomology sheaves.

By applying part (3) of the above result 13.3.6, we find

Corollary 13.3.9 (compare [B-B-D, 6.1.9]) Hypotheses and notations as in 13.3.4 above, suppose in addition that R is a strictly henselian local ring. Let s and $\overline{\eta}$ be geometric points of S := Spec(R) which lie over the special and generic points respectively. Then the natural inverse image functors

$$\mathcal{C}(S) \rightarrow \mathcal{C}(\overline{\eta}), \quad \mathcal{C}(S) \rightarrow \mathcal{C}(s)$$

are fully faithful.

By applying part (3) of the above result 13.3.6 to K and to all the shifts L[i] of L, we find the more pecise result:

Corollary 13.3.10 (compare [B-B-D, 6.1.9]) Hypotheses and notations as above, suppose in addition that R is a strictly henselian discrete valuation ring. Let s and $\overline{\eta}$ be geometric points of S := Spec(R) which lie over the special and generic points respectively. Then the natural inverse image functors

$$\mathcal{C}(S) \rightarrow \mathcal{C}(\overline{\eta}), \quad \mathcal{C}(S) \rightarrow \mathcal{C}(s)$$

are fully faithful, and the second is an equivalence of categories.

proof That the functors are fully faithful is the content of the previous corollary. It remains to explain why the second is essentially surjective. We proceed by induction on the amplitude

$$\mathrm{ampl}(K) := \sup\{i \mid \mathcal{H}^i(K) \neq 0\} - \inf\{i \mid \mathcal{H}^i(K) \neq 0\}.$$

If $\mathcal{H}^i(K) = 0$ for $i > n$, then we have a triangle

$$\tau_{<n}K \to K \to \mathcal{H}^n(K)[-n] \to (\tau_{<n}K)[1],$$

so K is a cone of the morphism $\mathcal{H}^n(K)[-n] \to (\tau_{<n}K)[1]$ between objects of lower amplitude.

It remains to treat the objects of amplitude zero. Any such is a shift of a sheaf $\mathcal{F}$ which is lisse and tame on $\mathbb{A}^1 - D$, and lisse on D. If we denote by $j: \mathbb{A}^1 - D \to \mathbb{A}^1$ and $i: D \to \mathbb{A}^1$ the inclusions, then $\mathcal{F}$ sits in the short exact sequence

$$0 \to j_!j^*\mathcal{F} \to \mathcal{F} \to i_*i^*\mathcal{F} \to 0,$$

so we recover $\mathcal{F}$ as the cone of the morphism $i_*i^*\mathcal{F} \to j_!j^*\mathcal{F}[1]$.

For objects of the form $i_*i^*\mathcal{F}$ the asserted equivalence is obvious: they are simply the constant sheaves on D, extended by zero. For objects of the form $j_!j^*\mathcal{F}$, the asserted equivalence results from Grothendieck's theory of the tame fundamental group (cf. [SGA I, XIII, Thm 2.4, 1] and the explication at the end the proof the Tame Specialization Theorem bis 8.17.14). QED

13.4 Location of the Singularities of a convolution

(13.4.1) In this section, we make precise the rough idea that "singularities add under convolution".

Proposition 13.4.2 Let R be a normal integral domain in which ℓ is invertible, whose fraction field has characteristic zero, and such that S := Spec(R) is a good scheme. Suppose that D_1, D_2, and D_3 are three divisors $D_i \subset \mathbb{A}^1{}_R$ each of which is finite etale over R of some degree $d_i \geq 1$, defined by a monic polynomial $f_i(x) \in R[x]$ of degree d_i whose discriminant Δ_i is a unit in R. Suppose that

$$(D_1 + D_2)^{\mathrm{red}} \subset D_3,$$

i.e., for every algebraically closed field k and every ring homomorphism $\varphi : R \to k$, if $\alpha_1, \alpha_2 \in k$ then

$$f_3(\alpha_1 + \alpha_2) = 0 \text{ if } f_1(\alpha_1) = 0 = f_2(\alpha_2) = 0.$$

Suppose that for $i = 1, 2$, K_i in $D^b{}_c(\mathbb{A}^1{}_R, \overline{\mathbb{Q}}_\ell)$ is adapted to the stratification $(\mathbb{A}^1{}_R - D_i, D_i)$. Then

(1) The additive ! convolution $K_1 *_{!+} K_2$ in $D^b{}_c(\mathbb{A}^1{}_R, \overline{\mathbb{Q}}_\ell)$ is adapted to the stratification $(\mathbb{A}^1{}_R - D_3, D_3)$, and its formation commutes with arbitrary change of base on Spec(R) to a good scheme.

(2) We have a canonical isomorphism $D(K_1 *_{!+} K_2) \approx DK_1 *_{*+} DK_2$, whose formation commutes with arbitrary change of base on Spec(R) to a good scheme.

(3) The additive * convolution $K_1 *_{*+} K_2$ in $D^b{}_c(\mathbb{A}^1{}_R, \overline{\mathbb{Q}}_\ell)$ is adapted to the stratification $(\mathbb{A}^1{}_R - D_3, D_3)$, and its formation commutes with arbitrary change of base on Spec(R) to a good scheme.

(4) We have a canonical isomorphism $D(K_1 *_{*+} K_2) \approx DK_1 *_{!+} DK_2$, whose formation commutes with arbitrary change of base on Spec(R) to a good scheme.

proof Making a finite etale base change on R, we may assume that all the divisors D_i are finite unions of disjoint sections, say $D_1 = \{a_i\}$, $D_2 = \{b_j\}$, $D_3 = \{c_k\}$. The hypothesis that $D_1 + D_2 \subset D_3$ is the the statement that each $a_i + b_j$ is a c_k.

We first prove that both convolutions are lisse on $U := \mathbb{A}^1{}_R - D_3$. Over U, the sum map becomes pr_2: $\mathbb{A}^1 \times_R U \to U$ with coordinates $(x,y) \mapsto y$, the source endowed with the object $K_1(x) \otimes K_2(y-x)$. This object is adapted to $(\mathbb{A}^1{}_U - D, D)$, for D the divisor $D_1 \amalg (y - D_2)$, which is finite etale over U of degree $d_1 + d_2$, defined by the $(\pm)$monic polynomial $f(x) := f_1(x)f_2(y - x)$. So this is just 13.3.4, (2) and (4).

To see that $K_1 *_{!+} K_2$ is lisse on D_3, we argue as follows. Since D_3 is the disjoint union of sections $\{c_k\}$, we may, by translation, suppose that $\{0\}$ is one of these sections. We must then show that $K_1 *_{!+} K_2$ is lisse along the zero section. By proper base change, $(K_1 *_{!+} K_2)|\{0\}$ is $R\pi_!(K_1 \otimes [x \mapsto -x]^* K_2)$. But on $\mathbb{A}^1{}_R$, $K_1 \otimes [x \mapsto -x]^* K_2$ is adapted to the stratification $(\mathbb{A}^1{}_R - D, D)$, for D union of the **disjoint** sections, each taken with multiplicity one, given by $\{a_i\} \cup \{-b_j\}$. So again we may apply 13.3.4 (2).

The base-change statement for $K_1 *_{!+} K_2$ is a special case of proper base change. This concludes the proof of (1).

Statement (2) is Verdier duality for the "sum" map, in the form $D \circ R\mathrm{sum}_! = R\mathrm{sum}_* \circ D$, together with 13.2.6 (2) applied to the situation $U = \mathbb{A}^1$, $X = \mathbb{P}^1$.

Once we have (2), apply it to the DK_i. Since $K_i \approx DDK_i$, this gives

$$D(DK_1 *_{!+} DK_2) \approx K_1 *_{*+} K_2,$$

which makes the (3) obvious. Assertion (4) is obtained by applying D to this, and recalling that $DK_1 *_{!+} DK_2$ is reflexive, because it is adapted to $(\mathbb{A}^1_R - D_3, D_3)$. QED

In the special case when D_2 is the zero section alone, i.e., when $f_2(x) = x$, then we can take $D_3 = D_1$, and we obtain:

Corollary 13.4.3 Let R be a normal integral domain in which ℓ is invertible, whose fraction field has characteristic zero, and such that $S := \mathrm{Spec}(R)$ is a good scheme. Suppose that $D \subset \mathbb{A}^1_R$ is a divisor which is finite etale over R of some degree $d \geq 1$, defined by a monic polynomial $f(x) \in R[x]$ of degree d whose discriminant Δ is a unit in R. Suppose that

K in $D^b_c(\mathbb{A}^1_R, \overline{\mathbb{Q}}_\ell)$ is adapted to the stratification $(\mathbb{A}^1_R - D, D)$,

L in $D^b_c(\mathbb{A}^1_R, \overline{\mathbb{Q}}_\ell)$ is adapted to the stratification $((\mathbb{G}_m)_R, \{0_R\})$.

Then

(1) The additive ! and * convolutions $K *_{!+} L$ and $K *_{*+} L$ in $D^b_c(\mathbb{A}^1_R, \overline{\mathbb{Q}}_\ell)$ are both adapted to the stratification $(\mathbb{A}^1_R - D, D)$, and of formation compatible with arbitrary change of base on Spec(R) to a good scheme.

(2) We have canonical isomorphisms

$$D(K *_{!+} L) \approx DK *_{*+} DL, \quad D(K *_{*+} L) \approx DK *_{!+} DL,$$

whose formation commutes with arbitrary change of base on Spec(R) to a good scheme.

13.5 The bookkeeping of iterated convolution

(13.5.1) Let R be a normal integral domain in which ℓ is invertible, whose fraction field has characteristic zero, and such that $S := \mathrm{Spec}(R)$

is a good scheme. Let $D \subset \mathbb{A}^1_R$ be a divisor which is finite etale over R of some degree $d \geq 1$, defined by a monic polynomial $f(x) \in R[x]$ of degree d whose discriminant Δ is a unit in R. At the expense of replacing R by a finite etale overring, we may suppose that D is a disjoint union of sections, i.e., that the polynomial f(x) splits completely in R, say $f(x) = \prod(x - a_i)$ with d distinct roots a_i. That Δ be a unit in R is the condition that for any two distinct roots $a_i \neq a_j$, the difference $a_i - a_j$ be a unit in R.

(13.5.2) Given any finite nonempty subset A of R, define its discriminant $\Delta(A) \in R - \{0\}$ by

$$\Delta(A) := \prod_{\alpha \neq \beta \text{ in } A}(\alpha - \beta),$$

with the empty product convention that if A consists of a single element, then $\Delta(A) = 1$. Denote by $f_A(x) \in R[x]$ the monic polynomial

$$f_A(x) := \prod_{\alpha \in A} (x - \alpha).$$

Then $f_A(x)$ defines a divisor D(A) in $\mathbb{A}^1_R$ which is finite flat over R of degree Card(A), and which is finite etale precisely over $R[1/\Delta(A)]$.

(13.5.3) Given two finite nonempty subsets A and B of R, we denote by A+B the finite nonempty subset of R consisting of all the sums a+b with a in A and b in B.

In this language, the previous proposition becomes

Proposition 13.5.4 Let R be a normal integral domain in which ℓ is invertible, whose fraction field has characteristic zero, and such that S := Spec(R) is a good scheme. Suppose that A and B are finite nonempty subsets of R, and that

$$\Delta(A)\Delta(B)\Delta(A+B) \text{ is a unit in R.}$$

Suppose that

K in $D^b_c(\mathbb{A}^1_R, \overline{\mathbb{Q}}_\ell)$ is adapted to $(\mathbb{A}^1_R - D(A), D(A))$,

L in $D^b_c(\mathbb{A}^1_R, \overline{\mathbb{Q}}_\ell)$ is adapted to $(\mathbb{A}^1_R - D(B), D(B))$.

Then

(1) The additive ! and * convolutions $K*_{!+}L$ and $K*_{*+}L$ in $D^b_c(\mathbb{A}^1_R, \overline{\mathbb{Q}}_\ell)$ are both adapted to the stratification $(\mathbb{A}^1_R - D(A+B), D(A+B))$, and of formation compatible with arbitrary change of base on Spec(R) to a good scheme.

(2) We have canonical isomorphisms

$$D(K_1*_{!+}K_2) \approx DK_1*_{*+}DK_2,\ D(K_1*_{*+}K_2) \approx DK_1*_{!+}DK_2,$$

whose formation commutes with arbitrary change of base on Spec(R) to a good scheme.

(13.5.5) To keep track of multiple convolutions, we formulate explicitly the following proposition, which follows immediately from the previous result by induction.

Proposition 13.5.6 Let R be a normal integral domain in which ℓ is invertible, whose fraction field has characteristic zero, and such that S := Spec(R) is a good scheme. Suppose that A is a finite nonempty subset of R. Define a sequence of subsets A_i, $i \geq 0$, of R as follows:

$$A_0 := \{0\}$$

$$A_1 := A$$

$$A_{i+1} := A_i \cup (A_1 + A_i) \text{ for } i \geq 1.$$

Their discriminants satisfy

$$1 = \Delta(A_0) \mid \Delta(A_1) \mid \Delta(A_2) \mid \Delta(A_3) \mid \Delta(A_4) \mid \dots .$$

Let $R_i := R[1/\Delta(A_i)]$, $U_i := \mathrm{Spec}(R_i)$. Thus

$$R = R_0 \subset R_1 \subset R_2 \subset R_3 \subset R_4 \subset \dots$$

$$\mathrm{Spec}(R) = U_0 \supset U_1 \supset U_2 \supset U_3 \supset U_4 \supset \dots .$$

Let $m \geq 0$ be an integer. Suppose given finitely many objects $K_1, \dots, K_r$ in $D^b{}_c(\mathbb{A}^1{}_{R_m}, \overline{\mathbb{Q}}_\ell)$, and for each i in [1, r] an integer $n(i) \geq 0$ such that

$\Sigma_i\, n(i) \leq m$,

for each i, K_i is adapted to $(\mathbb{A}^1{}_{R_m} - D(A_{n(i)}), D(A_{n(i)}))$.

Then

(1) The multiple ! and $*$ additive convolutions $K_1 *_{!+} K_2 *_{!+} \dots *_{!+} K_r$ and $K_1 *_{*+} K_2 *_{*+} \dots *_{*+} K_r$ are both adapted to $(\mathbb{A}^1{}_{R_m} - D(A_m), D(A_m))$, and of formation compatible with arbitrary change of base on $\mathrm{Spec}(R_m)$ to a good scheme.

(2) We have canonical isomorphisms

$$D(K_1 *_{!+} K_2 *_{!+} \dots *_{!+} K_r) \approx DK_1 *_{*+} DK_2 *_{*+} \dots *_{*+} DK_r,$$

$$D(K_1 *_{*+} K_2 *_{*+} \dots *_{*+} K_r) \approx DK_1 *_{!+} DK_2 *_{!+} \dots *_{!+} DK_r,$$

whose formation commutes with arbitrary change of base on $\mathrm{Spec}(R_m)$ to a good scheme.

13.5.7 Examples and remarks

(1) If A is the subset {0, 1}, then A_n is the subset {0, 1, ... , n}, and R_n is obtained from R by inverting all the prime numbers which are $\leq n$.

(2) If A is the subset {-1, 0, 1}, then A_n is the subset {-n, 1-n, ... , n-1,

n), and R_n is obtained from R by inverting all the prime numbers which are $\le 2n$.

(3) If R contains the d'th roots of unity for some $d \ge 3$, and A is μ_d, then $\Delta(A_n)$ is a nonzero integer (since the set A is $\mathbb{Q}$-rational). It is nontrivial to give a closed formula for the cardinality of A_n, much less to specify which primes divide $\Delta(A_n)$. Indeed the discussion of "exceptional primes" in 7.1 is basically the study of $\Delta(A_2)$.

(4) Suppose that, rather than starting with a finite nonempty subset A of R, we start with divisor D in $\mathbb{A}^1_R$ which is finite flat over R of some degree $d \ge 1$, defined by a monic polynomial f(x) in R[x] of degree d. Then in the integral closure R' of R in some finite galois extension E of the fraction field F of R, we can factor f(x) completely. Denote by $A \subset R'$ the finite subset consisting of the distinct roots of f. Each of the sets $A_n \subset R'$ is Galois-stable, so the polynomials

$$f_n(x) := \prod_{\alpha \text{ in } A_n} (x - \alpha)$$

have coefficients in R, and the quantities $\Delta(A_n)$ and their successive ratios $\Delta(A_{n+1})/\Delta(A_n)$ all lie in R. Therefore the successive localizations $R_n := R[1/\Delta(A_n)]$ make sense, and the divisors D_n defined by the polynomial f_n is finite etale over R_n. The previous proposition remains true if in its statement we replace "$D(A_i)$" by "D_i" throughout.

13.6 Various fibre-wise categories

(13.6.1) Let R be a normal integral domain in which ℓ is invertible, whose fraction field has characteristic zero, and such that S := Spec(R) is a good scheme. We denote by

$$D^b_{c,A!}(\mathbb{A}^1_R, \overline{\mathbb{Q}}_\ell) \subset D^b_c(\mathbb{A}^1_R, \overline{\mathbb{Q}}_\ell)$$

the full subcategory of $D^b_c(\mathbb{A}^1_R, \overline{\mathbb{Q}}_\ell)$ consisting of those objects which satisfy $R\pi_!K = 0$. We denote by

$$D^b_{c,\mathrm{tame}}(\mathbb{A}^1_R, \overline{\mathbb{Q}}_\ell) \subset D^b_c(\mathbb{A}^1_R, \overline{\mathbb{Q}}_\ell)$$

the full subcategory of $D^b_c(\mathbb{A}^1_R, \overline{\mathbb{Q}}_\ell)$ consisting of those objects whose restriction to every geometric fibre of $\pi : \mathbb{A}^1_R \to S = \mathrm{Spec}(R)$ is everywhere tame. We define

$$D^b_{c,A!,\mathrm{tame}}(\mathbb{A}^1_R, \overline{\mathbb{Q}}_\ell) := D^b_{c,A!}(\mathbb{A}^1_R, \overline{\mathbb{Q}}_\ell) \cap D^b_{c,\mathrm{tame}}(\mathbb{A}^1_R, \overline{\mathbb{Q}}_\ell),$$

as full subcategory of $D^b{}_c(\mathbb{A}^1{}_R, \overline{\mathbb{Q}}_\ell)$.

We denote by

$$\mathrm{Perv}(\mathbb{A}^1{}_R, \overline{\mathbb{Q}}_\ell) \subset D^b{}_c(\mathbb{A}^1{}_R, \overline{\mathbb{Q}}_\ell)$$

the full subcategory of $D^b{}_c(\mathbb{A}^1{}_R, \overline{\mathbb{Q}}_\ell)$ consisting of those objects whose restriction to every geometric fibre of $\pi : \mathbb{A}^1{}_R \to S = \mathrm{Spec}(R)$ is perverse. We define

$$\mathrm{Perv}_{A!}(\mathbb{A}^1{}_R, \overline{\mathbb{Q}}_\ell) := \mathrm{Perv}(\mathbb{A}^1{}_R, \overline{\mathbb{Q}}_\ell) \cap D^b{}_{c,A!}(\mathbb{A}^1{}_R, \overline{\mathbb{Q}}_\ell),$$
$$\mathrm{Perv}_{\mathrm{tame}}(\mathbb{A}^1{}_R, \overline{\mathbb{Q}}_\ell) := \mathrm{Perv}(\mathbb{A}^1{}_R, \overline{\mathbb{Q}}_\ell) \cap D^b{}_{c,\mathrm{tame}}(\mathbb{A}^1{}_R, \overline{\mathbb{Q}}_\ell),$$
$$\mathrm{Perv}_{A!,\mathrm{tame}}(\mathbb{A}^1{}_R, \overline{\mathbb{Q}}_\ell) := \mathrm{Perv}(\mathbb{A}^1{}_R, \overline{\mathbb{Q}}_\ell) \cap D^b{}_{c,A!,\mathrm{tame}}(\mathbb{A}^1{}_R, \overline{\mathbb{Q}}_\ell),$$

as full subcategories of $D^b{}_c(\mathbb{A}^1{}_R, \overline{\mathbb{Q}}_\ell)$.

(13.6.2) For $j : (\mathbb{G}_m)_R \to \mathbb{A}^1{}_R$ the inclusion, the object $Rj_*j^*\overline{\mathbb{Q}}_\ell(1)[1]$ of $D^b{}_c(\mathbb{A}^1{}_R, \overline{\mathbb{Q}}_\ell)$ sits in a canonical distinguished triangle

$$\overline{\mathbb{Q}}_\ell(1)[1] \to Rj_*j^*\overline{\mathbb{Q}}_\ell(1)[1] \to \delta_{0,\ell} \to.$$

So for any object K in $D^b{}_c(\mathbb{A}^1{}_R, \overline{\mathbb{Q}}_\ell)$, we have a distinguished triangle of additive ! convolutions

$$K*_{!+}\overline{\mathbb{Q}}_\ell(1)[1] \to K*_{!+}Rj_*j^*\overline{\mathbb{Q}}_\ell(1)[1] \to K \to.$$

On the other hand, by proper base change, we have a canonical isomorphism

$$\pi^*R\pi_!(K)(1)[1] \approx K*_{!+}\overline{\mathbb{Q}}_\ell(1)[1].$$

Combining these, we obtain a distinguished triangle

$$\pi^*R\pi_!(K)(1)[1] \to K*_{!+}Rj_*j^*\overline{\mathbb{Q}}_\ell(1)[1] \to K \to,$$

functorial in K.

Proposition 13.6.3 Notations as in 13.6.1 above, we have:

(1) The object $Rj_*j^*\overline{\mathbb{Q}}_\ell(1)[1]$ of $D^b{}_c(\mathbb{A}^1{}_R, \overline{\mathbb{Q}}_\ell)$ lies in $\mathrm{Perv}_{A!,\mathrm{tame}}(\mathbb{A}^1{}_R, \overline{\mathbb{Q}}_\ell)$.

(2) For any object K of $D^b{}_c(\mathbb{A}^1{}_R, \overline{\mathbb{Q}}_\ell)$, the additive ! convolution

$$K*_{!+}Rj_*j^*\overline{\mathbb{Q}}\ell(1)[1]$$

lies in $D^b{}_{c,A!}(\mathbb{A}^1{}_R, \overline{\mathbb{Q}}_\ell)$.

(3) The canonical map

$$K*_{!+}Rj_*j^*\overline{\mathbb{Q}}_\ell(1)[1] \to K$$

is an isomorphism if and only if K lies in $D^b{}_{c,A!}(\mathbb{A}^1{}_R, \overline{\mathbb{Q}}_\ell)$. The operator $K \mapsto K*_{!+}Rj_*j^*\overline{\mathbb{Q}}_\ell(1)[1]$ on $D^b{}_c(\mathbb{A}^1{}_R, \overline{\mathbb{Q}}_\ell)$ is idempotent; it is the projector onto $D^b{}_{c,A!}(\mathbb{A}^1{}_R, \overline{\mathbb{Q}}_\ell)$.

(4) $D^b{}_{c,A!}(\mathbb{A}^1{}_R, \overline{\mathbb{Q}}_\ell)$ is a $\overline{\mathbb{Q}}_\ell$-linear triangulated category. Using additive ! convolution as the "tensor" operation, $D^b{}_{c,A!}(\mathbb{A}^1{}_R, \overline{\mathbb{Q}}_\ell)$ is an ACU ⊗-category in the sense of [Saa], with $Rj_*j^**\overline{\mathbb{Q}}_\ell(1)[1]$ as unit object, in which ⊗ is $\overline{\mathbb{Q}}_\ell$-bilinear and bi-exact ("exact" in the sense of triangulated categories).

(5) The subcategory $D^b{}_{c,A!,tame}(\mathbb{A}^1{}_R, \overline{\mathbb{Q}}_\ell)$ of $D^b{}_{c,A!}(\mathbb{A}^1{}_R, \overline{\mathbb{Q}}_\ell)$ is a $\overline{\mathbb{Q}}_\ell$-linear triagulated subcategory, stable under additive ! convolution, so itself forms an ACU ⊗-category in the sense of [Saa], with $Rj_*j^**\overline{\mathbb{Q}}_\ell(1)[1]$ as unit object, in which ⊗ is $\overline{\mathbb{Q}}_\ell$-bilinear and bi-exact ("exact" in the sense of triangulated categories).

(6) The $\overline{\mathbb{Q}}_\ell$-linear additive category $\mathrm{Perv}_{A!,tame}(\mathbb{A}^1{}_R, \overline{\mathbb{Q}}_\ell)$ is stable under additive ! convolution. With this as ⊗-operation, it forms an ACU ⊗-category in the sense of [Saa], with $Rj_*j^*\overline{\mathbb{Q}}_\ell(1)[1]$ as unit object, in which ⊗ is $\overline{\mathbb{Q}}_\ell$-bilinear

(7)The $\overline{\mathbb{Q}}_\ell$-linear additive category $\mathrm{Perv}_{tame}(\mathbb{A}^1{}_R, \overline{\mathbb{Q}}_\ell)$ is a $\overline{\mathbb{Q}}_\ell$-linear additive category which is **not** stable under additive ! convolution, e.g., $\overline{\mathbb{Q}}_\ell[1]*_{!+}\overline{\mathbb{Q}}_\ell[1]$ has $\mathcal{H}^0 \approx \overline{\mathbb{Q}}_\ell(-1)$, so is not perverse. However, for K in $\mathrm{Perv}_{tame}(\mathbb{A}^1{}_R, \overline{\mathbb{Q}}_\ell)$, the additive ! convolution $K*_{!+}Rj_*j^*\overline{\mathbb{Q}}_\ell(1)[1]$ lies in $\mathrm{Perv}_{A!,tame}(\mathbb{A}^1{}_R, \overline{\mathbb{Q}}_\ell)$. The operator $K \mapsto K*_{!+}Rj_*j^*\overline{\mathbb{Q}}_\ell(1)[1]$ on $\mathrm{Perv}_{tame}(\mathbb{A}^1{}_R, \overline{\mathbb{Q}}_\ell)$ is idempotent; it is the projector onto $\mathrm{Perv}_{A!,tame}(\mathbb{A}^1{}_R, \overline{\mathbb{Q}}_\ell)$.

proof The formation of $Rj_*j^*\overline{\mathbb{Q}}_\ell(1)[1]$ commutes with arbitrary change of base on S = Spec(R) to a good scheme, so it suffices to check (1) when R is an algebraically closed field, in which case it is 12.10.4.

For (2), the Kunneth formula gives

$$R\pi_!(K*_{!+}Rj_*j^*\overline{\mathbb{Q}}_\ell(1)[1]) \approx R\pi_!(K)\otimes R\pi_!(Rj_*j^*\overline{\mathbb{Q}}_\ell(1)[1])$$

$$= R\pi_!(K)\otimes 0 = 0.$$

Assertion (3) is obvious from the exact triangle

$$\pi^* R\pi_!(K)(1)[1] \to K *_{!+} Rj_* j^* \overline{\mathbb{Q}}_\ell(1)[1] \to K \to.$$

Assertion (4) is simply the projection onto $D^b{}_{c,A!}(\mathbb{A}^1{}_R, \overline{\mathbb{Q}}_\ell)$ of the corresponding $\otimes$-structure on $D^b{}_c(\mathbb{A}^1{}_R, \overline{\mathbb{Q}}_\ell)$, with $\delta_{0,\ell}$ as unit object (cf. 13.1.4).

Assertion (5) amounts to the statement that over an algebraically closed field, additive ! convolution preserves tameness. This is vacuous in characteristic zero. In characteristic $p \neq \ell$ it is proven by Fourier Transform, where it becomes the stability under usual tensor product of objects M in $D^b{}_c(\mathbb{G}_m, \overline{\mathbb{Q}}_\ell)$ whose cohomology sheaves are all lisse on $\mathbb{G}_m$ and lie in $\mathcal{T}$ (cf. 12.11 for the perverse version).

Assertion (6) results formally from (5), once we show that $\mathrm{Perv}_{A!,\mathrm{tame}}(\mathbb{A}^1{}_R, \overline{\mathbb{Q}}_\ell)$ is stable under additive ! convolution. For this we are reduced to the case when R is an algebraically closed field k. In positive characteristic, this is 12.11.4. In characteristic zero, we argue as follows. If there exists an embedding of k into $\mathbb{C}$, then the required stability is 12.7.5. In the general case, we may reduce to this case because any finite collection of objects K_i in $\mathrm{Perv}_{A!,\mathrm{tame}}(\mathbb{A}^1{}_k, \overline{\mathbb{Q}}_\ell)$ is definable over an algebraically closed subfield k_0 of k of finite transcendence degree over $\mathbb{Q}$. Indeed, if we pick a single stratification $(\mathbb{A}^1{}_k - D, D)$, D a finite subet of $k = \mathbb{A}^1{}_k(k)$, to which all the K_i are adapted, then we may descend all the K_i to an algebraic closure of $\overline{\mathbb{Q}}(D)$.

To see this, we argue as follows. Since k has characteristic zero, "tame" is vacuous, and any descents of the K_i as perverse objects will automatically satisfy condition A! (proper base change for $R\pi_!$ under extension of algebraically closed field). So it suffices to descend a finite collection of objects K_i in $\mathrm{Perv}(\mathbb{A}^1{}_k, \overline{\mathbb{Q}}_\ell)$. Now these are successive extensions of perverse simple objects; because we are in characteristic zero, smooth base change assures us that the Ext groups in question are invariant under (necessarily separable) extension of algebraically closed field. So we are reduced to descending finitely many perverse simple objects, all adapted to a common stratification. In the case of a δ-module δ_α, we have a visible descent to $\overline{\mathbb{Q}}(\alpha)$. In the case of the middle extension of an irreducible local system on $\mathbb{A}^1{}_k - D$, the invariance of $\pi_1(\mathbb{A}^1{}_k - D)$ under extension of algebraically closed fields of characteristic zero shows that we have a descent to any algebraically closed overfield of $\overline{\mathbb{Q}}(D)$. This concludes the proof of (6).

Assertion (7) results from(3), once we know the result on the geometric fibres. In positive characteristic, this is 12.10.5. In characteristic zero, we reduce as above to the case when k embeds in $\mathbb{C}$, and apply 12.7.3. QED

CHAPTER 14
The Fundamental Comparison Theorems

14.1 The Basic Setting

(14.1.1) Fix a prime number ℓ, and an isomorphism of fields

$$\iota: \overline{\mathbb{Q}}_\ell \approx \mathbb{C}.$$

Let R be a subring of $\mathbb{C}$ which is a finitely generated $\mathbb{Z}[1/\ell]$-algebra. Let K be an object of $D^b_c(\mathbb{A}^1_R, \overline{\mathbb{Q}}_\ell)$. At the expense of replacing R by R[1/r] for some nonzero element $r \in R$, we may further assume that R is normal, and that K is adapted to the stratification $(\mathbb{A}^1_R - D, D)$ where $D \subset \mathbb{A}^1_R$ is a divisor which is finite etale over R of some degree $d \geq 1$, defined by a monic polynomial $f(x) \in R[x]$ of degree d whose discriminant Δ is a unit in R.

(14.1.2) In the following discussion, we view $R \subset \mathbb{C}$ by the given inclusion. This allows us to speak of the complex fibre $\mathbb{A}^1_\mathbb{C}$, and the object $K_\mathbb{C}$ in $D^b_c(\mathbb{A}^1_\mathbb{C}, \overline{\mathbb{Q}}_\ell)$.

Lemma 14.1.3 (cf. [Ka-PES, 1.7]) Hypotheses as above, the following conditions are equivalent:

(1) K is fibre-wise perverse, in the sense that for any algebraically closed field k, and any ring homomorphism $\varphi: R \to k$, the inverse image K_φ of K in $D^b_c(\mathbb{A}^1_k, \overline{\mathbb{Q}}_\ell)$ is perverse.

(2) There exists an algebraically closed field k, and a ring homomorphism $\varphi: R \to k$ such that the inverse image K_φ of K in $D^b_c(\mathbb{A}^1_k, \overline{\mathbb{Q}}_\ell)$ is perverse.

(3) The object $K_\mathbb{C}$ in $D^b_c(\mathbb{A}^1_\mathbb{C}, \overline{\mathbb{Q}}_\ell)$ is perverse.

(4) The cohomology sheaves $\mathcal{H}^i(K)$ and $\mathcal{H}^i(DK)$ both vanish for $i \neq 0, -1$, and for $i = 0$ both vanish on $\mathbb{A}^1_R - D$.

proof Each of the cohomology sheaves $\mathcal{H}^i(K)$, $\mathcal{H}^i(DK)$ is adapted to the stratification $(\mathbb{A}^1_R - D, D)$ of $\mathbb{A}^1_R$, so its vanishing on either stratum is detected on any geometric fibre. QED

(14.1.4) Suppose now that K is fibre-wise perverse. By 13.3.4 (5), we know that K is an object of $\mathrm{Perv}_{\mathrm{tame}}(\mathbb{A}^1_R, \overline{\mathbb{Q}}_\ell)$. We denote by $\mathcal{M}$ the holonomic RS $\mathcal{D}$-module on $\mathbb{A}^1_\mathbb{C}$ which corresponds to $K_\mathbb{C}$ via Riemann-Hilbert and the change of coefficients via ι. By 2.10.16, we know that $j^*FT(\mathcal{M})$ is a D.E. on $\mathbb{G}_{m,\mathbb{C}}$. We denote

$n(K)$:= the rank of the D.E. $j^*FT(\mathcal{M})$ on $\mathbb{G}_{m,\mathbb{C}}$,

G_{gal} := the differential galois group of $j^*FT(\mathcal{M})$ on $\mathbb{G}_{m,\mathbb{C}}$.

In this situation, we have

Proposition 14.1.5 For any algebraically closed field k of characteristic $p > 0$, any ring homomorphism $\varphi: R \to k$, and any nontrivial additive character ψ of any finite subfield of k, the perverse object $FT_\psi(K_\varphi)|\mathbb{G}_m$ is of the form $\mathcal{F}_\varphi[1]$, for $\mathcal{F}_\varphi$ a lisse sheaf on $(\mathbb{G}_m)_k$ of rank n(K).

proof Replacing K by $K *_{!+}(Rj_*j^*\overline{\mathbb{Q}}_\ell(1)[1])$ does not change either the D.E. $j^*FT(\mathcal{M})$ on $\mathbb{G}_{m,\mathbb{C}}$ or the lisse sheaves $\mathcal{F}_\varphi$ (by 12.5.8 and 12.10.5). Nor does it change the fact that K is adapted to the stratification $(\mathbb{A}^1_R - D, D)$ (cf 13.4.3). So we may assume in addition that K lies in $\mathrm{Perv}_{A!,tame}(\mathbb{A}^1_R, \overline{\mathbb{Q}}_\ell)$. In this case, we have seen (12.5.13) that n(K) is the generic rank of $\mathcal{M}$, i.e., it is the generic rank of $\mathcal{H}^{-1}(K_\mathbb{C})$. Because K is adapted to $(\mathbb{A}^1_R - D, D)$, this shows that the rank of the lisse sheaf $\mathcal{H}^{-1}(K) \mid \mathbb{A}^1_R - D$ is n(K).

Since K_φ is perverse, everywhere tame and satisfies condition (A!), we know (12.11.4) that $\mathcal{F}_\varphi$ is lisse of rank equal to the generic rank of $\mathcal{H}^{-1}(K_\varphi)$. But as K is adapted to $(\mathbb{A}^1_R - D, D)$, the generic rank of $\mathcal{H}^{-1}(K_\varphi)$ is equal to the rank on $\mathbb{A}^1_R - D$ of the lisse sheaf

$$\mathcal{H}^{-1}(K) \mid \mathbb{A}^1_R - D,$$

which as we have just seen is of rank n(K). QED

Reductive Comparison Theorem 14.2 Fix a prime number ℓ, and an isomorphism of fields $\iota: \overline{\mathbb{Q}}_\ell \approx \mathbb{C}$. Let R be a subring of $\mathbb{C}$ which is a finitely generated $\mathbb{Z}[1/\ell]$-algebra. Let K be an object of $D^b_c(\mathbb{A}^1_R, \overline{\mathbb{Q}}_\ell)$ which is adapted to a stratification $(\mathbb{A}^1_R - D, D)$ where $D \subset \mathbb{A}^1_R$ is a divisor which is finite etale over R of some degree $d \geq 1$, defined by a monic polynomial $f(x) \in R[x]$ of degree d whose discriminant Δ is a unit in R. Suppose that K is fibre-wise perverse, with $K_\mathbb{C}$ corresponding to the RS holonomic $\mathcal{D}$-module $\mathcal{M}$, whose $\mathcal{D}$-module Fourier Transform $FT(\mathcal{M})$ is a D.E. on $\mathbb{G}_{m,\mathbb{C}}$ of rank $n := n(K)$. Suppose that

(1) the differential galois group G_{gal} of $j^*FT(\mathcal{M})$ on $\mathbb{G}_{m,\mathbb{C}}$ is reductive.

(2) for any algebraically closed field k of characteristic $p > 0$, any ring homomorphism $\varphi: R \to k$, and any nontrivial additive character ψ of any finite subfield of k, denoting by $G_{geom,\varphi,\psi}$ the group G_{geom} for the lisse sheaf $\mathcal{F}_\varphi := j^*FT_\psi(K_\varphi)[-1]$ on $\mathbb{G}_{m,k}$ of rank $n := n(K)$ is reductive.

Then there exists a dense open set U of Spec(R), which depends only on

the original stratification $(\mathbb{A}^1_R - D, D)$ of $\mathbb{A}^1_R$ to which K was adapted, and

the conjugacy class of G_{gal} in $GL(n, \mathbb{C})$,

such that for any φ lying over a point of U, $G_{geom,\varphi,\psi}(\overline{\mathbb{Q}}_\ell)$ is conjugate in $GL(n, \overline{\mathbb{Q}}_\ell)$ to a subgroup of $\iota^{-1}G_{gal}(\mathbb{C})$.

proof Replacing K by $K *_{!+}(Rj_*j^*\overline{\mathbb{Q}}_\ell(1)[1])$ does not change either the D.E. $j^*FT(\mathcal{M})$ on $\mathbb{G}_{m,\mathbb{C}}$ or the lisse sheaves $\mathcal{F}_\varphi$. Nor does it change the fact that K is adapted to the stratification $(\mathbb{A}^1_R - D, D)$. So we may assume in addition that K lies in $Perv_{A!,tame}(\mathbb{A}^1_R, \overline{\mathbb{Q}}_\ell)$.

Consider the n-dimensional object $K_\mathbb{C}$ in the Tannakian category $Perv_{A!,tame}(\mathbb{A}^1_\mathbb{C}, \overline{\mathbb{Q}}_\ell)$, with convolution as the tensor operation. The Tannakian galois goup of $K_\mathbb{C}$ is $\iota^{-1}G_{gal}$ (by 12.8.3). By hypothesis, this group is reductive. So for some (D, N) with $1 \le D \le N$ and $N \ge 3$, $K_\mathbb{C}$ is (D, N)-determined.

For the dense open set U of Spec(R), we apply 13.5.6 and 13.5.7 (4), and take $U := Spec(R_N)$. Fix an algebraically closed field k of characteristic $p > 0$, a ring homomorphism $\varphi: R \to k$ which lies over a point u of U, and a nontrivial additive character ψ of a finite subfield of k. Consider the n-dimensional object K_φ in the Tannakian category $Perv_{A!,tame}(\mathbb{A}^1_k, \overline{\mathbb{Q}}_\ell)$, with convolution as the tensor operation. The Tannakian galois goup of K_φ is $G_{geom,\varphi,\psi}$.

Denote by $R_{u,hs}$ the strict henselization inside $\mathbb{C}$ of the local ring at u. Then φ is a geometric point of $Spec(R_{u,hs})$ lying over the closed point, and the inclusion of $R_{u,hs}$ into $\mathbb{C}$ is a geometric point lying over the generic point of $\mathrm{Spec}(R_{u,hs})$.

It remains only to apply to this situation the reductive specialization theorem 11.7.1. In the notations of that theorem, we take

$$\mathfrak{M} = \mathrm{Perv}_{A!,\mathrm{tame}}(\mathbb{A}^1{}_{R_{u,hs}}, \overline{\mathbb{Q}}_\ell),\ \mathbb{1} = Rj_*j^*\overline{\mathbb{Q}}_\ell(1)[1],\ \otimes = *_{!+},$$

$$V = K,$$

$$V^\vee = ([x \to -x]^* DK) *_{!+} (Rj_*j^*\overline{\mathbb{Q}}_\ell(1)[1]),$$

$$n = n(K),$$

$$\mathcal{C}_{\mathbb{C}} = \mathrm{Perv}_{A!}(\mathbb{A}^1{}_{\mathbb{C}}, \overline{\mathbb{Q}}_\ell),\ \mathbb{1} = Rj_*j^*\overline{\mathbb{Q}}_\ell(1)[1],\ \otimes = *_{!+},$$

$\mathfrak{M} \to \mathcal{C}_{\mathbb{C}}$ the functor "pullback to $\mathbb{A}^1{}_{\mathbb{C}}$"

$$\mathcal{C}_{\mathbb{F}} = \mathrm{Perv}_{A!,\mathrm{tame}}(\mathbb{A}^1{}_k, \overline{\mathbb{Q}}_\ell),\ \mathbb{1} = Rj_*j^*\overline{\mathbb{Q}}_\ell(1)[1],\ \otimes = *_{!+},$$

$\mathfrak{M} \to \mathcal{C}_{\mathbb{F}}$ the functor "pullback to $\mathbb{A}^1{}_k$".

The hypotheses (1$\mathbb{C}$) and (1$\mathbb{F}$) are satisfied, by 12.7.5 and 12.11.4. The reductivity hypotheses (2$\mathbb{C}$) and (2$\mathbb{F}$) are satisfied because we have assumed this. The hypotheses (3$\mathbb{C}$) and (3$\mathbb{F}$) are satisfied in virtue of 13.3.10. QED

The following corollary shows that among reductive subgroups of $GL(n, \mathbb{C})$, none smaller than G_{gal} "works" in the reductive comparison theorem.

Corollary 14.3 Hypotheses and notations as in 14.2, let H be a proper Zariski closed subgroup of G_{gal} which is reductive. Then there exists a dense open set U_1 of Spec(R), which depends only on

the original stratification ($\mathbb{A}^1{}_R$ - D, D) of $\mathbb{A}^1{}_R$ to which K was adapted, and

the conjugacy class of G_{gal} in $GL(n, \mathbb{C})$,

the conjugacy class of H in $GL(n, \mathbb{C})$,

such that for any φ lying over a point of U_1, $G_{geom,\varphi,\psi}(\overline{\mathbb{Q}}_\ell)$ is **not** conjugate in $GL(n, \overline{\mathbb{Q}}_\ell)$ to a subgroup of $\iota^{-1}H(\mathbb{C})$.

proof The idea is to exploit the fact that reductive subgroups of $GL(n, \mathbb{C}) = GL(V)$, are determined by their tensor invariants.

For each pair of nonnegative integers (a, b), we denote by $T^{a,b}(V)$ the representation $V^{\otimes a} \otimes (V^\vee)^{\otimes b}$. For any subgroup Γ of GL(V), we define

$$\mathrm{inv}(\Gamma;\, a,\, b) := \dim((T^{a,b}(V))^{\Gamma}) \in \mathbb{Z},$$

and the two-variable generating series

$$\mathrm{Inv}_{\Gamma}(x,\, y) := \Sigma_{a,b}\, \mathrm{inv}(\Gamma;\, a,\, b)x^a y^b \in \mathbb{Z}[[x,\, y]].$$

Notice that the dimensions $\mathrm{inv}(\Gamma;\, a,\, b)$ and the generating series $\mathrm{Inv}_{\Gamma}(x, y)$ depend only on the conjugacy class of Γ in $GL(V)$.

Since $H \subset G_{gal}$, we always have an inclusion of invariants

$$T^{a,b}(V)^{G_{gal}} \subset T^{a,b}(V)^{H},$$

so in particular an inequality of dimensions

$$\mathrm{inv}(H;\, a,\, b) \geq \mathrm{inv}(G_{gal};\, a,\, b).$$

Since H is a proper reductive subgroup of the reductive group G_{gal} inside $GL(n, \mathbb{C}) = GL(V)$, the two groups cannot have the same tensor invariants. Therefore there exists a pair of nonnegative integers (a, b) for which we have a strict inequality

$$\mathrm{inv}(H;\, a,\, b) > \mathrm{inv}(G_{gal};\, a,\, b).$$

Fix one such pair (a, b). In the notations of 13.5.6 and 13.5.7 (4), consider the open set $U := \mathrm{Spec}(R_{a+b})$ of Spec(R), and the object $T^{a,b}(K)$ of $\mathrm{Perv}_{A!,tame}(\mathbb{A}^1 R_{a+b},\, \overline{\mathbb{Q}}_\ell)$. In virtue of 13.3.6 (3), we have an equality of dimensions of spaces of invariants

$$\dim\mathrm{Hom}(\mathbb{1}_{\mathbb{C}},\, T^{a,b}(K_{\mathbb{C}})) = \dim\mathrm{Hom}(\mathbb{1}_{\varphi},\, T^{a,b}(K_{\varphi}))$$

for every geometric point $(k,\, \varphi: R_{a+b} \to k)$ of $\mathrm{Spec}(R_{a+b})$. For a geometric point of positive characteristic, we can read this in terms of the Fourier Transforms:

$$\dim\mathrm{Hom}_{D.E.(\mathbb{G}_m/\mathbb{C})}(\mathcal{O}_{\mathbb{G}_m},\, T^{a,b}(j^{*}FT(\mathcal{M}))) =$$

$$= \dim\mathrm{Hom}_{\text{lisse sheaves on } \mathbb{G}_{m,k}}(\overline{\mathbb{Q}}_\ell,\, T^{a,b}(\mathcal{F}_{\varphi})).$$

In other words, we have

$$\mathrm{inv}(G_{gal};\, a,\, b) = \mathrm{inv}(G_{geom,\varphi,\psi};\, a,\, b)$$

for every geometric point $(k,\, \varphi: R_{a+b} \to k)$ of $\mathrm{Spec}(R_{a+b})$ of positive characteristic. Therefore we cannot have $G_{geom,\varphi,\psi}$ conjugate to a subgroup of H, since it has too few invariants. QED

14.4 Remarks

(14.4.1) Indeed, over $\mathrm{Spec}(R_{a+b})$, 13.3.6 (3) shows that we have an equality of dimensions

$$\mathrm{inv}(G_{gal};\, c,\, d) = \mathrm{inv}(G_{geom,\varphi,\psi};\, c,\, d)$$

for every pair of nonnegative integers (c, d) with $c + d \leq a + b$. In other words, by shrinking on Spec(R), we can make **more and more** terms of the generating series $\mathrm{Inv}_{G_{\mathrm{geom},\varphi,\psi}}(x, y)$ for $G_{\mathrm{geom},\varphi,\psi}$ coincide with those of the generating series for G_{gal}. But we might never get all the terms right.

(14.4.2) Here is a simple example. Start over $\mathbb{Z}$, with the perverse object K given by the delta module δ_1 supported at 1. Over $\mathbb{C}$, the Fourier Transform is the D.E. for e^x, whose G_{gal} is $\mathbb{G}_m$. So for G_{gal} the generating series is

$$\mathrm{Inv}_{G_{\mathrm{gal}}}(x, y) = 1/(1 - xy).$$

In characteristic p, the Fourier Transform of δ_1 is $\mathcal{L}_\psi$, whose G_{geom} is the subgroup μ_p of $\mathbb{G}_m$, so for any φ of characteristic p we have

$$\mathrm{Inv}_{G_{\mathrm{geom},\varphi,\psi}}(x, y) = (1 - x^p y^p)/(1 - xy).$$

(14.4.3) This same example also illustrates the bookkeeping of iterated convolution (cf. 13.5). The natural stratification to which both $K = \delta_1$ and $[x \mapsto -x]^*DK = \delta_{-1}$ are simultaneously adapted is

$$(\mathbb{A}^1{}_{\mathbb{Z}[1/2]} - A, A) \text{ for } A := \{1, -1\}.$$

With this A, $n \geq 2$ convolutions bring us to the set

$$A_n := \{-n, 1-n, \ldots, n-1, n\},$$

so inverting $\Delta(A_n)$ requires the inverting of all primes which are $\leq 2n$.

(14.5) The following theorem shows that the above phenomenon of the generating series for $G_{\mathrm{geom},\varphi,\psi}$ never reaching that of G_{gal}, no matter how much we shrink, cannot occur if we require G_{gal} to be semisimple (and not just reductive). The proof is due to Ofer Gabber.

Semisimple Comparison Theorem 14.6 Hypotheses and notations as in the reductive comparison theorem 14.2, suppose in addition that the group G_{gal} is semisimple. Then there exists a dense open set U of Spec(R), which depends only on

the original stratification $(\mathbb{A}^1{}_R - D, D)$ of $\mathbb{A}^1{}_R$ to which K was adapted, and

the conjugacy class of G_{gal} in $GL(n, \mathbb{C})$,

such that for any φ lying over a point of U, $G_{\mathrm{geom},\varphi,\psi}(\overline{\mathbb{Q}}_\ell)$ is conjugate in $GL(n, \overline{\mathbb{Q}}_\ell)$ to $\iota^{-1}G_{\mathrm{gal}}(\mathbb{C})$.

proof Let us admit temporarily the following result, which I learned from Ofer Gabber.

Theorem 14.7 Let G be a Zariski closed subgroup of $GL(n)_{\mathbb{C}}$ which is semisimple. There exist finitely many proper reductive subgroups H_i of G such that **any** proper reductive subgroup H of G is G-conjugate to a subgroup of one of the listed subgroups H_i.

Granting this, the semisimple comparison theorem is immediate. First shrink on Spec(R) until the reductive comparison theorem comes into effect, say on the dense open U_0. Then apply the previous corollary to $G = G_{gal}$ and to each of the finitely many subgroups H_i of G; each application produces a dense open U_{H_i} of Spec(R) over which $G_{geom,\varphi,\psi}(\overline{\mathbb{Q}}_\ell)$ is **not** conjugate in $GL(n, \overline{\mathbb{Q}}_\ell)$ to a subgroup of $\iota^{-1}H_i(\mathbb{C})$. So over the intersection of U_0 and the finitely many U_{H_i}, $G_{geom,\varphi,\psi}(\overline{\mathbb{Q}}_\ell)$ must be conjugate in $GL(n, \overline{\mathbb{Q}}_\ell)$ to $\iota^{-1}G_{gal}(\mathbb{C})$. QED

(14.8) It remains to prove the group-theoretic theorem 14.7. This we will do by a combination of the unitarian trick and Jordan's theorem on finite subgroups of $GL(n, \mathbb{C})$. We first carry out the reduction to the compact case. Recall that reductive algebraic groups H over $\mathbb{C}$ have maximal compact subgroups K_H which are Zariski dense and all of which are H-conjugate. If H is a reductive subgroup of a reductive algebraic group G, then any choice of K_H is a compact subgroup of G, so contained in some maximal compact subgroup K_G of G. Moreover, any compact subgroup F of K_G is the maximal compact subgroup of a unique reductive subgroup H_F of G (namely H_F := the Zariski closure of F in G). So the theorem in question is equivalent to the following about compact Lie groups.

Theorem(bis) 14.9 Suppose that G is a compact Lie group which is semisimple. There exist finitely many proper compact subgroups H_i of G such that **any** proper compact subgroup H of G is G-conjugate to a subgroup of one of the listed subgroups H_i.

proof For H a proper compact subgroup of G, we have an obvious inclusion

$$H \subset N_G(H^0).$$

According to [Bour-L9, 9, exc 12, a)], as H^0 runs over the connected compact subgroups of G, all the groups $N_G(H^0)$ fall into finitely many G-conjugacy classes. So if H^0 is not a normal subgroup of G, we are done.

If H^0 is a normal subgroup of the semissimple G, then H^0 is necessarily semisimple (its connected center is normal in G). So Lie(H^0) is a semisimple ideal in Lie(G^0), of which there are only finitely many (cf. [Bour-L1, 6, exc 7]). Therefore there are only finitely many possible H^0 which are normal. For each of these finitely many, the group H corresponds to a finite subgoup of the quotient group G/H^0, which is itself semisimple. For each of these finitely many quotient groups, we must prove the theorem for all its finite subgroups.

Suppose now that H is a finite subgroup of G. In the proof of Jordan's theorem as given in [C-R-RT], one constructs a small neighborhood U of the identity in the compact group G (strictly speaking, in an ambient unitary group) and shows that if H is any finite subgroup of G, then any two elements of $H \cap U$ commute. Shrinking U, we may suppose that the log and exp maps are inverse bijections to a neighborhood of zero in Lie(G), and that U is stable by G-conjugation.

Suppose first that $H \cap U$ is reduced to the identity element. Then the order of H is bounded by the absolute constant c(G) := vol(G)/vol(U). By [Bour-L9, 9, exc 20], there are only finitely many conjugacy classes of finite subgroups of G of any given order.

Suppose now that $H \cap U$ is not reduced to the identity element. Consider the logarithms of the elements in $H \cap U$. This is a commuting set of elements in Lie(G), stable by H-conjugation, not all of which are zero, so their $\mathbb{R}$-span in Lie(G) is the Lie algebra of a nonzero connected torus T of G which is normalized by H. Therefore we have $H \subset N_G(T)$. Because G is semisimple, the torus T is **not** normal in G, and so by the earlier discussion the proper subgroup $N_G(T)$ lies in one of finitely many G-conjugacy classes. QED

Using this same group-theoretic result, we can also give a sharpening of the reductive comparison theorem, which includes the semisimple comparison theorem as a special case.

Sharpened Reductive Comparison Theorem 14.10 Hypotheses and notations as in the reductive comparison theorem 14.2, there exists a dense open set U of Spec(R), which depends only on

the original stratification $(\mathbb{A}^1_R - D, D)$ of $\mathbb{A}^1_R$ to which K was adapted, and

the conjugacy class of G_{gal} in $GL(n, \mathbb{C})$,

such that for any φ lying over a point of U, $G_{geom,\varphi,\psi}(\overline{\mathbb{Q}}_\ell)$ is conjugate in $GL(n, \overline{\mathbb{Q}}_\ell)$ to a subgroup of $\iota^{-1}G_{gal}(\mathbb{C})$. Via this conjugation the two groups have the same "semisimple connected parts",

$$G_{geom,\varphi,\psi}(\overline{\mathbb{Q}}_\ell)^{0,der} = \iota^{-1}G_{gal}(\mathbb{C})^{0,der},$$

and the composite map

$$G_{geom,\varphi,\psi} \subset G_{gal} \rightarrow G_{gal}/Z((G_{gal})^0)^0$$

is surjective.

proof For any reductive group G, one knows that $G^0 = G^{0,der}.Z(G^0)^0$, the quotient $G/Z(G^0)^0$ of G is the universal semisimple quotient of G, and the natural map

$$G^{0,der} \rightarrow G/Z(G^0)^0$$

is an isogeny of the source onto the identity component of the target. So if H is a closed reductive subgroup of G, then $H^{0,der} = G^{0,der}$ if and only if the composite

$$H \subset G \rightarrow G/Z(G^0)^0,$$

maps H onto a subgroup of finite index.

We now apply these remarks to the situation at hand. An initial shrinking on R allows us to assume that the conclusion of the reductive comparison theorem already holds. We fix choices of conjugations, and of ι, and view each $G_{geom,\varphi,\psi}$ as a subgroup of G_{gal}. Then each $(G_{geom,\varphi,\psi})^{0,der}$ is a subgroup of $(G_{gal})^{0,der}$. As explained above, it suffices to show that the canonical map

$$G_{geom,\varphi,\psi} \subset G_{gal} \rightarrow G_{gal}/Z((G_{gal})^0)^0$$

is surjective. If it is not surjective, its image is a proper reductive subgroup of the semisimple group $G_{gal}/Z((G_{gal})^0)^0$, and hence by the theorem 14.7 the image is conjugate to a subgroup of one of a **finite** list of proper subgroups H_i of $G_{gal}/Z((G_{gal})^0)^0$. Denote by $\tilde{H}_i \subset G_{gal}$ the inverse image of H_i in G_{gal}. Then the $\tilde{H}_i$ form a finite list of proper reductive subgroups of G_{gal}, and whenever the map

$$G_{geom,\varphi,\psi} \subset G_{gal} \to G_{gal}/Z((G_{gal})^0)^0$$

is not surjective, $G_{geom,\varphi,\psi}$ is conjugate in G_{gal} to a subgroup of one of the $\tilde{H}_i$. Now apply 14.3 to each of the finitely many $\tilde{H}_i$. QED

14.11 Interlude : a sufficient condition for the reductivity of G_{gal}

Reductivity Criterion 14.11.1 Fix a prime number ℓ, and an isomorphism of fields $\iota:\overline{\mathbb{Q}}_\ell \approx \mathbb{C}$. Let R be a subring of $\mathbb{C}$ which is a finitely generated $\mathbb{Z}[1/\ell]$-algebra. Let K be an object of $D^b{}_c(\mathbb{A}^1{}_R, \overline{\mathbb{Q}}_\ell)$ which is adapted to a stratification $(\mathbb{A}^1{}_R - D, D)$ where $D \subset \mathbb{A}^1{}_R$ is a divisor which is finite etale over R of some degree $d \geq 1$, defined by a monic polynomial $f(x) \in R[x]$ of degree d whose discriminant Δ is a unit in R. Suppose that K is fibre-wise perverse, with $K_\mathbb{C}$ corresponding to the RS holonomic $\mathcal{D}$-module $\mathcal{M}$, whose $\mathcal{D}$-module Fourier Transform $FT(\mathcal{M})$ is a D.E. on $\mathbb{G}_{m,\mathbb{C}}$ of rank $n := n(K)$. Suppose that for any algebraically closed field k of characteristic $p > 0$, any ring homomorphism $\varphi: R \to k$, and any nontrivial additive character ψ of any finite subfield of k, the group G_{geom} for the lisse sheaf $\mathcal{F}_{\varphi,\psi} := j^*FT_\psi(K_\varphi)[-1]$ on $\mathbb{G}_{m,k}$ of rank $n := n(K)$ is reductive.

Then the differential galois group G_{gal} of $j^*FT(\mathcal{M})$ on $\mathbb{G}_{m,\mathbb{C}}$ is reductive.

proof Replacing K by $K*_{!+}(Rj_*j^*\overline{\mathbb{Q}}_\ell(1)[1])$ does not change either the D.E. $j^*FT(\mathcal{M})$ on $\mathbb{G}_{m,\mathbb{C}}$ or the lisse sheaves $\mathcal{F}_{\varphi,\psi}$. Nor does it change the fact that K is adapted to the stratification $(\mathbb{A}^1{}_R - D, D)$. So we may assume in addition that K lies in $Perv_{A!,tame}(\mathbb{A}^1{}_R, \overline{\mathbb{Q}}_\ell)$.

We begin by explaining the idea of the proof. Suppose first that for for any algebraically closed field k of characteristic $p > 0$, any ring homomorphism $\varphi: R \to k$, and any nontrivial additive character ψ of any finite subfield of k, the lisse sheaf $\mathcal{F}_{\varphi,\psi}$ on $\mathbb{G}_{m,k}$ has no nonzero invariants under local inertia at zero:

$$(\mathcal{F}_{\varphi,\psi})^{I_0} = 0, \text{ i.e., } j_!\mathcal{F}_{\varphi,\psi} \approx j_*\mathcal{F}_{\varphi,\psi}.$$

Then $j_!\mathcal{F}_{\varphi,\psi}[1]$ is the middle extension of $\mathcal{F}_{\varphi,\psi}[1]$ to $\mathbb{A}^1{}_k$. By hypothesis, $\mathcal{F}_{\varphi,\psi}$ has G_{geom} reductive, which is to say that $\mathcal{F}_{\varphi,\psi}$ is geometrically the direct sum of irreducible lisse sheaves $\mathcal{F}_{\varphi,\psi,\alpha}$ on $\mathbb{G}_{m,k}$. Therefore the middle extension $j_!\mathcal{F}_{\varphi,\psi}[1]$ is the direct sum of the middle extensions of the $\mathcal{F}_{\varphi,\psi,\alpha}[1]$, each of which is perverse irreducible on $\mathbb{A}^1_k$. But $FT_\psi(K_\varphi) = j_!\mathcal{F}_{\varphi,\psi}[1]$, so by Fourier inversion and the fact that FT carries perverse irreducibles to perverse irreducibles, we deduce that K_φ is itself a direct sum of perverse irreducibles on $\mathbb{A}^1_k$, for every geometric fibre of $\mathbb{A}^1_R/\mathrm{Spec}(R)$ of positive characteristic. By [B-B-D, 6.1.9], it follows that the perverse object $K_{\mathbb{C}}$ on $\mathbb{A}^1_{\mathbb{C}}$ is a direct sum of perverse irreducibles. Then by 12.7.1.1 and Riemann-Hilbert, the corresponding $\mathcal{D}$-module $\mathcal{M}$ on $\mathbb{A}^1_{\mathbb{C}}$ is a direct sum of irreducible holonomic RS $\mathcal{D}$-modules. Therefore its $\mathcal{D}$-module Fourier Transform $FT(\mathcal{M})$ is a direct sum of irreducible holonomic $\mathcal{D}$-modules on $\mathbb{A}^1_{\mathbb{C}}$. Since the restriction of an irreducible holonomic to a nonvoid open set is (either zero or) irreducible holonomic, the restriction $j^*FT(\mathcal{M})$ to $\mathbb{G}_{m,\mathbb{C}}$ of $FT(\mathcal{M})$ is a direct sum of irreducible holonomics on $\mathbb{G}_{m,\mathbb{C}}$. Since $j^*FT(\mathcal{M})$ is itself a D.E., each of its irreducible holonomic summands is itself a D.E., so an irreducible D.E. on $\mathbb{G}_{m,\mathbb{C}}$. Therefore $j^*FT(\mathcal{M})$ is a semisimple object of the category $D.E.(\mathbb{G}_m/\mathbb{C})$, i.e., its G_{gal} is reductive.

We next explain how to reduce the general case to the one treated above. The idea is that if we convolve K with a suitable $j_!\mathcal{L}_{\overline{\chi}}[1]$, for χ a character of finite order, we replace $\mathcal{F}_{\varphi,\psi}$ by $\mathcal{L}_\chi \otimes \mathcal{F}_{\varphi,\psi}$, and if we take χ sufficiently general, then this sheaf has no nonzero inertial invariants at zero. On the other hand, the $\mathcal{D}$-module effect of the change is to replace $j^*FT(\mathcal{M})$ by an x^α twist $x^\alpha \otimes j^*FT(\mathcal{M})$ for some rational number α whose exact denominator is the order of χ. By the previous case, we conclude that $x^\alpha \otimes j^*FT(\mathcal{M})$ has its G_{gal} reductive, i.e., $x^\alpha \otimes j^*FT(\mathcal{M})$ is a semisimple object of the category $D.E.(\mathbb{G}_m/\mathbb{C})$. Twisting now by $x^{-\alpha}$, we find that $j^*FT(\mathcal{M})$ is itself a semisimple object of the category $D.E.(\mathbb{G}_m/\mathbb{C})$, as required.

To see that we can choose a single χ which "works" simultaneously in all the geometric fibres, recall that for any given object K of $D^b_c(\mathbb{A}^1_R, \overline{\mathbb{Q}}_\ell)$, there exists a finite extension E of $\mathbb{Q}_\ell$ inside $\overline{\mathbb{Q}}_\ell$ such that K "comes from" an object K_E of $D^b_c(\mathbb{A}^1_R, \overline{\mathbb{Q}}_\ell)$. The nontrivial characters of I_0 which occur in $\mathcal{F}_{\varphi,\psi}$ are, thanks to Laumon's analysis of the local monodromy of Fourier Transforms, among the characters of I_∞ which occur in the various cohomology sheaves $\mathcal{H}^i(K_\varphi)$. By the local monodromy theorem, the characters of I_∞ which occur in any particular $\mathcal{H}^i(K_\varphi)$ are all of finite order; because K lives over E, looking at characteristic polynomials shows that these characters take values which are algebraic over E of degree at most the generic rank of $\mathcal{H}^i(K)$. Since E has only finitely many extensions of any given degree, each of which contains only finitely many roots of unity, we see that the orders of the possible characters of of I_0 which occur in $\mathcal{F}_{\varphi,\psi}$ are uniformly bounded, say $\le$ N. So we have only to pick any integer M > N, replace R by the overring $R[1/M, \zeta_M]$ (inside $\mathbb{C}$), and pick for χ any character of exact order M. QED

Corollary 14.11.2 (criterion via purity) Fix a prime number ℓ, and an isomorphism of fields $\iota:\overline{\mathbb{Q}}_\ell \approx \mathbb{C}$. Let R be a subring of $\mathbb{C}$ which is a finitely generated $\mathbb{Z}[1/\ell]$-algebra. Let K be an object of $D^b{}_c(\mathbb{A}^1{}_R, \overline{\mathbb{Q}}_\ell)$ which is adapted to a stratification $(\mathbb{A}^1{}_R - D, D)$ where $D \subset \mathbb{A}^1{}_R$ is a divisor which is finite etale over R of some degree $d \ge 1$, defined by a monic polynomial $f(x) \in R[x]$ of degree d whose discriminant Δ is a unit in R. Suppose that K is fibre-wise perverse, with $K_\mathbb{C}$ corresponding to the RS holonomic $\mathcal{D}$-module $\mathcal{M}$, whose $\mathcal{D}$-module Fourier Transform $FT(\mathcal{M})$ is a D.E. on $\mathbb{G}_{m,\mathbb{C}}$ of rank n := n(K). Suppose that for any $\overline{\mathbb{F}}_p$ of characteristic p > 0, any ring homomorphism $\varphi: R \to \overline{\mathbb{F}}_p$, and any nontrivial additive character ψ of any finite subfield of $\overline{\mathbb{F}}_p$, the lisse sheaf $\mathcal{F}_{\varphi,\psi} := j^*FT_\psi(K_\varphi)[-1]$ on $\mathbb{G}_m/\overline{\mathbb{F}}_p$ is pure of some weight. Then the differential galois group G_{gal} of $j^*FT(\mathcal{M})$ on $\mathbb{G}_{m,\mathbb{C}}$ is reductive.

proof Indeed, by [De-WII, 3.4.1] the purity of the lisse sheaf $\mathcal{F}_{\varphi,\psi}$ implies that its G_{geom} is reductive (indeed, semisimple, in view of Grothendieck's theorem [De-WII, 1.3.8] that the radical of G_{geom} is unipotent for any lisse sheaf on on $\mathbb{G}_m/\overline{\mathbb{F}}_p$ which begins life over a finite field). QED

14.12 Application to hypergeometric sheaves

(14.12.1) We begin with a brief review of what we established earlier about hypergeometrics of type (n, n). Fix an isomorphism $\iota: \overline{\mathbb{Q}}_\ell \approx \mathbb{C}$.

For any integer $N \geq 1$, denote by

$\Phi_N(x)$:= the N'th cyclotomic polynomial,

R_N := the ring $\mathbb{Z}[1/N\ell, X]/(\Phi_N(X))$,

μ_N := the cyclic group $\mu_N(R_N)$ of order N,

S_N := $\mathrm{Spec}(R_N)$.

The group $\mathrm{Hom}(\mu_N, \mu_N)$ is canonically $(1/N)\mathbb{Z}/\mathbb{Z}$, with x in $(1/N)\mathbb{Z}/\mathbb{Z}$ corresponding to the character $\zeta \mapsto \zeta^{Nx}$.

(14.12.2) Once we fix an embedding

$$\iota_\ell : R_N \subset \overline{\mathbb{Q}}_\ell.$$

we have an induced isomorphism $\mu_N \approx \mu_N(\overline{\mathbb{Q}}_\ell)$; this allows us to identify $\overline{\mathbb{Q}}_\ell$-valued characters of μ_N with elements of $(1/N)\mathbb{Z}/\mathbb{Z}$.

(14.12.3) Suppose we are given an integer $n \geq 1$, a set $\{\chi_1, \ldots, \chi_n\}$ of n not necessarily distinct $\overline{\mathbb{Q}}_\ell$-valued characters χ_i of μ_N, and a **disjoint** set $\{\rho_1, \ldots, \rho_n\}$ of n not necessarily distinct $\overline{\mathbb{Q}}_\ell$-valued characters ρ_j of μ_N. Denote by x_i and the y_j be the unique elements of $(1/N)\mathbb{Z}/\mathbb{Z}$ to which the characters $\chi_1, \ldots, \chi_n; \rho_1, \ldots, \rho_n$, correspond. In (8.17.11), we constructed a $\overline{\mathbb{Q}}_\ell$-sheaf

$$\mathcal{H}(\chi\text{'s}; \rho\text{'s})$$

on $\mathbb{G}_m/S_N$ which is adapted to the stratification $(\mathbb{G}_m - \{1\}, \{1\})$, and such that

(1) the restriction of $\mathcal{H}(\chi\text{'s}; \rho\text{'s})$ to each geometric fibre $\mathbb{G}_m/\overline{\mathbb{F}}_p$ of positive chararacteristic is geometrically isomorphic to the hypergeometric sheaf $\mathcal{H}_1(!, \psi; \chi\text{'s}; \rho\text{'s})$.

(2) The restriction (via the composite inclusion $R_N \subset \overline{\mathbb{Q}}_\ell \approx \mathbb{C}$) of the sheaf $\mathcal{H}(\chi\text{'s}; \rho\text{'s})$ to the complex fibre corresponds (via passage to the analytic, the change of coefficients ι: $\overline{\mathbb{Q}}_\ell \approx \mathbb{C}$, and Riemann-Hilbert) to the hypergeometric $\mathcal{D}$-module $\mathcal{H}_1(x_1, \ldots, x_n; y_1, \ldots, y_n)$.

(14.12.4) We used the existence of such an "incarnation over $\mathbb{Z}$" in 8.17.12 to show that the differential galois group G_{gal} of the n'th order D.E. $\mathcal{H}_1(x_1, \ldots, x_n; y_1, \ldots, y_n)$ on $\mathbb{G}_m - \{1\}$ over $\mathbb{C}$ was "the same" as the group G_{geom} for the lisse, rank n sheaf $\mathcal{H}_1(!, \psi; \chi\text{'s}; \rho\text{'s})$ on $\mathbb{G}_m - \{1\}$ over $\overline{\mathbb{F}}_p$, for any p prime to $N\ell$.

In this section, we will apply the sharpened reductive comparison theorem 14.10 to the Kummer pullbacks of such "incarnations over $\mathbb{Z}$" of hypergeometrics of type (n, n) to get a comparison theorem for (Kummer pullbacks of) hypergeometrics of arbitrary **mixed** type.

Theorem 14.12.5 Fix an isomorphism $\iota: \overline{\mathbb{Q}}_\ell \approx \mathbb{C}$, an integer $N \geq 1$, and an embedding $\iota_\ell : R_N \subset \overline{\mathbb{Q}}_\ell$. Suppose we are given integers $n > m \geq 0$, a set $\{\chi_1, \ldots, \chi_n\}$ of n not necessarily distinct $\overline{\mathbb{Q}}_\ell$-valued characters χ_i of μ_N, and a **disjoint** set $\{\rho_1, \ldots, \rho_m\}$ of m not necessarily distinct $\overline{\mathbb{Q}}_\ell$-valued characters ρ_j of μ_N. Denote by x_i and the y_j be the unique elements of $(1/N)\mathbb{Z}/\mathbb{Z}$ to which the characters $\chi_1, \ldots, \chi_n; \rho_1, \ldots, \rho_m$, correspond. Define

$$d := n-m,$$

and fix a unit λ in the ring $R_N[1/d]$.

Denote by

$$G_{gal} \subset GL(n, \mathbb{C})$$

the differential galois group of $[d]^*\mathcal{H}_\lambda(x_1, \ldots, x_n; y_1, \ldots, y_m)$ on $\mathbb{G}_m/\mathbb{C}$. For any algebraically closed field k of characteristic $p > 0$, any ring homomorphism $\varphi: R_N[1/d] \to k$, and any nontrivial additive character ψ of any finite subfield of k, denote by

$$G_{geom,\varphi,\psi} \subset GL(n, \overline{\mathbb{Q}}_\ell)$$

the group G_{geom} for the lisse sheaf $[d]^*\mathcal{H}_\lambda(!, \psi; \chi_1, \ldots, \chi_n; \rho_1, \ldots, \rho_m)$ on $\mathbb{G}_{m,k}$.

The groups G_{gal} and $G_{geom,\varphi,\psi}$ are all reductive, and there exists a dense open set U of $\mathrm{Spec}(R_N[1/d])$, which depends only on the conjugacy class of G_{gal} in $GL(n, \mathbb{C})$, such that for any φ lying over a point of U, we have

(1) $G_{geom,\varphi,\psi}(\overline{\mathbb{Q}}_\ell)$ is conjugate in $GL(n, \overline{\mathbb{Q}}_\ell)$ to a subgroup of $\iota^{-1}G_{gal}(\mathbb{C})$.

(2) Using (1) and the isomorphism $\iota: \overline{\mathbb{Q}}_\ell \approx \mathbb{C}$ to identify $G_{geom,\varphi,\psi}$ with a subgroup of G_{gal}, we have in addition

(2a) $(G_{geom,\varphi,\psi})^{0,der} = (G_{gal})^{0,der}$.

(2b) $G_{geom,\varphi,\psi}$ maps onto the universal semisimple quotient $G_{gal}/Z((G_{gal})^0)^0$ of G_{gal}.

proof The groups in question are invariant under multiplicative translation of λ in $\mathbb{G}_m(R_N[1/d])$. The D.E. $\mathcal{H}_\lambda(x_1, \dots, x_n; y_1, \dots, y_m)$ on $\mathbb{G}_m/\mathbb{C}$ is irreducible, since the x_i and the y_j are disjoint mod $\mathbb{Z}$. Therefore any Kummer pullback is completely reducible as a D.E. on $\mathbb{G}_m/\mathbb{C}$, and hence G_{gal} is reductive. Similarly, $G_{geom,\varphi,\psi}$ is reductive.

It remains only to apply the reductive comparison theorem in its sharpened form (14.10). Recall that for k algebraically closed of characteristic p prime to $Nd\ell$ we have (9.4.2 (a) (2)) an isomorphism, for any λ in $k^\times$, of perverse sheaves on $\mathbb{A}^1{}_k$

$$j_*[d]^*\mathrm{Hyp}_\lambda(!, \psi; \chi_i\text{'s}; \rho_j\text{'s}) \approx$$

$$\approx \mathrm{FT}_\psi(j_*[d]^*\mathbf{Cancel}(\mathrm{Hyp}_{(-1)^{d+n-m}(d)^d/\lambda}(!, \psi; \Lambda_1, \dots, \Lambda_d, \overline{\rho}_j\text{'s}; \overline{\chi}_i\text{'s})).$$

And on the complex fibre, we have (6.4.2 (2)) an isomorphism, for any λ in $\mathbb{C}^\times$, of holonomic $\mathcal{D}$-modules on $\mathbb{A}^1{}_\mathbb{C}$

$$j_{!*}[d]^*\mathcal{H}_\lambda(x_i\text{'s}; y_j\text{'s}) \approx$$

$$\approx \mathrm{FT}(j_{!*}[d]^*\mathbf{Cancel}\mathcal{H}_{(-1)^{n+m+d}(d)^d/\lambda}(1/d, 2/d, \dots, d/d, -y_j\text{'s}; -x_i\text{'s})).$$

So for the particular choice

$$\lambda := (-1)^{n+m+d}(d)^d,$$

the asserted result is just the reductive comparison theorem in its sharpened form 14.10, applied to the object

$$K = j_*[d]^*\mathbf{Cancel}(\mathcal{H}(!, \psi; \Lambda_1, \dots, \Lambda_d, \overline{\rho}_j\text{'s}; \overline{\chi}_i\text{'s}))[1].$$

As the theorem is invariant under multiplicative translation of λ in $\mathbb{G}_m(R_N[1/d])$, it suffices to establish it in this case. QED

Corollary 14.12.6 (Hypergeomertic comparison) In the situation of the theorem 14.12.5, denote by

$$G_{gal} \subset GL(n, \mathbb{C})$$

the differential galois group of $\mathcal{H}_\lambda(x_1, \dots, x_n; y_1, \dots, y_m)$ on $\mathbb{G}_m/\mathbb{C}$. For any algebraically closed field k of characteristic $p > 0$, any ring homomorphism $\varphi: R_N[1/d] \to k$, and any nontrivial additive character ψ of any finite subfield of k, denote by

$$G_{geom,\varphi,\psi} \subset GL(n, \overline{\mathbb{Q}}_\ell)$$

the group G_{geom} for the lisse sheaf $\mathcal{H}_\lambda(!, \psi; \chi_1, \dots, \chi_n; \rho_1, \dots, \rho_m)$ on $\mathbb{G}_{m,k}$.

The groups G_{gal} and $G_{geom,\varphi,\psi}$ are all reductive, and there exists a dense open set U of $\mathrm{Spec}(R_N[1/d])$, which depends only on the conjugacy class of G_{gal} in $GL(n, \mathbb{C})$, such that for any φ lying over a point of U, we have

(1) $(G_{geom,\varphi,\psi})^0$ is conjugate in $GL(n, \overline{\mathbb{Q}}_\ell)$ to a subgroup of $\iota^{-1}(G_{gal})^0$.

(2) Using (1) and the isomorphism $\iota: \overline{\mathbb{Q}}_\ell \approx \mathbb{C}$ to identify $(G_{geom,\varphi,\psi})^0$ with a subgroup of $(G_{gal})^0$, we have in addition

(2a) $(G_{geom,\varphi,\psi})^{0,der} = (G_{gal})^{0,der}$.

(2b) $G(G_{geom,\varphi,\psi})^0$ maps onto the universal semisimple quotient $(G_{gal})^0/Z((G_{gal})^0)^0$ of $(G_{gal})^0$.

proof The groups G_{gal} and G_{geom} occuring here contain their homonyms occuring in the theorem as open normal subgroups of index dividing d with cyclic quotient. In particular they have the same identity components. QED

Remark 14.12.7 It is almost certainly true that the conclusions of the theorem 14.12.5 actually hold for the groups G_{gal} and G_{geom} of the corollary 14.12.6, and not just for their identity components. For instance, this is automatically the case if $d := n-m = 1$ (by the theorem itself), or if both groups G_{gal} and G_{geom} are connected, or... . In view of our detailed knowledge of both of these groups for irreducible hypergeometrics, one could envisage checking case by case.

14.13 Application to Fourier Transform of Cohomology along the fibres

(14.13.1) Fix a prime number ℓ, and an isomorphism of fields

$$\iota: \overline{\mathbb{Q}}_\ell \approx \mathbb{C}.$$

Let R be a subring of $\mathbb{C}$ which is a finitely generated $\mathbb{Z}[1/\ell]$-algebra. Let X/R be an affine R-scheme which is smooth over R, everywhere of relative dimension $d \geq 0$. Let $\mathcal{G}$ be a lisse $\overline{\mathbb{Q}}_\ell$-sheaf on X of rank $r \geq 1$. [Typically, $\mathcal{G}$ will be the constant sheaf $\overline{\mathbb{Q}}_\ell$, or R will contain the N'th roots of unity and $\mathcal{G}$ will be a sheaf of the form $\mathcal{L}_{\chi(g)}$ for some invertible function g on X, and some $\overline{\mathbb{Q}}_\ell$-valued character χ of the group $\mu_N(R)$.] Let

$$f : X \to \mathbb{A}^1_R$$

be a function on X, viewed as a morphism to $\mathbb{A}^1_R$.

(14.13.2) Now apply [Ka-Lau], 3.3.3] to the trivial stratification (X) of X and the morphism $f : X \to \mathbb{A}^1_R$. After shrinking on Spec(R), there exists a stratification of $\mathbb{A}^1_R$ of the form $(\mathbb{A}^1_R - D, D)$, where $D \subset \mathbb{A}^1_R$ is a divisor which is finite etale over R of some degree $d \geq 1$, defined by a monic polynomial $f(x) \in R[x]$ of degree d whose discriminant Δ is a unit in R, such that for any lisse lisse $\overline{\mathbb{Q}}_\ell$-sheaf $\mathcal{G}$ on X, the objects $Rf_!\mathcal{G}$ and $Rf_*\mathcal{G}$ of $D^b_c(\mathbb{A}^1_R, \overline{\mathbb{Q}}_\ell)$ are both adapted to $(\mathbb{A}^1_R - D, D)$, and their formation commutes with arbitrary change of base on Spec(R) to a good scheme.

Proposition 14.13.3 (Gabber) Let R be a subring of $\mathbb{C}$ which is a finitely generated $\mathbb{Z}[1/\ell]$-algebra. Let X/R be an affine R-scheme which is smooth over R, everywhere of relative dimension $d \geq 0$. Suppose given a stratification $(\mathbb{A}^1_R - D, D)$ of $\mathbb{A}^1_R$, where $D \subset \mathbb{A}^1_R$ is a divisor which is finite etale over R of some degree $d \geq 1$, defined by a monic polynomial $f(x) \in R[x]$ of degree d whose discriminant Δ is a unit in R, such that for any lisse lisse $\overline{\mathbb{Q}}_\ell$-sheaf $\mathcal{G}$ on X, the objects $Rf_!\mathcal{G}$ and $Rf_*\mathcal{G}$ of $D^b_c(\mathbb{A}^1_R, \overline{\mathbb{Q}}_\ell)$ are both adapted to $(\mathbb{A}^1_R - D, D)$, and their formation commutes with arbitrary change of base on Spec(R) to a good scheme.

For a given lisse $\overline{\mathbb{Q}}_\ell$-sheaf $\mathcal{G}$ on X, the following conditions are equivalent:

(1) For every triple (k, φ, ψ) consisting of an algebraically closed field k of positive characteristic, a ring homomorphism $\varphi : R \to k$, and a nontrivial $\overline{\mathbb{Q}}_\ell$-valued additive character ψ of a finite subfield of k, the natural "forget supports" maps

$$H^i_c(X\otimes_\varphi k, \mathcal{G}\otimes\mathcal{L}_{\psi(f)}) \to H^i(X\otimes_\varphi k, \mathcal{G}\otimes\mathcal{L}_{\psi(f)})$$

are all isomorphisms.

(1 bis) For every triple (k, φ, ψ) consisting of an algebraically closed field k of positive characteristic, a ring homomorphism $\varphi : R \to k$, and a nontrivial $\overline{\mathbb{Q}}_\ell$-valued additive character ψ of a finite subfield of k, we have

$$H^i_c(X\otimes_\varphi k, \mathcal{G}\otimes\mathcal{L}_{\psi(f)}) = 0 = H^i(X\otimes_\varphi k, \mathcal{G}\otimes\mathcal{L}_{\psi(f)}) \text{ for } i \neq d,$$

and the natural "forget supports" map

$$H^d_c(X\otimes_\varphi k, \mathcal{G}\otimes\mathcal{L}_{\psi(f)}) \to H^d(X\otimes_\varphi k, \mathcal{G}\otimes\mathcal{L}_{\psi(f)})$$

is an isomorphism.

(2) The "forget supports" mapping cone

$$K := K(\mathcal{G}) := [Rf_!\mathcal{G} \to Rf_*\mathcal{G}],$$

viewed as an object of $D^b{}_c(\mathbb{A}^1{}_R, \overline{\mathbb{Q}}_\ell)$, is lisse on $\mathbb{A}^1{}_R$, in the sense that all its cohomology sheaves are lisse on $\mathbb{A}^1{}_R$.

(2bis) $K_\mathbb{C}$ is lisse on $\mathbb{A}^1{}_\mathbb{C}$.

(2ter) The restriction of K to some geometric fibre of $\mathbb{A}^1{}_R/R$ is lisse.

proof Since $K(\mathcal{G})$ is adapted to the stratification $(\mathbb{A}^1{}_R - D, D)$, it is lisse on $\mathbb{A}^1{}_R$ if and only if it is lisse on any single geometric fibre $\mathbb{A}^1{}_k$, i.e., the conditions (2), (2bis), and (2ter) are all equivalent.

For any lisse $\mathcal{G}$ on X, we have a triangle on $\mathbb{A}^1{}_R$,

$$\to Rf_!\mathcal{G} \to Rf_*\mathcal{G} \to K(\mathcal{G}) \to,$$

whose formation commutes with arbitrary change of base on Spec(R) to a good scheme. Restrict to $\mathbb{A}^1{}_k$, and apply $FT_{\psi,!} \approx FT_{\psi,*}$. We get a triangle on $\mathbb{A}^1{}_k$,

$$\to FT_\psi(Rf_!\mathcal{G}) \to FT_\psi(Rf_*\mathcal{G}) \to FT_\psi(K(\mathcal{G})) \to .$$

Because $Rf_!\mathcal{G}$ is everywhere tame on $\mathbb{A}^1{}_k$, the object $FT_\psi(Rf_!\mathcal{G})$ is lisse on $\mathbb{G}_{m,k}$, and by proper base change it is automatically of formation compatible with arbitrary change of base on $\mathbb{G}_{m,k}$ to a good scheme. In terms of the morphism

$$pr_2 : X_k \times \mathbb{G}_{m,k} \to \mathbb{G}_{m,k}, \ (x, t) \mapsto t,$$

we have (by the very definition of $FT_{\psi,!}$)

$$FT_\psi(Rf_!\mathcal{G}) \mid \mathbb{G}_{m,k} = R(pr_2)_!(\mathcal{G} \otimes \mathcal{L}_{\psi(tf)})[1].$$

Hence its dual, which (up to a Tate twist and a shift) is

$$FT_{\overline{\psi},*}(Rf_*\mathcal{G}^\vee) \mid \mathbb{G}_{m,k} = R(pr_2)_*(\mathcal{G}^\vee \otimes \mathcal{L}_{\overline{\psi}(tf)})[1],$$

is itself lisse on $\mathbb{G}_{m,k}$, of formation compatible with arbitrary change of base on $\mathbb{G}_{m,k}$ to a good scheme. Applying this argument to $\mathcal{G}^\vee$ and $\overline{\psi}$, we see that

$$FT_\psi(Rf_*\mathcal{G}) \mid \mathbb{G}_{m,k} = R(pr_2)_*(\mathcal{G} \otimes \mathcal{L}_{\psi(tf)})[1]$$

is lisse on $\mathbb{G}_{m,k}$, of formation compatible with arbitrary change of base on $\mathbb{G}_{m,k}$ to a good scheme.

Passing to fibres, we see that condition (1) for $\mathcal{G}$ holds if and only if $FT_\psi(K(\mathcal{G})) \mid \mathbb{G}_{m,k}$ vanishes for every triple (k, φ, ψ), i.e., if and only if $FT_\psi(K(\mathcal{G}))$ is punctual, supported at the origin. By Fourier inversion, this is in turn equivalent to saying that $K(\mathcal{G})$ is geometrically constant, and hence lisse, on each geometric fibre $\mathbb{A}^1{}_k$ of positive characteristic, whence (2ter) holds.

Conversely, if $K(\mathcal{G})$ is lisse on $\mathbb{A}^1{}_R$, i.e., if condition (2) holds, then the restriction of $K(\mathcal{G})$ to each geometric fibre $\mathbb{A}^1{}_k$ of positive characteristic is both lisse and everywhere tamely ramified, and hence geometrically constant, whence $FT_\psi(K(\mathcal{G})) \mid \mathbb{G}_{m,k}$ vanishes. So looking fibre by fibre, we see that condition (1) holds.

The equivalence of (1) with (1bis), "mise pour memoire" results from the fact that each geometric fibre X_φ is affine and smooth, everywhere of dimension d, with $\mathcal{G}$ lisse, so by the Lefschetz affine theorem the H^i vanish for $i > d$, and dually the $H^i{}_c$ vanish for $i < d$. QED

Theorem 14.13.4 Hypotheses and notations as in 14.13.3, suppose in addition that the equivalent conditions (1) or (2) hold. Then

(1) There exists an integer n such that for any finite field k, any ring homomorphism $\varphi: R \to k$, and any nontrivial additive character ψ of k, the restriction to $\mathbb{G}_m$ of the Fourier Transform of $Rf_!\mathcal{G}[d]$ is of the form

$$\mathcal{F}_{\varphi,\psi}[1] := j^*FT_\psi(Rf_!\mathcal{G}[d])$$

with $\mathcal{F}_{\varphi,\psi}$ a single lisse sheaf of rank n on $\mathbb{G}_m/k$. For any finite extension E of k, and any point $\alpha \in E^\times = \mathbb{G}_m(E)$, the trace of $Frob_{E,\alpha}$ on $\mathcal{F}_{\varphi,\psi}$ is the sum

$$\operatorname{trace}(Frob_{E,\alpha} \mid \mathcal{F}_{\varphi,\psi}) = (-1)^{d-1}\sum_{x \text{ in } X_\varphi(E)} \psi_E(\alpha f(x))\operatorname{trace}(Frob_{E,x} \mid \mathcal{G}),$$

where ψ_E is the additive character $\psi\circ\operatorname{trace}_{E/k}$ of E.

(2) If $\mathcal{G}$ is pure of weight zero on X, then $\mathcal{F}_{\varphi,\psi}$ is pure of weight d on $\mathbb{G}_m$, and (consequently) the geometric monodromy group $G_{geom,\varphi,\psi}$ of $\mathcal{F}_{\varphi,\psi}$ is semisimple.

(3) The object K on $\mathbb{A}^1_R$ defined as the additive ! convolution

$$K := (Rf_!\mathcal{G}[d]) *_{!+} (Rj_* j^* \overline{\mathbb{Q}}_\ell(1)[1]),$$

is adapted to the stratification $(\mathbb{A}^1_R - D, D)$, is fibrewise perverse, and on each geometric fibre of positive characteristic its Fourier Transform is given by

$$FT_\psi(K_\varphi) = j_! \mathcal{F}_{\varphi,\psi}[1].$$

(4) Let $\mathcal{M}$ be the RS holonomic $\mathcal{D}$-module corresponding to $K_{\mathbb{C}}$. The $\mathcal{D}$-module Fourier Transform $FT(\mathcal{M})$ is a D.E. on $\mathbb{G}_{m,\mathbb{C}}$ of rank n. If $\mathcal{G}$ is pure of weight zero, then the differential galois group G_{gal} for $j^*FT(\mathcal{M})$ is reductive.

(5) If $\mathcal{G}$ is pure of weight zero, then there exists a dense open set U of Spec(R), which depends only on

the original stratification $(\mathbb{A}^1_R - D, D)$ of $\mathbb{A}^1_R$ of 14.13.3, and

the conjugacy class of G_{gal} in $GL(n, \mathbb{C})$,

such that for any φ lying over a point of U, $G_{geom,\varphi,\psi}(\overline{\mathbb{Q}}_\ell)$ is conjugate in $GL(n, \overline{\mathbb{Q}}_\ell)$ to a subgroup of $\iota^{-1}G_{gal}(\mathbb{C})$. Via this conjugation the two groups have the same "semisimple connected parts",

$$G_{geom,\varphi,\psi}(\overline{\mathbb{Q}}_\ell)^{0,der} = \iota^{-1}G_{gal}(\mathbb{C})^{0,der},$$

and the composite map

$$G_{geom,\varphi,\psi} \subset G_{gal} \rightarrow G_{gal}/Z((G_{gal})^0)^0$$

is surjective.

proof (1)The statification hypotheses show that there exists an integer n such that $j^*FT_\psi(Rf_!\mathcal{G}[d])$ has lisse cohomology sheaves on $\mathbb{G}_{m,k}$, the alternating sum of whose ranks is $-n$. Condition (1bis) then shows that $j^*FT_\psi(Rf_!\mathcal{G}[d])$ is of the form $\mathcal{F}_{\varphi,\psi}[1]$, with $\mathcal{F}_{\varphi,\psi}$ a single lisse sheaf of rank n on $\mathbb{G}_m/k$. The asserted trace formula is just a writing out of the Lefschetz Trace Formula in this case.

(2)If $\mathcal{G}$ is pure of weight zero, then condition (1bis) forces the purity, since H^d_c is mixed of weight $\leq$ d, while H^d is mixed of weight $\geq$ d. If $\mathcal{F}_{\varphi,\psi}$ is pure of some weight, then as recalled above (in the proof of 14.11.2) its G_{geom} is semisimple.

(3) That K is adapted to $(\mathbb{A}^1_R - D, D)$ has already been proven (13.4.3). To show that K is fibrewise perverse, it suffices (by 14.1.3) to show that

its restriction to any single geometric fibre is perverse. On a fibre in characteristic p, K_φ is perverse if and only if $FT_\psi(K_\varphi)$ is perverse. But

$$FT_\psi(K_\varphi) = j_!j^*FT_\psi(Rf_!\mathcal{G}[d]) = j_!\mathcal{F}_{\varphi,\psi}[1],$$

which is visibly perverse.

(4) That $j^*FT(\mathcal{M})$ is a D.E. of the same rank n has already been proven (14.1.5). That G_{gal} is reductive if $\mathcal{G}$ is pure results from part (2) above, via the purity criterion 14.11.2.

(5) This is the sharpened reductive comparison theorem 14.10, applied to K. QED

Here is the cohomological description of the D.E. $j^*FT(\mathcal{M})$ on $\mathbb{G}_m$.

Proposition 14.13.5 With the hypotheses and notations of theorem 14.13.4, denote by $\mathcal{N} \in D.E.(X_\mathbb{C}/\mathbb{C})$ the R.S. object on $X_\mathbb{C}$ which corresponds to the perverse object $\mathcal{G}[d]$ via Riemann-Hilbert. On the product $X_\mathbb{C} \times \mathbb{G}_{m,\mathbb{C}}$, with coordinates (x,t) consider the D.E. $pr_1{}^+(\mathcal{N}) \otimes e^{tf(x)}$, and take its $\mathcal{D}$-module direct images in the $*$ and ! sense via pr_2. The natural map

$$(pr_2)_!(pr_1{}^+(\mathcal{N}) \otimes e^{tf(x)}) \to (pr_2)_*(pr_1{}^+(\mathcal{N}) \otimes e^{tf(x)})$$

is an isomorphism, and the object

$$(pr_2)_*(pr_1{}^+(\mathcal{N}) \otimes e^{tf(x)}) \in D_b{}^{holo}(\mathbb{G}_{m,\mathbb{C}})$$

has a single nonzero cohomology sheaf, namely $j^*FT(\mathcal{M})$ in degree zero.

proof To see this, we argue as follows. The $\mathcal{D}$-module partners of $Rf_!\mathcal{G}[d]$ and of $Rf_*\mathcal{G}[d]$ are $f_!\mathcal{N}$ and $f_*\mathcal{N}$ respectively, and the mapping cylinder of

$$f_!\mathcal{N} \to f_*\mathcal{N}$$

is the $\mathcal{D}$-module partner of $K(\mathcal{G})$, so has all its cohomology sheaves constant (direct sums of $\mathcal{O}$) on $\mathbb{A}^1$. Therefore convolving on $\mathbb{A}^1$ with $j_*\mathcal{O}$ in the ! sense, gives, by 12.5.8, an isomorphism

$$(f_!\mathcal{N}) *_{!+} (j_*\mathcal{O}) \approx (f_*\mathcal{N}) *_{!+} (j_*\mathcal{O}).$$

The object $(f_!\mathcal{N}) *_{!+} (j_*\mathcal{O})$ is the Riemann-Hilbert partner of

$$K := (Rf_!\mathcal{G}[d]) *_{!+} (Rj_*j^*\overline{\mathbb{Q}}_\ell(1)[1]);$$

in other words, we have

$$\mathcal{M} \approx (f_!\mathcal{N}) *_{!+} (j_*\mathcal{O}) \approx (f_*\mathcal{N}) *_{!+} (j_*\mathcal{O}).$$

Taking the Fourier Transforms of these isomorphisms and restricting to

$\mathbb{G}_m$, we get isomorphisms in $D_b^{\mathrm{holo}}(\mathbb{G}_{m,\mathbb{C}})$

$$j^*FT(\mathcal{M}) \approx j^*FT(f_!\mathcal{N}) \approx j^*FT(f_*\mathcal{N}).$$

Visibly we have

$$j^*FT(f_!\mathcal{N}) \approx (pr_2)_!(pr_1{}^+(\mathcal{N})\otimes e^{tf(x)}),$$

$$j^*FT(f_*\mathcal{N}) \approx (pr_2)_*(pr_1{}^+(\mathcal{N})\otimes e^{tf(x)}),$$

as results from base change via the diagram

$$\begin{array}{ccc} X\times\mathbb{G}_m & \longrightarrow & X \\ f\times id \downarrow & & \downarrow f \\ \mathbb{A}^1\times\mathbb{G}_m & \longrightarrow & \mathbb{A}^1 \\ pr_2 \downarrow & & \\ \mathbb{G}_m. & & \end{array}$$

QED

14.14 Examples

(1) Suppose that $n \geq 1$ is an integer, and $(x_1, x_2, \ldots, x_n)$ are n functions on X which define a finite morphism from X to $\mathbb{A}^n{}_R$ (e.g., if X is given as a closed subscheme of $\mathbb{A}^n{}_R$, one might take for the x_i the coordinate functions in the ambient $\mathbb{A}^n{}_R$). Then by (the proof of) [Ka-Lau, 5.4], there exists a nonzero homogeneous polynomial in n variables Y_i,

$$F(Y_1, Y_2, \ldots, Y_n) \in R[Y_1, Y_2, \ldots, Y_n],$$

with the following property: for any n-tuple $(a_1, \ldots, a_n)$ of elements of R such that $F(a) \neq 0$, condition (1) of 14.13.3 holds for $X\otimes_R R[1/F(a)]$ over $R[1/F(a)]$, the function

$$f := \Sigma_i\, a_i x_i,$$

and any lisse $\mathcal{G}$ on $X\otimes_R R[1/F(a)]$.

(2) If X is $\mathbb{A}^n{}_R$, with coordinates $(x_1, x_2, \ldots, x_n)$, and f any polynomial in $R[x_1, x_2, \ldots, x_n]$ whose degree d is invertible in R and whose leading form $f_d(x_1, x_2, \ldots, x_n)$ defines a smooth hypersurface in $\mathbb{P}^{n-1}{}_R$, then condition (1) of 14.13.3 holds for the constant sheaf $\mathcal{G} = \overline{\mathbb{Q}}_\ell$. This example was given by Deligne in [De-WI]. See [Ka-SE, 5.1.1, 5.1.2] for

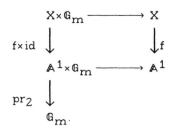

generalizations of this example, where X becomes the affine part of a smooth projective variety $\overline{X}/R$, and the function f on X has a particularly nice expression near $\overline{X}$ - X, but in which $\mathcal{G}$ remains the constant sheaf. The archetype of these generalizations is that of the finite part of a Lefschetz pencil. See [Ka-Lau, 5.6.1] for a discussion of the relations between these examples and those in (1) above.

(3) If the morphism $f : X \to \mathbb{A}^1_R$ is **finite**, then condition (2) of 14.13.3 is trivially satisfied for any $\mathcal{G}$, since $Rf_! = Rf_*$ and so $K(\mathcal{G}) = 0$.

(4) For $\mathcal{G}$ the constant sheaf $\overline{\mathbb{Q}}_\ell$, the condition (2bis) of 14.13.3 that $K_\mathbb{C}$ be lisse on $\mathbb{A}^1_\mathbb{C}$ is the purely topological condition on the complex morphism

$$(f_\mathbb{C})^{an} : (X_\mathbb{C})^{an} \to (\mathbb{A}^1_\mathbb{C})^{an},$$

that the mapping cylinder of

$$R(f_\mathbb{C})^{an}{}_!\mathbb{Q} \to R(f_\mathbb{C})^{an}{}_*\mathbb{Q}$$

have lisse cohomology sheaves on $(\mathbb{A}^1_\mathbb{C})^{an}$. With hindsight, many of the situations considered in [Ka-SE, Chapter 5] can be seen directly to satisfy this mapping cylinder condition.

5) Again for $\mathcal{G}$ the constant sheaf, Adolphson and Sperber (cf [Ad-Sp]) gives many "toroidal" examples where (1) of 14.13.3 holds. In their examples, X is $(\mathbb{G}_{m,R})^n$, and f is a Laurent polynomial whose "Newton polytope" is sufficiently nice.

Remark 14.15 In all of the above examples, one may need to do some initial shrinking on Spec(R) to arrange for the existence of the required stratification $(\mathbb{A}^1_R - D, D)$ of $\mathbb{A}^1_R$. To the extent that this initial shrinking is not very explicit, the apparent precision of the theorem in specifying upon what the final dense open U of Spec(R) may be taken to depend is somewhat illusory, at least from the point of view of effective calculation. This problem **did not arise** in comparing G_{gal} and G_{geom} for hypergeometrics $\mathcal{H}_1$, because in that case the initial good stratification of $\mathbb{A}^1$ is staring us in the face: $(\mathbb{A}^1_R - \{0,1\}, \{0,1\})$.

[Ad-Sp] Adolphson, A., and Sperber, S., Exponential sums and Newton polyhedra; cohomology and estimates, Ann. Math. 130 (1989), 367-406.

[Ax] Ax, J., Zeroes of polynomials over finite fields, Amer. J. Math. 86 (1964), 255-261.

[Bei] Beilinson, A. A., On the derived category of the category of perverse sheaves, in Manin, Yu. I. (ed), *K-Theory, Arithmetic and Geometry*, Springer Lecture Notes in Math. 1289, 1987, 27-41.

[B-B-D] Beilinson, A. A., Bernstein, I. N., and Deligne, P., Faisceaux Pervers, entire contents of *Analyse et Topologie sur les espaces singuliers I*, Conference de Luminy, Asterisque 100, 1982.

[Bel] Belyi, G.V., - On galois extensions of a maximal cyclotomic field, Math. U.S.S.R. Izv. 14 (1980), n°2, 247-256.

[Ber], Bernstein, J., Six lectures on the algebraic theory of $\mathcal{D}$-modules, xeroxed notes, E.T.H., Zurich, 1983.

[Bert] Berthelot, P., Cohomologie Cristalline des Schémas de Caractéristique p > 0, Springer Lecture Notes in Math. 407, 1974

[B-B-H] Beukers, F., Brownawell, D., and Heckman, G., Siegel Normality, Ann. Math. 127 (1988), 279-308.

[B-H] Beukers, F., and Heckman, G., Monodromy for the hypergeometric function ${}_nF_{n-1}$, Inv. Math. 95 (1989), 325-354.

[Bor] Borel, A. et al, *Algebraic D-Modules*, Academic Press, Boston, 1987.

[Bour-L1] Bourbaki, N., *Groupes et Algebres de Lie*, Chapitre 1, Diffusion CCLS, Paris, 1971.

[Bour-L6] Bourbaki, N., *Groupes et Algebres de Lie*, Chapitres 4, 5 et 6, Masson, Paris, 1981.

[Bour-L8] Bourbaki, N., *Groupes et Algebres de Lie*, Chapitres 7 et 8, Diffusion CCLS, Paris, 1975.

[Bour-L9] Bourbaki, N., *Groupes et Algebres de Lie*, Chapitre 9, Masson, Paris, 1982.

[Br] Brylinski, J.-L., Transformations canoniques, dualité projective, théorie de Lefschetz, transformations de Fourier et sommes trigonométriques, Astérisque 140-141 (1986), 3-134.

[C-R-RT] Curtis, C.W., and Reiner, I., *Representation Theory of Finite Groups and Associative Algebras*, Interscience Publ., New York and London, 1962.

[C-R-MRT] Curtis, C.W., and Reiner, I., *Methods of Representation Theory with applications to finite groups and orders, Vol. I*, John Wiley and Sons, New York, 1981.

[De-RFT] Deligne, P., Rapport sur la formule des traces, in *Cohomologie Etale (SGA 4 1/2)*, Springer Lecture Notes in Mathematics 569, 1977, 76-109.

[De-AFT] Deligne, P., Application de la formule des traces aux sommes trigonométriques, in *Cohomologie Etale (SGA 4 1/2)*, Springer Lecture Notes in Mathematics 569, 1977, 168-232.

[De-CT] Deligne, P., Catégories tannakiennes, I.A.S. preprint, 1987,to appear in *Grothendieck Festschrift*.

[Del-ED] Deligne, P., *Equations differentielles à points singuliers reguliers*, Springer Lecture Notes Math. 163, 1970.

[De-TF] Deligne, P., Theoremes de Finitude en Cohomologique ℓ-adique, in *Cohomologie Etale (SGA 4 1/2)*, Springer Lecture Notes in Mathematics 569, 1977, 233-251.

[De-WI] Deligne, P., La Conjecture de Weil I, Pub. Math. I.H.E.S. 48 (1974), 273-308.

[De-WII] Deligne, P., La conjecture de Weil II, Pub. Math. I.H.E.S. 52 (1981), 313-428.

[D-M] Deligne, P., and Milne, J., Tannakian categories, in Deligne, P., Milne, J., Ogus, A., and Shih, K.-y., *Hodge cycles, motives, and Shimura varieties*, Springer Lecture Notes in Math. 900, 1982, 101-228.

[Dw] Dwork, B., Bessel functions as p-adic functions of the argument, Duke Math. J. 41, 1974, 711-738.

[EGA] Grothendieck, A., rédigé avec la collaboration de Dieudonné, J., *Elements de Géométrie Algébrique*, Pub. Math. I.H.E.S. 4, 8, 11, 17, 20, 24, 28, 32, 1960-1967.

[Ek] Ekedahl, T., On the adic formalism, University of Stockholm preprint, 1988, to appear in *Grothendieck Festschrift*.

[Er] Erdelyi, A., *Higher Transcendental Functions, Vol. I (Bateman Manuscript Project)*, McGraw-Hill, New York, 1953.

[Ev] Evens, L., A generalization of the transfer map in the cohomology of groups, Trans. A. M. S. 108 (1963), 54-65.

[Gel] Gel'fand, I.M., The general theory of hypergeometric functions, Dokl. Akad. Nauk SSSR 288 No. 1 (1986), 14-18.

[Gre] Greene, J., Hypergeometric functions over finite fields, Trans. A.M.S. 301 (1987), 77-101.

[Gro-CDR] Grothendieck, A., Crsytals and the De Rham cohomology of schemes, reprinted in *Dix Exposés sur la cohomologie des schémas*, North-Holland, 1968.

[Gro-FL] Grothendieck, A., Formule de Lefschetz et rationalité des fonctions L, Seminaire Bourbaki 1964-65, Exposé 279, reprinted in *Dix Exposés sur la cohomologie des schémas*, North-Holland, 1968.

[Ill-ATF] Illusie, L., Appendix to Deligne, P., Theoremes de finitude en cohomologique ℓ-adique, in *Cohomologie Etale (SGA 4 1/2)*, Springer Lecture Notes in Mathematics 569, 1977, 233-251.

[Ill-DFT] Illusie, L., Deligne's ℓ-adic fourier transform, in Bloch, S.J. (ed), *Algebraic Geometry: Bowdoin 1985*, A.M.S., Providence, 1987

[Kash] Kashiwara, M., The Riemann-Hilbert problem for holonomic systems, RIMS, Kyoto Univ., 1983.

[Ka-AS] Katz, N., Algebraic solutions of differential equations; p-curvature and the Hodge filtration, Inv. Math. 18 (1972), 1-118.

[Ka-DGG] Katz, N., On the calculation of some differential galois groups, Inv. Math. 87 (1987), 13-61.

[Ka-GKM] Katz, N.,*Gauss sums, Kloosterman sums, and monodromy groups,* Annals of Math. Study 116, Princeton Univ. Press, 1988.

[Ka-LG] Katz, N., Local to global extensions of representations of fundamental groups, Ann. Inst. Fourier 36, 4 (1986), 59-106.

[Ka-MG] Katz, N., On the monodromy groups attached to certain families of exponential sums, Duke Math. J. 54 No. 1, (1987), 41-56.

[Ka-PES] Katz, N., Perversity and exponential sums, in *Algebraic Number Theory - in honor of K. Iwasawa,* Advanced Studies in Pure Mathematics 17, 1989, 209-259.

[Ka-ES] Katz, N., Exponential sums over finite fields and differential equations over the complex numbers: some interactions, xeroxed notes of Colloquium Lectures, A.M.S. Winter Meeting, Phoenix, 1989.

[Ka-CAT] Katz, N., A conjecture in the arithmetic theory of differential equations, Bull. Soc. Math. Fr. 110 (1982), 203-239.

[Ka-SE] Katz, N., *Sommes Exponentielles,* rédigé par G. Laumon, Asterisque 79, 1980.

[Ka-TL] Katz, N., Travaux de Laumon, Séminaire Bourbaki Exposé 691, in *Séminaire Bourbaki Volume 1987-88,* Asterisque 161-162 (1988), 105-132.

[Ka-Lau] Katz, N. and Laumon, G., Transformation de Fourier et majoration de sommes exponentielles, Pub. Math. I.H.E.S. 62 (1986), 361-418.

[Ka-Pi] Katz, N., and Pink, R., A note on pseudo-CM representations and differential galois groups, Duke Math. J. 54 No. 1 (1987), 57-65.

[Kob] Koblitz, N., The number of points on certain families of hypersurfaces over finite fields, Comp. Math. 48 (1983), 3-23.

[Kol] Kolchin, E., Algebraic groups and algebraic independence, Amer. J. Math. 90 (1968), 1151-1164.

[Kos] Kostant, B., A characterization of the classical groups, Duke Math. J. 25 (1958), 107-123.

[Lau-TF] Laumon, G., Transformation de Fourier, constantes d'équations fonctionelles et conjecture de Weil, Pub. Math. I.H.E.S. 65 (1987), 131-210.

[Lau-SCS] Laumon, G., Semi-continuité du conducteur de Swan (d'après P. Deligne), in *Caractéristique d'Euler-Poincaré, Seminaire E.N.S. 1978-79*, Astérisque 82-83 (1981), 173-219.

[Lev-JD] Levelt, A.H.M., Jordan decomposition for a class of singular differential operators, Arkiv. Math. 13 (1975), 1-27.

[Le-HF] Levelt, A.H.M., Hypergeometric functions, thesis, University of Amsterdam, 1961.

[Ma] Manin, J., Moduli fuchsiani, Ann. Sc. Norm. Pisa 19 (1965), 13-126.

[Me-HR] Mebkhout, Z., Sur le problème de Hilbert-Riemann, Springer Lecture Notes in Physics 129 (1980), 99-110.

[Me-EC] Mebkhout, Z., Une équivalence de catégories et une autre équivalence de catégories, Comp. Math. 51 (1984), 55-69.

[Me-SO] Mebkhout, Z., *Le formalisme des six operations de Grothendieck pour les D_X-modules cohérents*, Travaux en Cours 35, Hermann, 1989.

[Ri] Ribet, K., Galois action on division points of abelian varieties with real multiplication, Amer. J. Math. 98 (1976), 751-805.

[Saa] Saaveedra Rivano, N., *Catgories tannakiennes*, Springer Lecture Notes in Math. 265, 1972.

[SGA] A. Grothendieck et al - *Séminaire de Géométrie Algébrique du Bois-Marie*, SGA 1, SGA 4 Parts I, II, and III, SGA 4½, SGA 5, SGA 7 Parts I and II, Springer Lecture Notes in Math. 224, 269-270-305, 569, 589, 288-340, 1971 to 1977.

[Se-Ta] Serre, J.-P., and Tate, J., Good reduction of abelian varieties, Ann. Math. 88 (1968), 492-517.

[Ver] Verdier, J.-L., Specialization de faisceaux et monodromie moderé, in *Analyse et topologie sur les espaces singuliers*, vol. II and III, Astérisque 101-102, 1983, 332-364.

[Za-WS] Zarkhin, Yu. G., Weights of simple Lie algebras in the cohomology of algebraic varieties, Izv. Akad. Nauk. SSSR, Ser. Math. 48 (1984), 264-304, English translation in Math. USSR Izvestiya 24 (1985), 245-281

[Za-LS] Zarkhin, Yu. G., Linear simple Lie algebras and ranks of operators, preprint Orsay, 1988, to appear in *Grothendieck Festschrift*.

Milton Keynes UK
Ingram Content Group UK Ltd.
UKHW040739121224
452420UK00001B/59